THE GREAT NEW WILDERNESS DEBATE

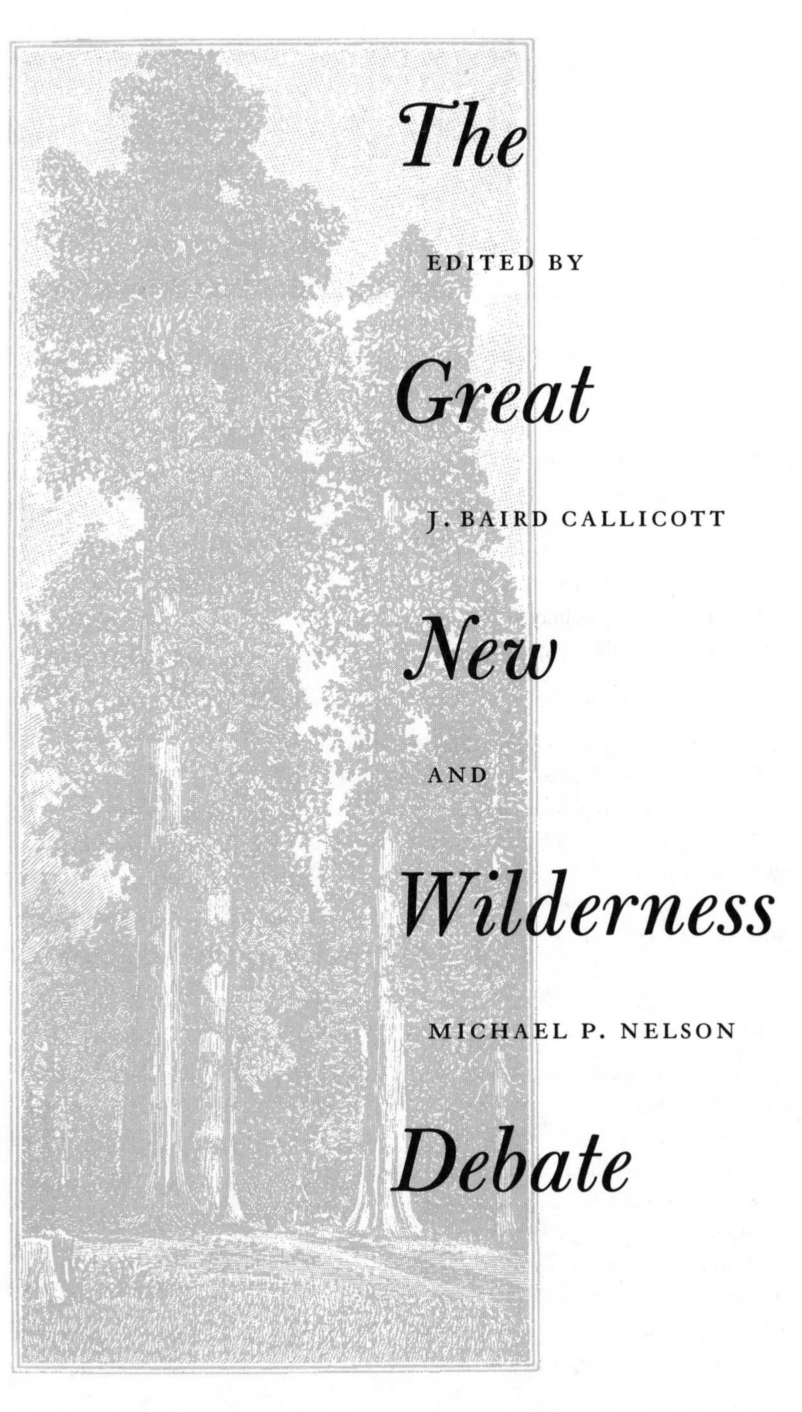

The Great New Wilderness Debate

EDITED BY
J. BAIRD CALLICOTT
AND
MICHAEL P. NELSON

THE UNIVERSITY OF GEORGIA PRESS ATHENS AND LONDON

Designed by Christine Taylor
Set in Granjon text with Bulmer display
by Wilsted & Taylor Publishing Services

The paper in this book meets the guidelines for permanence
and durability of the Committee on Production Guidelines
for Book Longevity of the Council on Library Resources.

Printed in the United States of America

02 01 00 99 98 C 5 4 3 2 1
02 01 00 P 5 4 3 2

LIBRARY OF CONGRESS CATALOGING IN PUBLICATION DATA

The great new wilderness debate / edited by J. Baird Callicott and
Michael P. Nelson.
p. cm.
Includes bibliographical references and index.
ISBN 0-8203-1983-X (alk. paper). — ISBN 0-8203-1984-8 (pbk. :
alk. paper)
1. Nature conservation. 2. Nature conservation—Philosophy.
I. Callicott, J. Baird. II. Nelson, Michael P., 1966– .
QH75.G69 1998
333.78′216—dc21 97-47023

British Library Cataloging in Publication Data available

We dedicate this book

to a wonderful scholar,

teacher, colleague, and friend—

ARTHUR L. HERMAN

CONTENTS

PART THREE *The Wilderness Idea Roundly Criticized and Defended*

PART FOUR *Beyond the Wilderness Idea*

ACKNOWLEDGMENTS

First and foremost, we would like to thank all authors—living and dead—who contributed to this volume. For secretarial assistance in scanning, proofreading, and retyping the individual essays we owe a debt of gratitude to Carolee Cote. For proofreading individual essays we thank Julie Sparhawk. And we thank Nancy Evans of Wilsted & Taylor Publishing Services for expert and careful copyediting.

Essays in the public domain include "Images or Shadows of Divine Things," "Christian Doctrine of Original Sin Defended," and "Sinners in the Hands of an Angry God" by Jonathan Edwards (1758); Chapter 1 of Ralph Waldo Emerson's essay "Nature" (1836); selections from the essay "Walking" by Henry David Thoreau (1862); "The Wild Parks and Forest Reservations of the West" and "The American Forests" from John Muir's *Our National Parks* (1901); and Theodore Roosevelt's "The American Wilderness: Wilderness Hunters and Wilderness Game" (originally published in *The Complete Works of Theodore Roosevelt* [New York: P.F. Collier, 1897]).

Selections from Henry David Thoreau's essay "Huckleberries" (1861) are reprinted by permission of the Henry W. and Albert A. Berg Collection, The New York Public Library, and the Astor, Lenox and Tilden Foundations.

Aldo Leopold's essays "Wilderness as a Form of Land Use" (originally published in *The Journal of Land and Public Utility Economics* 1[4] [October 1925]: 398–404), "Threatened Species: A Proposal to the Wildlife Conference for an Inventory of the Needs of Near-Extinct Birds and Animals" (originally published in *American Forests* 42[3] [March 1936]: 116–19), and "Wilderness" (originally written in 1935) were also published (the latter for the first time) in *The River of the Mother of God and Other Essays by Aldo Leopold,* edited by Susan L. Flader and J. Baird Callicott (Madison: University of Wisconsin Press, 1991); they are all reprinted here by permission of the Aldo Leopold Foundation and the University of Wisconsin Press.

Robert Marshall's essay "The Problem of the Wilderness" is reprinted with permission from *The Scientific Monthly* 30 (February 1930): 141–48. © 1930 American Association for the Advancement of Science.

"Why Wilderness?" by Sigurd Olson was originally published in *American Forests* (September 1938): 395–430.

"Wildlife Management in the National Parks" (or, The Leopold Report) by A. Starker Leopold et al. was originally published in *Transactions of the Twenty-Eighth North American Wildlife and Natural Resources Conference,* James B. Trefethen, ed. (Washington, D.C.: Wildlife Management Institute, 1963), pp. 29–44.

The Wilderness Act of 1964 is United States of America Public Law 88-577, 88th Congress, September 3, 1964.

"Indian Wisdom" by Chief Luther Standing Bear was originally published in *Land of the Spotted Eagle* (Boston: Houghton Mifflin, 1933), pp. 38, 192–97.

Roderick Nash's "The International Perspective" was originally published in *Wilderness and the American Mind,* 3d ed. (New Haven: Yale University Press, 1982), pp. 342–78.

David Harmon's "Cultural Diversity, Human Subsistence, and the National Park Ideal" was originally published in *Environmental Ethics* 9 (Summer 1987): 147–58.

Ramachandra Guha's "Radical American Environmentalism and Wilderness Preservation: A Third World Critique" was originally published in *Environmental Ethics* 11 (Spring 1989): 71–83.

"The Relevance of Deep Ecology to the Third World: Some Preliminary Comments" by David M. Johns was originally published in *Environmental Ethics* 12 (Fall 1990): 233–52.

Arne Naess's "The Third World, Wilderness, and Deep Ecology" was originally published in *Deep Ecology for the 21st Century: Readings on the Philosophy and Practice of the New Environmentalism,* edited by George Sessions, © 1995. Reprinted by arrangement with Shambhala Publications, Inc., 300 Massachusetts Avenue, Boston, Mass., 02115.

"Taming the Wilderness Myth" by Arturo Gómez-Pompa and Andrea Kaus was originally published in *BioScience* 42(4) (April 1992): 271–79. © 1992 American Institute of Biological Sciences.

Fabienne Bayet's "Overturning the Doctrine: Indigenous People and Wilderness—Being Aboriginal in the Environmental Movement" was originally published in *Social Alternatives* 13(2) (July 1994): 27–32.

J. Baird Callicott's "The Wilderness Idea Revisited: The Sustainable Development Alternative" was originally published in *The Environmental Professional* 13 (1991): 235–47. Holmes Rolston III's "The Wilderness Idea Reaffirmed" was originally published in *The Environmental Professional* 13 (1991): 370–77. J. Baird Callicott's "That Good Old-Time Wilderness Religion" was originally published in *The Environmental Professional* 13 (1991): 378–79.

"Sustainability and Wilderness" by Reed F. Noss was originally published in *Conservation Biology* 5(1) (1991): 120–22.

William M. Denevan's "The Pristine Myth: The Landscape of the Americas in 1492" was originally published in *Annals of the Association of American Geographers* 82(3) (1992): 369–85.

Thomas H. Birch's "The Incarceration of Wildness: Wilderness Areas as Prisons" was originally published in *Environmental Ethics* 12 (Spring 1990): 3–26.

"The Trouble with Wilderness, or, Getting Back to the Wrong Nature" by William Cronon was originally published in *Uncommon Ground: Toward Reinventing Nature* (New York: W. W. Norton, 1995), pp. 69–90. © 1995 by William Cronon. Reprinted by permission of W. W. Norton & Company, Inc.

Marvin Henberg's "Wilderness, Myth, and American Character" was originally published in *The Key Reporter* 59(3) (Spring 1994): 7–11, and is used by permission of the Phi Beta Kappa Society.

Reed F. Noss's "Wilderness Recovery: Thinking Big in Restoration Ecology" was originally published in *The Environmental Professional* 13 (1991): 225–34.

Dave Foreman's "Wilderness: From Scenery to Nature" was originally published in *Wild Earth* 5(4) (Winter 1995/96): 9–16.

J. Baird Callicott's "Should Wilderness Areas Become Biodiversity Reserves?" was first published in *The George Wright Forum* 13(2) (1996): 32–38.

"Using Biodiversity as a Justification for Nature Protection in the US" by R. Edward Grumbine was originally published in *Wild Earth* 6(4) (Winter 1996/97): 71–80.

"In Wildness Is the Preservation of the World" by Jack Turner was originally published in *Northern Lights* 6(4) (1991): 22–25.

Gary Paul Nabhan's "Cultural Parallax in Viewing North American Habitats" was originally published in *Reinventing Nature? Responses to Postmodern Deconstruction,* edited by Michael E. Soulé and Gary Lease (Washington, D.C.: Island Press, 1995), pp. 87–101.

Gary Snyder's "The Rediscovery of Turtle Island" was originally published in *A Place in Space: Ethics, Aesthetics, and Watersheds* (Washington, D.C.: Counterpoint, 1995), pp. 236–51. Reprinted with permission.

We also thank Mark Woods ("Federal Wilderness Preservation in the United States: The Preservation of Wilderness?"), Michael P. Nelson ("An Amalgamation of Wilderness Preservation Arguments"), Ramachandra Guha ("Deep Ecology Revisited"), Carl Talbot ("The Wilderness Narrative and the Cultural Logic of Capitalism"), Dave Foreman ("Wilderness Areas for Real"), Donald M. Waller ("Getting Back to the Right Nature: A Reply to Cronon's 'The Trouble with Wilderness'"), and Val Plumwood ("Wilderness Skepticism and Wilderness Dualism"), for making original contributions to this book.

J. Baird Callicott and Michael P. Nelson

Introduction

ITS TITLE MAY SUGGEST to some that this volume exposes and systematizes the diatribes against wilderness preservation forthcoming from talk radio demagogues such as Rush Limbaugh, and documents the defenses against such assaults mounted by beleaguered environmentalists. But that is *not* what this book is about. Well-heeled special interests—off-road recreational vehicle manufacturers, corporate cattle ranchers, mining companies, oil companies, and timber companies—have funded a new and reactionary so-called Wise Use Movement, dedicated to the undoing of local, state, and especially federal environmental legislation and regulation. The name itself is a dissembling perversion of the credo of Gifford Pinchot, the high-minded, well-intentioned chief architect of the Progressive conservation movement—which was born a century ago and which began the tradition of public commitment to environmental protection in North America. Typically, these special interests (some of them foreign-owned) —who stand to profit from the opening of wilderness areas to motorized recreation, grazing, mining, drilling, and clear-cutting—wrap their greed in the flag of individual freedom and private property rights. Their wealthy propagandists cynically claim to represent the ordinary, middle-class American in his or her mythic, Manichaean struggle against big government and cryptic socialism. Ron Arnold, Rush Limbaugh, and other irresponsible critics of wilderness preservation have nothing of intellectual interest to say, though the damage that their vituperative disinformation campaigns can do to the cause of conservation is real enough. Talk-radio-type wilderness bashing will, therefore, not be represented in this book; it will not be included alongside sincere and honest critical discussion of the wilderness idea nor be dignified as worthy of serious consideration. (In Part IV of this collection of essays, Gary Snyder more fully discusses the Wise Use Movement and its nefarious agenda.)

What then is this book about? It is about a *concept,* the "received wilderness idea"—that is, the notion of wilderness that we have inherited from our forebears. And it is an anthology; it includes, with a few notable exceptions, previously published work by many authors, who approach the concept of wilderness from many different points of departure, and who write in a wide variety of styles, addressed to a wide variety of primary audiences. The theme that binds these otherwise disparate writings into a coherent whole is the concept of wilderness. Some of the authors contributing to this anthology are academics, and some are not; those academics who formally cite sources do so in various ways—the ways typical of their several disciplines—and in many specific styles. Except for a few excerpts from longer works, most of the items in this anthology are complete, freestanding essays. The received wilderness idea is currently the subject of intense attack and impassioned defense on several fronts at once. The wilderness idea is alleged to be ethnocentric, androcentric, phallocentric, unscientific, unphilosophic, impolitic, outmoded, even genocidal. Defenders of the wilderness idea insist that it is none of these things. The received wilderness idea, has, in short, recently been the subject of heated debate. In sum, then, this anthology documents the current debate about the received wilderness idea. Before we go on to introduce this great new wilderness debate, however, we should first contextualize it.

Just who are "we" who have inherited the wilderness idea? And who are "our" forebears? Most immediately, *we,* the editors of this anthology, are Euro-American men, and our own cultural legacy is patriarchal Western civilization in its current postcolonial, globally hegemonic form. Though often resented and sometimes resisted, Americanized Western civilization (*civilization* here not in its congratulatory, but in its descriptive sense, ultimately from the Latin *civitas,* city) has—for better or worse, like it or not

—come to dominate the planet. Therefore, to one degree or another, the "we" in the first sentence of the previous paragraph also, most generally, comprises everyone on Earth and our (the editors') cultural legacy is also everyone's cultural legacy. However, we (the editors) are also academic philosophers; and from its Socratic beginnings, Western philosophy has involved, among other things, self-examination. Though our cultural legacy may be postcolonial, patriarchal, hegemonic Western civilization, we (the editors) believe that we can be, if not objective, then at least critically self-aware and, accordingly, sincerely strive not to privilege the discourse of the ethnic and gender groups to which, as an accident of birth, we happen to belong.

In the "conversation of the West," the voices giving shape to the concept of wilderness, those from whom we have received the wilderness idea, are mostly the colonial and postcolonial male writers represented in Part I of this anthology—Jonathan Edwards, Ralph Waldo Emerson, Henry David Thoreau, John Muir, Theodore Roosevelt, Aldo Leopold, Robert Marshall, and Sigurd Olson. With the possible exception of Robert Marshall, who, like Aldo Leopold, was a well-born wilderness-minded employee of the United States Forest Service, and Sigurd Olson, a northwoods nature writer, these men are all well-known figures in American letters and need no introduction by us. The work of these writers is included in this anthology, not because, like us, they are male and Euro-American, nor because they are the only historical writers on the subject of wilderness, nor because they had the most profound, sophisticated, or even interesting things to say about wilderness, but because their writings on wilderness, fairly sampled here, most influenced the popular wilderness idea. They articulated the concept of wilderness that is variously criticized (often as being both androcentric and ethnocentric) and defended in Parts II and III of this book.

Also included in Part I are a 1963 document, conventionally known as "The Leopold Report," that has exerted considerable influence on public wilderness policy in the United States, and the oft-quoted text of the legislation enacted by the Congress of the United States, conventionally known as "The Wilderness Act of 1964" (ghost-written by the pro-wilderness Washington lobbyist Howard Zahnizer), that established a national system of wilderness preserves. In Zahnizer's now standard definition of *wilderness* in the Wilderness Act—"a wilderness, in contrast with those areas where man and his own works dominate the landscape, is hereby recognized as an area where the earth and its community of life are untrammeled by man, where man is a visitor who does not remain"—the received wilderness idea is crystallized. Concluding Part I are two essays published here for the first time. The first is a conceptual analysis of the Wilderness Act by environmental philosopher Mark Woods. The second, by environmental philosopher Michael Nelson (one of the editors of this anthology), is a collection, summary, and evaluation of the many and various arguments for wilderness preservation advanced by those who have taken the concept of wilderness at face value, who have innocently believed that the word *wilderness,* like the word *mountain,* was the innocuous and unproblematic English name for something that exists in the world independently of any socially constructed skein of ideas.

Logically enough, that's how Part I ends, but why does it begin with an excerpt from the writings of Jonathan Edwards, a Puritan preacher? Edwards certainly has a widely acknowledged preeminent place in Euro-American intellectual history, but his name does not spring to mind—as do the names of Thoreau, Muir, Leopold, and Marshall—in connection with the wilderness idea. As environmental historian Roderick Nash demonstrates, in his now classic 1967 study of the wilderness idea, *Wilderness and the American Mind,* the "wilderness condition" of North America was certainly a preoccupation of the Puritans. For the first generation of Puritan colonists, it was, understandably, a wholly negative condition, something to be feared, loathed, and ultimately eradicated—something to be replaced by fair farms and shining cities on hills. The very success of their immigrant forebears in transforming the New England landscape into something resembling the landscape of the mother country, however, bequeathed prosperity to subsequent generations of Puritans. A cornerstone of Puritan Presbyterianism is the doctrine of original sin, which seemed to Edwards to express itself, in his own time, less in the reduced and pacified Native American population and the thoroughly domesticated countryside, and more in the prosperous populace of New England towns and cities. By contrast, the wild remnants of pre-settlement America that could be found here and there in Connecticut and Massachusetts appeared innocent and pure; to Edwards they seemed even to portend the divine.

We (the editors) are convinced that the originally colonial and eventually postcolonial received concept of wilderness is first and foremost an artifact of the sharp dichotomy, in Puritan thinking, between humanity, on the one hand—exclusively created in the image of God, but also fallen and depraved—and nature, on the other. The first generation of Puritans thought of themselves primarily as God's emissaries in the New World—which, to their perfervid, religion-besotted imaginations, was the wild, unruly stronghold of Satan. In the no less vivid imagination of Jonathan Edwards, the true stronghold of Satan had become the sinful human heart in the breast of his Euro-American neighbors, and the pristine American landscape had become Edenic. Interestingly, many of the most notable and most passionate subsequent defenders of the wilderness faith have a direct connection to Calvinism. Two stand out especially. Muir, famously, was brought up in a strict and austere Presbyterian household; and environmental philosopher Holmes Rolston III, among the most stalwart contemporary defenders of the wilderness idea, is an ordained Presbyterian minister.

The first criticism of the wilderness idea was voiced by those upon whom it was imposed and those whom it dispossessed. As documented by "Indian Wisdom," from *Land of the Spotted Eagle* by Chief Luther Standing Bear (like the famous Black Elk, an Oglala Lakota), published in 1933 and reprinted here, the wilderness idea was directly challenged by Native Americans, who were its first victims. But until very recently, the voices of Native Americans on this matter—as on almost all others—were muffled and ignored. In the third edition of *Wilderness and the American Mind,* published in 1982, Nash duly though belatedly noted Standing Bear's protest against the wilderness idea, but did not accordingly revise his sympathetic, even celebratory account of how the concept of wilderness gradually shed its largely negative connotations in mainstream Euro-American culture, and acquired positive connotations.

Since World War II, American economic, political, and cultural influence has spread inexorably. Along with many other things American, the wilderness idea and the public policies it inspired were adopted by empowered elements in postcolonial nation-states throughout the world, but especially in Africa and India. To set up Part II of this anthology, we have included an excerpt from Nash's chapter, "The International Perspective," newly written for the third edition of *Wilderness and the American Mind,* in which he details, without much critical reflection or comment, the emergence of an international trade in the wilderness experience. In that discussion, Nash focuses mainly on African national parks. We (the editors) were unable to find any essay by a native African spokesperson that specifically discusses the wilderness idea. However, a sympathetic Euro-American writer, David Harmon, tried to indicate the impact of the wilderness idea when it was exported to Africa in "Cultural Diversity, Human Subsistence, and the National Park Ideal," originally published in the journal *Environmental Ethics* in 1987. In *The Mountain People,* British anthropologist Colin Turnbull painted a morbidly fascinating portrait of the infamous Ik, a tribe of people who seemed inhumanly indifferent to one another. Harmon reveals that these unfortunate people had once been isolated gatherer-hunters, happily, successfully, sustainably, and humanly living by traditional means in the Kidepo highlands of Uganda. They seem literally to have abandoned their humanity, in their abject despair over having been evicted from their homeland and forced to live in sedentary villages, so that when in Kidepo did President Milton "Apollo" Abote a stately national park decree, it might measure up to the American ideal of a wilderness park, a place "where man is a visitor who does not remain."

Harmon's critique of the wilderness idea from a mostly fourth-world perspective—that is, from the perspective not of the "progressive" elite in developing countries, but of disempowered traditional tribal groups whose way of life is threatened by "progress" and "development" in the third world—went mostly unnoticed and unanswered. In 1989, "Radical American Environmentalism and Wilderness Preservation: A Third World Critique," by Ramachandra Guha, an Indian sociologist, appeared in *Environmental Ethics.* Perhaps because the words *wilderness, critique,* and *third world* were right in its title, and Guha's name was recognizably non-Western, his article attracted the attention of the community of Western environmental philosophers, most of whom innocently thought of wilderness as nature's sanctum sanctorum. Two proponents of the wilderness idea—the distinguished Norwegian philosopher Arne Naess, founder of the Deep Ecology school of thought, and political scientist David Johns, a wilderness activist—replied in subsequent issues of the same journal. Thus did the great new wilderness debate commence. This anthology brings together in one volume the most notable contributions to this debate. Because a debate is by nature dialectical—proceeding by point and counterpoint, thesis and antithesis—immediately following Guha's third-world critique of the wilderness idea we have included the rejoinder by Johns, "The Relevance of Deep Ecology to the Third World." Following that we have placed "Deep Ecology Revisited," Guha's response to Johns and other apologists for Deep Ecology's embrace of the received wilderness idea. Coming last in this Third World–Deep Ecology exchange, Arne Naess's "The Third World, Wilderness, and Deep Ecology" strives to integrate—in the concept of "free nature"—Guha's evident concern for peoples subsisting by traditional means with the concern of Deep Ecologists for nonhuman species.

Also representing third- and fourth-world perspectives on the wilderness idea are the renowned Mexican ethnobotanist Arturo Gómez-Pompa and his collaborator Andrea Kaus, a Euro-American anthropologist. Their aggressively titled scientific critique of the wilderness idea, "Taming the Wilderness Myth" (originally published in *BioScience*), is written from a Latin American point of view. In addition, we (the editors) feel very fortunate indeed to have found and to be able to reprint Fabienne Bayet's article "Overturning the Doctrine: Indigenous People and Wilderness—Being Aboriginal in the Environmental Movement." Bayet, as she explains in her essay, is an Australian Aboriginal woman of color *and* a committed environmentalist, for whom criticizing the wilderness idea is—as it is for us the editors of this anthology, who are also committed environmentalists—a painful thing to have to do. Her poignant piece records her struggle to reconcile conflicting loyalties.

Bayet's essay exposes a more sinister aspect of the received wilderness idea. The European conquest and settlement of Australia occurred more recently and in a less desultory way than did the European conquest and settlement of the Americas. To think of Australia before European settlement as a wilderness of continental proportions—as a *terra nullius* (an empty land) in the jargon of Anglo-Australian jurisprudence—made the dispossession and extermination of its Aboriginal human inhabitants morally more palatable. The wilderness idea, in effect, erased those inhabitants from Western consciousness—and thus from conscience. While the wilderness idea served colonial Anglo-Australians by concealing from themselves and from the rest of the Western world their systematic policy of genocide, it now serves contemporary postcolonial Australian environmentalists, no less than it does postcolonial American environmentalists, as a means of checking industrial environmental rapine. This then is Bayet's dilemma. The wilderness idea is unhistorical. Australia was not a wilderness, a *terra nullius,* before British conquest and settlement. It was, as Bayet notes, fully settled and actively managed by its Aboriginal inhabitants. But designated wilderness areas are now a vital element in Australian nature conservation, just as they are in the United States. To deconstruct the wilderness idea therefore risks undermining nature conservation. Part II ends with an essay, "The Wilderness Narrative and the Cultural Logic of Capitalism," by British environmental philosopher Carl Talbot, written from a Marxist point of view and published here for the first time. Talbot provides the critical antidote to Nash's enthusiastic endorsement of an international trade in the wilderness experience, thus bringing Part II full circle.

Environmental philosopher J. Baird Callicott (one of the editors of this anthology) was as surprised and troubled as any other Western environmental philosopher by Guha's third-world critique of the received wilderness idea. His response was different, however, from that of Johns and Naess. Rather than chiming in with other apologists, Callicott attempted to widen and deepen Guha's critique of the concept of wilderness. In "The Wilderness Idea Revisited: The Sustainable Development Alternative," he points out that in addition to the untoward social consequences of exporting the wilderness idea to (in Guha's words) "long-settled, densely populated" regions of the world, the received wilderness idea might actually be conceptually incoherent. First, and most generally, it perpetuates the pre-Darwinian separation of "man" from nature, while from an evolutionary point of view, Homo sapiens is a part of nature. Second, in serving its colonial purpose of erasing from mind indigenous peoples, whose existence, if acknowledged and honestly confronted, might morally impede the march of empire, it also blinds those it enthralls to the considerable impact of such peoples on the biotic communities that they inhabit(ed). Only Antarctica would qualify as a wilderness area of continental proportion, according to the definition of the Wilderness Act. Most of North and South America and Australia certainly would not, as these areas were thoroughly inhabited by indigenous peoples—Australia for more than 40,000 years, the Americas for more than 11,000—who were, of course, not visitors in their own homelands. Moreover, the works of these peoples did (and often still do) dominate the landscape ecologically, though not in the same way, or as evidently to the untutored eye, as do the works of industrial Homo sapiens.

Callicott's paper was originally published in a journal called *The Environmental Professional* and was followed in the next issue of the same volume (1991) by environmental philosopher Holmes Rolston III's rejoinder, "The Wilderness Idea Reaffirmed," and by a brief response to Rolston by Callicott. Rolston reaffirms the separation of Homo sapiens from nature, but not on such fanciful traditional grounds as the biblical doctrine that "man" is unique because "he" is created in the image of God or the classical philosophical doctrine that "man" is unique because "he" is uniquely rational. Rather, Rolston argues, Homo sapiens uniquely possesses culture, a means of adapting to the environment (and adapting the environment to the species) so disproportionate to that of other species that Homo sapiens has literally transcended nature. However, pre-Columbian Native Americans and Australian Aboriginals before the advent of James Cook, while fully cultural Homo sapiens, had such ineffectual, largely Stone Age cultures, Rolston believes, that they little impacted their environments—which remained, therefore, largely "untrammeled," as per the definition of *wilderness* in the Wilderness Act, not "areas where man and his own works dominate the landscape." Wilderness preservation, Rolston therefore believes, is a laudable and eminently coherent effort to prevent the sphere of the natural from being wholly reduced to the cultural. To all of which Callicott replies in "That Good Old-Time Wilderness Religion" that Rolston is simply reasserting the old Puritan dichotomous distinctions between "man" and nature and civilization and savagery in more acceptable secular terminology. The Callicott-Rolston-Callicott exchange opens Part III of this anthology.

In keeping with the dialectical organization of this anthology, Callicott's reply to Rolston is followed by two essays written in defense of the wilderness idea. The first is by Dave Foreman, one of the most daring contemporary captains of that army putatively commanded by Aldo Leopold, to which Rolston alludes, fighting the war for wilderness preservation declared by Robert Marshall. For a Euro-American environmental activist like Dave Foreman, what a ruminating academic philosopher such as Callicott dares to utter in polite discourse addressed to his peers, should it fail to be confined to the ivory tower, can have dire political consequences.

Foreman essays to refute Callicott's critique of the wilderness idea point for point, but his evident irritation, directed at Callicott personally, stems from his concern that the currently fashionable academic deconstruction of the wilderness *idea* can be abused by those implacable foes of wilderness *preservation* whose dissembling and unreasoned voices we began this introduction by excluding from the great new wilderness debate. The second is by leading Euro-American conservation biologist Reed Noss. The subtitle of Callicott's "The Wilderness Idea Revisited" is "The Sustainable Development Alternative." In "Sustainability and Wilderness," Noss resists what he perceives as a shift from what he believes to be the socially more demanding, but more effective, wilderness preservation paradigm to what he believes to be the socially more agreeable, but less effective, sustainable development paradigm in conservation policy. Noss identifies four values of designated and de facto wilderness areas (assuming that any such actually exist): their scientific value (by which he means a scientific "control" or "base datum of normality," against which the ecological performance of humanly inhabited and exploited areas can be measured); their biological value (by which he means habitat for species, especially the large predators, which do not coexist well with industrialized Homo sapiens); their value as a source of humility; and their intrinsic value.

One of the principal bones of contention between Callicott and Rolston is the extent and intensity of the environmental transformation of the Americas effected by Native Americans. This is an empirical question, which neither Callicott nor Rolston is qualified to answer with authority. Cultural geographer William Denevan is, however, eminently qualified to do so. In "The Pristine Myth: The Landscape of the Americas in 1492," originally published in "The Americas Before and After 1492: Current Geographical Research," a special issue of the *Annals of the Association of American Geographers,* observing the quincentennial of Columbus's first transatlantic voyage, Denevan reviews the evidence supporting his contention that the pre-Columbian New World was a humanized landscape almost everywhere. But to this argument, Denevan adds a novel twist. Old World diseases—which originated with the domestication of animals in Europe, Asia, and Africa—inversely decimated (to decimate literally means "to select by lot and kill one in ten," from the Latin *decimus,* tenth) Native American populations. With the exception of dogs, pre-Columbian Native Americans associated with no domesticated animals, and hence had evolved no resistances to those diseases, such as smallpox, which leapt from livestock to their human masters in the Old World. Only one Native American in ten, or, perhaps, one in twenty, survived the cycle of pathogen pandemics that swept through the Americas during the century after contact. Thus between 1492 and 1607, when the first permanent English settlement on the Atlantic coast was established, the Americas had been depopulated—the diseases, once arrived, having been communicated from Native to Native. Consequently, the landscape had begun to recover something of "its primeval character and influence" (in the words of the Wilderness Act). So, if Denevan's interpretation of the evidence is correct, the Puritans did find a wilderness condition, after all, in the New World, but it was—ironically, even oxymoronically—an artificial wilderness condition, a condition that was produced, albeit inadvertently and indirectly, by human agency.

"The Incarceration of Wildness: Wilderness Areas as Prisons," by Euro-American environmental philosopher Thomas Birch, was originally published in *Environmental Ethics* in 1990, the year after Guha's third-world critique and the year before the Callicott-Rolston-Callicott exchange in the *Environmental Professional.* Birch expresses discomfort with the wilderness idea in a North American context, its home turf. Adapting a style of analysis pioneered by French philosopher Michel Foucault, Birch articulates an image of the designated wilderness areas in the national forests and parks of the United States and Canada as being like prisons or mental institutions, places in which the nonhuman "Other"—the wild and untamable forms and forces of nature—can be isolated from the polite (from the Greek *polis,* city) "imperium," confined—and thus after a fashion, controlled and mastered. Birch nevertheless stops short of calling for a repeal of the Wilderness Act, though such a recommendation would seem to follow if one traces the practical implication of his argument to its logical conclusion.

William Cronon's widely read 1983 book, *Changes in the Land,* was instrumental in precipitating the great new wilderness debate. For in that book Cronon, a Euro-American environmental historian, detailed the manner in which the New England landscape had been humanly inhabited, exploited, and transformed by Native American peoples as a prelude to his description of how the English colonists differently inhabited, exploited, and transformed it. While Cronon's *Changes in the Land* is regularly cited by the current critics of the concept of wilderness, the author himself was slow to directly criticize the wilderness idea. When he got around to doing so, he seemed unaware of the raging academic brouhaha for which *Changes in the Land* was partly responsible. "The Trouble with Wilderness" was originally published as the lead essay in a 1995 book, *Uncommon Ground,* edited by Cronon, that was one of several forthcoming from a series of seminars entitled Reinventing Nature, organized and sponsored by the University of California's Humanities Research Institute. This essay is a fitting climax to the general critique of the wilderness idea in Part III of this book because it is so sweeping. It begins by summarizing the intellectual history of the wilderness idea set forth in a more leisurely way by Roderick Nash in *Wilderness and the American Mind* thirty years ago (though Cronon cites Nash's classic only once and only in passing). From there it moves on to recapitulate, on behalf of Native Americans, the third- and fourth-world critique of the received wilderness idea offered by Guha (on behalf of rural Indians) and Gómez-Pompa and Kaus (on behalf of Central American peasants), and then to make the points here severally made by Talbot, Callicott, Birch, and others (all without citing these path-breaking authors, as if Cronon were articulating these thoughts for the first time). In short, Cronon's essay, though largely unoriginal, is a forcefully written summary and crystallization of the case against the received wilderness idea made piecemeal by the authors of the foregoing essays in Parts II and III of this anthology. With the publication of a condensed version of "The Trouble with Wilderness" in the *New York Times Sunday Magazine* on August 13, 1995, the great new wilderness debate finally burst out of the ivory tower and came to the attention of the general public. In keeping with the dialectical spirit of the debate format of this anthology and in scrupulous observance of the title of Part III, The Wilderness Idea Roundly Criticized *and* Defended, we close this section with a quiet but elegant defense of the received wilderness idea by Euro-American philosopher Marvin Henberg.

We (the editors) believe that the received wilderness idea has been mortally wounded by the withering critique to which it has been lately subjected. Even its most indignant and impassioned apologist, Dave Foreman, seems now to have capitulated, as a side-by-side comparison of his two contributions to this anthology will bear witness. The first, "Wilderness Areas for

Real"—his implacable response to Callicott's "The Wilderness Idea Revisited" in Part III—categorically defends the received wilderness idea and the classic nineteenth- and twentieth-century wilderness preservation movement associated with it. The second, "Wilderness: From Scenery to Nature" in Part IV, concedes that the historic wilderness preservation movement, though well intentioned, was, from the point of view of biological conservation, misguided. Nevertheless, however flawed, the wilderness idea has been indispensable to the twentieth-century nature conservation and environmental movements. Its reluctant critics cannot, in good conscience, just turn their attention to some other enticing intellectual puzzle and leave nature more vulnerable to exploitation than ever, since the wilderness idea has, by all accounts, been the most powerful antidote to such exploitation in the environmentalists' cognitive arsenal. As we enter the twenty-first century, we must carefully gather up and embrace the proverbial baby before we throw out the proverbial bath water.

We see two alternatives to the received wilderness idea currently taking shape. One alternative would deanthropocentrize the classic wilderness idea; the other would replace the received wilderness idea with the obviously related, but very different, concept of wildness and the concepts of free nature, sustainability, and reinhabitation that are allied with it.

As the seminal items in Part I amply indicate, wilderness was classically conceived to be a resource for human use—for nonconsumptive human use, to be sure, but for human use nevertheless: for human recreation, aesthetic gratification, spiritual communion, character building, scientific study, and so on. While John Muir and certainly Aldo Leopold, in writings other than those reprinted here, adumbrated a less anthropocentric point of view, it was not until the emergence of academic environmental philosophy in the 1970s that a fully and self-consciously nonanthropocentric environmental ethic was articulated, at least not in the conversation of the West. Thus, at last, we can dare to think about and argue for the preservation of areas of the Earth for the primary (if not sole) use and enjoyment of nonhuman beings. In terminology now standard in conservation biology, we can at last readily conceive of biodiversity reserves.

Wilderness areas, though originally set aside for purposes of virile recreation, scenery, and solitude, now have a new, nonanthropocentric raison d'être. They are habitat for intrinsically valuable rare and endangered species, especially such large carnivores as the brown bear and gray wolf—

which have been and continue to be persecuted by modern Homo sapiens. By this trope wilderness areas become not the playgrounds of wilderness recreationalists, the art galleries of natural esthetes, and the cathedrals of solitude seekers, but refugia for nonhuman forms of life. Conservation biology, the science of biological scarcity and diversity, should guide the selection, design, and management of these refugia.

The old system of wilderness areas in the United States and elsewhere represents only a point of departure, a cornerstone, for a new system of biodiversity reserves. As Dave Foreman here notes, under the present system, designated wilderness areas were not selected for preservation because they were either particularly rich in species or because they were the preferred habitat of threatened species. They were selected because they appeared to be untrammeled, had little foreseeable commercial value, and contained monumental scenery or opportunities for a primitive and unconfined type of recreation. Nor were their boundaries drawn with the habitat requirements of threatened species in mind.

Reconceiving wilderness areas as biodiversity reserves is a forward and proactive, not a backward and defensive step for the neo-Puritan cause of nature preservation. It provides a scientific mandate for expanding, not shrinking, existing wilderness set-asides and for connecting them with wild corridors. Moreover, it provides a scientific mandate for conferring biodiversity-reserve status on many other habitats—lowland forests, level grasslands, deserts, and wetlands, especially—that were despised by twentieth-century wilderness preservationists because they were not rugged enough to present a challenge to hikers and climbers or because they were not sufficiently grand and picturesque. But they too are biologically rich and diverse; and they too harbor endangered species.

The other alternative to the received wilderness idea is less obvious, less well-defined, and less easily identified. Arne Naess calls it "free nature"; Baird Callicott calls it "sustainability"; Euro-American nature poet and bioregionalist Gary Snyder calls it "reinhabitation." As Naess documents in his contribution to Part II of this anthology, Snyder frequently reminds us that the places we like to think of as humanly uninhabited before Europeans "discovered" them did indeed have their human denizens. Even the most forbidding places had names and were traversed by trails.

The biodiversity-reserve reconstruction of the thoroughly deconstructed received wilderness idea is essentially neo-Puritan because it segregates people from nature—not, however, on the basis of religious metaphysics (the image of God and original sin), philosophical metaphysics (rationality), or even their contemporary scientific successor (non-natural culture), but on the basis of necessity. If we are to preserve threatened species we have to provide them with habitat. That implies that we exclude incompatible human inhabitation and use (including, *nota bene,* recreational use)—which, as things now stand, is most human residence and use—from their habitat. So, reconceiving wilderness areas as biodiversity reserves effectively partitions the ecologically degraded human sphere from the remnant and recovering natural sphere. Substituting the concept of wildness for wilderness, we can envision (re)inhabiting nature symbiotically. In contrast, the basic free nature/sustainability/reinhabitation idea does not deanthropocentrize the classic preservation approach to conservation, but tries to maintain or reestablish, as the case may be, a human harmony with nature, a mutually beneficial relationship between Homo sapiens and the ecosystems human beings inhabit. The biodiversity reserve alternative perpetuates, indeed even exaggerates, the dualistic separation of people from nature implicit in the classic wilderness idea. The free nature/sustainability/reinhabitation alternative, to the contrary, rests on the premise that people are a part of nature. Some peoples still live sustainably and symbiotically with their nonhuman neighbors. But if Homo sapiens is a part of nature, all peoples can, in principle, either rediscover or reinvent a way of living sustainably and symbiotically with their nonhuman neighbors.

We begin Part IV with two short essays by Aldo Leopold, "Threatened Species" and "Wilderness." That would seem to be anachronistic. Leopold was an architect of the received wilderness idea and, to the very end of his life, he was a vocal partisan of classic wilderness preservation. But Leopold was, as is often said of him, a prophet, a person who thought far ahead of his time and foresaw the shape of things to come. In "Threatened Species," Leopold advocates setting aside habitat for those species threatened by human encroachment. And in "Wilderness," Leopold attends less to wilderness than to the potential for wildness in the middle landscape, as it is sometimes called, of North America—the rural landscape between densely settled urban areas and the largely unsettled designated and de facto wilderness areas.

In the next two items in Part IV, "Wilderness Recovery: Thinking Big in Restoration Ecology" and "Getting Back to the Right Nature: A Reply to Cronon's 'The Trouble with Wilderness,'" contemporary Euro-American conservation biologists Reed Noss and Donald Waller, respectively, advocate the establishment of extensive biodiversity reserves—but, somewhat confusingly, in the name of wilderness preservation. Indeed, Waller expressly argues that there is perfect continuity between the old wilderness idea and the new biodiversity reserve idea—despite the fact that, by his own account, the former serves primarily anthropocentric and the latter primarily nonanthropocentric values. For Waller, it seems, the operative distinction is not that between anthropocentric and nonanthropocentric values, but that between consumptive and nonconsumptive values (the latter including both the intrinsic value of biodiversity *and* its instrumental value as, say, a photo opportunity). Noss's essay originally appeared in the same issue of the *Environmental Professional* as Callicott's "The Wilderness Idea Revisited," a special issue on ecological restoration. Waller's essay is published here for the first time. In "Wilderness: From Scenery to Nature" and "Should Wilderness Areas Become Biodiversity Reserves?," Dave Foreman and J. Baird Callicott, respectively, detail the ways in which the new concept of biodiversity reserves differs from the received wilderness idea. Like Noss and Waller, Foreman regards the provision of habitat for threatened species as a new and potent rationale for the same old thing—"wilderness preservation"—in contemporary conservation policy, while Callicott believes that the biodiversity reserve concept and its rationale are sufficiently different, though evolved out of and continuous with the classic concept of wilderness, to warrant a different name. Whatever the name, the main idea is not to preserve, in the famous phrase of the Leopold Report, "vignettes of primitive America," in order to entertain, edify, or inspire human *visitors,* but to provide living space for species threatened by residential, commercial, and industrial development. In "Using Biodiversity as a Justification for Nature Protection in the US," R. Edward Grumbine traces the history of thought about the preservation of biodiversity back to the early twentieth century (though, of course, the term *biodiversity* itself has only recently been coined). Unfortunately, the voices of those visionaries whom Grumbine identifies—Joseph Grinnell and Victor Shelford are notable among them—who conceived of wilderness as biodiversity reserves were drowned out by those who mainly conceived of it as scenic areas principally dedicated to virile recreation.

The Puritan roots of the received wilderness idea are the source of some of its biggest present problems. Calvinist theology sharply divides humanity per se from nature. Hence wilderness areas were defined not in contrast to domesticated or civilized regions of the Earth, but in contrast to human inhabitation and human influence in general. Had wilderness been defined not in contrast to areas where "man and his own works dominate the landscape," and especially not in contrast to humanly inhabited areas—such that wilderness is, by definition, an area "where man himself is a visitor who does not remain"—but in contrast to cities and their pastoral-agrarian hinterlands, then there might be no great wilderness debate going on right now. At least several main problems with the received wilderness idea would have been obviated.

First, in 1492, as noted, the only continent measuring up to the definition of wilderness in the Wilderness Act was Antarctica. The Americas were humanly inhabited from the Bering Strait to the Strait of Magellan and from San Francisco Bay to Guanabara Bay; and they were, overall, radically transformed by their human inhabitants. Much of their most magnificent fauna—horses, camels, elephants, for example—was exterminated by the original discoverers of the New World, ten thousand or more years before Columbus stumbled on it. The European latecomers hardly found a "virgin" hemisphere. The pre-Columbian flora, moreover, was modified by anthropogenic fires; and those animal species populations—bison and deer, for example—not reduced to extinction by the immigrant Siberian big-game hunters and their immediate descendants were affected by the anthropogenic modification of the flora. If wilderness had been defined in contrast to civilization, not in contrast to human inhabitation and impact, then all of Australia and vast parts of the Americas—central Mexico, parts of the Andes, and the central Mississippi Valley are among the exceptions—would incontestably have been in a wilderness condition upon discovery by civilized Europeans. So there would be no plausibility to the claim that the alleged wilderness condition of the Americas and Australia is a "myth" made up by European colonists in order historically to "erase" indigenous peoples and assuage any residual guilt the colonizers might have felt about slaughtering the majority and dispossessing the rest.

Second, were wilderness areas defined in contrast to civilization, not in contrast to human inhabitation and impact, then, in establishing wilderness reserves in Africa and India, it would not have been necessary to forcibly remove the human residents living there—provided such residents were subsisting sustainably in what Arne Naess calls free nature by means of foraging, horticulture, or some combination of the two. The international socioeconomic problem—so forcefully stated by Guha and several other authors in Part II of this anthology—with the wilderness idea might have been obviated, in other words, had a wilderness condition, all along, been understood to mean the absence of cities, surplus agriculture, and domesticated livestock, not the absence of people per se and their sustainable subsistence economies.

Third, the more abstract, philosophical problem with the received wilderness idea—that it perpetuates the pre-Darwinian metaphysical separation of man from nature—would, of course, have been obviated were wilderness defined in contrast to civilization, not in contrast to human inhabitation and use.

"In Wildness Is the Preservation of the World," Jack Turner, a Euro-American who holds a Ph.D. in philosophy and works as a backcountry guide, begins, in the context of this anthology, to develop the free nature/sustainability/reinhabitation alternative. Turner points out that Thoreau's famous dictum is often misquoted as "In wild*er*ness is the preservation of the world." Echoing Birch, Turner claims that formally designating areas as wilderness often, ironically, tames the wildness in them. To reclaim the wildness in wilderness and in ourselves, we must, Turner believes, live and work in the wild world. How to do so without destroying that wildness, is, however, a big problem; and though he broaches it, Turner does not directly essay to solve it. One approach to solving that problem is to ask how those indigenous peoples who live(d) and work(ed) in the wild world, without destroying its wildness, manage(d) to do so. In "Cultural Parallax in Viewing North American Habitats," Euro-American ethnobotanist Gary Nabhan explores the way his neighbors in southern Arizona, the O'odham (formerly called Papago), managed to do so. For contemporary Euro-Americans or Anglo-Australians to go native, as it were, is, of course, impossible; it is even difficult, as Nabhan points out, for contemporary Native peoples to sustain their own adaptive cultures. But from study of the way longtime residents in a place have symbiotically adapted to it, the general principles of reinhabitation may be learned and applied in fresh and creative ways. That is just what Gary Snyder's essay, "The Rediscovery of Turtle Island," is all about. Snyder provides examples, from his own extensive experience, of how to go about gently reinhabiting free nature in a manner mindful of how the earlier inhabitants did so, but in a manner that is thoroughly autochthonous.

This anthology ends with a comprehensive philosophical reflection on the great new wilderness debate by Anglo-Australian ecofeminist Val Plumwood. In "Wilderness Skepticism and Wilderness Dualism," published here for the first time, Plumwood exposes the androcentrism of the received wilderness idea as well as its ethnocentrism. The received wilderness idea—of wilderness as virgin, unsullied territory—expresses, she suggests, an essentially male point of view, as well as an essentially colonial point of view. After detailing the slight differences in the postcolonial Anglo-Australian and Euro-American wilderness movements, Plumwood takes up the deeper conundrum presented by the concept of wilderness, identifying three views about the culture-nature relationship, all of which she believes to be faulty. First, in Holmes Rolston's view, the acquisition of culture divorced Homo sapiens from nature. Human culture is biologically revolutionary, as Rolston sees it, providing Homo sapiens with a means of adaptation many orders of magnitude more rapid than adaptation through genetic mutation and natural selection, to which all other species are limited. Rolston is a modern classic dualist on the nature-culture question; human beings transcend nature. Second, in Baird Callicott's view, human culture is not unique; many other species transmit cultural information as well as genetic information from generation to generation. The difference between Homo sapiens and other species is, in this regard, a matter of degree, not of kind. In Plumwood's opinion, Callicott reduces culture to nature. Third, William Cronon seems to adopt the poststructuralist view that "nature" is a cultural construct, varying across history, gender, and society. In Plumwood's opinion, his reduction runs in a direction opposite to Callicott's: Cronon reduces nature to culture.

Plumwood's resolution of this triangular nature-culture affair is subtle, but it seems that she is saying that the classic modern dualistic view of the nature-culture relationship, so forthrightly represented by Rolston, is not correct, but neither are Callicott's reduction of culture to nature nor Cronon's reduction of nature to culture good ways to solve the problem. Both terms of the old nature-culture dichotomy need to be maintained, but not opposed. If one were to try to put their point graphically and succinctly, one might say that nature and culture can be united as the yin and yang. They are opposites, yet not opposed. They are two, yet together form one whole, neither complete without the other. Nature and culture—like male and female or self and other—are, in a word, complementary.

Plumwood thus provides analytic depth to Callicott's sketch, in "Should Wilderness Areas Become Biodiversity Reserves?," of a complementary twenty-first-century philosophy of conservation that transcends the dichotomous twentieth-century philosophies of conservation classically articulated by Gifford Pinchot and John Muir—wise use of natural resources versus wilderness preservation. We have long known that utilitarian resource management, as envisioned by Gifford Pinchot and his successors, is flawed—because it ignores the relationship of resources to nonresources; that is, because it is ecologically uninformed. We have just found out that wilderness preservation, as envisioned by Muir and his successors, is equally flawed—for all the reasons elaborated in this volume. Conservation philosophy is presently, therefore, in a state of doubt and confusion. But out of such states can arise something new and more refined. And what might that be? We (the editors) envision, first, the creation of a global system of scientifically selected, designed, managed, and interconnected biodiversity reserves; complemented, second, by support for those traditional peoples who are doing so to continue living symbiotically with their nonhuman neighbors in free nature, and also support for people like Turner and Snyder who are attempting to harmoniously reinhabit free nature; and, third and finally, a commitment on the part of everyone else to the development of ecologically sustainable economies. This represents, of course, a utopian prospect—an ideal. Utopian thinking has been a mainstay of Western philosophy from Plato on. In the context of human ethics, Aristotle, Plato's successor in the Western tradition, observed that we are more likely to go right if we have a target at which to aim. The same seems true to us in our effort to get beyond the received wilderness idea.

PART ONE

The Received Wilderness Idea

SELECTIONS FROM Jonathan Edwards

The Images or Shadows of Divine Things (1758)

THE BEAUTY OF THE WORLD

THE BEAUTY OF THE WORLD consists wholly of sweet mutual consents, either within itself or with the supreme being. As to the corporeal world, though there are many other sorts of consents, yet the sweetest and most charming beauty of it is its resemblance of spiritual beauties. The reason is that spiritual beauties are infinitely the greatest, and bodies being but the shadows of being, they must be so much the more charming as they shadow forth spiritual beauties. This beauty is peculiar to natural things, it surpassing the art of man.

Thus there is the resemblance of a decent trust, dependence and acknowledgment in the planets continually moving around the sun, receiving his influences by which they are made happy, bright and beautiful: a decent attendance in the secondary planets, an image of majesty, power, and glory, and beneficence in the sun in the midst of all, and so in terrestrial things, as I have shown in another place.

It is very probable that the wonderful suitableness of green for the grass and plants, the blue of the skie, the white of the clouds, the colours of flowers, consists in a complicated proportion that these colours make one with another, either in their magnitude of the rays, the number of vibrations that are caused in the atmosphere, or some other way. So there is a great suitableness between the objects of different senses, as between sounds, colours, and smells; as between colours of the woods and flowers and the smells and the singing of birds, which it is probable consist in a certain proportion of the vibrations that are made in the different organs. So there are innumerable other agreeablenesses of motions, figures, etc. The gentle motions of waves, of the lily, etc., as it is agreeable to other things that represent calmness, gentleness and benevolence, etc., the fields and woods seem to rejoice, and how joyfull do the birds seem to be in it. How much a resemblance is there of every grace in the field covered with plants and flowers when the sun shines serenely and undisturbedly upon them, how a resemblance, I say, of every grace and beautiful disposition of mind, of an inferiour towards a superiour cause, preserver, benevolent benefactor, and a fountain of happiness.

How great a resemblance of a holy and virtuous soul is a calm, serene day. What an infinite number of such like beauties is there in that one thing, the light, and how complicated an harmony and proportion it is probable belongs to it.

There are beauties that are more palpable and explicable, and there are hidden and secret beauties. The former pleases, and we can tell why; we can explain the particular point for the agreement that renders the thing pleasing. Such are all artificial regularities; we can tell wherein the regularity lies that affects us. [The] latter sort are those beauties that delight us and we cannot tell why. Thus, we find ourselves pleased in beholding the colour of the violets, but we know not what secret regularity or harmony it is that creates that pleasure in our minds. These hidden beauties are commonly by far the greatest, because the more complex a beauty is, the more hidden is it. In this latter fact consists principally the beauty of the world, and very much in light and colours. Thus mere light is pleasing to the mind. If it be to the degree of effulgence, it is very sensible, and mankind have agreed in it: they all represent glory and extraordinary beauty by brightness. The reason of it is either that light or our organ of seeing is so contrived that an harmonious motion is excited in the animal spirits and propagated to the brain. That mixture we call white is a proportionate mixture that is harmonious, as Sir Isaac Newton has shown, to each particular simple colour, and contains in some harmony or other that is delightfull. And each sort of rays play a distinct tune to the soul, besides those lovely mixtures that are found in nature. Those beauties, how lovely is the green of the face of the earth in all manner of colours, in flowers, the colour of the skies, and lovely tinctures of the morning and evening.

Corollary: Hence the reason why almost all men, and those that seem to be very miserable, love life, because they cannot bear to lose sight of such a beautiful and lovely world. The ideas, that every moment whilst we live have a beauty that we take no distinct notice of, brings a pleasure that, when we come to the trial, we had rather live in such pain and misery than lose.

Christian Doctrine of Original Sin Defended

All mankind constantly, in all Ages, without Fail in any one Instance, run into that moral Evil, which is in effect their own utter and eternal Perdition in a total privation of GOD's Favour, and suffering of his Vengeance and Wrath.

Sinners in the Hands of an Angry God

THE USE OF THIS AWFUL SUBJECT may be for awakening unconverted persons in this congregation. This that you have heard is the case of every one of you that are out of Christ.—That world of misery, that lake of burning brimstone, is extended abroad under you. There is the dreadful pit of the glowing flames of the wrath of God; there is hell's wide gaping mouth open; and you have nothing to stand upon, nor any thing to take hold of; there is nothing between you and hell but the air; it is only the power and mere pleasure of God that holds you up.

You probably are not sensible of this; you find you are kept out of hell, but do not see the hand of God in it; but look at other things, as the good state of your bodily constitution, your care of your own life, and the means you use for your own preservation. But indeed these things are nothing; if God should withdraw his hand, they would avail no more to keep you from falling, than the thin air to hold up a person that is suspended in it.

Your wickedness makes you as it were heavy as lead, and to tend downwards with great weight and pressure towards hell; and if God should let you go, you would immediately sink and swiftly descend and plunge into the bottomless gulf, and your healthy constitution, and your own care and prudence, and best contrivance, and all your righteousness, would have no more influence to uphold you and keep you out of hell, than a spider's web would have to stop a fallen rock. Were it not for the sovereign pleasure of God, the earth would not bear you one moment; for you are a burden to it; the creation groans with you; the creature is made subject to the bondage of your corruption, not willingly; the sun does not willingly shine upon you to give you light to serve sin and Satan; the earth does not willingly yield her increase to satisfy your lusts; nor is it willingly a stage for your wickedness to be acted upon; the air does not willingly serve you for breath to maintain the flame of life in your vitals, while you spend your life in the service of God's enemies. God's creatures are good, and were made for men to serve God with, and do not willingly subserve to any other purpose, and groan when they are abused to purposes so directly contrary to their nature and end. And the world would spew you out, were it not for the sovereign hand of him who hath subjected it in hope. There are black clouds of God's wrath now hanging directly over your heads, full of the dreadful storm, and big with thunder; and were it not for the restraining hand of God, it would immediately burst forth upon you. The sovereign pleasure of God, for the present, stays his rough wind; otherwise it would come with fury, and your destruction would come like a whirlwind, and you would be like the chaff of the summer threshing floor. . . .

The God that holds you over the pit of hell, much as one holds a spider, or some loathsome insect over the fire, abhors you, and is dreadfully provoked: his wrath towards you burns like fire; he looks upon you as worthy of nothing else, but to be cast into the fire; he is of purer eyes than to bear to have you in his sight; you are ten thousand times more abominable in his eyes, than the most hateful venomous serpent is in ours. You have offended him infinitely more than ever a stubborn rebel did his prince; and yet it is nothing but his hand that holds you from falling into the fire every moment. It is to be ascribed to nothing else, that you did not go to hell the last night; that you was suffered to awake again in this world, after you closed your eyes to sleep. And there is no other reason to be given, why you have not dropped into hell since you rose in the morning, but that God's hand has held you up. There is no other reason to be given why you have not gone to hell, since you have sat here in the house of God, provoking his pure eyes by your sinful wicked manner of attending his solemn worship. Yea, there is nothing else that is to be given as a reason why you do not this very moment drop down into hell.

O sinner! Consider the fearful danger you are in: it is a great furnace of wrath, a wide and bottomless pit, full of the fire of wrath, that you are held over in the hand of that God, whose wrath is provoked and incensed as much against you, as against many of the damned in hell. You hang by a slender thread, with the flames of divine wrath flashing about it, and ready every moment to singe it, and burn it asunder; and you have no interest in any Mediator, and nothing to lay hold of to save yourself, nothing to keep off the flames of wrath, nothing of your own, nothing that you ever have done, nothing that you can do, to induce God to spare you one moment.

Ralph Waldo Emerson

SELECTIONS FROM *Nature* (1836)

CHAPTER I.

To go into solitude, a man needs to retire as much from his chamber as from society. I am not solitary whilst I read and write, though nobody is with me. But if a man would be alone, let him look at the stars. The rays that come from those heavenly worlds, will separate between him and vulgar things. One might think the atmosphere was made transparent with this design, to give man, in the heavenly bodies, the perpetual presence of the sublime. Seen in the streets of cities, how great they are! If the stars should appear one night in a thousand years, how would men believe and adore; and preserve for many generations the remembrance of the city of God which had been shown! But every night come out these preachers of beauty, and light the universe with their admonishing smile.

The stars awaken a certain reverence, because though always present, they are always inaccessible; but all natural objects make a kindred impression, when the mind is open to their influence. Nature never wears a mean appearance. Neither does the wisest man extort all her secret, and lose his curiosity by finding out all her perfection. Nature never became a toy to a wise spirit. The flowers, the animals, the mountains, reflected all the wisdom of his best hour, as much as they had delighted the simplicity of his childhood.

When we speak of nature in this manner, we have a distinct but most poetical sense in the mind. We mean the integrity of impression made by manifold natural objects. It is this which distinguishes the stick of timber of the wood-cutter, from the tree of the poet. The charming landscape which I saw this morning, is indubitably made up of some twenty or thirty farms. Miller owns this field, Locke that, and Manning the woodland beyond. But none of them owns the landscape. There is a property in the horizon which no man has but he whose eye can integrate all the parts, that is, the poet. This is the best part of these men's farms, yet to this their land-deeds give them no title.

To speak truly, few adult persons can see nature. Most persons do not see the sun. At least they have a very superficial seeing. The sun illuminates only the eye of the man, but shines into the eye and the heart of the child. The lover of nature is he whose inward and outward senses are still truly adjusted to each other; who has retained the spirit of infancy even into the era of manhood. His intercourse with heaven and earth, becomes part of his daily food. In the presence of nature, a wild delight runs through the man, in spite of real sorrows. Nature says,—he is my creature, and maugre all his impertinent griefs, he shall be glad with me. Not the sun or the summer alone, but every hour and season yields its tribute of delight; for every hour and change corresponds to and authorizes a different state of the mind, from breathless noon to grimmest midnight. Nature is a setting that fits equally well a comic or a mourning piece. In good health, the air is a cordial of incredible virtue. Crossing a bare common, in snow puddles, at twilight, under a clouded sky, without having in my thoughts any occurrence of special good fortune, I have enjoyed a perfect exhilaration. Almost I fear to think how glad I am. In the woods too, a man casts off his years, as the snake his slough, and at what period soever of life, is always a child. In the woods, is perpetual youth. Within these plantations of God, a decorum and sanctity reign, a perennial festival is dressed, and the guest sees not how he should tire of them in a thousand years. In the woods, we return to reason and faith. There I feel that nothing can befal me in life,—no disgrace, no calamity, (leaving me my eyes,) which nature cannot repair. Standing on the bare ground,—my head bathed by the blithe air, and uplifted into infinite space,—all mean egotism vanishes. I become a transparent eye-ball. I am nothing. I see all. The currents of the Universal Being circulate through me; I am part or particle of God. The name of the nearest friend sounds then foreign and accidental. To be brothers, to be acquaintances,—master or servant, is then a trifle and a disturbance. I am the lover of uncontained and immortal beauty. In the wilderness, I find something more dear and connate than in streets or villages. In the tranquil landscape, and especially in the distant line of the horizon, man beholds somewhat as beautiful as his own nature.

The greatest delight which the fields and woods minister, is the suggestion of an occult relation between man and the vegetable. I am not alone and unacknowledged. They nod to me and I to them. The waving of the boughs in the storm is new to me and old. It takes me by surprise, and yet is not unknown. Its effect is like that of a higher thought or a better emotion coming over me, when I deemed I was thinking justly or doing right.

Yet it is certain that the power to produce this delight, does not reside in nature, but in man, or in a harmony of both. It is necessary to use these pleasures with great temperance. For, nature is not always tricked in holiday attire, but the same scene which yesterday breathed perfume and glittered as for the frolic of the nymphs, is overspread with melancholy today. Nature always wears the colors of the spirit. To a man laboring under calamity, the heat of his own fire hath sadness in it. Then, there is a kind of contempt of the landscape felt by him who has just lost by death a dear friend. The sky is less grand as it shuts down over less worth in the population.

SELECTIONS FROM Henry David Thoreau

Walking (1862)

I WISH TO SPEAK A WORD for Nature, for absolute freedom and wildness, as contrasted with a freedom and culture merely civil,—to regard man as an inhabitant, or a part and parcel of Nature, rather than a member of society. I wish to make an extreme statement, if so I may make an emphatic one, for there are enough champions of civilization: the minister and the school-committee, and every one of you will take care of that. . . .

Living much out of doors, in the sun and wind, will no doubt produce a certain roughness of character,—will cause a thicker cuticle to grow over some of the finer qualities of our nature, as on the face and hands, or as severe manual labor robs the hands of some of their delicacy of touch. So staying in the house, on the other hand, may produce a softness and smoothness, not to say thinness of skin, accompanied by an increased sensibility to certain impressions. Perhaps we should be more susceptible to some influences important to our intellectual and moral growth, if the sun had shone and the wind blown on us a little less; and no doubt it is a nice matter to proportion rightly the thick and thin skin. But methinks that is a scurf that will fall off fast enough,—that the natural remedy is to be found in the proportion which the night bears to the day, the winter to the summer, thought to experience. There will be so much the more air and sunshine in our thoughts. The callous palms of the laborer are conversant with finer tissues of self-respect and heroism, whose touch thrills the heart, than the languid fingers of idleness. That is mere sentimentality that lies abed by day and thinks itself white, far from the tan and callus of experience....

Nowadays almost all man's improvements, so called, as the building of houses, and the cutting down of the forest and of all large trees, simply deform the landscape, and make it more and more tame and cheap. A people who would begin by burning the fences and let the forest stand! I saw the fences half consumed, their ends lost in the middle of the prairie, and some worldly miser with a surveyor looking after his bounds, while heaven had taken place around him, and he did not see the angels going to and fro, but was looking for an old post-hole in the midst of paradise. I looked again, and saw him standing in the middle of a boggy, stygian fen, surrounded by devils, and he had found his bounds without a doubt three little stones, where a stake had been driven, and looking nearer, I saw that the Prince of Darkness was his surveyor.

I can easily walk ten, fifteen, twenty, any number of miles, commencing at my own door, without going by any house, without crossing a road except where the fox and the mink do: first along by the river, and then the brook, and then the meadow and the wood-side. There are square miles in my vicinity which have no inhabitant. From many a hill I can see civilization and the abodes of man afar. The farmers and their works are scarcely more obvious than woodchucks and their burrows. Man and his affairs, church and state and school, trade and commerce, and manufactures and agriculture, even politics, the most alarming of them all,—I am pleased to see how little space they occupy in the landscape. Politics is but a narrow field, and that still narrower highway yonder leads to it. I sometimes direct the traveller thither. If you would go to the political world, follow the great road,—follow that market-man, keep his dust in your eyes, and it will lead you straight to it; for it, too, has its place merely, and does not occupy all space. I pass from it as from a bean-field into the forest, and it is forgotten. In one half-hour I can walk off to some portion of the earth's surface where a man does not stand from one year's end to another, and there consequently, politics are not, for they are but as the cigar-smoke of a man.

The village is the place to which the roads tend, a sort of expansion of the highway, as a lake of a river. It is the body of which roads are the arms

I know not what there is of joyous and smooth in the aspect of Amer-
ɔlants;" and I think that in this country there are no, or at most very
Africanæ bestiæ, African beasts, as the Romans called them, and that
s respect also it is peculiarly fitted for the habitation of man. We are
hat within three miles of the centre of the East-Indian city of Singa-
some of the inhabitants are annually carried off by tigers; but the trav-
:an lie down in the woods at night almost anywhere in North Amer-
ithout fear of wild beasts.

ese are encouraging testimonies. If the moon looks larger here than
rope, probably the sun looks larger also. If the heavens of America
r infinitely higher, and the stars brighter, I trust that these facts are
olical of the height to which the philosophy and poetry and religion
r inhabitants may one day soar. At length, perchance, the immaterial
n will appear as much higher to the American mind, and the intima-
that star it as much brighter. For I believe that climate does thus react
an,—as there is something in the mountain-air that feeds the spirit
ıspires. Will not man grow to greater perfection intellectually as well
ysically under these influences? Or is it unimportant how many foggy
here are in his life? I trust that we shall be more imaginative, that our
ghts will be clearer, fresher, and more ethereal, as our sky,—our un-
ınding more comprehensive and broader, like our plains,—our intel-
enerally on a grander scale, like our thunder and lightning, our rivers
nountains and forests,—and our hearts shall even correspond in
th and depth and grandeur to our inland seas. Perchance there will
r to the traveller something, he knows not what, of *laeta* and *glabra,*
ous and serene, in our very faces. Else to what end does the world go
ıd why was America discovered?

Americans I hardly need to say,—

"Westward the star of empire takes its way."

:rue patriot, I should be ashamed to think that Adam in paradise was
favorably situated on the whole than the backwoodsman in this
:ry.

r sympathies in Massachusetts are not confined to New England;
gh we may be estranged from the South, we sympathize with the
. There is the home of the younger sons, as among the Scandinavians

and legs,—a trivial or quadrivial place, the thoroughfare and ordinary of travellers. The word is from the Latin *villa,* which, together with *via,* a way, or more anciently *ved* and *vella,* Varro derives from *veho,* to carry, because the villa is the place to and from which things are carried. They who got their living by teaming were said *vellaturam facere.* Hence, too, apparently, the Latin word *vilis* and our vile; also *villain.* This suggests what kind of degeneracy villagers are liable to. They are wayworn by the travel that goes by and over them, without travelling themselves. . . .

At present, in this vicinity, the best part of the land is not private property; the landscape is not owned, and the walker enjoys comparative freedom. But possibly the day will come when it will be partitioned off into so-called pleasure-grounds, in which a few will take a narrow and exclusive pleasure only,—when fences shall be multiplied, and man-traps and other engines invented to confine men to the *public* road, and walking over the surface of God's earth shall be construed to mean trespassing on some gentleman's grounds. To enjoy a thing exclusively is commonly to exclude yourself from the true enjoyment of it. Let us improve our opportunities, then, before the evil days come.

What is it that makes it so hard sometimes to determine whither we will walk? I believe that there is a subtile magnetism in Nature, which, if we unconsciously yield to it, will direct us aright. It is not indifferent to us which way we walk. There is a right way; but we are very liable from heedlessness and stupidity to take the wrong one. We would fain take that walk, never yet taken by us through this actual world, which is perfectly symbolical of the path which we love to travel in the interior and ideal world; and sometimes, no doubt, we find it difficult to choose our direction, because it does not yet exist distinctly in our idea.

When I go out of the house for a walk, uncertain as yet whither I will bend my steps, and submit myself to my instinct to decide for me, I find, strange and whimsical as it may seem, that I finally and inevitably settle southwest, toward some particular wood or meadow or deserted pasture or hill in that direction. My needle is slow to settle,—varies a few degrees, and does not always point due southwest, it is true, and it has good authority for this variation, but it always settled between west and south-southwest. The future lies that way to me, and the earth seems more unexhausted and richer on that side. The outline which would bound my walks would be, not a circle, but a parabola, or rather like one of those cometary orbits which have been thought to be non-returning curves, in this case opening westward, in which my house occupies the place of the sun. I turn round and round irresolute sometimes for a quarter of an hour, until I decide, for a thousandth time, that I will walk into the southwest or west. Eastward I go only by force; but westward I go free. Thither no business leads me. It is hard for me to believe that I shall find fair landscapes or sufficient wildness and freedom behind the eastern horizon. I am not excited by the prospect of a walk thither; but I believe that the forest which I see in the western horizon stretches uninterruptedly toward the setting sun, and there are no towns nor cities in it of enough consequence to disturb me. Let me live where I will, on this side is the city, on that the wilderness, and ever I am leaving the city more and more, and withdrawing into the wilderness. I should not lay so much stress on this fact, if I did not believe that something like this is the prevailing tendency of my countrymen. I must walk toward Oregon, and not toward Europe. And that way the nation is moving, and I may say that mankind progress from east to west. Within a few years we have witnessed the phenomenon of a southeastward migration, in the settlement of Australia; but this affects us as a retrograde movement, and, judging from the moral and physical character of the first generation of Australians, has not yet proved a successful experiment. The eastern Tartars think that there is nothing west beyond Thibet. "The world ends there," say they, "beyond there is nothing but a shoreless sea." It is unmitigated East where they live.

We go eastward to realize history and study the works of art and literature, retracing the steps of the race; we go westward as into the future, with a spirit of enterprise and adventure. The Atlantic is a Lethean stream, in our passage over which we have had an opportunity to forget the Old World and its institutions. If we do not succeed this time, there is perhaps one more chance for the race left before it arrives on the banks of the Styx; and that is in the Lethe of the Pacific, which is three times as wide. . . .

Where on the globe can there be found an area of equal extent with that occupied by the bulk of our States, so fertile and so rich and varied in its productions, and at the same time so habitable by the European, as this is? Michaux, who knew but part of them, says that "the species of large trees are much more numerous in North America than in Europe; in the United States there are more than one hundred and forty species that exceed thirty feet in height; in France there are but thirty that attain this size." Later bot-

anists more than confirm his observations. Humbo
realize his youthful dreams of a tropical vegetatio
greatest perfection in the primitive forests of the A
tic wilderness on the earth, which he has so eloque
ographer Guyot, himself a European, goes farth
ready to follow him; yet not when he says,—"As t
animal, as the vegetable world is made for the an
made for the man of the Old World. . . . The man o
upon his way. Leaving the highlands of Asia, he c
station towards Europe. Each of his steps is mark
superior to the preceding, by a greater power of c
the Atlantic, he pauses on the shore of this unkno
which he knows not, and turns upon his footprin
he has exhausted the rich soil of Europe, and rein
recommences his adventurous career westward as
far Guyot.

From this western impulse coming in contact w
lantic sprang the commerce and enterprise of moc
Michaux, in his "Travels West of the Alleghanie
common inquiry in the newly settled West was, '
world have you come?' As if these vast and fertile
be the place of meeting and common country of
globe."

To use an obsolete Latin word, I might say, *Ex*
FRUX. From the East light; from the West fruit.

Sir Francis Head, an English traveller and a G
ada, tells us that "in both the northern and sout
New World, Nature has not only outlined her wo
has painted the whole picture with brighter and m
used in delineating and in beautifying the Old W
America appear infinitely higher, the sky is bluer,
is intenser, the moon looks larger, the stars are
louder, the lightning is vivider, the wind is strong
mountains are higher, the rivers longer, the f
broader." This statement will do at least to set ag
this part of the world and its productions.

Linnaeus said long ago, "Nescio quae facies *lae*

they took to the sea for their inheritance. It is too late to be studying Hebrew; it is more important to understand even the slang of to-day.

Some months ago I went to see a panorama of the Rhine. It was like a dream of the Middle Ages. I floated down its historic stream in something more than imagination, under bridges built by the Romans, and repaired by later heroes, past cities and castles whose very names were magic to my ears, and each of which was the subject of a legend. There were Ehrenbreitstein and Rolandseck and Coblentz, which I knew only in history. They were ruins that interested me chiefly. There seemed to come up from its waters and its vine-clad hills and valleys a hushed music as of Crusaders departing for the Holy Land. I floated along under the spell of enchantment, as if I had been transported to an heroic age, and breathed an atmosphere of chivalry.

Soon after, I went to see a panorama of the Mississippi, and as I worked my way up the river in the light of to-day, and saw the steamboats wooding up, counted the rising cities, gazed on the fresh ruins of Nauvoo, beheld the Indians moving west across the stream, and, as before I had looked up the Moselle now looked up the Ohio and the Missouri, and heard the legends of Dubuque and of Wenona's Cliff,—still thinking more of the future than of the past or present,—I saw that this was a Rhine stream of a different kind; that the foundations of castles were yet to be laid, and the famous bridges were yet to be thrown over the river; and I felt that *this was the heroic age itself,* though we know it not, for the hero is commonly the simplest and obscurest of men.

The West of which I speak is but another name for the Wild; and what I have been preparing to say is, that in Wildness is the preservation of the World. Every tree sends its fibres forth in search of the Wild. The cities import it at any price. Men plough and sail for it. From the forest and wilderness come the tonics and barks which brace mankind. Our ancestors were savages. The story of Romulus and Remus being suckled by a wolf is not a meaningless fable. The founders of every State which has risen to eminence have drawn their nourishment and vigor from a similar wild source. It was because the children of the Empire were not suckled by the wolf that they were conquered and displaced by the children of the Northern forests who were.

I believe in the forest, and in the meadow, and in the night in which the corn grows. We require an infusion of hemlock-spruce or arborvitæ in our tea. There is a difference between eating and drinking for strength and from mere gluttony. The Hottentots eagerly devour the marrow of the koodoo and other antelopes raw, as a matter of course. Some of our Northern Indians eat raw marrow of the Arctic reindeer, as well as various other parts, including the summits of the antlers, as long as they are soft. And herein, perchance, they have stolen a march on the cooks of Paris. They get what usually goes to feed the fire. This is probably better than stall-fed beef and slaughter-house pork to make a man of. Give me a wildness whose glance no civilization can endure,—as if we lived on the marrow of koodoos devoured raw.

There are some intervals which border the strain of the wood-thrush, to which I would migrate,—wild lands where no settler has squatted; to which, methinks, I am already acclimated.

The African hunter Cummings tells us that the skin of the eland, as well as that of most other antelopes just killed, emits the most delicious perfume of trees and grass. I would have every man so much like a wild antelope, so much a part and parcel of Nature, that his very person should thus sweetly advertise our senses of his presence, and remind us of those parts of Nature which he most haunts. I feel no disposition to be satirical, when the trapper's coat emits the odor of musquash eve; it is a sweeter scent to me than that which commonly exhales from the merchant's or the scholar's garments. When I go into their wardrobes and handle their vestments, I am reminded of no grassy plains and flowery meads which they have frequented, but of dusty merchants' exchanges and libraries rather.

A tanned skin is something more than respectable, and perhaps olive is a fitter color than white for a man,—a denizen of the woods. "The pale white man!" I do not wonder that the African pitied him. Darwin the naturalist says, "A white man bathing by the side of a Tahitian was like a plant bleached by the gardener's art, compared with a fine, dark green one, growing vigorously in the open fields."

Ben Jonson exclaims,—

> "How near to good is what is fair!"

So I would say,—

> How near to good is what is *wild!*

Life consists with wildness. The most alive is the wildest. Not yet subdued to man, its presence refreshes him. One who pressed forward incessantly and never rested from his labors, who grew fast and made infinite demands on life, would always find himself in a new country or wilderness, and surrounded by the raw material of life. He would be climbing over the prostrate stems of primitive forest-trees. . . .

In Literature it is only the wild that attracts us. Dulness is but another name for tameness. It is the uncivilized free and wild thinking in "Hamlet" and "Iliad," in all the Scriptures and Mythologies, not learned in the schools, that delights us. As the wild duck is more swift and beautiful than the tame, so is the wild—the mallard—thought, which 'mid falling dews wings its way above the fens. A truly good book is something as natural, and as unexpectedly and unaccountably fair and perfect, as a wild flower discovered on the prairies of the West or in the jungles of the East. Genius is a light which makes the darkness visible, like the lightning's flash, which perchance shatters the temple of knowledge itself,—and not a taper lighted at the hearth-stone of the race, which pales before the light of common day. . . .

I do not know of any poetry to quote which adequately expresses this yearning for the Wild. Approached from this side, the best poetry is tame. I do not know where to find in any literature, ancient or modern, any account which contents me of that Nature with which even I am acquainted. You will perceive that I demand something which no Augustan nor Elizabethan age, which no *culture,* in short, can give. Mythology comes nearer to it than anything. How much more fertile a Nature, at least, has Grecian mythology its root in than English literature! Mythology is the crop which the Old World bore before its soil was exhausted, before the fancy and imagination were affected with blight; and which it still bears, wherever its pristine vigor is unabated. All other literatures endure only as the elms which overshadow our houses; but this is like the great dragon-tree of the Western Isles, as old as mankind, and, whether that does or not, will endure as long; for the decay of other literatures makes the soil in which it thrives.

The West is preparing to add its fables to those of the East. The valleys of the Ganges, the Nile, and the Rhine, having yielded their crop, it remains to be seen what the valleys of the Amazon, the Platte, the Orinoco, the St. Lawrence, and the Mississippi will produce. Perchance, when, in the course of ages, American liberty has become a fiction of the past,—as it is to some extent a fiction of the present,—the poets of the world will be inspired by American mythology. . . .

In short, all good things are wild and free. There is something in a strain of music, whether produced by an instrument or by the human voice,—take the sound of a bugle in a summer night, for instance,—which by its wildness, to speak without satire, reminds me of the cries emitted by wild beasts in their native forests. It is so much of their wildness as I can understand. Give me for my friends and neighbors wild men, not tame ones. The wildness of the savage is but a faint symbol of the awful ferity with which good men and lovers meet. . . .

Here is this vast, savage, howling mother of ours, Nature, lying all around, with such beauty, and such affection for her children, as the leopard; and yet we are so early weaned from her breast to society, to that culture which is exclusively an interaction of man on man,—a sort of breeding in and in, which produces at most a merely English nobility, a civilization destined to have a speedy limit.

In society, in the best institutions of men, it is easy to detect a certain precocity. When we should still be growing children, we are already little men. Give me a culture which imports much muck from the meadows, and deepens the soil,—not that which trusts to heating manures, and improved implements and modes of culture only! . . .

I would not have every man nor every part of a man cultivated, any more than I would have every acre of earth cultivated: part will be tillage, but the greater part will be meadow and forest, not only serving an immediate use, but preparing a mould against a distant future, by the annual decay of the vegetation which it supports.

There are other letters for the child to learn than those which Cadmus invented. The Spaniards have a good term to express this wild and dusky knowledge,—*Gramatica parda,* tawny grammar,—a kind of mother-wit derived from that same leopard to which I have referred. . . .

For my part, I feel that with regard to Nature I live a sort of border life, on the confines of a world into which I make occasional and transional and transient forays only, and my patriotism and allegiance to the State into whose territories I seem to retreat are those of a mosstrooper. Unto a life which I call natural I would gladly follow even a will-o'-the-wisp through bogs and sloughs unimaginable, but no moon nor fire-fly has shown me the causeway to it. Nature is a personality so vast and universal that we have never seen one of her features. The walker in the familiar fields which stretch around my native town sometimes finds himself in another land than is described in their owners' deeds, as it were in some far-away field on the confines of the actual Concord, where her jurisdiction ceases, and the idea which the word Concord suggests ceases to be suggested. These farms which I have myself surveyed, these bounds which I have set up, appear dimly still as through a mist; but they have no chemistry to fix them; they fade from the surface of the glass; and the picture which the painter painted stands out dimly from beneath. The world with which we are commonly acquainted leaves no trace, and it will have no anniversary. . . .

Huckleberries (1861)

. . . AMONG THE INDIANS, the earth and its productions generally were common and free to all the tribe, like the air and water—but among us who have supplanted the Indians, the public retain only a small yard or common in the middle of the village, with perhaps a graveyard beside it, and the right of way, by sufferance, by a particular narrow route, which is annually becoming narrower, from one such yard to another. I doubt if you can ride out five miles in any direction without coming to where some individual is tolling in the road—and he expects the time when it will all revert to him or his heirs. This is the way we civilized men have arranged it.

I am not overflowing with respect and gratitude to the fathers who thus laid out our New England villages, whatever precedent they were influenced by, for I think that a 'prentice hand liberated from Old English prejudices could have done much better in this new world. If they were in earnest seeking thus far away 'freedom to worship God,' as some assure us—why did they not secure a little more of it, when it was so cheap and they were about it? At the same time that they built meeting-houses why did they not preserve from desecration and destruction far grander temples not made with hands?

What are the natural features which make a township handsome—and worth going far to dwell in? A river with its water-falls—meadows, lakes—hills, cliffs or individual rocks, a forest and single ancient trees—such things are beautiful. They have a high use which dollars and cents never represent. If the inhabitants of a town were wise they would seek to preserve these things though at a considerable expense. For such things educate far more than any hired teachers or preachers, or any at present recognized system of school education.

I do not think him fit to be the founder of a state or even of a town who does not foresee the use of these things, but legislates as it were, for oxen chiefly.

It would be worth the while if in each town there were a committee appointed, to see that the beauty of the town received no detriment. If here is the largest boulder in the country, then it should not belong to an individual nor be made into door-steps. In some countries precious metals belong to the crown—so here more precious objects of great natural beauty should belong to the public.

Let us try to keep the new world new, and while we make a wary use of the city, preserve as far as possible the advantages of living in the country.

I think of no natural feature which is a greater ornament and treasure to this town than the river. It is one of the things which determine whether a man will live here or in another place, and it is one of the first objects which we show to a stranger. In this respect we enjoy a great advantage over those neighboring towns which have no river. Yet the town, as a corporation, has never turned any but the most purely utilitarian eyes upon it—and has done nothing to preserve its natural beauty.

They who laid out the town should have made the river available as a common possession forever. The town collectively should at least have done as much as an individual of taste who owns an equal area commonly does in England. Indeed I think that not only the channel but one or both banks of every river should be a public highway—for a river is not useful merely to float on. In this case, one bank might have been reserved as a public walk and the trees that adorned it have been protected, and frequent avenues have been provided leading to it from the main street. This would have cost but few acres of land and but little wood, and we should all have been gainers by it. Now it is accessible only at the bridges at points comparatively distant from the town, and there is not a foot of shore to stand on unless you trespass on somebody's lot—and if you attempt a quiet stroll down the bank—you soon meet with fences built at right angles with the stream and projecting far over the water—where individuals, naturally enough, under the present arrangement—seek to monopolize the shore. At last we shall get our only view of the stream from the meeting-house belfry.

As for the trees which fringed the shore within my remembrance—where are they? and where will the remnant of them be after ten years more?

So if there is any central and commanding hill-top, it should be reserved for the public use. Think of a mountain top in the township—even to the Indians a sacred place—only accessible through private grounds. A temple as it were which you cannot enter without trespassing—nay the temple itself private property and standing in a man's cow yard—for such is commonly the case. New Hampshire courts have lately been deciding, as if it was for them to decide, whether the top of Mount Washington belonged to A or B—and it being decided in favor of B, I hear that he went up one winter with the proper officers and took formal possession. That area should be left unappropriated for modesty and reverence's sake—if only to suggest that the traveller who climbs thither in a degree rises above himself, as well as his native valley, and leaves some of his grovelling habits behind.

I know it is a mere figure of speech to talk about temples nowadays, when men recognize none, and associate the word with heathenism. Most men, it appears to me, do not care for Nature, and would sell their share in all her beauty, for as long as they may live, for a stated and not very large sum. Thank God they cannot yet fly and lay waste the sky as well as the earth. We are safe on that side for the present. It is for the very reason that some do not care for these things that we need to combine to protect all from the vandalism of a few.

It is true, we as yet take liberties and go across lots in most directions but we naturally take fewer and fewer liberties every year, as we meet with more resistance, and we shall soon be reduced to the same straights they are in England, where going across lots is out of the question—and we must ask leave to walk in some lady's park.

There are a few hopeful signs. There is the growing *library*—and then the town does set trees along the highways. But does not the broad landscape itself deserve attention?

We cut down the few old oaks which witnessed the transfer of the township from the Indian to the white man, and perchance commence our museum with a cartridge box taken from a British soldier in 1775. How little we insist on truly grand and beautiful natural features. There may be the most beautiful landscapes in the world within a dozen miles of us, for aught we know—for their inhabitants do not value nor perceive them—and so have not made them known to others—but if a grain of gold were picked up there, or a pearl found in a fresh-water clam, the whole state would resound with the news.

Thousands annually seek the White Mountains to be refreshed by their wild and primitive beauty—but when the country was discovered a similar kind of beauty prevailed all over it—and much of this might have been preserved for our present refreshment if a little foresight and taste had been used.

I do not believe that there is a town in this country which realizes in what its true wealth consists.

I visited the town of Boxboro only eight miles west of us last fall—and far the handsomest and most memorable thing which I saw there, was its noble oak wood. I doubt if there is a finer one in Massachusetts. Let it stand fifty years longer and men will make pilgrimages to it from all parts of the country, and for a worthier object than to shoot squirrels in it—and yet I said to myself, Boxboro would be very like the rest of New England, if she were ashamed of that wood-land. Probably, if the history of this town is written, the historian will have omitted to say a word about this forest—the most interesting thing in it—and lay all the stress on the history of the parish.

It turned out that I was not far from right—for not long after I came across a very brief historical notice of Stow—which then included Boxboro—written by the Reverend John Gardiner in the *Massachusetts Historical Collections,* nearly a hundred years ago. In which Mr. Gardiner, after telling us who was his predecessor in the ministry, and when he himself was settled, goes on to say, 'As for any remarkables, I am of mind there have been the fewest of any town of our standing in the Province. . . . I can't call to mind above one thing worthy of public notice, and that is the grave of Mr. John Green' who, it appears, when in England, 'was made clerk of the exchequer' by Cromwell. 'Whether he was excluded the act of oblivion or not I cannot tell,' says Mr. Gardiner. At any rate he returned to New England and as Gardiner tells us 'lived and died, and lies buried in this place.'

I can assure Mr. Gardiner that he was not excluded from the act of oblivion.

It is true Boxboro was less peculiar for its woods at that date—but they were not less interesting absolutely.

I remember talking a few years ago with a young man who had undertaken to write the history of his native town—a wild and mountainous town far up country, whose very name suggested a hundred things to me, and I almost wished I had the task to do myself—so few of the original settlers had been driven out—and not a single clerk of the exchequer buried in it. But to my chagrin I found that the author was complaining of want of materials, and that the crowning fact of his story was that the town had been the residence of General C—and the family mansion was still standing.

I have since heard, however, that Boxboro is content to have that forest stand, instead of the houses and farms that might supplant it—not because of its beauty—but because the land pays a much larger tax now than it would then.

Nevertheless it is likely to be cut off within a few years for ship-timber and the like. It is too precious to be thus disposed of. I think that it would be wise for the state to purchase and preserve a few such forests.

If the people of Massachusetts are ready to found a professorship of Natural History—so they must see the importance of preserving some portions of nature herself unimpaired.

I find that the rising generation in this town do not know what an oak or a pine is, having seen only inferior specimens. Shall we hire a man to lecture on botany, on oaks for instance, our noblest plants—while we permit others to cut down the few best specimens of these trees that are left? It is like teaching children Latin and Greek while we burn the books printed in those languages.

I think that each town should have a park, or rather a primitive forest, of five hundred or a thousand acres, either in one body or several—where a stick should never be cut for fuel—nor for the navy, nor to make wagons, but stand and decay for higher uses—a common possession forever, for instruction and recreation.

All Walden wood might have been reserved, with Walden in the midst of it, and the Easterbrooks country, an uncultivated area of some four square miles in the north of the town, might have been our huckleberry field. If any owners of these tracts are about to leave the world without natural heirs who need or deserve to be specially remembered, they will do wisely to abandon the possession to all mankind, and not will them to some individual who perhaps has enough already—and so correct the error that was made when the town was laid out. As some give to Harvard College or another Institution, so one might give a forest or a huckleberry field to Concord. This town surely is an institution which deserves to be remembered. Forget the heathen in foreign parts, and remember the pagans and savages here.

We hear of cow commons and ministerial lots, but we want *men* commons and *lay* lots as well. There is meadow and pasture and woodlot for the town's poor, why not a forest and huckleberry field for the town's rich?

We boast of our system of education, but why stop at schoolmasters and schoolhouses? We are all schoolmasters and our schoolhouse is the universe. To attend chiefly to the desk or schoolhouse, while we neglect the scenery in which it is placed, is absurd. If we do not look out we shall find our fine schoolhouse standing in a cow yard at last.

It frequently happens that what the city prides itself on most is its park—those acres which require to be the least altered from their original condition.

Live in each season as it passes; breathe the air, drink the drink, taste the fruit, and resign yourself to the influences of each. Let these be your only diet-drink and botanical medicines.

In August live on berries, not dried meats and pemmican as if you were on shipboard making your way through a waste ocean, or in the Darien Grounds, and so die of ship-fever and scurvy. Some will die of ship-fever and scurvy in an Illinois prairie, they lead such stifled and scurvy lives.

Be blown on by all the winds. Open all your pores and bathe in all the tides of nature, in all her streams and oceans, at all seasons. Miasma and infection are from within, not without. The invalid brought to the brink of the grave by an unnatural life, instead of imbibing the great influence that nature is—drinks only of the tea made of a particular herb—while he still continues his unnatural life—saves at the spile and wastes at the bung. He does not love nature or his life and so sickens and dies and no doctor can save him.

Grow green with spring—yellow and ripe with autumn. Drink of each season's influence as a vial, a true panacea of all remedies mixed for your especial use. The vials of summer never made a man sick, only those which he had stored in his cellar. Drink the wines not of your own but of nature's bottling—not kept in a goat- or pig-skin, but in the skins of a myriad fair berries.

Let Nature do your bottling, as also your pickling and preserving.

For all nature is doing her best each moment to make us well. She exists for no other end. Do not resist her. With the least inclination to be well we should not be sick. Men have discovered, or think that they have discovered the salutariness of a few wild things only, and not of all nature. Why nature is but another name for health. Some men think that they are not well in Spring or Summer or Autumn or Winter, (if you will excuse the pun) it is only because they are not indeed *well,* that is fairly *in* those seasons.

John Muir

SELECTIONS FROM *Our National Parks* (1901)

THE WILD PARKS AND FOREST RESERVATIONS OF THE WEST

. . . THE TENDENCY NOWADAYS TO WANDER in wilderness is delightful to see. Thousands of tired, nerve-shaken, over-civilized people are beginning to find out that going to the mountains is going home; that wildness is a necessity; and that mountain parks and reservations are useful not only as fountains of timber and irrigating rivers, but as fountains of life. Awakening from the stupefying effects of the vice of over-industry and the deadly apathy of luxury, they are trying as best they can to mix and enrich their own little ongoings with those of Nature, and to get rid of rust and disease. Briskly venturing and roaming, some are washing off sins and cobweb cares of the devil's spinning in all-day storms on mountains; sauntering in rosiny pinewoods or in gentian meadows, brushing through chaparral, bending down and parting sweet, flowery sprays; tracing rivers to their sources, getting in touch with the nerves of Mother Earth; jumping from rock to rock, feeling the life of them, learning the songs of them, panting in whole-souled exercise, and rejoicing in deep, long-drawn breaths of pure wildness. This is fine and natural and full of promise. So also is the growing interest in the care and preservation of forests and wild places in general, and in the half wild parks and gardens of towns. Even the scenery habit in its most artificial forms, mixed with spectacles, silliness, and kodaks; its devotees arrayed more gorgeously than scarlet tanagers, frightening the wild game with red umbrellas—even this is encouraging, and may well be regarded as a hopeful sign of the times.

All the Western mountains are still rich in wildness, and by means of good roads are being brought nearer civilization every year. To the sane and free it will hardly seem necessary to cross the continent in search of wild beauty, however easy the way, for they find it in abundance wherever they chance to be. Like Thoreau they see forests in orchards and patches of huckleberry brush, and oceans in ponds and drops of dew. Few in these hot, dim, strenuous times are quite sane or free; choked with care like clocks full of dust, laboriously doing so much good and making so much money—or so little—they are no longer good for themselves.

When, like a merchant taking a list of his goods, we take stock of our wildness, we are glad to see how much of even the most destructible kind is still unspoiled. Looking at our continent as scenery when it was all wild, lying between beautiful seas, the starry sky above it, the starry rocks beneath it, to compare its sides, the East and the West, would be like comparing the sides of a rainbow. But it is no longer equally beautiful. The rainbows of today are, I suppose, as bright as those that first spanned the sky; and some of our landscapes are growing more beautiful from year to year, notwithstanding the clearing, trampling work of civilization. New plants and animals are enriching woods and gardens, and many landscapes wholly new, with divine sculpture and architecture, are just now coming to the light of day as the mantling folds of creative glaciers are being withdrawn, and life in a thousand cheerful, beautiful forms is pushing into them, and new-born rivers are beginning to sing and shine in them. The old rivers, too, are growing longer, like healthy trees, gaining new branches and lakes as the residual glaciers at their highest sources on the mountains recede, while the rootlike branches in their flat deltas are at the same time spreading farther and wider into the seas and making new lands.

Under the control of the vast mysterious forces of the interior of the earth all the continents and islands are slowly rising or sinking. Most of the mountains are diminishing in size under the wearing action of the weather, though a few are increasing in height and girth, especially the volcanic ones, as fresh floods of molten rocks are piled on their summits and spread in successive layers, like the wood-rings of trees, on their sides. New mountains, also, are being created from time to time as islands in lakes and seas, or as subordinate cones on the slopes of old ones, thus in some measure balancing the waste of old beauty with new. Man, too, is making many far-reaching changes. This most influential half animal, half angel is rapidly multiplying and spreading, covering the seas and lakes with ships, the land with huts, hotels, cathedrals, and clustered city shops and homes, so that soon, it would seem, we may have to go farther than Nansen to find a good sound solitude. None of Nature's landscapes are ugly so long as they are wild; and much, we can say comfortingly, must always be in great part wild, particularly the sea and the sky, the floods of light from the stars, and the warm, unspoilable heart of the earth, infinitely beautiful, though only dimly visible to the eye of imagination. The geysers, too, spouting from the hot underworld; the steady, long-lasting glaciers on the mountains, obedient only to the sun; Yosemite domes and the tremendous grandeur of rocky cañons and mountains in general—these must always be wild, for man can change them and mar them hardly more than can the butterflies that hover above them. But the continent's outer beauty is fast passing away, especially the plant part of it, the most destructible and most universally charming of all.

Only thirty years ago, the great Central Valley of California, five hundred miles long and fifty miles wide, was one bed of golden and purple flowers. Now it is ploughed and pastured out of existence, gone forever—scarce a memory of it left in fence corners and along the bluffs of the streams. The gardens of the Sierra, also, and the noble forests in both the reserved and unreserved portions are sadly hacked and trampled, notwithstanding the ruggedness of the topography—all excepting those of the parks guarded by a few soldiers. In the noblest forests of the world, the ground, once divinely beautiful, is desolate and repulsive, like a face ravaged by disease. This is true also of many other Pacific Coast and Rocky Mountain valleys and forests. The same fate, sooner or later, is awaiting them all, unless awakening public opinion comes forward to stop it. Even the great deserts in Arizona, Nevada, Utah, and New Mexico, which offer so little to attract settlers, and which a few years ago pioneers were afraid of, as places of desolation and death, are now taken as pastures at the rate of one or two square miles per cow, and of course their plant treasures are passing away—the delicate abronias, phloxes, gilias, etc. Only a few of the bitter, thorny, unbitable shrubs are left, and the sturdy cactuses that defend themselves with bayonets and spears.

Most of the wild plant wealth of the East also has vanished—gone into dusty history. Only vestiges of its glorious prairie and woodland wealth remain to bless humanity in boggy, rocky, unploughable places. Fortunately, some of these are purely wild, and go far to keep Nature's love visible. White water-lilies, with rootstocks deep and safe in mud, still send up every summer a Milky Way of starry, fragrant flowers around a thousand lakes, and many a tuft of wild grass waves its panicles on mossy rocks, beyond reach of trampling feet, in company with saxifrages, bluebells, and ferns. Even in the midst of farmers' fields, precious sphagnum bogs, too soft for the feet of cattle, are preserved with their charming plants unchanged—chiogenes, Andromeda, Kalmia, Linnæa, Arethusa, etc. Calypso borealis still hides in the arbor vitæ swamps of Canada, and away to the southward there are a few unspoiled swamps, big ones, where miasma, snakes, and alligators, like guardian angels, defend their treasures and keep them as pure as paradise. And beside a' that and a' that, the East is blessed with good winters and blossoming clouds that shed white flowers over all the land, covering every scar and making the saddest landscape divine at least once a year.

The most extensive, least spoiled, and most unspoilable of the gardens of the continent are the vast tundras of Alaska. In summer they extend smooth, even, undulating, continuous beds of flowers and leaves from about lat. 62° to the shores of the Arctic Ocean; and in winter sheets of snowflowers make all the country shine, one mass of white radiance like a star. Nor are these Arctic plant people the pitiful frost-pinched unfortunates they are guessed to be by those who have never seen them. Though lowly in stature, keeping near the frozen ground as if loving it, they are bright and cheery, and speak Nature's love as plainly as their big relatives of the South. Tenderly happed and tucked in beneath downy snow to sleep through the long, white winter, they make haste to bloom in the spring without trying to grow tall, though some rise high enough to ripple and wave in the wind, and display masses of color—yellow, purple, and blue—so rich that they look like beds of rainbows, and are visible miles and miles away.

As early as June one may find the showy Geum glaciale in flower, and the dwarf willows putting forth myriads of fuzzy catkins, to be followed quickly, especially on the dryer ground, by mertensia, eritrichium, polemonium, oxytropis, astragalus, lathyrus, lupinus, myosotis, dodecatheon, arnica, chrysanthemum, nardosmia, saussurea, senecio, erigeron, matrecaria, caltha, valeriana, stellaria, Tofieldia, polygonum, papaver, phlox, lychnis, cheiranthus, Linnæa, and a host of drabas, saxifrages, and heathworts, with bright stars and bells in glorious profusion, particularly Cassiope, Andromeda, ledum, pyrola, and vaccinium—Cassiope the most abundant and beautiful of them all. Many grasses also grow here, and wave fine purple spikes and panicles over the other flowers—poa, aira, calamagrostis, alopecurus, trisetum, elymus, festuca, glyceria, etc. Even ferns are found thus far north, carefully and comfortably unrolling their precious fronds—aspidium, cystopteris, and woodsia, all growing on a sumptuous bed of mosses and lichens; not the scaly lichens seen on rails and trees and fallen logs to the southward, but massive, round-headed, finely colored plants like corals, wonderfully beautiful, worth going round the world to see. I should like to mention all the plant friends I found in a summer's wanderings in this cool reserve, but I fear few would care to read their names, although everybody, I am sure, would love them could they see them blooming and rejoicing at home.

On my last visit to the region about Kotzebue Sound, near the middle of September, 1881, the weather was so fine and mellow that it suggested the Indian summer of the Eastern States. The winds were hushed, the tundra glowed in creamy golden sunshine, and the colors of the ripe foliage of the heathworts, willows, and birch—red, purple, and yellow, in pure bright tones—were enriched with those of berries which were scattered everywhere, as if they had been showered from the clouds like hail. When I was back a mile or two from the shore, reveling in this color-glory, and thinking how fine it would be could I cut a square of the tundra sod of conventional picture size, frame it, and hang it among the paintings on my study walls at home, saying to myself, "Such a Nature painting taken at random from any part of the thousand-mile bog would make the other pictures look dim and coarse," I heard merry shouting, and, looking round, saw a band of Eskimos—men, women, and children, loose and hairy like wild animals—running towards me. I could not guess at first what they were seeking, for they seldom leave the shore; but soon they told me, as they threw themselves down, sprawling and laughing, on the mellow bog, and began to feast on the berries. A lively picture they made, and a pleasant one, as they frightened the whirring ptarmigans, and surprised their oily stomachs with the beautiful acid berries of many kinds, and filled sealskin bags with them to carry away for festive days in winter.

Nowhere else on my travels have I seen so much warmblooded, rejoicing life as in this grand Arctic reservation, by so many regarded as desolate. Not only are there whales in abundance along the shores, and innumerable seals, walruses, and white bears, but on the tundras great herds of fat reindeer and wild sheep, foxes, hares, mice, piping marmots, and birds. Perhaps more birds are born here than in any other region of equal extent on the continent. Not only do strong-winged hawks, eagles, and water-fowl, to whom the length of the continent is merely a pleasant excursion, come up here every summer in great numbers, but also many short-winged warblers, thrushes, and finches, repairing hither to rear their young in safety, reinforce the plant bloom with their plumage, and sweeten the wilderness with song; flying all the way, some of them, from Florida, Mexico, and Central America. In coming north they are coming home, for they were born here, and they go south only to spend the winter months, as New Englanders go to Florida. Sweet-voiced troubadours, they sing in orange groves and vine-clad magnolia woods in winter, in thickets of dwarf birch and alder in summer, and sing and chatter more or less all the way back and forth, keeping the whole country glad. Oftentimes, in New England, just as the last snow-patches are melting and the sap in the maples begins to flow, the blessed wanderers may be heard about orchards and the edges of fields where they have stopped to glean a scanty meal, not tarrying long, knowing they have far to go. Tracing the footsteps of spring, they arrive in their tundra homes in June or July, and set out on their return journey in September, or as soon as their families are able to fly well.

This is Nature's own reservation, and every lover of wildness will rejoice with me that by kindly frost it is so well defended. The discovery lately made that it is sprinkled with gold may cause some alarm; for the strangely exciting stuff makes the timid bold enough for anything, and the lazy destructively industrious. Thousands at least half insane are now pushing their way into it, some by the southern passes over the mountains, perchance the first mountains they have ever seen—sprawling, struggling, gasping for breath, as, laden with awkward, merciless burdens of provisions and tools, they climb over rough-angled boulders and cross thin miry bogs. Some are going by the mountains and rivers to the eastward through Canada, tracing the old romantic ways of the Hudson Bay traders; others by Bering Sea and the Yukon, sailing all the way, getting glimpses perhaps of the famous fur-seals, the ice-floes, and the innumerable islands and bars of the great Alaska river. In spite of frowning hardships and the frozen ground, the Klondike gold will increase the crusading crowds for years to come, but comparatively little harm will be done. Holes will be burned and dug into the hard ground here and there, and into the quartz-ribbed mountains and hills; ragged towns like beaver and muskrat villages will be built, and mills and locomotives will make rumbling, screeching, disenchanting noises; but the miner's pick will not be followed far by the plough, at least not until Nature is ready to unlock the frozen soil-beds with her slow-turning climate key. On the other hand, the roads of the pioneer miners will lead many a lover of wildness into the heart of the reserve, who without them would never see it.

In the meantime, the wildest health and pleasure grounds accessible and available to tourists seeking escape from care and dust and early death are the parks and reservations of the West. There are four national parks[1]—the Yellowstone, Yosemite, General Grant, and Sequoia—all within easy reach, and thirty forest reservations, a magnificent realm of woods, most of which, by railroads and trails and open ridges, is also fairly accessible, not only to the determined traveler rejoicing in difficulties, but to those (may their tribe increase) who, not tired, not sick, just naturally take wing every summer in search of wildness. The forty million acres of these reserves are in the main unspoiled as yet, though sadly wasted and threatened on their more open margins by the axe and fire of the lumberman and prospector, and by hoofed locusts, which, like the winged ones, devour every leaf within reach, while the shepherds and owners set fires with the intention of making a blade of grass grow in the place of every tree, but with the result of killing both the grass and the trees.

In the million acre Black Hills Reserve of South Dakota, the easternmost of the great forest reserves, made for the sake of the farmers and miners, there are delightful, reviving sauntering-grounds in open parks of yellow pine, planted well apart, allowing plenty of sunshine to warm the ground. This tree is one of the most variable and most widely distributed of American pines. It grows sturdily on all kinds of soil and rocks, and, protected by a mail of thick bark, defies frost and fire and disease alike, daring every danger in firm, calm beauty and strength. It occurs here mostly on the outer hills and slopes where no other tree can grow. The ground beneath it is yellow most of the summer with showy Wythia, arnica, applopappus, solidago, and other sun-loving plants, which, though they form no heavy entangling growth, yet give abundance of color and make all the woods a garden. Beyond the yellow pine woods there lies a world of rocks of wildest architecture, broken, splintery, and spiky, not very high, but the strangest in form and style of grouping imaginable. Countless towers and spires, pinnacles and slender domed columns, are crowded together, and feathered with sharp-pointed Engelmann spruces, making curiously mixed forests—half trees, half rocks. Level gardens here and there in the midst of them offer charming surprises, and so do the many small lakes with lilies on their meadowy borders, and bluebells, anemones, daisies, castilleias, comandras, etc., together forming landscapes delightfully novel, and made still wilder by many interesting animals—elk, deer, beavers, wolves, squirrels, and birds. Not very long ago this was the richest of all the red man's hunting-grounds hereabout. After the season's buffalo hunts were over—as described by Parkman, who, with a picturesque cavalcade of Sioux savages, passed through these famous hills in 1846—every winter deficiency was here made good, and hunger was unknown until, in spite of most determined, fighting, killing opposition, the white gold-hunters entered the fat game reserve and spoiled it. The Indians are dead now, and so are most of the hardly less striking free trappers of the early romantic Rocky Mountain times. Arrows, bullets, scalping-knives, need no longer be feared; and all the wilderness is peacefully open.

The Rocky Mountain reserves are the Teton, Yellowstone, Lewis and Clark, Bitter Root, Priest River and Flathead, comprehending more than twelve million acres of mostly unclaimed, rough, forest-covered mountains in which the great rivers of the country take their rise. The commonest tree in most of them is the brave, indomitable, and altogether admirable Pinus contorta, widely distributed in all kinds of climate and soil, growing cheerily in frosty Alaska, breathing the damp salt air of the sea as well as the dry biting blasts of the Arctic interior, and making itself at home on the most dangerous flame-swept slopes and ridges of the Rocky Mountains in immeasurable abundance and variety of forms. Thousands of acres of this species are destroyed by running fires nearly every summer, but a new growth springs quickly from the ashes. It is generally small, and yields few sawlogs of commercial value, but is of incalculable importance to the farmer and miner; supplying fencing, mine timbers, and firewood, holding the porous soil on steep slopes, preventing landslips and avalanches, and giving kindly, nourishing shelter to animals and the widely outspread sources of the life-giving rivers. The other trees are mostly spruce, mountain pine, cedar, juniper, larch, and balsam fir; some of them, especially on the western slopes of the mountains, attaining grand size and furnishing abundance of fine timber.

Perhaps the least known of all this grand group of reserves is the Bitter Root, of more than four million acres. It is the wildest, shaggiest block of forest wildness in the Rocky Mountains, full of happy, healthy, storm-loving trees, full of streams that dance and sing in glorious array, and full of Nature's animals—elk, deer, wild sheep, bears, cats, and innumerable smaller people.

In calm Indian summer, when the heavy winds are hushed, the vast forests covering hill and dale, rising and falling over the rough topography and vanishing in the distance, seem lifeless. No moving thing is seen as we climb the peaks, and only the low, mellow murmur of falling water is heard, which seems to thicken the silence. Nevertheless, how many hearts with warm red blood in them are beating under cover of the woods, and how many teeth and eyes are shining! A multitude of animal people, intimately related to us, but of whose lives we know almost nothing are as busy about their own affairs as we are about ours: beavers are building and mending dams and huts for winter, and storing them with food; bears are studying winter quarters as they stand thoughtful in open spaces, while the gentle breeze ruffles the long hair on their backs; elk and deer, assembling on the heights, are considering cold pastures where they will be farthest away from the wolves; squirrels and marmots are busily laying up provisions and lining their nests against coming frost and snow foreseen; and countless thousands of birds are forming parties and gathering their young about them for flight to the southlands; while butterflies and bees, apparently with no thought of hard times to come, are hovering above the late-blooming goldenrods, and, with countless other insect folk, are dancing and humming right merrily in the sunbeams and shaking all the air into music.

Wander here a whole summer, if you can. Thousands of God's wild blessings will search you and soak you as if you were a sponge, and the big days will go by uncounted. If you are business-tangled, and so burdened with duty that only weeks can be got out of the heavy-laden year, then go to the Flathead Reserve; for it is easily and quickly reached by the Great Northern Railroad. Get off the track at Belton Station, and in a few minutes you will find yourself in the midst of what you are sure to say is the best care-killing scenery on the continent—beautiful lakes derived straight from glaciers, lofty mountains steeped in lovely nemophila-blue skies and clad with forests and glaciers, mossy, ferny waterfalls in their hollows, nameless and numberless, and meadowy gardens abounding in the best of everything. When you are calm enough for discriminating observation, you will find the king of the larches, one of the best of the Western giants, beautiful, picturesque, and regal in port, easily the grandest of all the larches in the world. It grows to a height of one hundred and fifty to two hundred feet, with a diameter at the ground of five to eight feet, throwing out its branches into the light as no other tree does. To those who before have seen only the European larch or the Lyall species of the eastern Rocky Mountains, or the little tamarack or hackmatack of the Eastern States and Canada, this Western king must be a revelation. . . .

These grand reservations should draw thousands of admiring visitors at least in summer, yet they are neglected as if of no account, and spoilers are allowed to ruin them as fast as they like.[2] A few peeled spars cut here were set up in London, Philadelphia, and Chicago, where they excited wondering attention; but the countless hosts of living trees rejoicing at home on the mountains are scarce considered at all. Most travelers here are content with what they can see from car windows or the verandas of hotels, and in going from place to place cling to their precious trains and stages like wrecked sailors to rafts. When an excursion into the woods is proposed, all sorts of dangers are imagined—snakes, bears, Indians. Yet it is far safer to wander in God's woods than to travel on black highways or to stay at home. The snake danger is so slight it is hardly worth mentioning. Bears are a peaceable people, and mind their own business, instead of going about like the devil seeking whom they may devour. Poor fellows, they have been poisoned, trapped, and shot at until they have lost confidence in brother man, and it is not now easy to make their acquaintance. As to Indians, most of them are dead or civilized into useless innocence. No American wilderness that I know of is so dangerous as a city home "with all the modern improvements." One should go to the woods for safety, if for nothing else.

Lewis and Clark, in their famous trip across the continent in 1804–1805, did not lose a single man by Indians or animals, though all the West was then wild. Captain Clark was bitten on the hand as he lay asleep. That was one bite among more than a hundred men while traveling nine thousand miles. Loggers are far more likely to be met than Indians or bears in the reserves or about their boundaries, brown weather-tanned men with faces furrowed like bark, tired-looking, moving slowly, swaying like the trees they chop. A little of everything in the woods is fastened to their clothing, rosiny and smeared with balsam, and rubbed into it, so that their scanty outer garments grow thicker with use and never wear out. Many a forest giant have these old woodmen felled, but, round-shouldered and stooping, they too are leaning over and tottering to their fall. Others, however, stand ready to take their places, stout young fellows, erect as saplings; and always the foes of trees outnumber their friends. Far up the white peaks one can hardly fail to meet the wild goat, or American chamois—an admirable mountaineer, familiar with woods and glaciers as well as rocks—and in leafy thickets deer will be found; while gliding about unseen there are many sleek furred animals enjoying their beautiful lives, and birds also, notwithstanding few are noticed in hasty walks. The ousel sweetens the glens and gorges where the streams flow fastest, and every grove has its singers, however silent it seems—thrushes, linnets, warblers; humming-birds glint about the fringing bloom of the meadows and peaks, and the lakes are stirred into lively pictures by water-fowl. . . .

The Sierra of California is the most openly beautiful and useful of all the forest reserves, and the largest excepting the Cascade Reserve of Oregon and the Bitter Root of Montana and Idaho. It embraces over four million acres of the grandest scenery and grandest trees on the continent, and its forests are planted just where they do the most good, not only for beauty, but for farming in the great San Joaquin Valley beneath them. It extends southward from the Yosemite National Park to the end of the range, a distance of nearly two hundred miles. No other coniferous forest in the world contains so many species or so many large and beautiful trees—Sequoia gigantea, king of conifers, "the noblest of a noble race," as Sir Joseph Hooker well says; the sugar pine, king of all the world's pines, living or extinct; the yellow pine, next in rank, which here reaches most perfect development, forming noble towers of verdure two hundred feet high; the mountain pine, which braves the coldest blasts far up the mountains on grim, rocky slopes; and five others, flourishing each in its place, making eight species of pine in one forest, which is still further enriched by the great Douglas spruce, libocedrus, two species of silver fir, large trees and exquisitely beautiful, the Paton hemlock, the most graceful of evergreens, the curious tumion, oaks of many species, maples, alders, poplars, and flowering dogwood, all fringed with flowery underbrush, manzanita, ceanothus, wild rose, cherry, chestnut, and rhododendron. Wandering at random through these friendly, approachable woods, one comes here and there to the loveliest lily gardens, some of the lilies ten feet high, and the smoothest gentian meadows, and Yosemite valleys known only to mountaineers. Once I spent a night by a camp-fire on Mount Shasta with Asa Gray and Sir Joseph Hooker, and, knowing that they were acquainted with all the great forests of the world, I asked whether they knew any coniferous forest that rivaled that of the Sierra. They unhesitatingly said: "No. In the beauty and grandeur of individual trees, and in number and variety of species, the Sierra forests surpass all others."

This Sierra Reserve, proclaimed by the President of the United States in September, 1893, is worth the most thoughtful care of the government for its own sake, without considering its value as the fountain of the rivers on which the fertility of the great San Joaquin Valley depends. Yet it gets no care at all. In the fog of tariff, silver, and annexation politics it is left wholly unguarded, though the management of the adjacent national parks by a few soldiers shows how well and how easily it can be preserved. In the meantime, lumbermen are allowed to spoil it at their will, and sheep in uncountable ravenous hordes to trample it and devour every green leaf within reach; while the shepherds, like destroying angels, set innumerable fires, which burn not only the undergrowth of seedlings on which the permanence of the forest depends, but countless thousands of the venerable giants. If every citizen could take one walk through this reserve, there would be no more trouble about its care; for only in darkness does vandalism flourish. . . .

THE AMERICAN FORESTS

. . . Even in Congress a sizable chunk of gold, carefully concealed, will outtalk and outfight all the nation on a subject like forestry, well smothered in ignorance, and in which the money interests of only a few are conspicuously involved. Under these circumstances, the bawling, blethering oratorical stuff drowns the voice of God himself. Yet the dawn of a new day in forestry is breaking. Honest citizens see that only the rights of the government are being trampled, not those of the settlers. Only what belongs to all alike is reserved, and every acre that is left should be held together under the federal government as a basis for a general policy of administration for the public good. The people will not always be deceived by selfish opposition, whether from lumber and mining corporations or from sheepmen and prospectors, however cunningly brought forward underneath fables and gold.

Emerson says that things refuse to be mismanaged long. An exception would seem to be found in the case of our forests, which have been mismanaged rather long, and now come desperately near being like smashed eggs and spilt milk. Still, in the long run the world does not move backward. The wonderful advance made in the last few years, in creating four national parks in the West, and thirty forest reservations, embracing nearly forty million acres; and in the planting of the borders of streets and highways and spacious parks in all the great cities, to satisfy the natural taste and hunger for landscape beauty and righteousness that God has put, in some measure, into every human being and animal, shows the trend of awakening public opinion. The making of the far-famed New York Central Park was opposed by even good men, with misguided pluck, perseverance, and ingenuity; but straight right won its way, and now that park is appreciated. So we confidently believe it will be with our great national parks and forest reservations. There will be a period of indifference on the part of the rich, sleepy with wealth, and of the toiling millions, sleepy with poverty, most of whom never saw a forest; a period of screaming protest and objection from the plunderers, who are as unconscionable and enterprising as Satan. But light is surely coming, and the friends of destruction will preach and bewail in vain.

The United States government has always been proud of the welcome it has extended to good men of every nation, seeking freedom and homes and bread. Let them be welcomed still as Nature welcomes them, to the woods as well as to the prairies and plains. No place is too good for good men, and still there is room. They are invited to heaven, and may well be allowed in America. Every place is made better by them. Let them be as free to pick gold and gems from the hills, to cut and hew, dig and plant, for homes and bread, as the birds are to pick berries from the wild bushes, and moss and leaves for nests. The ground will be glad to feed them, and the pines will come down from the mountains for their homes as willingly as the cedars came from Lebanon for Solomon's temple. Nor will the woods be the worse for this use, or their benign influences be diminished any more than the sun is diminished by shining. Mere destroyers, however, tree-killers, wool and mutton men, spreading death and confusion in the fairest groves and gardens ever planted—let the government hasten to cast them out and make an end of them. For it must be told again and again, and be burningly borne in mind, that just now, while protective measures are being deliberated languidly, destruction and use are speeding on faster and farther every day. The axe and saw are insanely busy, chips are flying thick as snowflakes, and every summer thousands of acres of priceless forests, with their underbrush, soil, springs, climate, scenery, and religion, are vanishing away in clouds of smoke, while, except in the national parks, not one forest guard is employed.

All sorts of local laws and regulations have been tried and found wanting, and the costly lessons of our own experience, as well as that of every civilized nation, show conclusively that the fate of the remnant of our forests is in the hands of the federal government, and that if the remnant is to be saved at all, it must be saved quickly.

Any fool can destroy trees. They cannot run away; and if they could, they would still be destroyed—chased and hunted down as long as fun or a dollar could be got out of their bark hides, branching horns, or magnificent bole backbones. Few that fell trees plant them; nor would planting avail much towards getting back anything like the noble primeval forests. During a man's life only saplings can be grown, in the place of the old trees—tens of centuries old—that have been destroyed. It took more than three thousand years to make some of the trees in these Western woods—trees that are still standing in perfect strength and beauty, waving and singing in the mighty forests of the Sierra. Through all the wonderful, eventful centuries since Christ's time—and long before that—God has cared for these trees, saved them from drought, disease, avalanches, and a thousand straining, leveling tempests and floods; but he cannot save them from fools—only Uncle Sam can do that.

NOTES

1. There are now five parks and thirty-eight reservations.

2. The outlook over forest affairs is now encouraging. Popular interest, more practical than sentimental in whatever touches the welfare of the country's forests, is growing rapidly, and a hopeful beginning has been made by the Government in real protection for the reservations as well as for the parks. From July 1, 1900, there have been 9 superintendents, 39 supervisors, and from 330 to 445 rangers of reservations.

Theodore Roosevelt

The American Wilderness (1897)

Wilderness Hunters and Wilderness Game

MANIFOLD ARE THE SHAPES taken by the American wilderness. In the east, from the Atlantic coast to the Mississippi valley, lies a land of magnificent hardwood forest. In endless variety and beauty, the trees cover the ground, save only where they have been cleared away by man, or where toward the west the expanse of the forest is broken by fertile prairies. Toward the north, this region of hardwood-trees merges insensibly into the southern extension of the great subarctic forest; here the silver stems of birches gleam against the sombre background of coniferous evergreens. In the southeast again, by the hot, oozy coasts of the South Atlantic and the Gulf, the forest becomes semitropical; palms wave their feathery fronds, and the tepid swamps teem with reptile life.

Some distance beyond the Mississippi, stretching from Texas to North Dakota, and westward to the Rocky Mountains, lies the plains country. This is a region of light rainfall, where the ground is clad with short grass, while cottonwood-trees fringe the courses of the winding plains streams; streams that are alternately turbid torrents and mere dwindling threads of water. The great stretches of natural pasture are broken by gray sage-brush plains and tracts of strangely shaped and colored Bad Lands; sun-scorched wastes in summer, and in winter arctic in their iron desolation. Beyond the plains rise the Rocky Mountains, their flanks covered with coniferous woods; but the trees are small, and do not ordinarily grow very closely together. Toward the north the forest becomes denser, and the peaks higher; and glaciers creep down toward the valleys from the fields of everlasting snow. The brooks are brawling, trout-filled torrents; the swift rivers foam over rapid and cataract on their way to one or the other of the two great oceans.

Southwest of the Rockies evil and terrible deserts stretch for leagues and leagues, mere waterless wastes of sandy plain and barren mountain, broken here and there by narrow strips of fertile ground. Rain rarely falls, and there are no clouds to dim the brazen sun. The rivers run in deep canyons, or are swallowed by the burning sand; the smaller watercourses are dry throughout the greater part of the year.

Beyond this desert region rise the sunny Sierras of California, with their flower-clad slopes and groves of giant trees; and north of them, along the coast, the rain-shrouded mountain chains of Oregon and Washington, matted with the towering growth of the mighty evergreen forest.

The white hunters, who from time to time first penetrated the different parts of this wilderness, found themselves in such hunting-grounds as those wherein, long ages before, their Old World forefathers had dwelled; and the game they chased was much the same as that their lusty barbarian ancestors followed, with weapons of bronze and of iron, in the dim years before history dawned. As late as the end of the seventeenth century the turbulent village nobles of Lithuania and Livonia hunted the bear, the bison, the elk, the wolf, and the stag, and hung the spoils in their smoky wooden palaces; and so, two hundred years later, the free hunters of Montana, in the interludes between hazardous mining quests and bloody Indian campaigns, hunted game almost or quite the same in kind, through the cold mountain forests surrounding the Yellowstone and Flathead lakes, and decked their log cabins and ranch-houses with the hides and horns of the slaughtered beasts.

Zoologically speaking, the north temperate zones of the Old and New Worlds are very similar, differing from one another much less than they do from the various regions south of them, or than these regions differ among themselves. The untrodden American wilderness resembles both in game and physical character the forests, the mountains, and the steppes of the Old World as it was at the beginning of our era. Great woods of pine and fir, birch and beech, oak and chestnut; streams where the chief game-fish are spotted trout and silvery salmon; grouse of various kinds as the most common game-birds; all these the hunter finds as characteristic of the New World as of the Old. So it is with most of the beasts of the chase, and so also with the fur-bearing animals that furnish to the trapper alike his life-work and his means of livelihood. The bear, wolf, bison, moose, caribou, wapiti, deer, and bighorn, the lynx, fox, wolverene, sable, mink, ermine, beaver, badger, and otter of both worlds are either identical or more or less closely kin to one another. Sometimes of the two forms, that found in the Old World is the larger. Perhaps more often the reverse is true, the American beast being superior in size. This is markedly the case with the wapiti, which is merely a giant brother of the European stag, exactly as the fisher is merely a very large cousin of the European sable or marten. The extraordinary prongbuck, the only hollow-horned ruminant which sheds its horns annually, is a distant representative of the Old World antelopes of the steppes; the queer white antelope-goat has for its nearest kinsfolk certain Himalayan species. Of the animals commonly known to our hunters and trappers, only a few, such as the cougar, peccary, raccoon, possum (and among birds the wild turkey), find their nearest representatives and type forms in tropical America.

Of course this general resemblance does not mean identity. The differences in plant life and animal life, no less than in the physical features of the land, are sufficiently marked to give the American wilderness a character distinctly its own. Some of the most characteristic of the woodland animals, some of those which have most vividly impressed themselves on the imagination of the hunters and pioneer settlers, are the very ones which have no Old World representatives. The wild turkey is in every way the king of American game-birds. Among the small beasts the coon and the possum are those which have left the deepest traces in the humbler lore of the frontier; exactly as the cougar—usually under the name of panther or mountain-lion—is a favorite figure in the wilder hunting tales. Nowhere else is there anything to match the wealth of the eastern hardwood forests, in number, variety, and beauty of trees; nowhere else is it possible to find conifers approaching in size the giant redwoods and sequoias of the Pacific slope. Nature here is generally on a larger scale than in the Old World home of our race. The lakes are like inland seas, the rivers like arms of the sea. Among stupendous mountain chains there are valleys and canyons of fathomless depth and incredible beauty and majesty. There are tropical swamps, and sad, frozen marshes; deserts and Death Valleys, weird and evil, and the strange wonderland of the Wyoming geyser region. The waterfalls are rivers rushing over precipices; the prairies seem without limit, and the forest never-ending.

At the time when we first became a nation, nine-tenths of the territory now included within the limits of the United States was wilderness. It was during the stirring and troubled years immediately preceding the outbreak of the Revolution that the most adventurous hunters, the vanguard of the hardy army of pioneer settlers, first crossed the Alleghanies, and roamed far and wide through the lonely, danger-haunted forests which filled the No-man's-land lying between the Tennessee and the Ohio. They waged ferocious warfare with Shawnee and Wyandot and wrought huge havoc among the herds of game with which the forest teemed. While the first Continental Congress was still sitting, Daniel Boone, the archetype of the American hunter, was leading his bands of tall backwoods riflemen to settle in the beautiful country of Kentucky, where the red and the white warriors strove with such obstinate rage that both races alike grew to know it as "the dark and bloody ground."

Boone and his fellow hunters were the heralds of the oncoming civilization, the pioneers in that conquest of the wilderness which has at last been practically achieved in our own day. Where they pitched their camps and built their log huts or stockaded hamlets towns grew up, and men who were tillers of the soil, not mere wilderness wanderers, thronged in to take and hold the land. Then, ill at ease among the settlements for which they had themselves made ready the way, and fretted even by the slight restraints of the rude and uncouth semicivilization of the border, the restless hunters moved onward into the yet unbroken wilds where the game dwelt and the red tribes marched forever to war and hunting. Their untamable souls ever found something congenial and beyond measure attractive in the lawless freedom of the lives of the very savages against whom they warred so bitterly.

Step by step, often leap by leap, the frontier of settlement was pushed westward; and ever from before its advance fled the warrior tribes of the red men and the scarcely less intractable array of white Indian fighters and game-hunters. When the Revolutionary War was at its height, George Rogers Clark, himself a mighty hunter of the old backwoods type, led his handful of hunter-soldiers to the conquest of the French towns of the Illinois. This was but one of the many notable feats of arms performed by the wild soldiery of the backwoods. Clad in their fringed and tasselled hunting-shirt of buckskin or homespun, with coonskin caps and deer-hide leggings and moccasins, with tomahawk and scalping-knife thrust into their bead-worked belts, and long rifles in hand, they fought battle after battle of the most bloody character, both against the Indians, as at the Great Kanawha, at the Fallen Timbers, and at Tippecanoe, and against more civilized foes, as at King's Mountain, New Orleans, and the River Thames.

Soon after the beginning of the present century Louisiana fell into our hands, and the most daring hunters and explorers pushed through the forests of the Mississippi valley to the great plains, steered across these vast seas of grass to the Rocky Mountains, and then through their rugged defiles onward to the Pacific Ocean. In every work of exploration, and in all the earlier battles with the original lords of the western and southwestern lands, whether Indian or Mexican, the adventurous hunters played the leading part; while close behind came the swarm of hard, dogged, border farmers—a masterful race, good fighters and good breeders, as all masterful races must be.

Very characteristic in its way was the career of quaint, honest, fearless Davy Crockett, the Tennessee rifleman and Whig Congressman, perhaps the best shot in all our country, whose skill in the use of his favorite weapon passed into a proverb, and who ended his days by a hero's death in the ruins of the Alamo. An even more notable man was another mighty hunter, Houston, who when a boy ran away to the Indians; who while still a lad returned to his own people to serve under Andrew Jackson in the campaigns which that greatest of all the backwoods leaders waged against the Creeks, the Spaniards, and the British. He was wounded at the storming of one of the strongholds of Red Eagle's doomed warriors, and returned to his Tennessee home to rise to high civil honor, and become the foremost man of his State. Then, while Governor of Tennessee, in a sudden fit of moody anger, and of mad longing for the unfettered life of the wilderness, he abandoned his office, his people, and his race, and fled to the Cherokees beyond the Mississippi. For years he lived as one of their chiefs; until one day, as he lay in ignoble ease and sloth, a rider from the south, from the rolling plains of the San Antonio and Brazos, brought word that the Texans were up, and in doubtful struggle striving to wrest their freedom from the lancers and carabineers of Santa Anna. Then his dark soul flamed again into burning life; riding by night and day he joined the risen Texans, was hailed by them as a heaven-sent leader, and at the San Jacinto led them on to the overthrow of the Mexican host. Thus the stark hunter, who had been alternately Indian fighter and Indian chief, became the President of the new republic, and, after its admission into the United States, a Senator at Washington; and, to his high honor, he remained to the end of his days stanchly loyal to the flag of the Union.

By the time that Crockett fell, and Houston became the darling leader of the Texans, the typical hunter and Indian fighter had ceased to be a backwoodsman; he had become a plainsman or mountain-man; for the frontier, east of which he never willingly went, had been pushed beyond the Mississippi. Restless, reckless, and hardy, he spent years of his life in lonely wanderings through the Rockies as a trapper; he guarded the slowly moving caravans which for purposes of trade journeyed over the dangerous Santa Fe trail; he guided the large parties of frontier settlers who, driving before them their cattle, with all their household goods in their white-topped wagons, spent perilous months and seasons on their weary way to Oregon or California. Joining in bands, the stalwart, skin-clad riflemen waged ferocious war on the Indians, scarcely more savage than themselves, or made long raids for plunder and horses against the outlying Mexican settlements. The best, the bravest, the most modest of them all was the renowned Kit Carson. He was not only a mighty hunter, a daring fighter, a finder of trails, and maker of roads through the unknown, untrodden wilderness, but also a real leader of men. Again and again he crossed and recrossed the continent, from the Mississippi to the Pacific; he guided many of the earliest military and exploring expeditions of the United States Government; he himself led the troops in victorious campaigns against Apache and Navajo; and in the Civil War he was made a colonel of the Federal Army.

After him came many other hunters. Most were pure-blooded Americans, but many were creole Frenchmen, Mexicans, or even members of the so-called civilized Indian tribes, notably the Delawares. Wide were their wanderings, many their strange adventures in the chase, bitter their unending warfare with the red lords of the land. Hither and thither they roamed, from the desolate, burning deserts of the Colorado to the grassy plains of the Upper Missouri; from the rolling Texas prairies, bright beneath their sunny skies, to the high snow peaks of the northern Rockies, or the giant pine forests and soft rainy weather of the coasts of Puget Sound. Their main business was trapping, furs being the only articles yielded by the wilderness, as they knew it, which were both valuable and portable. These early hunters were all trappers likewise, and, indeed, used their rifles only to procure meat or repel attacks. The chief of the fur-bearing animals they followed was the beaver, which abounded in the streams of the plains and mountains; in the far north they also trapped otter, mink, sable, and fisher. They married squaws from among the Indian tribes with which they happened for the moment to be at peace; they acted as scouts for the United States troops in their campaigns against the tribes with which they happened to be at war.

Soon after the Civil War the life of these hunters, taken as a class, entered on its final stage. The Pacific coast was already fairly well settled, and there were few mining-camps in the Rockies; but most of this Rocky Mountain region, and the entire stretch of plains country proper, the vast belt of level or rolling grassland lying between the Rio Grande and the Saskatchewan, still remained primeval wilderness, inhabited only by roving hunters and formidable tribes of Indian nomads, and by the huge herds of game on which they preyed. Beaver swarmed in the streams and yielded a rich harvest to the trapper; but trapping was no longer the mainstay of the adventurous plainsmen. Foremost among the beasts of the chase, on account of its numbers, its size, and its economic importance, was the bison, or American buffalo; its innumerable multitudes darkened the limitless prairies. As the transcontinental railroads were pushed toward completion, and the tide of settlement rolled onward with ever increasing rapidity, buffalo-robes became of great value. The hunters forthwith turned their attention mainly to the chase of the great clumsy beasts, slaughtering them by the hundreds of thousands for their hides; sometimes killing them on horseback, but more often on foot, by still-hunting, with the heavy, long-range Sharp's rifle. Throughout the fifteen years during which this slaughter lasted, a succession of desperate wars was waged with the banded tribes of the Horse Indians. All the time, in unending succession, long trains of big white-topped wagons crept slowly westward across the prairies, marking the steady oncoming of the frontier settlers.

By the close of 1883 the last buffalo-herd was destroyed. The beaver were trapped out of all the streams, or their numbers so thinned that it no longer paid to follow them. The last formidable Indian war had been brought to a successful close. The flood of the incoming whites had risen over the land; tongues of settlement reached from the Mississippi to the Rocky Mountains, and from the Rocky Mountains to the Pacific. The frontier had come to an end; it had vanished. With it vanished also the old race of wilderness hunters, the men who spent all their days in the lonely wilds, and who killed game as their sole means of livelihood. Great stretches of wilderness still remained in the Rocky Mountains, and here and there in the plains country, exactly as much smaller tracts of wild land are to be found in the Alleghanies and northern New York and New England; and on these tracts occasional hunters and trappers still linger; but as a distinctive class, with a peculiar and important position in American life, they no longer exist.

There were other men besides the professional hunters who lived on the borders of the wilderness and followed hunting, not only as a pastime but also as yielding an important portion of their subsistence. The frontier farmers were all hunters. In the Eastern backwoods, and in certain places in the West, as in Oregon, these adventurous tillers of the soil were the pioneers among the actual settlers; in the Rockies their places were taken by the miners, and on the great plains by the ranchmen and cowboys, the men who lived in the saddle, guarding their branded herds of horses and horned stock. Almost all of the miners and cowboys were obliged on occasions to turn hunters.

Moreover, the regular army, which played so important a part in all the later stages of the winning of the West, produced its full share of mighty hunters. The later Indian wars were fought principally by the regulars. The West Point officer and his little company of trained soldiers appeared abreast of the first hardy cattlemen and miners. The ordinary settlers rarely made their appearance until, in campaign after campaign, always inconceivably wearing and harassing and often very bloody in character, the scarred and tattered troops had broken and overthrown the most formidable among the Indian tribes. Faithful, uncomplaining, unflinching, the soldiers wearing the national uniform lived for many weary years at their lonely little posts, facing unending toil and danger with quiet endurance, surrounded by the desolation of vast solitudes, and menaced by the most merciless of foes. Hunting was followed not only as a sport but also as the only means of keeping the posts and the expeditionary trains in meat.

Many of the officers became equally proficient as marksmen and hunters. The three most famous Indian fighters since the Civil War, Generals Custer, Miles, and Crook, were all keen and successful followers of the chase.

Of American big game the bison, almost always known as the buffalo, was the largest and most important to man. When the first white settlers landed in Virginia the bison ranged east of the Alleghanies almost to the seacoast, westward to the dry deserts lying beyond the Rocky Mountains, northward to the Great Slave Lake and southward to Chihuahua. It was a beast of the forests and mountains, in the Alleghanies no less than in the Rockies; but its true home was on the prairies and the high plains. Across these it roamed, hither and thither, in herds of enormous, of incredible, magnitude; herds so large that they covered the waving grass-land for hundreds of square leagues, and when on the march occupied days and days in passing a given point. But the seething myriads of shaggy-maned wild cattle vanished with remarkable and melancholy rapidity before the inroads of the white hunters, and the steady march of the oncoming settlers. Now they are on the point of extinction. Two or three hundred are left in that great national game-preserve the Yellowstone Park; and it is said that others still remain in the wintry desolation of Athabasca. Elsewhere only a few individuals exist—probably considerably less than half a hundred all told—scattered in small parties in the wildest and most remote and inaccessible portions of the Rocky Mountains. A bison bull is the largest American animal. His huge bulk, his short, curved black horns, the shaggy mane clothing his great neck and shoulders, give him a look of ferocity which his conduct belies. Yet he is truly a grand and noble beast, and his loss from our prairies and forests is as keenly regretted by the lover of nature and of wild life as by the hunter.

Next to the bison in size, and much superior in height to it and to all other American game—for it is taller than the tallest horse—comes the moose, or broad-horned elk. It is a strange, uncouth-looking beast, with very long legs, short, thick neck, a big, ungainly head, a swollen nose, and huge shovel horns. Its home is in the cold, wet pine and spruce forests which stretch from the subarctic region of Canada southward in certain places across our frontier. Two centuries ago it was found as far south as Massachusetts. It has now been exterminated from its former haunts in northern New York and Vermont, and is on the point of vanishing from northern Michigan. It is still found in northern Maine and northeastern

Minnesota and in portions of northern Idaho and Washington; while along the Rockies it extends its range southward through western Montana to northwestern Wyoming, south of the Tetons. In 1884 I saw the fresh hide of one that was killed in the Bighorn Mountains.

The wapiti, or round-horned elk, like the bison, and unlike the moose, had its centre of abundance in the United States, though extending northward into Canada. Originally its range reached from ocean to ocean and it went in herds of thousands of individuals; but it has suffered more from the persecution of hunters than any other game except the bison. By the beginning of this century it had been exterminated in most localities east of the Mississippi; but a few lingered on for many years in the Alleghanies. Colonel Cecil Clay informs me that an Indian whom he knew killed one in Pennsylvania in 1869. A very few still exist here and there in northern Michigan and Minnesota, and in one or two spots on the western boundary of Nebraska and the Dakotas; but it is now properly a beast of the wooded Western mountains. It is still plentiful in western Colorado, Wyoming, and Montana, and in parts of Idaho, Washington, and Oregon. Though not as large as the moose, it is the most beautiful and stately of all animals of the deer kind, and its antlers are marvels of symmetrical grandeur.

The woodland caribou is inferior to the wapiti both in size and symmetry. The tips of the many branches of its long, irregular antlers are slightly palmated. Its range is the same as that of the moose, save that it does not go so far southward. Its hoofs are long and round; even larger than the long, oval hoofs of the moose, and much larger than those of the wapiti. The tracks of all three can be told apart at a glance, and cannot be mistaken for the footprints of other game. Wapiti tracks, however, look much like those of yearling and two-year-old cattle, unless the ground is steep or muddy, in which case the marks of the false hoofs appear, the joints of wapiti being more flexible than those of domestic stock.

The whitetail deer is now, as it always has been, the best known and most abundant of American big game, and though its numbers have been greatly thinned it is still found in almost every State of the Union. The common blacktail or mule deer, which has likewise been sadly thinned in numbers, though once extraordinarily abundant, extends from the great plains to the Pacific; but is supplanted on the Puget Sound coast by the Columbian blacktail. The delicate, heart-shaped footprints of all three are nearly indistinguishable; when the animal is running the hoof points are, of course, separated. The track of the antelope is more oval, growing squarer with age. Mountain-sheep leave footmarks of a squarer shape, the points of the hoof making little indentations in the soil, well apart, even when the animal is only walking; and a yearling's track is not unlike that made by a big prongbuck when striding rapidly with the toes well apart. White-goat tracks are also square, and as large as those of the sheep; but there is less indentation of the hoof points, which come nearer together.

The antelope, or prongbuck, was once found in abundance from the eastern edge of the great plains to the Pacific, but it has everywhere diminished in numbers, and has been exterminated along the eastern and western borders of its former range. The bighorn, or mountain-sheep, is found in the Rocky Mountains from northern Mexico to Alaska; and in the United States from the Coast and Cascade ranges to the Bad Lands of the western edges of the Dakotas, wherever there are mountain chains or tracts of rugged hills. It was never very abundant, and, though it has become less so, it has held its own better than most game. The white goat, however, alone among our game animals, has positively increased in numbers since the advent of settlers; because white hunters rarely follow it, and the Indians who once sought its skin for robes now use blankets instead. Its true home is in Alaska and Canada, but it crosses our borders along the lines of the Rockies and Cascades, and a few small isolated colonies are found here and there southward to California and New Mexico.

The cougar and wolf, once common throughout the United States, have now completely disappeared from all save the wildest regions. The black bear holds its own better; it was never found on the great plains. The huge grizzly ranges from the great plains to the Pacific. The little peccary, or Mexican wild hog, merely crosses our southern border.

The finest hunting-ground in America was, and indeed is, the mountainous region of western Montana and northwestern Wyoming. In this high, cold land of lofty mountains, deep forests, and open prairies, with its beautiful lakes and rapid rivers, all the species of big game mentioned above, except the peccary and Columbian blacktail, are to be found. Until 1880 they were very abundant, and they are still, with the exception of the bison, fairly plentiful. On most of the long hunting expeditions which I made away from my ranch, I went into this region.

The bulk of my hunting has been done in the cattle country, near my ranch on the Little Missouri, and in the adjoining lands round the lower

Powder and Yellowstone. Until 1881 the valley of the Little Missouri was fairly thronged with game, and was absolutely unchanged in any respect from its original condition of primeval wildness. With the incoming of the stockmen all this changed, and the game was wofully slaughtered; but plenty of deer and antelope, a few sheep and bear, and an occasional elk are still left.

Since the professional hunters have vanished with the vast herds of game on which they preyed, the life of the ranchman is that which yields most chance of hunting. Life on a cattle-ranch, on the great plains or among the foot-hills of the high mountains, has a peculiar attraction for those hardy, adventurous spirits who take most kindly to a vigorous out-of-door existence, and who are therefore most apt to care passionately for the chase of big game. The free ranchman lives in a wild, lonely country, and exactly as he breaks and tames his own horses and guards and tends his own branded herds, so he takes the keenest enjoyment in the chase, which is to him not merely the pleasantest of sports, but also a means of adding materially to his comforts, and often his only method of providing himself with fresh meat.

Hunting in the wilderness is of all pastimes the most attractive, and it is doubly so when not carried on merely as a pastime. Shooting over a private game-preserve is of course in no way to be compared to it. The wilderness hunter must not only show skill in the use of the rifle and address in finding and approaching game, but he must also show the qualities of hardihood, self-reliance, and resolution needed for effectively grappling with his wild surroundings. The fact that the hunter needs the game, both for its meat and for its hide, undoubtedly adds a zest to the pursuit. Among the hunts which I have most enjoyed were those made when I was engaged in getting in the winter's stock of meat for the ranch, or was keeping some party of cowboys supplied with game from day to day.

Aldo Leopold

Wilderness as a Form of Land Use (1925)

FROM THE EARLIEST TIMES one of the principal criteria of civilization has been the ability to conquer the wilderness and convert it to economic use. To deny the validity of this criterion would be to deny history. But because the conquest of wilderness has produced beneficial reactions on social, political, and economic development, we have set up, more or less unconsciously, the converse assumption that the ultimate social, political, and economic development will be produced by conquering the wilderness entirely—that is, by eliminating it from our environment.

My purpose is to challenge the validity of such an assumption and to show how it is inconsistent with certain cultural ideas which we regard as most distinctly American.

Our system of land use is full of phenomena which are sound as tendencies but become unsound as ultimates. It is sound for a city to grow but unsound for it to cover its entire site with buildings. It was sound to cut down our forests but unsound to run out of wood. It was sound to expand our agriculture, but unsound to allow the momentum of that expansion to result in the present overproduction. To multiply examples of an obvious truth would be tedious. The question, in brief, is whether the benefits of wilderness-conquest will extend to ultimate wilderness-elimination.

The question is new because in America the point of elimination has only recently appeared upon the horizon of foreseeable events. During our four centuries of wilderness-conquest the possibility of disappearance has been too remote to register in the national consciousness. Hence we have no mental language in which to discuss the matter. We must first set up some ideas and definitions.

WHAT IS A WILDERNESS AREA?

The term wilderness, as here used, means a wild, roadless area where those who are so inclined may enjoy primitive modes of travel and subsistence, such as exploration trips by pack-train or canoe.

The first idea is that wilderness is a resource, not only in the physical sense of the raw materials it contains, but also in the sense of a distinctive environment which may, if rightly used, yield certain social values. Such a conception ought not to be difficult, because we have lately learned to think of other forms of land use in the same way. We no longer think of a municipal golf links, for instance, as merely soil and grass.

The second idea is that the value of wilderness varies enormously with location. As with other resources, it is impossible to dissociate value from location. There are wilderness areas in Siberia which are probably very similar in character to parts of our Lake states, but their value to us is negligible, compared with what the value of a similar area in the Lake states would be, just as the value of a golf links would be negligible if located so as to be out of reach of golfers.

The third idea is that wilderness, in the sense of an environment as distinguished from a quantity of physical materials, lies somewhere between the class of non-reproducible resources like minerals, and the reproducible resources like forests. It does not disappear proportionately to use, as minerals do, because we can conceive of a wild area which, if properly administered, could be traveled indefinitely and still be as good as ever. On the other hand, wilderness certainly cannot be built at will, like a city park or a tennis court. If we should tear down improvements already made in order to build a wilderness, not only would the cost be prohibitive, but the result would probably be highly dissatisfying. Neither can a wilderness be grown like timber, because it is something more than trees. The practical point is that if we want wilderness, we must foresee our want and preserve the proper areas against the encroachment of inimical uses.

Fourth, wilderness exists in all degrees, from the little accidental wild spot at the head of a ravine in a Corn Belt woodlot to vast expanses of virgin country—

> Where nameless men by nameless rivers wander
> And in strange valleys die strange deaths alone.

What degree of wilderness, then, are we discussing? The answer is, *all degrees*. Wilderness is a relative condition. As a form of land use it cannot be a rigid entity of unchanging content, exclusive of all other forms. On the contrary, it must be a flexible thing, accommodating itself to other forms and blending with them in that highly localized give-and-take scheme of land-planning which employs the criterion of "highest use." By skilfully adjusting one use to another, the land planner builds a balanced whole without undue sacrifice of any function, and thus attains a maximum net utility of land.

Just as the application of the park idea in civic planning varies in degree from the provision of a public bench on a street corner to the establishment of a municipal forest playground as large as the city itself, so should the application of the wilderness idea vary in degree from the wild, roadless spot of a few acres left in the rougher parts of public forest devoted to timber-growing, to wild, roadless regions approaching in size a whole national forest or a whole national park. For it is not to be supposed that a public wilderness area is a new kind of public land reservation, distinct from public forests and public parks. It is rather a new kind of land-dedication within our system of public forests and parks, to be duly correlated with dedications to the other uses which that system is already obligated to accommodate.

Lastly, to round out our definitions, let us exclude from practical consideration any degree of wilderness so absolute as to forbid reasonable protection. It would be idle to discuss wilderness areas if they are to be left subject to destruction by forest fires, or wide open to abuse. Experience has demonstrated, however, that a very modest and unobtrusive framework of trails, telephone line and lookout stations will suffice for protective purposes. Such improvements do not destroy the wild flavor of the area, and are necessary if it is to be kept in usable condition.

WILDERNESS AREAS IN A BALANCED LAND SYSTEM

What kind of case, then, can be made for wilderness as a form of land use?

To preserve any land in a wild condition is, of course, a reversal of economic tendency, but that fact alone should not condemn the proposal. A study of the history of land utilization shows that good use is largely a matter of good balance—of wise adjustment between opposing tendencies. The modern movements toward diversified crops and live stock on the farm, conservation of eroding soils, forestry, range management, game management, public parks—all these are attempts to balance opposing tendencies that have swung out of counterpoise.

One noteworthy thing about good balance is the nature of the opposing tendencies. In its more utilitarian aspect, as seen in modern agriculture, the needed adjustment is between economic uses. But in the public park movement the adjustment is between an economic use, on the one hand, and a purely social use on the other. Yet, after a century of actual experience, even the most rigid economic determinists have ceased to challenge the wisdom of a reasonable reversal of economic tendency in favor of public parks.

I submit that the wilderness is a parallel case. The parallelism is not yet generally recognized because we do not yet conceive of the wilderness environment as a resource. The accessible supply has heretofore been unlimited, like the supply of air-power, or tide-power, or sunsets, and we do not recognize anything as a resource until the demand becomes commensurable with the supply.

Now after three centuries of overabundance, and before we have even realized that we are dealing with a non-reproducible resource, we have come to the end of our pioneer environment and are about to push its remnants into the Pacific. For three centuries that environment has determined the character of our development; it may, in fact, be said that, coupled with the character of our racial stocks, it is the very stuff America is made of. Shall we now exterminate this thing that made us American?

Ouspensky says that, biologically speaking, the determining characteristic of rational beings is that their evolution is self-directed. John Burroughs cites the opposite example of the potato bug, which, blindly obedient to the law of increase, exterminates the potato and thereby exterminates itself. Which are we?

WHAT THE WILDERNESS HAS CONTRIBUTED TO AMERICAN CULTURE

Our wilderness environment cannot, of course, be preserved on any considerable scale as an economic fact. But, like many other receding economic facts, it can be preserved for the ends of sport. But what is the justification of sport, as the word is here used?

Physical combat between men, for instance, for unnumbered centuries was an economic fact. When it disappeared as such, a sound instinct led us to preserve it in the form of athletic sports and games. Physical combat between men and beasts since first the flight of years began was an economic fact, but when it disappeared as such, the instinct of the race led us to hunt and fish for sport. The transition of these tests of skill from an economic to a social basis has in no way destroyed their efficacy as human experiences—in fact, the change may be regarded in some respects as an improvement.

Football requires the same kind of back-bone as battle but avoids its moral and physical retrogressions. Hunting for sport in its highest form is an improvement on hunting for food in that there has been added, to the test of skill, an ethical code which the hunter formulates for himself and must often execute without the moral support of bystanders.

In these cases the surviving sport is actually an improvement on the receding economic fact. Public wilderness areas are essentially a means for allowing the more virile and primitive forms of outdoor recreation to survive the receding economic fact of pioneering. These forms should survive because they likewise are an improvement on pioneering itself.

There is little question that many of the attributes most distinctive of America and Americans are the impress of the wilderness and the life that accompanied it. If we have any such thing as an American culture (and I think we have), its distinguishing marks are a certain vigorous individualism combined with ability to organize, a certain intellectual curiosity bent to practical ends, a lack of subservience to stiff social forms, and an intolerance of drones, all of which are the distinctive characteristics of successful pioneers. These, if anything, are the indigenous part of our Americanism, the qualities that set it apart as a new rather than an imitative contribution to civilization. Many observers see these qualities not only bred into our people, but built into our institutions. Is it not a bit beside the point for us to be so solicitous about preserving those institutions without giving so much as a thought to preserving the environment which produced them and which may now be one of our effective means of keeping them alive?

WILDERNESS LOCATIONS

But the proposal to establish wilderness areas is idle unless acted on before the wilderness has disappeared. Just what is the present status of wilderness remnants in the United States?

Large areas of half a million acres and upward are disappearing very rapidly, not so much by reason of economic need, as by extension of motor roads. Smaller areas are still relatively abundant in the mountainous parts of the country, and will so continue for a long time.

The disappearance of large areas is illustrated by the following instance: In 1910 there were six roadless regions in Arizona and New Mexico, ranging in size from half a million to a million acres, where the finest type of mountain wilderness pack trips could be enjoyed. Today roads have eliminated all but one area of about half a million acres.

In California there were seven large areas ten years ago, but today there are only two left unmotorized.

In the Lake states no large unmotorized playgrounds remain. The motor launch, as well as the motor road, is rapidly wiping out the remnants of canoe country.

In the Northwest large roadless areas are still relatively numerous. The land plans of the Forest Service call for exclusion of roads from several areas of moderate size.

Unless the present attempts to preserve such areas are greatly strengthened and extended, however, it may be predicted with certainty that, except in the Northwest, all of the large areas already in public ownership will be invaded by motors in another decade.

In selecting areas for retention as wilderness, the vital factor of location must be more decisively recognized. A few areas in the national forests of Idaho or Montana are better than none, but, after all, they will be of limited usefulness to the citizen of Chicago or New Orleans who has a great desire but a small purse and a short vacation. Wild areas in the poor lands of the Ozarks and the Lake states would be within his reach. For the great urban populations concentrated on the Atlantic seaboards, wild areas in both ends of the Appalachians would be especially valuable.

Are the remaining large wilderness areas disappearing so rapidly because they contain agricultural lands suitable for settlement? No; most of them are entirely devoid of either existing or potential agriculture. Is it because they contain timber which should be cut? It is true that some of them do contain valuable timber, and in a few cases this fact is leading to a legitimate extension of logging operations; but in most of the remaining wilderness the timber is either too thin and scattered for exploitation, or else the topography is too difficult for the timber alone to carry the cost of roads or railroads. In view of the general belief that lumber is being overproduced in relation to the growing scarcity of stumpage, and will probably so continue for several decades, the sacrifice of wilderness for timber can hardly be justified on grounds of necessity.

Generally speaking, it is not timber, and certainly not agriculture, which is causing the decimation of wilderness areas, but rather the desire to attract tourists. The accumulated momentum of the good-roads movement constitutes a mighty force, which, skilfully manipulated by every little mountain village possessed of a chamber of commerce and a desire to become a metropolis, is bringing about the extension of motor roads into every remaining bit of wild country, whether or not there is economic justification for the extension.

Our remaining wild lands are wild because they are poor. But this poverty does not deter the booster from building expensive roads through them as bait for motor tourists.

I am not without admiration for this spirit of enterprise in backwoods villages, nor am I attempting a censorious pose toward the subsidization of their ambitions from the public treasuries; nor yet am I asserting that the resulting roads are devoid of any economic utility. I do maintain, (1) that such extensions of our road systems into the wilderness are seldom yielding a return sufficient to amortize the public investment; (2) that even where they do yield such a return, their construction is not necessarily in the public interest, any more than obtaining an economic return from the last vacant lot in a parkless city would be in the public interest. On the contrary, the public interest demands the careful planning of a system of wilderness areas and the permanent reversal of the ordinary economic process within their borders.

To be sure, to the extent that the motor-tourist business is the cause of invasion of these wilderness playgrounds, one kind of recreational use is merely substituted for another. But this substitution is a vitally serious matter from the point of view of good balance. It is just as unwise to devote 100% of the recreational resources of our public parks and forests to motorists as it would be to devote 100% of our city parks to merry-go-rounds. It would be just as unreasonable to ask the aged to indorse a park with only swings and trapezes, or the children a park with only benches, or the motorists a park with only bridlepaths, as to ask the wilderness recreationist to indorse a universal priority for motor roads. Yet that is what our land plans—or rather lack of them—are now doing; and so sacred is our dogma of "development" that there is no effective protest. The inexorable molding of the individual American to a standardized pattern in his economic activities makes all the more undesirable this unnecessary standardization of his recreational tastes.

PRACTICAL ASPECTS OF ESTABLISHING WILDERNESS AREAS

Public wilderness playgrounds differ from all other public areas in that both their establishment and maintenance would entail very low costs. The wilderness is the one kind of public land that requires no improvements. To be sure, a simple system of fire protection and administrative patrol would be required, but the cost would not exceed two or three cents per acre per year. Even that would not usually be a new cost, since the greater part of the needed areas are already under administration in the rougher parts of the national forests and parks. The action needed is the permanent differentiation of a suitable system of wild areas within our national park and forest system.

In regions such as the Lake states, where the public domain has largely disappeared, lands would have to be purchased; but that will have to be done, in any event, to round out our park and forest system. In such cases a lesser degree of wilderness may have to suffice, the only ordinary utilities practicable to exclude being cottages, hotels, roads, and motor boats.

The retention of certain wild areas in both national forests and national parks will introduce a healthy variety into the wilderness idea itself, the forest areas serving as public hunting grounds, the park areas as public wild-life sanctuaries, and both kinds as public playgrounds in which the wilderness environments and modes of travel may be preserved and enjoyed.

THE CULTURAL VALUE OF WILDERNESS

Are these things worth preserving? This is the vital question. I cannot give an unbiased answer. I can only picture the day that is almost upon us when canoe travel will consist in paddling in the noisy wake of a motor launch and portaging through the back yard of a summer cottage. When that day comes, canoe travel will be dead, and dead, too, will be a part of our Americanism. Joliet and LaSalle will be words in a book, Champlain will be a blue spot on a map, and canoes will be merely things of wood and canvas, with a connotation of white duck pants and bathing "beauties."

The day is almost upon us when a pack-train must wind its way up a graveled highway and turn out its bell-mare in the pasture of a summer hotel. When that day comes the pack-train will be dead, the diamond hitch will be merely rope, and Kit Carson and Jim Bridger will be names in a history lesson. Rendezvous will be French for "date," and Forty-Nine will be the number preceding fifty. And thenceforth the march of empire will be a matter of gasoline and four-wheel brakes.

European outdoor recreation is largely devoid of the thing that wilderness areas would be the means of preserving in this country. Europeans do not camp, cook, or pack in the woods for pleasure. They hunt and fish when they can afford it, but their hunting and fishing is merely hunting and fishing, staged in a setting of ready-made hunting lodges, elaborate fare, and hired beaters. The whole thing carries the atmosphere of a picnic rather than that of a pack trip. The test of skill is confined almost entirely to the act of killing, itself. Its value as a human experience is reduced accordingly.

There is a strong movement in this country to preserve the distinctive democracy of our field sports by preserving free hunting and fishing, as distinguished from the European condition of commercialized hunting and fishing privileges. Public shooting grounds and organized cooperative relations between sportsmen and landowners are the means proposed for keeping these sports within reach of the American of moderate means. Free hunting and fishing is a most worthy objective, but it deals with only one of the two distinctive characteristics of American sport. The other characteristic is that our test of skill is primarily the act of living in the open, and only secondarily the act of killing game. It is to preserve this primary characteristic that public wilderness playgrounds are necessary.

Herbert Hoover aptly says that there is no point in increasing the average American's leisure by perfecting the organization of industry, if the expansion of industry is allowed to destroy the recreational resources on which leisure may be beneficially employed. Surely the wilderness is one of the most valuable of these resources, and surely the building of unproductive roads in the wrong places at public expense is one of the least valuable of industries. If we are unable to steer the Juggernaut of our own prosperity, then surely there is an impotence in our vaunted Americanism that augurs ill for our future. The self-directed evolution of rational beings does not apply to us until we become collectively, as well as individually, rational and self-directing.

Wilderness as a form of land use is, of course, premised on a qualitative conception of progress. It is premised on the assumption that enlarging the range of individual experience is as important as enlarging the number of individuals; that the expansion of commerce is a means, not an end; that the environment of the American pioneers had values of its own, and was not merely a punishment which they endured in order that we might ride in motors. It is premised on the assumption that the rocks and rills and templed hills of this America are something more than economic materials, and should not be dedicated exclusively to economic use.

The vanguard of American thought on the use of land has already recognized all this, in theory. Are we too poor in spirit, in pocket, or in idle acres to recognize it likewise in fact?

Robert Marshall

The Problem of the Wilderness (1930)

I

IT IS APPALLING TO REFLECT how much useless energy has been expended in arguments which would have been inconceivable had the terminology been defined. In order to avoid such futile controversy I shall undertake at the start to delimit the meaning of the principal term with which this paper is concerned. According to Dr. Johnson a *wilderness* is "a tract of solitude and savageness," a definition more poetic than explicit. Modern lexicographers do better with "a tract of land, whether a forest or a wide barren plain, uncultivated and uninhabited by human beings."[1] This definition gives a rather good foundation, but it still leaves a penumbra of partially shaded connotation.

For the ensuing discussion I shall use the word *wilderness* to denote a region which contains no permanent inhabitants, possesses no possibility of conveyance by any mechanical means and is sufficiently spacious that a person in crossing it must have the experience of sleeping out. The dominant attributes of such an area are: first, that it requires any one who exists in it to depend exclusively on his own effort for survival; and second, that it preserves as nearly as possible the primitive environment. This means that all roads, power transportation and settlements are barred. But trails and temporary shelters, which were common long before the advent of the white race, are entirely permissible.

When Columbus effected his immortal debarkation, he touched upon a wilderness which embraced virtually a hemisphere. The philosophy that progress is proportional to the amount of alteration imposed upon nature never seemed to have occurred to the Indians. Even such tribes as the Incas, Aztecs and Pueblos made few changes in the environment in which they were born. "The land and all that it bore they treated with consideration; not attempting to improve it, they never desecrated it."[2] Consequently, over billions of acres the aboriginal wanderers still spun out their peripatetic careers, the wild animals still browsed in unmolested meadows and the forests still grew and moldered and grew again precisely as they had done for undeterminable centuries.

It was not until the settlement of Jamestown in 1607 that there appeared the germ for that unabated disruption of natural conditions which has characterized all subsequent American history. At first expansion was very slow. The most intrepid seldom advanced further from their neighbors than the next drainage. At the time of the Revolution the zone of civilization was still practically confined to a narrow belt lying between the Atlantic Ocean and the Appalachian valleys. But a quarter of a century later, when the Louisiana Purchase was consummated, the outposts of civilization had reached the Mississippi, and there were foci of colonization in half a dozen localities west of the Appalachians, though the unbroken line of the frontier was east of the mountains.[3]

It was yet possible as recently as 1804 and 1805 for the Lewis and Clark Expedition to cross two thirds of a continent without seeing any culture more advanced than that of the Middle Stone Age. The only routes of travel were the uncharted rivers and the almost impassable Indian trails. And continually the expedition was breaking upon some "truly magnificent and sublimely grand object, which has from the commencement of time been concealed from the view of civilized man."[4]

This exploration inaugurated a century of constantly accelerating emigration such as the world had never known. Throughout this frenzied period the only serious thought ever devoted to the wilderness was how it might be demolished. To the pioneers pushing westward it was an enemy of diabolical cruelty and danger, standing as the great obstacle to industry and development. Since these seemed to constitute the essentials for felicity, the obvious step was to excoriate the devil which interfered. And so the path of empire proceeded to substitute for the undisturbed seclusion of nature the conquering accomplishments of man. Highways wound up valleys which had known only the footsteps of the wild animals; neatly planted gardens and orchards replaced the tangled confusion of the primeval forest; factories belched up great clouds of smoke where for centuries trees had transpired toward the sky, and the ground-cover of fresh sorrel and twinflower was transformed to asphalt spotted with chewing-gum, coal dust and gasoline.

To-day there remain less than twenty wilderness areas of a million acres, and annually even these shrunken remnants of an undefiled continent are being despoiled. Aldo Leopold has truly said:

> The day is almost upon us when canoe travel will consist in paddling up the noisy wake of a motor launch and portaging through the back yard of a summer cottage. When that day comes canoe travel will be dead, and dead too will be a part of our Americanism. . . . The day is almost upon us when a pack train must wind its way up a graveled highway and turn out its bell mare in the pasture of a summer hotel. When that day comes the pack train will be dead, the diamond hitch will be merely a rope and Kit Carson and Jim Bridger will be names in a history lesson.[5]

Within the next few years the fate of the wilderness must be decided. This is a problem to be settled by deliberate rationality and not by personal prejudice. Fundamentally, the question is one of balancing the total happiness which will be obtainable if the few undesecrated areas are perpetuated against that which will prevail if they are destroyed. For this purpose it will be necessary: first, to consider the extra-ordinary benefits of the wilderness; second, to enumerate the drawbacks to undeveloped areas; third, to evaluate the relative importance of these conflicting factors, and finally, to formulate a plan of action.

II

The benefits which accrue from the wilderness may be separated into three broad divisions: the physical, the mental and the esthetic.

Most obvious in the first category is the contribution which the wilderness makes to health. This involves something more than pure air and quiet, which are also attainable in almost any rural situation. But toting a fifty-pound pack over an abominable trail, snowshoeing across a blizzard-swept plateau or scaling some jagged pinnacle which juts far above timber all develop a body distinguished by a soundness, stamina and élan unknown amid normal surroundings.

More than mere heartiness is the character of physical independence which can be nurtured only away from the coddling of civilization. In a true wilderness if a person is not qualified to satisfy all the requirements of existence, then he is bound to perish. As long as we prize individuality and competence it is imperative to provide the opportunity for complete self-sufficiency. This is inconceivable under the effete superstructure of urbanity; it demands the harsh environment of untrammeled expanses.

Closely allied is the longing for physical exploration which bursts through all the chains with which society fetters it. Thus we find Lindbergh, Amundsen, Byrd gaily daring the unknown, partly to increase knowledge, but largely to satisfy the craving for adventure. Adventure, whether physical or mental, implies breaking into unpenetrated ground, venturing beyond the boundary of normal aptitude, extending oneself to the limit of capacity, courageously facing peril. Life without the chance for such exertions would be for many persons a dreary game, scarcely bearable in its horrible banality.

It is true that certain people of great erudition "come inevitably to feel that if life has any value at all, then that value comes in thought,"[6] and so they regard mere physical pleasures as puerile inconsequences. But there are others, perfectly capable of comprehending relativity and the quantum theory, who find equal ecstasy in non-intellectual adventure. It is entirely irrelevant which view-point is correct; each is applicable to whoever entertains it. The important consideration is that both groups are entitled to indulge their penchant, and in the second instance this is scarcely possible without the freedom of the wilderness.

III

One of the greatest advantages of the wilderness is its incentive to independent cogitation. This is partly a reflection of physical stimulation, but more inherently due to the fact that original ideas require an objectivity and perspective seldom possible in the distracting propinquity of one's fellow men. It is necessary to "have gone behind the world of humanity, seen its institutions like toadstools by the wayside."[7] This theorizing is justified empirically by the number of America's most virile minds, including Thomas Jefferson, Henry Thoreau, Louis Agassiz, Herman Melville, Mark Twain, John Muir and William James, who have felt the compulsion of periodical retirements into the solitudes. Withdrawn from the contaminating notions of their neighbors, these thinkers have been able to meditate, unprejudiced by the immuring civilization.

Another mental value of an opposite sort is concerned not with incitement but with repose. In a civilization which requires most lives to be passed amid inordinate dissonance, pressure and intrusion, the chance of retiring now and then to the quietude and privacy of sylvan haunts becomes for some people a psychic necessity. It is only the possibility of convalescing in the wilderness which saves them from being destroyed by the terrible neural tension of modern existence.

There is also a psychological bearing of the wilderness which affects, in contrast to the minority who find it indispensable for relaxation, the whole of human kind. One of the most profound discoveries of psychology has been the demonstration of the terrific harm caused by suppressed desires. To most of mankind a very powerful desire is the appetite for adventure. But in an age of machinery only the extremely fortunate have any occasion to satiate this hankering, except vicariously. As a result people become so choked by the monotony of their lives that they are readily amenable to the suggestion of any lurid diversion. Especially in battle, they imagine, will be found the glorious romance of futile dreams. And so they endorse war with enthusiasm and march away to stirring music, only to find their adventure a chimera, and the whole world miserable. It is all tragically ridiculous, and yet there is a passion there which can not be dismissed with a contemptuous reference to childish quixotism. William James has said that "militarism is the great preserver of ideals of hardihood, and human life with no use for hardihood would be contemptible."[8] The problem, as he points out, is to find a "moral equivalent of war," a peaceful stimulation for the hardihood and competence instigated in bloodshed. This equivalent may be realized if we make available to every one the harmless excitement of the wilderness. Bertrand Russell has skilfully amplified this idea in his essay on "Machines and the Emotions." He expresses the significant conclusion that "many men would cease to desire war if they had opportunities to risk their lives in Alpine climbing."[9]

IV

In examining the esthetic importance of the wilderness I will not engage in the unprofitable task of evaluating the preciousness of different sorts of beauty, as, for instance, whether an acronical view over the Grand Canyon is worth more than the Apollo of Praxiteles. For such a rating would always have to be based on a subjective standard, whereas the essential for any measure is impersonality. Instead of such useless metaphysics I shall call attention to several respects in which the undisputed beauty of the primeval, whatever its relative merit, is distinctly unique.

Of the myriad manifestations of beauty, only natural phenomena like the wilderness are detached from all temporal relationship. All the beauties in the creation of alteration of which man has played even the slightest role are firmly anchored in the historic stream. They are temples of Egypt, oratory of Rome, painting of the Renaissance or music of the Classicists. But in the wild places nothing is moored more closely than to geologic ages. The silent wanderer crawling up the rocky shore of the turbulent river could be a savage from some prehistoric epoch or a fugitive from twentieth century mechanization.

The sheer stupendousness of the wilderness gives it a quality of intangibility which is unknown in ordinary manifestations of ocular beauty. These are always very definite two or three dimensional objects which can be physically grasped and circumscribed in a few moments. But "the beauty that shimmers in the yellow afternoons of October, who ever could clutch it."[10] Any one who has looked across a ghostly valley at midnight, when moonlight makes a formless silver unity out of the drifting fog, knows how impossible it often is in nature to distinguish mass from hallucination. Any one who has stood upon a lofty summit and gazed over an inchoate tangle of deep canyons and cragged mountains, of sunlit lakelets and black expanses of forest, has become aware of a certain giddy sensation that there are no distances, no measures, simply unrelated matter rising and falling without any analogy to the banal geometry of breadth, thickness and height. A fourth dimension of immensity is added which makes the location of some dim elevation outlined against the sunset as incommensurable to the figures of the topographer as life itself is to the quantitative table of elements which the analytic chemist proclaims to constitute vitality.

Because of its size the wilderness also has a physical ambiency about it which most forms of beauty lack. One looks from outside at works of art and architecture, listens from outside to music or poetry. But when one looks at and listens to the wilderness he is encompassed by his experience of beauty, lives in the midst of his esthetic universe.

A fourth peculiarity about the wilderness is that it exhibits a dynamic beauty. A Beethoven symphony or a Shakespearean drama, a landscape by Corot or a Gothic cathedral, once they are finished become virtually static. But the wilderness is in constant flux. A seed germinates, and a stunted seedling battles for decades against the dense shade of the virgin forest. Then some ancient tree blows down and the long-suppressed plant suddenly enters into the full vigor of delayed youth, grows rapidly from sapling to maturity, declines into the conky senility of many centuries, dropping millions of seeds to start a new forest upon the rotting debris of its own ancestors, and eventually topples over to admit the sunlight which ripens another woodland generation.

Another singular aspect of the wilderness is that it gratifies every one of the senses. There is unanimity in venerating the sights and sounds of the forest. But what are generally esteemed to be the minor senses should not be slighted. No one who has ever strolled in springtime through seas of blooming violets, or lain at night on boughs of fresh balsam, or walked across dank holms in early morning can omit odor from the joys of the primordial environment. No one who has felt the stiff wind of mountaintops or the softness of untrodden sphagnum will forget the exhilaration experienced through touch. "Nothing ever tastes as good as when it's cooked in the woods" is a trite tribute to another sense. Even equilibrium causes a blithe exultation during many a river crossing on tenuous foot log and many a perilous conquest of precipice.

Finally, it is well to reflect that the wilderness furnishes perhaps the best opportunity for pure esthetic enjoyment. This requires that beauty be observed as a unity, and that for the brief duration of any pure esthetic experience the cognition of the observed object must completely fill the spectator's cosmos. There can be no extraneous thoughts—no question about the creator of the phenomenon, its structure, what it resembles or what vanity in the beholder it gratifies. "The purely esthetic observer has for the moment forgotten his own soul";[11] he has only one sensation left and that is exquisiteness. In the wilderness, with its entire freedom from the manifestations of human will, that perfect objectivity which is essential for pure esthetic rapture can probably be achieved more readily than among any other forms of beauty.

V

But the problem is not all one-sided. Having discussed the tremendous benefits of the wilderness, it is now proper to ponder upon the disadvantages which uninhabited territory entails.

In the first place, there is the immoderate danger that a wilderness without developments for fire protection will sooner or later go up in smoke and down in ashes.

A second drawback is concerned with the direct economic loss. By locking up wilderness areas we as much as remove from the earth all the lumber, minerals, range land, water-power and agricultural possibilities which they contain. In the face of the tremendous demand for these resources it seems unpardonable to many to render nugatory this potential material wealth.

A third difficulty inherent in undeveloped districts is that they automatically preclude the bulk of the population from enjoying them. For it is admitted that at present only a minority of the genus *Homo* cares for wilderness recreation, and only a fraction of this minority possesses the requisite virility for the indulgence of this desire. Far more people can enjoy the woods by automobile. Far more would prefer to spend their vacations in luxurious summer hotels set on well-groomed lawns than in leaky, fly-infested shelters bundled away in the brush. Why then should this majority have to give up its rights?

VI

As a result of these last considerations the irreplaceable values of the wilderness are generally ignored, and a fatalistic attitude is adopted in regard to the ultimate disappearance of all unmolested localities. It is my contention that this outlook is entirely unjustified, and that almost all the disadvantages of the wilderness can be minimized by forethought and some compromise.

The problem of protection dictates the elimination of undeveloped areas of great fire hazard. Furthermore, certain infringements on the concept of an unsullied wilderness will be unavoidable in almost all instances. Trails, telephone lines and lookout cabins will have to be constructed, for without such precaution most forests in the west would be gutted. But even with these improvements the basic primitive quality still exists: dependence on personal effort for survival.

Economic loss could be greatly reduced by reserving inaccessible and unproductive terrain. Inasmuch as most of the highly valuable lands have already been exploited, it should be easy to confine a great share of the wilderness tracts to those lofty mountain regions where the possibility of material profit is unimportant. Under these circumstances it seems like the grossest illogicality for any one to object to the withdrawal of a few million acres of low-grade timber for recreational purposes when one hundred million acres of potential forest lie devastated.[12] If one tenth portion of this denuded land were put to its maximum productivity, it could grow more wood than all the proposed wilderness areas put together. Or if our forests, instead of attaining only 22 per cent of their possible production,[13] were made to yield up to capacity, we could refrain from using three quarters of the timber in the country and still be better off than we are to-day. The way to meet our commercial demands is not to thwart legitimate divertisement, but to eliminate the unmitigated evils of fire and destructive logging. It is time we appreciated that the real economic problem is to see how little land need be employed for timber production, so that the remainder of the forest may be devoted to those other vital uses incompatible with industrial exploitation.

Even if there should be an underproduction of timber, it is well to recall that it is much cheaper to import lumber for industry than to export people for pastime. The freight rate from Siberia is not nearly as high as the passenger rate to Switzerland.

What small financial loss ultimately results from the establishment of wilderness areas must be accepted as a fair price to pay for their unaccessible preciousness. We spend about twenty-one billion dollars a year for entertainment of all sorts.[14] Compared with this there is no significance to the forfeiture of a couple of million dollars of annual income, which is all that our maximum wilderness requirements would involve. Think what an enormously greater sum New York City alone sacrifices in the maintenance of Central Park.

But the automobilists argue that a wilderness domain precludes the huge majority of recreation-seekers from deriving any amusement whatever from it. This is almost as irrational as contending that because more people enjoy bathing than art exhibits therefore we should change our picture galleries into swimming pools. It is undeniable that the automobilist has more roads than he can cover in a lifetime. There are upward of 3,000,000[15] miles of public highways in the United States, traversing many of the finest scenic features in the nation. Nor would the votaries of the wilderness object to the construction of as many more miles in the vicinity of the old roads, where they would not be molesting the few remaining vestiges of the primeval. But when the motorists also demand for their particular diversion the insignificant wilderness residue, it makes even a Midas appear philanthropic.

> Such are the differences among human beings in their sources of pleasure, that unless there is a corresponding diversity in their modes of life, they neither obtain their fair share of happiness, nor grow up to the mental, moral and esthetic stature of which their nature is capable. Why then should tolerance extend only to tastes and modes of life which extort acquiescence by the multitude of their adherents?[16]

It is of the utmost importance to concede the right of happiness also to people who find their delight in unaccustomed ways. This prerogative is valid even though its exercise may encroach slightly on the fun of the majority, for there is a point where an increase in the joy of the many causes a decrease in the joy of the few out of all proportion to the gain of the former. This has been fully recognized not only by such philosophers of democracy as Paine, Jefferson and Mill, but also in the practical administration of governments which spend prodigious sums of money to satisfy the expensive wants of only a fragment of the community. Public funds which could bring small additional happiness to the majority are diverted to support museums, art galleries, concerts, botanical gardens, menageries and golf-links. While these, like wilderness areas, are open to the use of every one, they are vital to only a fraction of the entire population. Nevertheless, they are almost universally approved, and the appropriations to maintain them are growing phenomenally.

VII

These steps of reasoning lead up to the conclusion that the preservation of a few samples of undeveloped territory is one of the most clamant issues before us today. Just a few years more of hesitation and the only trace of that wilderness which has exerted such a fundamental influence in molding American character will lie in the musty pages of pioneer books and the mumbled memories of tottering antiquarians. To avoid this catastrophe demands immediate action.

A step in the right direction has already been initiated by the National Conference on Outdoor Recreation,[17] which has proposed twenty-one possible wilderness areas. Several of these have already been set aside in a tentative way by the Forest Service; others are undergoing more careful scrutiny. But this only represents the incipiency of what ought to be done.

A thorough study should forthwith be undertaken to determine the probable wilderness needs of the country. Of course, no precise reckoning could be attempted, but a radical calculation would be feasible. It ought to be radical for three reasons: because it is easy to convert a natural area to industrial or motor usage, impossible to do the reverse; because the population which covets wilderness recreation is rapidly enlarging and because the higher standard of living which may be anticipated should give millions the economic power to gratify what is to-day merely a pathetic yearning. Once the estimate is formulated, immediate steps should be taken to establish enough tracts to insure every one who hungers for it a generous opportunity of enjoying wilderness isolation.

To carry out this program it is exigent that all friends of the wilderness ideal should unite. If they do not present the urgency of their view-point the other side will certainly capture popular support. Then it will only be a few years until the last escape from society will be barricaded. If that day arrives there will be countless souls born to live in strangulation, countless human beings who will be crushed under the artificial edifice raised by man. There is just one hope of repulsing the tyrannical ambition of civilization to conquer every niche on the whole earth. That hope is the organization of spirited people who will fight for the freedom of the wilderness.

NOTES

1. Webster's New International Dictionary.
2. Willa Cather, "Death Comes for the Archbishop."
3. Frederic L. Paxson, "History of the American Frontier."
4. Reuben G. Thwaites, "Original Journals of the Lewis and Clark Expedition, 1804–1806," June 13, 1805.
5. Aldo Leopold, "The Last Stand of the Wilderness," *American Forests and Forest Life,* October, 1925.
6. Joseph Wood Krutch, "The Modern Temper."
7. Henry David Thoreau, "Journals," April 2, 1852.
8. William James, "The Moral Equivalent of War."
9. Bertrand Russell, "Essays in Scepticism."
10. Ralph Waldo Emerson, "Nature."
11. Irwin Edman, "The World, the Arts and the Artist."
12. George P. Ahern, "Deforested America," Washington, D.C.
13. U.S. Department of Agriculture, "Timber, Mine or Crop?"
14. Stuart Chase, "Whither Mankind?"
15. "The World Almanac," 1929.
16. John Stuart Mill, "On Liberty."
17. National Conference on Outdoor Recreation, "Recreation Resources of Federal Lands," Washington, D.C.

Sigurd Olson

Why Wilderness? (1938)

IN SOME MEN, the need of unbroken country, primitive conditions and intimate contact with the earth is a deeply rooted cancer gnawing forever at the illusion of contentment with things as they are. For months or years this hidden longing may go unnoticed and then, without warning, flare forth in an all consuming passion that will not bear denial. Perhaps it is the passing of a flock of wild geese in the spring, perhaps the sound of running water, or the smell of thawing earth that brings the transformation. Whatever it is, the need is more than can be borne with fortitude and for the good of their families and friends, and their own particular restless souls, they head toward the last frontiers and escape.

I have seen them come to the "jumping off places" of the North, these men whereof I speak. I have seen the hunger in their eyes, the torturing hunger for action, distance and solitude, and a chance to live as they will. I know these men and the craving that is theirs; I know also that in the world today there are only two types of experience which can put their minds at peace, the way of wilderness or the way of war.

As a guide in the primitive lake regions of the Hudson's Bay watershed, I have lived with men from every walk of life, have learned to know them more intimately than their closest friends at home, their dreams, their hopes, their aspirations. I have seen them come from the cities down below, worried and sick at heart, and have watched them change under the stimulus of wilderness living into happy, carefree, joyous men, to whom the successful taking of a trout or the running of a rapids meant far more than the rise and fall of stocks and bonds. Ask these men what it is they have found and it would be difficult for them to say. This they do know, that hidden back there in the country beyond the steel and the traffic of towns is something real, something as definite as life itself, that for some reason or other is an answer and a challenge to civilization.

At first, I accepted the change that was wrought with the matter of factness of any woodsman, but as the years went by I began to marvel at the infallibility of the wilderness formula. I came to see that here was a way of life as necessary and as deeply rooted in some men as the love of home and family, a vital cultural aspect of life which brought happiness and lasting content.

The idea of wilderness enjoyment is not new. Through our literature we find abundant reference to it, but seldom of the virile, masculine type of experience men need today. Since the beginning of time poets have sung of the healing power of solitude and of communion with nature, but for them the wilderness meant the joys of contemplation. Typical of this tone of interpretation is Thoreau with his "tonic of wildness," but to the men I have come to know his was an understanding that did not begin to cover what they feel. To him, the wild meant the pastoral meadows of Concord and Walden Pond, and the joy he had, though unmistakably genuine, did not approach the fierce, unquenchable desire of my men of today. For them the out-of-doors is not enough; nor are the delights of meditation. They need the sense of actual struggle and accomplishment, where the odds are real and where they know that they are no longer playing make believe. These men need more than picnics, purling streams, or fields of daffodils to stifle their discontent, more than mere solitude and contemplation to give them peace.

Burroughs, another lover of the out-of-doors, spoke often of the wilderness, but knew it not at all. When he regretted having to leave Old Slabsides on the Hudson for the wilds of Alaska and the West, we knew there was little of the primitive urge in his nature. The birds, the common phenomena of the passing seasons, and work in his vineyard satisfied abundantly his need of reality and physical contact with the earth. For him the wild had little charm. As we explore our literature for men who have felt deeply about wilderness we find them few indeed, perhaps because in the past there was wilderness enough and men had not learned to wean themselves so completely away from its influence. Invariably men wrote of the struggle and the dominating effect of wilderness as a mighty unconquered force and everywhere we find evidence of the part it played in molding the lives of those it touched. Fear was the keynote of the past, fear of the brooding monster of the unknown, and little of the joy of adventure and freedom is ever in evidence. Were it not for a few such daring souls as Joseph Conrad and Jack London, we would know little of the feeling some men have for the far places of the earth.

With the rapid elimination of the frontiers, due to increased facility of transportation and huge development programs, the opportunity to see and know real wilderness has become increasingly difficult. As it approaches the status of rarity for the first time in history, we see it not as something to be feared and subdued, not as an encumbrance to the advance of civilization, but instead as a distinctly cultural asset which contributes to spiritual satisfaction. The greater part of the old wilderness is gone, but during the centuries in which we fought our way through it we unconsciously absorbed its influence. Now as conquering invaders, we feel the need of the very elements which a short time ago we fought to eradicate. The wild has left its mark upon us and now that we have succeeded in surrounding ourselves with a complexity of new and often unnatural habits of daily living, we long for the old stimulus which only the unknown could give.

Why wilderness? No two men would have the same explanation. Something definite does happen to most men, however, when they hit the out trails of our last frontiers, and though they react in various ways there is a certain uniformity noticeable to one who has often seen them make the break with civilization. Whatever it is, they are changed almost overnight from the prosaic conformists they may have been, who dress, think and act like all the rest, to adventurers ready to die with their boots on, explorers pushing into the blue, once more members of a pioneering band.

It is surprising how quickly a man sheds the habiliments of civilization and how soon he feels at home in the wilds. Before many days have passed, he feels that the life he has been living was merely an interruption in a long wilderness existence and that now again he is back at the real business of living. And when we think of the comparatively short time that we have been living and working as we do now, when we recall that many of us are hardly a generation removed from the soil, and that a scant few thousand years ago our ancestors roamed and hunted the fastnesses of Europe, it is not strange that the smell of woodsmoke and the lure of the primitive is with us yet. Racial memory is a tenacious thing, and for some it is always easy to slip back into the deep grooves of the past. What we feel most deeply are those things which as a race we have been doing the longest, and the hunger men feel for the wilds and a roving life is natural evidence of the need of repeating a plan of existence that for untold centuries was common practice. It is still in our blood and many more centuries must pass before we lose much of its hold.

Civilized living in the great towns, with all their devices for comfort and convenience, is far too sudden a violation of slowly changing racial habit and we find that what gave men pleasure in the past—simple, primitive tasks and the ordinary phenomena of life in the open—today give the same satisfaction. Men have found at last that there is a penalty for too much comfort and ease, a penalty of lassitude and inertia and the frustrated feeling that goes with unreality. Certainly the adjustment for many has been difficult and it is those who must ever so often break their bonds and hie themselves away.

All do not feel the need and there are many perfectly content with life as they find it. They will always be the picnickers and the strollers, and for them are highways, gravelled trails and country clubs. For them scenic vistas of the wild from the shelter of broad and cool verandas. The others, those who cannot rest, are of a different breed. For them is sweat and toil, hunger and thirst, and the fierce satisfaction that only comes with hardship.

While wilderness means escape from the perplexing problems of everyday life and freedom from the tyranny of wires, bells, schedules and pressing responsibility, nevertheless, it may be at first a decided shock and days and even weeks may pass before men are finally aware that the tension is gone. When the realization does come, they experience a peace of mind and relaxation which a short time before would have seemed impossible. With this dramatic change of atmosphere comes an equally dramatic change in individual reactions as they feel that the need of front and reserve is gone.

I have seen staid educators, dignified surgeons, congressmen and admirals tie up their heads in gaudy bandannas, go shirtless to bring on the tan of the northern sun, and wear bowie knives in their belts. I have seen them glory in the muck of portages, fight the crashing combers on stormy lakes with the abandon of boys on their first adventure. I have heard them laugh as they haven't laughed for years and bellow old songs in the teeth of a gale. With their new found freedom and release many things become important that were half forgotten—sunsets, the coloring of clouds and leaves, reflections in the water. I can honestly say, that I have heard more laughter in a week out there than in any month in town. Men laugh and sing as naturally as breathing once the strain is gone.

With escape comes perspective. Far from the towns and all they denote, engrossed in their return to the old habits of wilderness living, men begin to wonder if the speed and pressure they have left are not a little senseless. Here where matters of food, shelter, rest and new horizons are all important, they begin to question the worthwhileness of their old objectives. Now they have long days with nothing to clutter their minds but the simple problems of wilderness living, and at last they have time to think. Then comes the transformation and, of a sudden, they are back to earth. Things move slowly, majestically in the wilds and the coming of the full moon in itself becomes of major importance. Countless natural phenomena begin to show themselves, things long forgotten and needing only the rejuvenating experience of actual contact to bring them back. With this, some of the old primitive philosophy works itself into their thinking, and in their new calm they forget to worry. Their own affairs seem trivial. Perspective? I sometimes think that men go to the wilds for that alone. Finding it means equilibrium, the long-time point of view so often lost in towns.

Ernest Holt, one-time guard to the late Colonel Fawcett on his first Amazon expedition, told me that in the depths of the jungle he experienced a spiritual uplift and sense of oneness with life that he could find nowhere else. I believe that here is a sensation born of perspective that most men know in any wilderness. Whenever it comes, men are conscious of a unity with the primal forces of creation and all life that swiftly annihilates the feeling of futility, frustration and unreality. When men realize that they are on their own, that if they are to be sheltered and fed and, what is more, return to civilization, they must depend entirely on their own ingenuity, everything they do assumes tremendous importance. Back home, mistakes can be made and easily excused or remedied, but here mistakes might cause discomfort or catastrophe. Knowing this makes all the difference in the world in a man's attitude toward the commonplace activities of daily life. Simple duties like the preparation of food, the taking of a fish, or the caching of supplies become fraught with import. Life soon develops a new and fascinating angle and days which to the uninitiated may seem humdrum or commonplace are filled with the adventure of living for its own sake. There is no make-believe here, but reality in the strictest sense of the word.

Men who have shared campfires together, who have known the pinch of hunger and what it means to cut a final cigarette in half two hundred miles from town, enjoy a comradeship that others never know. Only at war or on wilderness expeditions can this type of association be found, and I believe that it is this that men miss as much in civilized living as contact with the wild itself. I know a busy surgeon who once left his hospital operating room and traveled without thought of compensation a thousand miles through the bitter cold of midwinter to save the life of his guide, stricken with pneumonia. Nothing could have made him consider deserting his practice to take such a long hazardous trip but a call from a comrade in need. I stood at the bedside of that woodsman as he babbled incoherently of rapids and lakes and wilderness camps they had known together, and I knew then that here was a bond between men that could only be forged in the wilds, something deep and fine, something based on loyalty to open skies and distance and a way of life men need.

I do not advocate that the men of whom I speak allow the wilderness idea to claim all of their energy or enthusiasm. I do believe, however, that if for a short time each year it were possible for them to get away, not necessarily to the great wildernesses of the Arctic or the Canadian lakes, but to some wild part of the country which has not as yet been entirely caught up in some scheme of exploitation or development, that they would return to their friends and families strengthened and rejuvenated.

Why wilderness? Ask the men who have known it and who have made it part of their lives. They might not be able to explain, but your very question will kindle a light in eyes that have reflected the campfires of a continent, eyes that have known the glory of dawns and sunsets and nights under the stars. Wilderness to them is real and this they do know: when the pressure becomes more than they can stand, somewhere back of beyond, where roads and steel and towns are still forgotten, they will find release.

A. Starker Leopold (Chairman), S. A. Cain,
C. M. Cottam, I. N. Gabrielson, T. L. Kimball

Wildlife Management in the National Parks (1963) (or, The Leopold Report)

HISTORICAL

In the Congressional Act of 1916 which created the National Park Service, preservation of native animal life was clearly specified as one of the purposes of the parks. A frequently quoted passage of the Act states "... which purpose is to conserve the scenery and the natural historic objects and the wild life therein and to provide for the enjoyment of the same in such manner and by such means as will leave them unimpaired for the enjoyment of future generations."

In implementing this Act, the newly formed Park Service developed a philosophy of wildlife *protection,* which in that era was indeed the most obvious and immediate need in wildlife conservation. Thus the parks were established as refuges, the animal populations were protected from hunting and their habitats were protected from wildfire. For a time predators were controlled to protect the "good" animals from the "bad" ones, but this endeavor mercifully ceased in the 1930's. On the whole, there was little major change in the Park Service practice of wildlife management during the first 40 years of its existence.

During the same era, the concept of wildlife management evolved rapidly among other agencies and groups concerned with the production of wildlife for recreational hunting. It is now an accepted truism that maintenance of suitable habitat is the key to sustaining animal populations, and that protection, though it is important, is not of itself a substitute for habitat. Moreover, habitat is not a fixed or stable entity that can be set aside and preserved behind a fence, like a cliff dwelling or a petrified tree. Biotic communities change through natural stages of succession. They can be changed deliberately through manipulation of plant and animal populations. In recent years the National Park Service has broadened its concept of wildlife conservation to provide for purposeful management of plant and animal communities as an essential step in preserving wildlife resources "... unimpaired for the enjoyment of future generations." In a few parks active manipulation of habitat is being tested, as for example in the Everglades where controlled burning is now used experimentally to maintain the open glades and piney woods with their interesting animal and plant life. Excess populations of grazing ungulates are being controlled in a number of parks to preserve the forage plants on which the animals depend. The question already has been posed—how far should the National Park Service go in utilizing the tools of management to maintain wildlife populations?

THE CONCEPT OF PARK MANAGEMENT

The present report proposes to discuss wildlife management in the national parks in terms of three questions which shift emphasis progressively from the general to the specific:

1. What should be the *goals* of wildlife management in the national parks?
2. What general *policies* of management are best adapted to achieve the pre-determined goals?
3. What are some of the *methods* suitable for on-the-ground implementation of policies?

It is acknowledged that this Advisory Board was requested by the Secretary of the Interior to consider particularly one of the methods of management, namely, the procedure of removing excess ungulates from some of the parks. We feel that this specific question can only be viewed objectively in the light of goals and operational policies, and our report is framed accordingly. In speaking of national parks we refer to the whole system of parks and monuments; national recreation areas are discussed briefly near the end of the report.

As a prelude to presenting our thoughts on the goals, policies, and methods of managing wildlife in the parks of the United States we wish to quote in full a brief report on "Management of National Parks and Equivalent Areas" which was formulated by a committee of the First World Conference on National Parks that convened in Seattle in July, 1962. The committee consisted of 15 members of the Conference, representing eight nations; the chairman was Francois Bourliere of France. In our judgment this report suggests a firm basis for park management. The statement of the committee follows:

"1. Management is defined as an activity directed toward achieving or maintaining a given condition in plant and/or animal populations and/or habitats in accordance with the conservation plan for the area. A prior definition of the purposes and objectives of each park is assumed.

"Management may involve active manipulation of the plant and animal communities, or protection from modification or external influences.

"2. Few of the world's parks are large enough to be in fact self-regulatory ecological units; rather, most are ecological islands subject to direct or indirect modification by activities and conditions in the surrounding areas. These influences may involve such factors as immigration and/or emigration of animal and plant life, changes in the fire regime, and alterations in the surface or subsurface water.

"3. There is no need for active modification to maintain large examples of the relatively stable 'climax' communities which under protection perpetuate themselves indefinitely. Examples of such communities include large tracts of undisturbed rain-forest, tropical mountain paramos, and arctic tundra.

"4. However, most biotic communities are in a constant state of change due to natural or man-caused processes of ecological succession. In these 'successional' communities it is necessary to manage the habitat to achieve or stabilize it at a desired stage. For example, fire is an essential management tool to maintain East African open savanna or American prairie.

"5. Where animal populations get out of balance with their habitat and threaten the continued existence of a desired environment, population control becomes essential. This principle applies, for example, in situations where ungulate populations have exceeded the carrying capacity of their habitat through loss of predators, immigration from surrounding areas, or compression of normal migratory patterns. Specific examples include excess populations of elephants in some African parks and of ungulates in some mountain parks.

"6. The need for management, the feasibility of management methods, and evaluation of results must be based upon current and continuing scientific research. Both the research and management itself should be undertaken only by qualified personnel. Research, management planning, and execution must take into account, and if necessary regulate, the human uses for which the park is intended.

"7. Management based on scientific research is, therefore, not only desirable but often essential to maintain some biotic communities in accordance with the conservation plan of a national park or equivalent area."

THE GOAL OF PARK MANAGEMENT IN THE UNITED STATES

Item 1 in the report just quoted specifies that "a prior definition of the purposes and objectives of each park is assumed." In other words, the goal must first be defined.

As a primary goal, we would recommend that the biotic associations within each park be maintained, or where necessary recreated, as nearly as possible in the condition that prevailed when the area was first visited by the white man. A national park should represent a vignette of primitive America.

The implications of this seemingly simple aspiration are stupendous. Many of our national parks—in fact most of them—went through periods of indiscriminate logging, burning, livestock grazing, hunting and predator control. Then they entered the park system and shifted abruptly to a regime of equally unnatural protection from lightning fires, from insect outbreaks, absence of natural controls of ungulates, and in some areas elimination of normal fluctuations in water levels. Exotic vertebrates, insects, plants, and plant diseases have inadvertently been introduced. And of course lastly there is the factor of human use—or roads and trampling and camp grounds and pack stock. The resultant biotic associations in many of our parks are artifacts, pure and simple. They represent a complex ecologic history but they do not necessarily represent primitive America.

Restoring the primitive scene is not done easily nor can it be done completely. Some species are extinct. Given time, an eastern hardwood forest can be regrown to maturity but the chestnut will be missing and so will the roar of pigeon wings. The colorful drapanid finches are not to be heard again in the lowland forests of Hawaii, nor will the jack-hammer of the ivory-bill ring in southern swamps. The wolf and grizzly bear cannot readily be reintroduced into ranching communities, and the factor of human use of the parks is subject only to regulation, not elimination. Exotic plants, animals, and diseases are here to stay. All these limitations we full realize. Yet if the goal cannot be fully achieved it can be approached. A reasonable illusion of primitive America could be recreated, using the utmost in skill, judgment, and ecologic sensitivity. This in our opinion should be the objective of every national park and monument.

To illustrate the goal more specifically, let us cite some cases. A visitor entering Grand Teton National Park from the south drives across Antelope Flats. But there are no antelope. No one seems to be asking the question—why aren't there? If the mountain men who gathered here in rendezvous fed their squaws on antelope, a 20th century tourist at least should be able to see a band of these animals. Finding out what aspect of the range needs rectifying, and doing so, would appear to be a primary function of park management.

When the forty-niners poured over the Sierra Nevada into California, those that kept diaries spoke almost to a man of the wide-spaced columns of mature trees that grew on the lower western slope in gigantic magnificence. The ground was a grass parkland, in springtime carpeted with wildflowers. Deer and bears were abundant. Today much of the west slope is a dog-hair thicket of young pines, white fir, incense cedar, and mature brush—a direct function of overprotection from natural ground fires. Within the four national parks—Lassen, Yosemite, Sequoia, and Kings Canyon—the thickets are even more impenetrable than elsewhere. Not only is this accumulation of fuel dangerous to the giant sequoias and other mature trees but the animal life is meager, wildflowers are sparse, and to some at least the vegetative tangle is depressing, not uplifting. Is it possible that the primitive open forest could be restored, at least on a local scale?

And if so, how? We cannot offer an answer. But we are posing a question to which there should be an answer of immense concern to the National Park Service.

The scarcity of bighorn sheep in the Sierra Nevada represents another type of management problem. Though they have been effectively protected for nearly half a century, there are fewer than 400 bighorns in the Sierra. Two-thirds of them are found in summer along the crest which lies within the eastern border of Sequoia and Kings Canyon National Parks. Obviously, there is some shortcoming of habitat that precludes further increase in the population. The high country is still recovering slowly from the devastation of early domestic sheep grazing so graphically described by John Muir. But the present limitation may not be in the high summer range at all but rather along the eastern slope of the Sierra where the bighorns winter on lands in the jurisdiction of the Forest Service. These areas are grazed in summer by domestic livestock and large numbers of mule deer, and it is possible that such competitive use is adversely affecting the bighorns. It would seem to us that the National Park Service might well take the lead in studying this problem and in formulating cooperative management plans with other agencies even though the management problem lies outside the park boundary. The goal, after all, is to restore the Sierra bighorn. If restoration is achieved in the Sequoia–Kings Canyon region, there might follow a program of re-introduction and restoration of bighorns in Yosemite and Lassen National Parks, and Lava Beds National Monument, within which areas this magnificent native animal is presently extinct.

We hope that these examples clarify what we mean by the goal of park management.

POLICIES OF PARK MANAGEMENT

The major policy change which we would recommend to the National Park Service is that it recognize the enormous complexity of ecologic communities and the diversity of management procedures required to preserve them. The traditional, simple formula of protection may be exactly what is needed to maintain such climax associations as arctic-alpine heath, the rain forests of Olympic peninsula, or the Joshua trees and saguaros of southwestern deserts. On the other hand, grasslands, savannas, aspen, and other successional shrub and tree associations may call for very different treatment. Reluctance to undertake biotic management can never lead to a realistic presentation of primitive America, much of which supported successional communities that were maintained by fires, floods, hurricanes, and other natural forces.

A second statement of policy that we would reiterate—and this one conforms with present Park Service standards—is that management be limited to native plants and animals. Exotics have intruded into nearly all of the parks but they need not be encouraged, even those that have interest or ecologic values of their own. Restoration of antelope in Jackson Hole, for example, should be done by managing native forage plants, not by planting crested wheat grass or plots of irrigated alfalfa. Gambel quail in a desert wash should be observed in the shade of a mesquite, not a tamarisk. A visitor who climbs a volcano in Hawaii ought to see mamane trees and silverswords, not goats.

Carrying this point further, observable artificiality in any form must be minimized and obscured in every possible way. Wildlife should not be displayed in fenced enclosures; this is the function of a zoo, not a national park. In the same category is artificial feeding of wildlife. Fed bears become bums, and dangerous. Fed elk deplete natural ranges. Forage relationships in wild animals should be natural. Management may at times call for the use of the tractor, chain-saw, rifle, or flame-thrower but the signs and sounds of such activity should be hidden from visitors insofar as possible. In this regard, perhaps the most dangerous tool of all is the roadgrader. Although the American public demands automotive access to the parks, road systems must be rigidly prescribed as to extent and design. Roadless wilderness area should be permanently zoned. The goal, we repeat, is to maintain or create the mood of wild America. We are speaking here of restoring wildlife to enhance this mood, but the whole effect can be lost if the parks are overdeveloped for motorized travel. If too many tourists crowd the roadways, then we should ration the tourists rather than expand the roadways.

Additionally in this connection, it seems incongruous that there should exist in the national parks mass recreation facilities such as golf courses, ski lifts, motorboat marinas, and other extraneous developments which completely contradict the management goal. We urge the National Park Service to reverse its policy of permitting these nonconforming uses, and to liquidate them as expeditiously as possible (painful as this will be to concessionaires). Above all other policies, the maintenance of naturalness should prevail.

Another major policy matter concerns the research which must form the basis for all management programs. The agency best fitted to study park management problems is the National Park Service itself. Much help and guidance can be obtained from ecologic research conducted by other agencies, but the objectives of park management are so different from those of state fish and game departments, the Forest Service, etc., as to demand highly skilled studies of a very specialized nature. Management without knowledge would be a dangerous policy indeed. Most of the research now conducted by the National Park Service is oriented largely to interpretive functions rather than to management. We urge the expansion of the research activity in the Service to prepare for future management and restoration programs. As models of the type of investigation that should be greatly accelerated we cite some of the recent studies of elk in Yellowstone and of bighorn sheep in Death Valley. Additionally, however, there are needed equally critical appraisals of ecologic relationships in various plant associations and of many lesser organisms such as azaleas, lupines, chipmunks, towhees, and other non-economic species.

In consonance with the above policy statements, it follows logically that every phase of management itself be under the full jurisdiction of biologically trained personnel of the Park Service. This applies not only to habitat manipulation but to all facets of regulating animal populations. Reducing the numbers of elk in Yellowstone or of goats on Haleakala Crater is part of an overall scheme to preserve or restore a natural biotic scene. The purpose is single-minded. We cannot endorse the view that responsibility for removing excess game animals be shared with state fish and game departments whose primary interest would be to capitalize on the recreational value of the public hunting that could thus be supplied. Such a proposal imputes a multiple use concept of park management which was never intended, which is not legally permitted, nor for which we can find any impelling justification today.

Purely from the standpoint of how best to achieve the goal of park management, as here defined, unilateral administration directed to a single objective is obviously superior to divided responsibility in which secondary goals, such as recreational hunting, are introduced. Additionally, uncontrolled public hunting might well operate in opposition to the goal, by removing roadside animals and frightening the survivors, to the end that public viewing of wildlife would be materially impaired. In one national park, namely Grand Teton, public hunting was specified by Congress as the method to be used in controlling elk. Extended trial suggests this to be an awkward administrative tool at best.

Since this whole matter is of particular current interest it will be elaborated in a subsequent section on methods.

METHODS OF HABITAT MANAGEMENT

It is obviously impossible to mention in this brief report all the possible techniques that might be used by the National Park Service in manipulating plant and animal populations. We can, however, single out a few examples. In so doing, it should be kept in mind that the total area of any one park, or of the parks collectively, that may be managed intensively is a very modest part indeed. This is so for two reasons. First, critical areas which may determine animal abundance are often a small fraction of total range. One deer study on the west slope of the Sierra Nevada, for example, showed that important winter range, which could be manipulated to support the deer, constituted less than two per cent of the year-long herd range. Roadside areas that might be managed to display a more varied and natural flora and fauna can be rather narrow strips. Intensive management, in short, need not be extensive to be effective. Secondly, manipulation of vegetation is often exorbitantly expensive. Especially will this be true when the objective is to manage "invisibly"—that is, to conceal the signs of management. Controlled burning is the only method that may have extensive application.

The first step in park management is historical research, to ascertain as accurately as possible what plants and animals and biotic associations existed originally in each locality. Much of this has been done already.

A second step should be ecologic research on plant-animal relationships leading to formulation of a management hypothesis.

Next should come small-scale experimentation to test the hypothesis in practice. Experimental plots can be situated out of sight of roads and visitor centers.

Lastly, application of tested management methods can be undertaken on critical areas.

By this process of study and pre-testing, mistakes can be minimized. Likewise, public groups vitally interested in park management can be shown the results of research and testing before general application, thereby eliminating possible misunderstanding and friction.

Some management methods now in use by the National Park Service seem to us potentially dangerous. For example, we wish to raise a serious question about the mass application of insecticides in the control of forest insects. Such application may (or may not) be justified in commercial timber stands, but in a national park the ecologic impact can have unanticipated effects on the biotic community that might defeat the overall management objective. It would seem wise to curtail this activity, at least until research and small-scale testing have been conducted.

Of the various methods of manipulating vegetation, the controlled use of fire is the most "natural" and much the cheapest and easiest to apply. Unfortunately, however, forest and chaparral areas that have been completely protected from fire for long periods may require careful advance treatment before even the first experimental blaze is set. Trees and mature brush may have to be cut, piled, and burned before a creeping ground fire can be risked. Once fuel is reduced, periodic burning can be conducted safely and at low expense. On the other hand, some situations may call for a hot burn. On Isle Royale, moose range is created by periodic holocausts that open the forest canopy. Maintenance of the moose population is surely one goal of management on Isle Royale.

Other situations may call for the use of the bulldozer, the disc harrow, or the spring-tooth harrow to initiate desirable changes in plant succession. Buffalo wallows on the American prairie were the propagation sites of a host of native flowers and forbs that fed the antelope and the prairie chicken. In the absence of the great herds, wallows can be simulated.

Artificial reintroduction of rare native plants is often feasible. Overgrazing in years past led to local extermination of many delicate perennials such as some of the orchids. Where these are not reappearing naturally they can be transplanted or cultured in a nursery. A native plant, however small and inconspicuous, is as much a part of the biota as a redwood tree or a forage species for elk.

In essence, we are calling for a set of ecologic skills unknown in this country today. Americans have shown a great capacity for degrading and fragmenting native biotas. So far we have not exercised much imagination or ingenuity in rebuilding damaged biotas. It will not be done by passive protection alone.

CONTROL OF ANIMAL POPULATIONS

Good park management requires that ungulate populations be reduced to the level that the range will carry in good health and without impairment to the soil, the vegetation, or to habitats of other animals. This problem is world-wide in scope, and includes non-park as well as park lands. Balance may be achieved in several ways.

(a) *Natural predation.*—Insofar as possible, control through natural predation should be encouraged. Predators are now protected in the parks of the United States, although unfortunately they were not in the early years and the wolf, grizzly bear, and mountain lion became extinct in many of the national parks. Even today populations of large predators, where they still occur in the parks, are kept below optimal level by programs of predator control applied outside the park boundaries. Although the National Park Service has attempted to negotiate with control agencies of federal and local governments for the maintenance of buffer zones around the parks where predators are not subject to systematic control, these negotiations have been only partially successful. The effort to protect large predators in and around the parks should be greatly intensified. At the same time, it must be recognized that predation alone can seldom be relied upon to control ungulate numbers, particularly the larger species such as bison, moose, elk, and deer; additional artificial controls frequently are called for.

(b) *Trapping and transplanting.*—Traditionally in the past the National Park Service has attempted to dispose of excess ungulates by trapping and transplanting. Since 1892, for example, Yellowstone National Park alone has supplied 10,478 elk for restocking purposes. Many of the elk ranges in the western United States have been restocked from this source. Thousands of deer and lesser numbers of antelope, bighorns, mountain goats, and bison also have been moved from the parks. This program is fully justified so long as breeding stocks are needed. However, most big game ranges of the United States are essentially filled to carrying capacity, and the cost of a continuing program of trapping and transplanting cannot be sustained solely on the basis of controlling populations within the parks. Trapping and handling of a big game animal usually costs $50 to $150 and in some situations much more. Since annual surpluses will be produced indefinitely into the future, it is patently impossible to look upon trapping as a practical plan of disposal.

(c) *Shooting excess animals that migrate outside the parks.*—Many park herds are migratory and can be controlled by public hunting outside the park boundaries. Especially is this true in mountain parks which usually consist largely of summer game range with relatively little winter range. Effective application of this form of control frequently calls for special regulations, since migration usually occurs after normal hunting dates. Most of the western states have cooperated with the National Park Service in scheduling late hunts for the specific purpose of reducing park game herds, and in fact most excess game produced in the parks is so utilized. This is by far the best and most widely applied method of controlling park populations of ungulates. The only danger is that migratory habits may be eliminated from a herd by differential removal, which would favor survival of non-migratory individuals. With care to preserve, not eliminate, migratory traditions, this plan of control will continue to be the major form of herd regulation in national parks.

(d) *Control by shooting within the parks.*—Where other methods of control are inapplicable or impractical, excess park ungulates must be removed by killing. As stated above in the discussion of park policy, it is the unanimous recommendation of this Board that such shooting be conducted by competent personnel, under the sole jurisdiction of the National Park Service, and for the sole purpose of animal removal, not recreational hunting. If the magnitude of a given removal program requires the services of additional shooters beyond regular Park Service personnel, the selection, employment, training, deputization, and supervision of such additional personnel should be entirely the responsibility of the National Park Service. Only in this manner can the primary goal of wildlife management in the parks be realized. A limited number of expert riflemen, properly equipped and working under centralized direction, can selectively cull a herd with a minimum of disturbance to the surviving animals or to the environment. General public hunting by comparison is often non-selective and grossly disturbing.

Moreover, the numbers of game animals that must be removed annually from the parks by shooting is so small in relation to normally hunted populations outside the parks as to constitute a minor contribution to the public bag, even if it were so utilized. All of these points can be illustrated in the example of the north Yellowstone elk population which has been a focal point of argument about possible public hunting in national parks.

(e) *The case of Yellowstone.*—Elk summer in all parts of Yellowstone Park and migrate out in nearly all directions, where they are subject to hunting on adjoining public and private lands. One herd, the so-called Northern Elk Herd, moves only to the vicinity of the park border where it may winter largely inside or outside the park, depending on the severity of the winter. This herd was estimated to number 35,000 animals in 1914 which was far in excess of the carrying capacity of the range. Following a massive die-off in 1919–20 the herd has steadily decreased. Over a period of 27 years, the National Park Service removed 8,825 animals by shooting and 5,765 by live-trapping; concurrently, hunters took 40,745 elk from this herd outside the park. Yet the range continued to deteriorate. In the winter of 1961–62 there were approximately 10,000 elk in the herd and carrying capacity of the winter range was estimated at 5,000. So the National Park Service at last undertook a definitive reduction program, killing 4,283 elk by shooting, which along with 850 animals removed in other ways (hunting outside the park, trapping, winter kill) brought the herd down to 5,725 as censused from helicopter. The carcasses of the elk were carefully processed and distributed to Indian communities throughout Montana and Wyoming; so they were well used. The point at issue is whether this same reduction could or should have been accomplished by public hunting.

In autumn during normal hunting season the elk are widely scattered through rough inaccessible mountains in the park. Comparable areas, well stocked with elk, are heavily hunted in adjoining national forests. Applying the kill statistics from the forests to the park, a kill of 200–400 elk might be achieved if most of the available pack stock in the area were used to transport hunters within the park. Autumn hunting could not have accomplished the necessary reduction.

In mid-winter when deep snow and bitter cold forced the elk into lower country along the north border of the park, the National Park Service undertook its reduction program. With snow vehicles, trucks, and helicopters they accomplished the unpleasant job in temperatures that went as low as

−40° F. Public hunting was out of the question. Thus, in the case most bitterly argued in the press and in legislative halls, reduction of the herd by recreational hunting would have been a practical impossibility, even if it had been in full conformance with park management objectives.

From now on, the annual removal from this herd may be in the neighborhood of 1,000 to 1,800 head. By January 31, 1963, removals had totalled 1,300 (300 shot outside the park by hunters, 600 trapped and shipped, and 406 killed by park rangers). Continued special hunts in Montana and other forms of removal will yield the desired reduction by spring. The required yearly maintenance kill is not a large operation when one considers that approximately 100,000 head of big game are taken annually by hunters in Wyoming and Montana.

(f) *Game control in other parks.*—In 1961–62, excluding Yellowstone elk, there were approximately 870 native animals transplanted and 827 killed on 18 national parks and monuments. Additionally, about 2,500 feral goats, pigs and burros were removed from three areas. Animal control in the park system as a whole is still a small operation. It should be emphasized, however, that removal programs have not in the past been adequate to control ungulates in many of the parks. Future removals will have to be larger and in many cases repeated annually. Better management of wildlife habitat will naturally produce larger annual surpluses. But the scope of this phase of park operation will never be such as to constitute a large facet of management. On the whole, reductions will be small in relation to game harvests outside the parks. For example, from 50 to 200 deer a year are removed from a problem area in Sequoia National Park; the deer kill in California is 75,000 and should be much larger. In Rocky Mountain National Park 59 elk were removed in 1961–62 and the trim should perhaps be 100 per year in the future; Colorado kills over 10,000 elk per year in open hunting ranges. In part, this relates to the small area of the national park system, which constitutes only 3.9 per cent of the public domain; hunting ranges under the jurisdiction of the Forest Service and Bureau of Land Management make up approximately 70 per cent.

In summary, control of animal populations in the national parks would appear to us to be an integral part of park management, best handled by the National Park Service itself. In this manner excess ungulates have been controlled in the national parks of Canada since 1943, and the same principle is being applied in the parks of many African countries. Selection of personnel to do the shooting likewise is a function of the Park Service. In most small operations this would logically mean skilled rangers. In larger removal programs, there might be included additional personnel, selected from the general public, hired and deputized by the Service or otherwise engaged, but with a view to accomplishing a task, under strict supervision and solely for the protection of park values. Examples of some potentially large removal programs where expanded crews may be needed are mule deer populations on plateaus fringing Dinosaur National Monument and Zion National Park (west side), and white-tailed deer in Acadia National Park.

WILDLIFE MANAGEMENT ON NATIONAL RECREATION AREAS

By precedent and logic, the management of wildlife resources on the national recreation areas can be viewed in a very different light than in the park system proper. National recreation areas are by definition multiple use in character as regards allowable types of recreation. Wildlife management can be incorporated into the operational plans of these areas with public hunting as one objective. Obviously, hunting must be regulated in time and place to minimize conflict with other uses, but it would be a mistake for the National Park Service to be unduly restrictive of legitimate hunting in these areas. Most of the existing national recreation areas are federal holdings surrounding large water impoundments; there is little potentiality for hunting. Three national seashore recreational areas on the East Coast (Hatteras, Cape Cod, and Padre Island) offer limited waterfowl shooting. But some of the new areas being acquired or proposed for acquisition will offer substantial hunting opportunity for a variety of game species. This opportunity should be developed with skill, imagination, and (we would hopefully suggest) with enthusiasm.

On these areas as elsewhere, the key to wildlife abundance is a favorable habitat. The skills and techniques of habitat manipulation applicable to parks are equally applicable on the recreation areas. The regulation of hunting, on such areas as are deemed appropriate to open for such use, should be in accord with prevailing state regulations.

NEW NATIONAL PARKS

A number of new national parks are under consideration. One of the critical issues in the establishment of new parks will be the manner in which the wildlife resources are to be handled. It is our recommendation that the basic objectives and operating procedures of new parks be identical with those of established parks. It would seem awkward indeed to operate a national park system under two sets of ground rules. On the other hand, portions of several proposed parks are so firmly established as traditional hunting ground that impending closure of hunting may preclude public acceptance of park status. In such cases it may be necessary to designate core areas as national parks in every sense of the word, establishing protective buffer zones in the form of national recreation areas where hunting is permitted. Perhaps only through compromises of this sort will the park system be rounded out.

SUMMARY

The goal of managing the national parks and monuments should be to preserve, or where necessary to recreate, the ecologic scene as viewed by the first European visitors. As part of this scene, native species of wild animals should be present in maximum variety and reasonable abundance. Protection alone, which has been the core of Park Service wildlife policy, is not adequate to achieve this goal. Habitat manipulation is helpful and often essential to restore or maintain animal numbers. Likewise, populations of the animals themselves must sometimes be regulated to prevent habitat damage; this is especially true of ungulates.

Active management aimed at restoration of natural communities of plants and animals demands skills and knowledge not now in existence. A greatly expanded research program, oriented to management needs, must be developed within the National Park Service itself. Both research and the application of management methods should be in the hands of skilled park personnel.

Insofar as possible, animal populations should be regulated by predation and other natural means. However, predation cannot be relied upon to control the populations of larger ungulates, which sometimes must be reduced artificially.

Most ungulate populations within the parks migrate seasonally outside the park boundaries where excess numbers can be removed by public hunting. In such circumstances the National Park Service should work closely with state fish and game departments and other interested agencies in conducting the research required for management and in devising cooperative management programs.

Excess game that does not leave a park must be removed. Trapping and transplanting has not proven to be a practical method of control, though it is an appropriate source of breeding stock as needed elsewhere.

Direct removal by killing is the most economical and effective way of regulating ungulates within a park. Game removal by shooting should be conducted under the complete jurisdiction of qualified park personnel and solely for the purpose of reducing animals to preserve park values. Recreational hunting is an inappropriate and non-conforming use of the national parks and monuments.

Most game reduction programs can best be accomplished by regular park employees. But as removal programs increase in size and scope, as well may happen under better wildlife management, the National Park Service may find it advantageous to employ or otherwise engage additional shooters from the general public. No objection to this procedure is foreseen so long as the selection, training, and supervision of shooting crews is under rigid control of the Service and the culling operation is made to conform to primary park goals.

Recreational hunting is a valid and potentially important use of national recreation areas, which are also under jurisdiction of the National Park Service. Full development of hunting opportunities on these areas should be provided by the Service.

The Wilderness Act of 1964

Public Law 88-577
88th Congress, S.4
September 3, 1964

AN ACT

To establish a National Wilderness Preservation System
for the permanent good of the whole people,
and for other purposes.

Be it enacted by the Senate and House of Representatives of the United States of America in Congress assembled,

SHORT TITLE

Section 1. This Act may be cited as the "Wilderness Act".

WILDERNESS SYSTEM ESTABLISHED
STATEMENT OF POLICY

Sec. 2. (a) In order to assure that an increasing population, accompanied by expanding settlement and growing mechanization, does not occupy and modify all areas within the United States and its possessions, leaving no lands designated for preservation and protection in their natural conditions, it is hereby declared to be the policy of the Congress to secure for the American people of present and future generations the benefits of an enduring resource of wilderness. For this purpose there is hereby established a National Wilderness Preservation System to be composed of federally owned areas designated by Congress as "wilderness areas", and these shall be administered for the use and enjoyment of the

American people in such manner as will leave them unimpaired for future use and enjoyment as wilderness, and so as to provide for the protection of these areas, the preservation of their wilderness character, and for the gathering and dissemination of information regarding their use and enjoyment as wilderness; and no Federal lands shall be designated as "wilderness areas" except as provided for in this Act or by a subsequent Act.

(b) The inclusion of an area in the National Wilderness Preservation System notwithstanding, the area shall continue to be managed by the Department and agency having jurisdiction thereover immediately before its inclusion in the national Wilderness Preservation System unless otherwise provided by Act of Congress. No appropriation shall be available for the payment of expenses or salaries for the administration of the national Wilderness Preservation System as a separate unit nor shall any appropriations be available for additional personnel stated as being required solely for the purpose of managing or administering areas solely because they are included within the National Wilderness Preservation System.

DEFINITION OF WILDERNESS

(c) A wilderness, in contrast with those areas where man and his own works dominate the landscape, is hereby recognized as an area where the earth and its community of life are untrammeled by man, where man himself is a visitor who does not remain. An area of wilderness is further defined to mean in this Act an area of undeveloped Federal land retaining its primeval character and influence, without permanent improvements or human habitation, which is protected and managed so as to preserve its natural conditions and which (1) generally appears to have been affected primarily by the forces of nature, with the imprint of man's work substantially unnoticeable; (2) has outstanding opportunities for solitude or a primitive and unconfined type of recreation; (3) has at least five thousand acres of land or is of sufficient size as to make practicable its preservation and use in an unimpaired condition; and (4) may also contain ecological, geological, or other features of scientific, educational, scenic, or historical value.

NATIONAL WILDERNESS PRESERVATION SYSTEM—EXTENT OF SYSTEM

Sec. 3. (a) All areas within the national forests classified at least 30 days before the effective date of this Act by the Secretary of Agriculture or the Chief of the Forest Service as "wilderness", "wild", or "canoe" are hereby designated as wilderness areas. The Secretary of Agriculture shall—

1. Within one year after the effective date of this Act, file a map and legal description of each wilderness area with the Interior and Insular Affairs Committees of the United States Senate and the House of Representatives, and such descriptions shall have the same force and effect as if included in this Act: *Provided, however,* That correction of clerical and typographical errors in such legal descriptions and maps be made.
2. Maintain, available to the public, records pertaining to said wilderness areas, including maps and legal descriptions, copies of regulations governing them, copies of public notices of, and reports submitted to Congress regarding pending additions, eliminations, or modifications. Maps, legal descriptions, and regulations pertaining to wilderness areas within their respective jurisdictions also shall be available to the public in the offices of regional foresters, national forest supervisors, and forest rangers.

(b) The Secretary of Agriculture shall, within ten years after the enactment of this Act, review, as to its suitability or nonsuitability for preservations as wilderness, each area in the national forests classified on the effective date of this Act by the Secretary of Agriculture or the Chief of the Forest Service as "primitive" and report his findings to the President. The President shall advise the United States Senate and House of Representatives of his recommendations with respect to the designation as "wilderness" or other reclassification of each area on which review has been completed, together with maps and a definition of boundaries. Such advice shall be given with respect to not less than one-third of all the areas now classified as "primitive" within three years after the enactment of this Act, not less than two-thirds within seven years after the enactment of this Act, and the remaining areas within ten years after the enactment of this Act. Each recommendation of the President for designation as "wilderness" shall become effective only if so provided by an Act of Congress. Areas classified as "primitive" on the effective date of this Act shall continue to be administered under the rules and regulations affecting such areas on the effective date of this Act until Congress has determined otherwise. Any such area may be increased in size by the President at the time he submits his recommendations to the Congress by not more than five thousand acres with no more than one thousand two hundred and eighty acres of such increase in any one compact unit; if it is proposed to increase the size of any such area by more than five thousand acres or by more than one thousand two hundred and eighty acres in any one compact unit the increase in size shall not become effective until acted upon by Congress. Nothing herein contained shall limit the President in proposing, as part of his recommendations to Congress, the alteration of existing boundaries of primitive areas or recommending the addition of any contiguous area of national forest lands predominantly of wilderness value. Notwithstanding any other provisions of this Act, the Secretary of Agriculture may complete his review and delete such area as may be necessary, but not to exceed seven thousand acres, from the southern tip of the Gore Range–Eagles Nest Primitive Area, Colorado, if the Secretary determines that such action is in the public interest.

(c) Within ten years after the effective date of this Act the Secretary of the Interior shall review every roadless area of five thousand contiguous acres or more in the national parks, monuments and other units of the national park system and every such area of, and every roadless island within, the national wildlife refuges and game ranges, under his jurisdiction on the effective date of this Act and shall report to the President his recommendation as to the suitability or nonsuitability of each such area or island for preservation as wilderness. The President shall advise the President of the Senate and the Speaker of the House of Representatives of his recommendation with respect to the designation as wilderness of each such area or island on which review has been completed, together with a map thereof and a definition of its boundaries. Such advice shall be given with respect to not less than one-third of the areas and islands to be reviewed under this subsection within three years after enactment of this Act, not less than two-thirds within seven years of enactment of this Act, and the remainder within ten years of enactment of this Act. A recommendation of the President for designation as wilderness shall become effective only if so provided by an Act of Congress. Nothing contained herein shall, by implication or otherwise, be construed to lessen the present statutory authority of the Secretary of the Interior with respect to the maintenance of roadless areas within units of the national park system.

(d) (1) The Secretary of Agriculture and the Secretary of the Interior shall, prior to submitting any recommendations to the President with respect to the suitability of any area for preservation as wilderness—

A. give such public notice of the proposed action as they deem appropriate, including publication in the Federal Register and in a newspaper having general circulation in the area or areas in the vicinity of the affected land;
B. hold a public hearing or hearings at a location or locations convenient to the area affected. The hearings shall be announced through such means as the respective Secretaries involved deem appropriate, including notices in the Federal Register and in newspapers of general circulation in the area: *Provided,* that if the lands involved are located in more than one State, at least one hearing shall be held in each State in which a portion of the land lies;
C. at least thirty days before the date of a hearing advise the Governor of each State and the governing board of each county, or in Alaska the borough, in which the lands are located, and Federal departments and agencies concerned, and invite such officials and Federal agencies to submit their views on the proposed action at the hearing or by no later than thirty days following the date of the hearing.

(2) Any views submitted to the appropriate Secretary under the provisions of (1) of this subsection with respect to any area shall be included with any recommendations to the President and to Congress with respect to such area.

(e) Any modification or adjustment of boundaries of any wilderness area shall be recommended by the appropriate Secretary after public notice of such proposal and public hearing or hearings as provided in subsection (d) of this section. The proposed modification or adjustment shall then be recommended with map and description thereof to the President. The President shall advise the United States Senate and the House of Representatives of his recommendations with respect to such modification or adjustment and such recommendations shall become effective only in the same manner as provided for in subsections (b) and (c) of this section.

USE OF WILDERNESS AREAS

Sec. 4. (a) The purposes of this Act are hereby declared to be within and supplemental to the purposes for which national forests and units of the national park and national wildlife refuge systems are established and administered and—

1. Nothing in this Act shall be deemed to be in interference with the purpose for which national forests are established as set forth in the Act of June 4, 1897 (30 Stat. 11), and the Multiple-Use Sustained-Yield Act of June 12, 1960 (74 Stat. 215).
2. Nothing in this Act shall modify the restrictions and provisions of the Shipstead-Nolan Act (Public Law 539, Seventy-first Congress, July 10, 1930; 46 Stat. 1020), the Thye-Blatnik Act (Public Law 733, Eightieth Congress, June 22, 1948; 62 Stat. 568), and the Humphrey-Thye-Blatnik-Andersen Act (Public Law 607, Eighty-fourth Congress, June 22, 1956; 70 Stat. 326), as applying to the Superior National Forest or the regulations of the Secretary of Agriculture.
3. Nothing in this Act shall modify the statutory authority under which units of the national park system are created. Further, the designation of any area of any park, monument, or other unit of the national park system as a wilderness area pursuant to this Act shall in no manner lower the standards evolved for the use and preservation of such park, monument, or other unit of the national park system in accordance with the Act of August 25, 1916, the statutory authority under which the area was created, or any other Act of Congress which might pertain to or affect such area, including, but not limited to, the Act of June 8, 1906 (34 Stat. 225; 16 U.S.C. 432 et seq.); section 3(2) of the Federal Power Act (16 U.S.C. 796(2)); and the Act of August 21, 1935 (49 Stat. 666; 16 U.S.C. 461 et seq.).

(b) Except as otherwise provided in this Act, each agency administering any area designated as wilderness shall be responsible for preserving the wilderness character of the area and shall so administer such area for such other purposes for which it may have been established as also to preserve its wilderness character. Except as otherwise provided in this Act, wilderness areas shall be devoted to the public purposes of recreational, scenic, scientific, educational, conservation, and historical use.

PROHIBITION OF CERTAIN USES

(c) Except as specifically provided for in this Act, and subject to existing private rights, there shall be no commercial enterprise and no permanent road within any wilderness area designated by this Act and, except as necessary to meet minimum requirements for the administration of the area for the purpose of this Act (including measures required in emergencies involving the health and safety of persons within the area), there shall be no temporary road, no use of motor vehicles, motorized equipment or motorboats, no landing of aircraft, no other form of mechanical transport, and no structure or installation within any such area.

SPECIAL PROVISIONS

(d) The following special provisions are hereby made:

(1) Within wilderness areas designated by this Act the use of aircraft or motorboats, where these uses have already become established, may be permitted to continue subject to such restrictions as the Secretary of Agriculture deems desirable. In addition, such measures may be taken as may be necessary in the control of fire, insects and diseases, subject to such conditions as the Secretary deems desirable.

(2) Nothing in this Act shall prevent within national forest wilderness areas any activity, including prospecting, for the purpose of gathering information about mineral or other resources, if such activity is carried on in a manner compatible with the preservation of the wilderness environment. Furthermore, in accordance with such program as the Secretary of the Interior shall develop and conduct in consultation with the Secretary of Agriculture, such areas shall be surveyed on a planned, recurring basis consistent with the concept of wilderness preservation by the Geological Survey and the Bureau of Mines to determine the mineral values, if any, that may

be present; and the results of such surveys shall be made available to the public and submitted to the President and Congress.

(3) Notwithstanding any other provisions of this Act, until midnight December 31, 1983, the United States mining laws and all laws pertaining to mineral leasing shall, to the same extent as applicable prior to the effective date of this Act, extend to those national forest lands designated by this Act as "wilderness areas"; subject, however, to such reasonable regulations governing ingress and egress as may be prescribed by the Secretary of Agriculture consistent with the use of the land for mineral location and development and exploration, drilling, and production, and use of land for transmission lines, waterlines, telephone lines, or facilities necessary in exploring, drilling, producing, mining, and processing operations, including where essential the use of mechanized ground or air equipment and restoration as near as practicable of the surface of the land disturbed in performing prospecting, location, and, in oil and gas leasing, discovery work, exploration, drilling, and production, as soon as they have served their purpose. Mining locations lying within the boundaries of said wilderness areas shall be held and used solely for mining or processing operations and uses reasonably incident thereto; and hereafter, subject to valid existing rights, all patents issued under the mining laws of the United States affecting national forest lands designated by this Act as wilderness areas shall convey title to the mineral deposits within the claim, together with the right to cut and use so much of the mature timber therefrom as may be needed in the extraction, removal, and beneficiation of the mineral deposits, if needed timber is not otherwise reasonably available, and if the timber is cut under sound principles of forest management as defined by the national forest rules and regulations, but each such patent shall reserve to the United States all title in or to the surface of the lands and products thereof, and no use of the surface of the claim or the resources therefrom not reasonably required for carrying on mining or prospecting shall be allowed except as otherwise expressly provided in this Act: *Provided,* That, unless hereafter specifically authorized, no patent within wilderness areas designated by this Act shall issue after December 31, 1983, except for the valid claims existing on or before December 31, 1983. Mining claims located after the effective date of this Act within the boundaries of wilderness areas designated by this Act shall create no rights in excess of those rights which may be patented under the provisions of this

subsection. Mineral leases, permits, and licenses covering lands within national forest wilderness areas designated by this Act shall contain such reasonable stipulations as may be prescribed by the Secretary of Agriculture for the protection of the wilderness character of the land consistent with the use of the land for the purposes for which they are leased, permitted, or licensed. Subject to valid rights then existing, effective January 1, 1984, the minerals in lands designated by this Act as wilderness areas are withdrawn from all forms of appropriation under the mining laws and from disposition under all laws pertaining to mineral leasing and all amendments thereto.

(4) Within wilderness areas in the national forests designated by the Act, (1) the President may, within a specific area and in accordance with such regulations as he may deem desirable, authorize prospecting for water resources, the establishment and maintenance of reservoirs, water-conservation works, power projects, transmission lines, and other facilities needed in the public interest, including the road construction and maintenance essential to development and use thereof, upon his determination that such use or uses in the specific area will better serve the interests of the United States and the people thereof than will its denial; and (2) the grazing of livestock, where established prior to the effective date of this Act, shall be permitted to continue subject to such reasonable regulations as are deemed necessary by the Secretary of Agriculture.

(5) Other provisions of this Act to the contrary notwithstanding, the management of the Boundary Waters Canoe Area, formerly designated as the Superior, Little Indian Sioux, and Caribou Roadless Areas, in the Superior National Forest, Minnesota, shall be in accordance with regulations established by the Secretary of Agriculture in accordance with the general purpose of maintaining, without unnecessary restrictions on other uses, including that of timber, the primitive character of the area, particularly in the vicinity of lakes, streams, and portages: *Provided,* That nothing in this Act shall preclude the continuance within the area of any already established use of motorboats.

(6) Commercial services may be performed within the wilderness areas designated by this Act to the extent necessary for activities which are proper for realizing the recreational or other wilderness purposes of the areas.

(7) Nothing in this Act shall constitute an express or implied claim or

denial on the part of the Federal Government as to exemption from State water laws.

(8) Nothing in this Act shall be construed as affecting the jurisdiction or responsibilities of the several States with respect to wildlife and fish in the national forests.

STATE AND PRIVATE LANDS WITHIN WILDERNESS AREAS

Sec. 5. (a) In any case where State-owned or privately owned land is completely surrounded by national forest lands within areas designated by this Act as wilderness, such State or private owner shall be given such rights as may be necessary to assure adequate access to such State-owned or privately owned land by such State or private owner and their successors in interest, or the State-owned land or privately owned land shall be exchanged for federally owned land in the same State of approximately equal value under authorities available to the Secretary of Agriculture: *Provided, however,* That the United States shall not transfer to a State or private owner any mineral interests unless the State or private owner relinquishes or causes to be relinquished to the United States the mineral interest in the surrounded land.

(b) In any case where valid mining claims or other valid occupancies are wholly within a designated national forest wilderness area, the Secretary of Agriculture shall, by reasonable regulations consistent with the preservation of the area as wilderness, permit ingress and egress to such surrounded areas by means which have been or are being customarily enjoyed with respect to other such areas similarly situated.

(c) Subject to the appropriation of funds by Congress, the Secretary of Agriculture is authorized to acquire privately owned land within the perimeter of any area designated by this Act as wilderness if (1) the owner concurs in such acquisition or (2) the acquisition is specifically authorized by Congress.

GIFTS, BEQUESTS, AND CONTRIBUTIONS

Sec. 6. (a) The Secretary of Agriculture may accept gifts or bequests of land within wilderness areas designated by this Act for preservation as wilderness. The Secretary of Agriculture may also accept gifts or bequests of land adjacent to wilderness areas designated by this Act for preservation as wilderness if he has given sixty days advance notice thereof to the President of the Senate and the Speaker of the House of Representatives. Land accepted by the Secretary of Agriculture under this section shall become part of the wilderness area involved. Regulations with regard to any such land may be in accordance with such agreements, consistent with the policy of this Act, as are made at the time of such gift, or such conditions, consistent with such policy, as may be included in, and accepted with, such bequest.

(b) The Secretary of Agriculture or the Secretary of the Interior is authorized to accept private contributions and gifts to be used to further the purposes of this Act.

ANNUAL REPORTS

Sec. 7. At the opening of each session of Congress, the Secretaries of Agriculture and Interior shall jointly report to the President for transmission to Congress on the status of the wilderness system, including a list and descriptions of the areas in the system, regulations in effect, and other pertinent information, together with any recommendations they may care to make.

Approved September 3, 1964.

LEGISLATIVE HISTORY

House Reports: No. 1538 accompanying H.R. 9070 (Comm. on Interior & Insular Affairs) and No. 1829 (Comm. of Conference).

Senate Report No. 109 (Comm. on Interior & Insular Affairs).

Congressional Record:

Vol. 109 (1963): Apr. 4, 8, considered in Senate.
Apr. 9, considered and passed Senate.

Vol. 110 (1964): July 28, considered in House.
July 30, considered and passed House, amended, in lieu of H.R. 9070.
Aug. 20, House and Senate agreed to conference report.

Mark Woods

Federal Wilderness Preservation in the United States (1998)

The Preservation of Wilderness?

MUCH OF ENVIRONMENTAL PHILOSOPHY has been concerned with articulating and defending the value of wilderness. Federal wilderness preservation is a legal umbrella for protecting wilderness and its value, but in the past philosophers have paid relatively little attention to this legal umbrella. The following is an attempt to direct philosophical attention to how federal wilderness is preserved in the United States. Within federal wilderness preservation are tensions I identify as paradoxes that should force us to question the role and purpose of legal wilderness preservation.

THE RATIONALES AND COMPROMISE OF THE WILDERNESS ACT OF 1964

The Wilderness Act of 1964 (hereafter referred to as the "Act") is the federal statute that created the public land category of "wilderness area" and a national system of wilderness areas known today as the "National Wilderness Preservation System" (NWPS). It led to mandates for four federal agencies to preserve and manage wilderness: the United States Forest Service (USFS), the National Park Service (NPS), the Fish and Wildlife Service (FWS), and the Bureau of Land Management (BLM).[1] Today it serves as the legal basis from which all decisions to preserve and manage wilderness are made.

Two separate rationales supporting legal wilderness preservation and a legal compromise that limits wilderness preservation are expressed in the Act. Because these rationales and this compromise express the values of different interest groups who lobbied for and against the Act, they are essentially political.[2]

Concern that some federal public lands remain in an undeveloped condition defines one of the two explicit rationales for wilderness preservation in the Wilderness Act of 1964. The Act's preamble, Section 2(a), articulates a preservation rationale and states that certain tracts of land with "wilderness character" be set aside from and closed to future development and human habitation so such lands can be preserved and protected in their undeveloped state or "natural condition." Section 2(a) expresses a concern that continued economic and population growth could lead to the inhabitation and cultivation of all areas of the United States *unless* certain areas are set aside as wilderness. Because the preservation rationale justifies a policy of preserving wilderness areas in their undeveloped state (McCloskey 1966, 309; Rohlf and Honnold 1988, 255–56), a strong case can be made for saying that the preservation rationale appeals to the noninstrumental value (i.e., intrinsic value) of wilderness. Thus, this rationale resonates with biocentric and ecocentric arguments that philosophers have variously articulated about the value of wilderness.[3] Those who write about wilderness management concur: "management should emphasize the natural integrity of wilderness ecosystems. This reflects a biocentric management philosophy" (Hendee et al. 1990, 20).

A careful reading of the Act's preamble reveals the second explicit rationale for the Act. Wilderness is declared to be an "enduring resource" for present and future generations of Americans and is to be administered "for the use and enjoyment of the American people." This "use and enjoyment rationale" justifies preserving wilderness because of its instrumental value: the main use of wilderness is as a setting for certain outdoor recreational activities (and for solitude) that cannot be enjoyed in non-wilderness settings (McCloskey 1966, 309; Rohlf and Honnold 1988, 256–57). This is a recreation rationale:[4] one of the primary political motivations behind the Act was to create a system of federal wilderness areas that would be available for all wilderness recreationists to use. Preserving wilderness for use and enjoyment by people resonates with anthropocentric arguments that philosophers have variously articulated about the value of wilderness.

A group of pro-development-motivated clauses that limits wilderness preservation can be found in Section 4 of the Act—which defines what humans can and cannot do in wilderness areas. This section stipulates several important "exceptions" to wilderness preservation by permitting continued use of aircraft and motorboats, hardrock mining, mineral leasing, grazing, access road construction, and water development projects in wilderness areas. Wilderness acts that have been passed by Congress since 1964 (there are well over 100) mention many other "exceptions" to wilderness preservation that further limit wilderness preservation as defined and mandated in the original Act. These exceptions indicate that wilderness preservation—within legally designated wilderness areas—can be overridden by various commercial and local interests in using wilderness areas for motorized recreation, mining, grazing, water development, and the like (Rohlf and Honnold 1988, 257–58). Further, Sections 4(a) (1) and (2) stipulate that the purposes of wilderness areas indicated by Congress are *supplemental*—that is, ancillary—to the purposes of the agencies responsible for managing wilderness areas. Because the USFS, responsible for managing all initial fifty-four wilderness areas, was charged by Congress to manage the national forests in accordance with the Multiple-Use, Sustained-Yield Act of 1960, the management of forest lands for wilderness preservation was deemed to be "supplemental" to the management of forest lands for recreation, logging, mining, grazing, and water development.[5] The Wilderness Act of 1964 explicitly charges wilderness preservation be made compatible with various local and commercial activities that spoil the "wilderness character" of legally designated wilderness areas. Thus, non-wilderness or anti-wilderness preservation values are reflected in the Act.

LEGAL WILDERNESS AREAS: HOW ARE THEY DEFINED?

The Act's anthropocentric and non-anthropocentric rationales legally indicate *why* we should preserve wilderness, and its exceptions indicate what competing values preempt preservation values. The Act famously tells us *what* wilderness is:

> A wilderness, in contrast with those areas where man and his own works dominate the landscape, is hereby recognized as an area where the earth and its community of life are untrammeled by man, where man himself is a visitor who does not remain. An area of wilderness is further defined to mean in this Act an area of undeveloped Federal land retaining its primeval character and influence, without permanent improvements or human habitation, which is protected and managed so as to preserve its natural conditions and which (1) generally appears to have been affected primarily by the forces of nature, with the imprint of man's work substantially unnoticeable; (2) has outstanding opportunities for solitude or a primitive and unconfined type of recreation; (3) has at least five thousand acres of land or is of sufficient size as to make practicable its preservation and use in an unimpaired condition; and (4) may also contain ecological, geological, or other features of scientific, educational, scenic, or historical value. (Wilderness Act of 1964, Section 2[c])

There is a lot packed into this passage that must be analyzed in order to understand what makes a tract of de facto wilderness qualify to become de jure wilderness.[6] What are the legal requirements that a natural area must meet in order to qualify for inclusion within the NWPS?

THE NATURALNESS REQUIREMENT

First there is a naturalness requirement. When we combine what the Act says about naturalness with the past thirty-two years of legal wilderness preservation, we see several different interpretations of the meaning of "naturalness."

UNTRAMMELED AND PRIMEVAL

The first sentence from Section 2(c) of the Act states that wilderness is an area that is "untrammeled by man." "Untrammeled" should not be confused with "untrampled"; the latter means not walked upon or treated harshly, while the former means "not subject to human controls and manipulations that hamper the free play of natural forces" (Hendee et al. 1990, 108). The use of the term "untrammeled" then denotes the opposite of a human-dominated landscape: the nonhuman forces of nature are to be given free rein in wilderness. As an example, consider a series of ridges that geographically define the continental divide in the San Juan Mountains of southwest Colorado buried deep within the Weminuche Wilderness Area.

These ridges are "untrammeled" because they generally appear to have been affected primarily by the forces of nature.

Many legal commentators and wilderness managers claim that this untrammeled requirement of wilderness is an ideal (i.e., it is too strong and not practical for legal purposes) and that the next sentence of Section 2(c) goes on to qualify wilderness in a legal sense as lands that are "without permanent improvements or human habitation" (Coggins et al. 1993, 1015; Hendee et al. 1990, 108). Such lands are labeled as "primeval."

For the sake of brevity, I shall call the existence of naturalness in wilderness, as legally interpreted, the untrammeled condition of wilderness. The term "untrammeled" is a less precise way to say that wilderness areas are undeveloped "without permanent improvements or human habitation" that retain a "primeval character and influence."

After the passage of the Act in 1964 a number of industry groups opposed to wilderness and the USFS argued that the Act required an untrammeled condition of wilderness in the strictest sense possible: lands previously impacted by humans in any way could no longer be considered as candidates for wilderness designation. Defining wilderness in this way came to be known as the purity definition of wilderness. Nine years after the passage of the Act, Jeffrey Foote made the following comment about the debate over the classification of wilderness:

> The Forest Service, industry, and their Congressional supporters have taken a "purist" stance concerning the classification of wilderness. By "purist," it is meant that they accept only the strictest definitional criteria. The "purist" policy has been to oppose application of the wilderness designation to areas where *any* man-made intrusions exist.
>
> (Foote 1973, 255; emphasis added)

Although the USFS itself cannot designate federal wilderness areas (only Congress can do this), it was required to inventory its lands and to recommend to Congress tracts of these lands that met legal qualifications for wilderness designation. The USFS disqualified areas for wilderness recommendation that did not contain a strict, "pure" naturalness condition (Worf 1980). The USFS interpreted the Act as "defining away" areas as wilderness candidates that did not meet this requirement.

Most lands within national forests generally east of the Mississippi River were roaded, logged, farmed, or otherwise impacted by humans at one time or another in the past. The USFS voluntarily conducted an inventory of all of its roadless lands for possible inclusion in the NWPS between 1967 and 1972. The results of this inventory, known as RARE I (Roadless Area Review and Evaluation I; a RARE II occurred later), contained recommendations of 235 USFS roadless areas (12.3 million acres) for wilderness suitability out of 1,448 USFS roadless areas (56 million acres). Only 3 roadless areas in the eastern United States were included in this recommendation. The USFS argued that inclusion of previously impacted eastern areas (beyond the three they recommended for wilderness designation) would fall afoul of the naturalness requirement for wilderness designation and thus compromise the entire purpose of wilderness preservation (Rickart 1980, 893). Wilderness advocates in turn argued that the previous human impacts in many more eastern roadless areas were substantially unnoticeable because these areas had "recovered" due to the free play of natural forces.[7]

This difference of opinion about how unnoticeable human impacts had to be culminated in the so-called Eastern Wilderness Act of 1975 that established sixteen new wilderness areas in national forests in the eastern United States.[8] Most of these wilderness areas contained land that was previously trammeled and impacted by humans. In effect, therefore, Congress rejected the purity definition of wilderness.[9]

The legal history of "untrammeled" wilderness reveals an important dimension of the naturalness requirement of wilderness. The USFS interpreted the terms "untrammeled" and "primeval" in a backward-looking sense: lands that had been trammeled by humans *in the past* and *had lost* their "primeval character" had lost a condition of naturalness required for wilderness designation. The Congressional and judicial rejection of the USFS's purity definition of wilderness suggested that the untrammeled character of naturalness was more forward-looking: wilderness lands were to be managed in such a way that they *would be* untrammeled and *return* to primeval conditions *in the future.* As Senator Buckley testified:

> I take [the Act's language] to mean that the primitive area in question will remain untrammeled and undisturbed by man's activities in the future. If an area has recovered from man's past activities and nature's healing processes have restored its character, so that it is impossible to distinguish it

> from a pristine area, I believe it is fully consistent with the intent of the Wilderness Act to include the area in the national wilderness preservation system.[10]

A distinction between *synchronic* naturalness and *diachronic* naturalness may be helpful here. The former refers to naturalness that exists at the time we look at an area, while the latter refers to the naturalness that exists over an extended period of time that includes both the present and the future. Legal wilderness preservation enshrines a diachronic sense of present *and* future naturalness.

One sustained philosophical discussion of naturalness is Robert Elliot's "Faking Nature." Wilderness embodies naturalness, and "[w]hat is significant about wilderness is its causal continuity with the past" (Elliot 1982, 87). Because Elliot thinks that causal continuity with the past is a necessary condition of naturalness, he is skeptical of restoring a trammeled area to a wilderness condition, and he seems to be advocating a purity definition of wilderness similar to that assumed by the USFS. The history of legal wilderness preservation suggests that we should follow Elliot part way and argue that the naturalness of wilderness must have *some* causal continuity with untrammeled naturalness. For example, it is difficult to argue for the eventual designation of the Wall Street Wilderness Area because Manhattan has been so extensively trammeled in the past that its recovery of naturalness anytime in the foreseeable future is unimaginable. A strictly past-oriented form of naturalness, however, is too strong for legal purposes. If a tract of wilderness loses some *essential* feature of wilderness because it has been impacted by humans in the past, then it seems difficult to claim that this same tract of land (now non-wilderness) could ever revert back to a wilderness condition in the future. If an essential naturalness condition of wilderness is *strictly* past-oriented, then future ascription of the term "wilderness" to any tract of land previously impacted by humans seems impossible. In light of evidence that the ancestors of American Indians ubiquitously inhabited the Americas and left their imprint on the lands they inhabited, it may be impossible to find any wilderness at all in the United States if we accept a strong purity definition of wilderness (Callicott 1991; Denevan 1992).[11] But a forward-looking form of naturalness legally circumvents the problems raised by interpreting naturalness strictly in terms of the past and suggests that naturalness is a matter of degree, rather than a sharp category distinction between pristine wilderness and previously trammeled tracts of nature. Future naturalness should have substantial causal continuity with past naturalness.[12] For legal purposes, however, we must reject a strong purity definition of wilderness and argue that naturalness is conditioned by future possibilities as well as past causalities.

ROADLESSNESS

Many people believe that the lack of roads—the condition of roadlessness—is a central component of the naturalness requirement of wilderness. Dave Foreman and Howie Wolke use the term "primitive" (which is close in meaning to the term "primeval") to define wilderness, and they gloss this term with the condition of roadlessness: road-building is one of the defining characteristics of modern humans, and roadless areas that existed before modern humans came on the scene are primitive—are de facto wilderness—and thus qualify for legal wilderness preservation (Foreman and Wolke 1992, 1–13).[13] Section 4(c) of the Wilderness Act of 1964 states that there shall be "no permanent road within any wilderness area designated by this Act." As just indicated, natural areas that contain roads may qualify for wilderness designation if the roads will disappear under the influence of the forces of nature. However, this does not mean that Congress will routinely ignore the presence of existing roads in an area that is under consideration for wilderness designation. In fact, the presence of existing roads will *usually* disqualify an area for possible inclusion within the NWPS.

But what is a road? The term is not defined in the Act. In 1972, the Southwest Region of the USFS defined a road as "a parallel set of tire tracks that remained visible into the next season" during their RARE I wilderness inventory (Foreman and Wolke 1992, 12). This insured that any passage of a vehicle that left tire tracks from the previous year disqualified an area for wilderness recommendation. In many of the fragile lands of New Mexico and Arizona (where this definition of "road" applied), ranging from soft alpine mountain meadows to arid desert lands, the passage of one vehicle in 1971 could disqualify an area from the USFS's wilderness inventory.

Beyond federal agency whims, agreement over what constituted a "road" emerged in 1976. In the legislative history of legal wilderness preservation, the only definition pertaining to roads is the following: "The word 'roadless' refers to the absence of roads which have been improved *and* maintained by mechanical means to insure relatively regular and continuous use. A way maintained solely by the passage of vehicles does not constitute a road" (emphasis added).[14] This definition of the term "roadless" was officially adopted by the BLM in 1976,[15] and most legal commentators and wilderness managers point to this as the meaning of the term "roadlessness" today. The key feature of this definition is that roads require mechanical improvements and maintenance, and the mere passage of a four-wheel-drive vehicle over a route that is not mechanically improved and maintained does not establish a road. Is this noncontroversial?

No. There has been debate over what constitutes a road in the arid, desert lands of the southwestern United States administered by the BLM. In 1983, the Interior Board of Land Appeals (IBLA—the agency within the Department of the Interior that resolves appeals from decisions relating to the use and disposition of federal public lands and resources) affirmed that a road on Nevada BLM land existed that had been mechanically improved *but not* mechanically maintained, reasoning that the arid climate of Nevada had eliminated the need for mechanized road maintenance (Torrey 1992, 485). This example is indicative of a general complaint that the BLM and the IBLA have not been consistent in applying a definition of what constitutes a road in the BLM's wilderness inventory processes (Torrey 1992, 483–87). Anyone who has spent time in the desert Southwest can appreciate this problem of identifying roads. The arid climate, dry washes, and open expanses of land can in many places allow easy access for traversing the landscape in four-wheel-drive vehicles. On a strict reading of the above definition of the term "roadless," such a passage of vehicles would not constitute a "road." However, the passage of one or more vehicles can leave marks that many people would identify as a road. Because governmental agencies, especially the BLM, are given wide latitude over whether or not they will include a tract of land in a wilderness inventory based on whether or not they "see" roadless naturalness, the issue of what constitutes roadlessness can be very important in establishing the wilderness potential of an area.

A common practice to keep a road out of a wilderness area is to "cherrystem" a road that penetrates but does not cross a wilderness area. This involves drawing the boundary of a wilderness area around both sides of a road; thus, the road appears as a "cherrystem" penetrating the wilderness area (the "cherry" is the adjacent non-wilderness land). The obvious problem with this procedure is that it can eliminate large areas of land for possible wilderness designation due to the presence of roads. Many wilderness advocates argue that the obvious solution to the controversy over what constitutes a "road" is simply to close existing roads and let natural forces overgrow and eliminate the offending problem (Foreman and Wolke 1992). The recommended future-oriented interpretation of untrammeled naturalness supports such a solution.

PERCEPTUAL NATURALNESS

The conditions of naturalness (as untrammeled and roadless) are qualified by Section 2(c) (1) of the Act which says that wilderness "*generally appears* to have been affected primarily by the forces of nature, with the imprint of man's work *substantially unnoticeable*" (emphasis added). This denotes a perceptual condition of naturalness: the naturalness that can qualify an area for wilderness designation need only be apparent; how a tract of land appears determines whether it is natural or not. Is the naturalness of wilderness grounded in nothing more than this?

The history of legal wilderness preservation seems to suggest an affirmative answer to this question. The purity definition of wilderness was thrown out on grounds that empirical evidence of previous human impact could be ignored when the forces of nature could eventually erase such evidence. Untrammeledness can re-emerge in spite of past and current human interferences. And what counts as a road depends upon both one's chosen definition of a "road" and what "appears" to be a road. Most legal commentators readily acknowledge the perceptual nature of the naturalness of wilderness, but discussion of what this means is surprisingly absent.

Many believe that accepting the perceptual condition of wilderness (or something broader) is unavoidable (Nash 1982).[16] Human impact to one degree or another exists virtually everywhere upon the Earth. If we are to have any legally designated wilderness areas, then we must choose among the types and degrees of actual human imprints that will disqualify an area for wilderness designation and those that will be ignored and regarded as consistent with a naturalness requirement of wilderness. However, qualifying the naturalness of wilderness *strictly* with a perceptual condition leads us into a problem of relativism: I may see little human impact and a condition of naturalness when I visit the 19,130 roadless acres on and around Bullion Butte in the Little Missouri National Grasslands in the North Dakota Badlands, but USFS officials who manage this land may see substantial human impacts and little evidence of naturalness in this area. Whose perceptual beliefs should Congress listen to when trying to resolve the issue of wilderness designation in the Badlands?

The forward-looking interpretation of naturalness discussed above may help us answer this question. Naturalness is diachronic, and, given some continuity with the past, it can recrudesce over time. When we talk about a perceptual condition of naturalness, we are referring both to what we see now and to what we could see in the future. Thus, if we can see some degree of naturalness today, and if we can see the potential for more naturalness in the future, we could be seeing wilderness.

Because a perceptual definition of naturalness is written into the Wilderness Act of 1964, we have to live with it for now.[17] But in order to address the problem of relativism, for legal purposes we can insist that a particular perception of naturalness or nonnaturalness be backed with supporting evidence. Thus, the debate over the existence of naturalness on and around Bullion Butte can be resolved by looking at evidence of the degree of human impact offered by wilderness advocates, anti-wilderness advocates, and the USFS.[18] To the extent that evidence can be evaluated in a legal fashion, perceived naturalness need not be wholly arbitrary or relative.

THE HUMAN SOLITUDE/RECREATION REQUIREMENT

Beyond naturalness the Act mandates a second requirement for wilderness: "[Wilderness] has outstanding opportunities for solitude or a primitive and unconfined type of recreation" (Section 2[c][2]). I shall call this the human solitude/recreation requirement of wilderness. This requirement suggests that humans must be able to find solitude or be able to recreate primitively without confinement in wilderness. The lack of outstanding opportunities within an area to find solitude or recreate primitively could lead to the conclusion that such an area could not qualify as legal wilderness.

When we think of wilderness as a place of solitude, we commonly think of it as a place that is secluded or unfrequented by other people. Jay Hansford Vest (1987) argues that wilderness solitude is a state of mind that combines an aesthetic contemplation of what we find beautiful and sublime in a wilderness setting with some kind of a religious and/or spiritual experience one has in this setting. But because different people can have different states of mind, we run into another problem of relativism or subjectivism when we think of wilderness solitude: I may find outstanding solitude in the Bob Marshall Wilderness Area in Montana; someone else may not. Who is to judge what wilderness solitude is? How do we measure it?

The presence of other humans or of nearby human development and/or habitation can play an important role in determining outstanding wilderness solitude. Thus, wilderness solitude may be more than a mere state of mind. The BLM attempted to capture this physical component of wilderness solitude when inventorying several areas on the Kaiparowits Plateau in south-central Utah for possible wilderness recommendation (BLM 1986). The BLM used what they called a topographic screening standard and a vegetation screening standard to help determine whether wilderness solitude was present in several Wilderness Study Areas (WSAs). If topography or vegetation screened one from other humans present in the WSA, then the WSA was said to meet some wilderness solitude standards. Because of numerous open spaces and arid landscapes with little vegetational cover (much of the area is dominated by desert scrubs and piñon-juniper flora), the BLM had a difficult time finding wilderness solitude (UWC 1990). The BLM went on to use another screening standard to find wilderness solitude: "the effects of outside sights and sounds." Using this standard the glow from nearby Wahweap Marina and the Glen Canyon Dam area, visible for many miles in places, eliminated the possibility of finding wilderness solitude in several large areas.[19] Finally, the BLM used the standard of the ability to find a secluded spot to find wilderness solitude. Because of a lack of topographic and vegetational screening and the presence of nearby human development and habitation, the BLM again had difficulty finding wilderness solitude. Out of 650,500 roadless acres in the Kaiparowits, the BLM found only 91,361 acres suitable for wilderness designation (UWC 1990). To be sure, the BLM did not disqualify 558,639 acres of roadless land simply because of an inability to find wilderness solitude.[20] However, when one reads the BLM Draft Environmental Impact Statements for the five WSAs of the Kaiparowits Plateau (BLM 1986), the inability to find wilderness solitude is cited again and again to justify recommending against wilderness designation of these areas. As Vest tells us, the BLM officials looking for solitude in the Kaiparowits may have had better luck had they stayed home:[21] "The simple-mindedness of agency approach is shown by the remarks of a Bureau of Land Management district manager, who once stated that he could find solitude in his own closet and rhetorically asked, 'Why are we looking for it in all these wide-open spaces?'" (Vest 1987, 305).

The BLM in the above instances claimed to rely upon some physical criteria to find wilderness solitude. If one can believe their reports, the Kaiparowits lacked these criteria, and they could not find wilderness solitude. However, this is the wrong conclusion. First, BLM officials were out looking for outstanding wilderness solitude on the Kaiparowits. The human solitude/recreation requirement of wilderness stipulates that wilderness should have *opportunities* for outstanding solitude. BLM officials were looking for the wrong thing. Second, the opportunities for solitude in this case are qualified as *wilderness* solitude. If we are to take the naturalness requirement of wilderness seriously, then wilderness solitude must be tied to the opportunity to be free from substantially noticeable human impacts and the opportunity to be alone in a primitive region. Naturalness, even perceptual naturalness, must make reference to something in wilderness that is untrammeled and roadless. This helps circumvent the problem of subjectivism. Although solitude is a state of mind, outstanding opportunities for wilderness solitude necessarily rely upon naturalness. And similar to perceptual naturalness, what counts as an opportunity for wilderness solitude can be backed with a reasoned argument. Although the BLM tried to connect solitude with naturalness on the Kaiparowits, in many cases they failed to give good reasons for why WSAs on the Kaiparowits lacked opportunities for wilderness solitude.

We see similar problems when we turn to the issue of outstanding opportunities for primitive recreation in wilderness. Because the Act stipulates that there can be no use of motor vehicles, motorized equipment, or mechanized transport (Section 4[c]), wilderness recreation is qualified by the term "primitive" which essentially means non-motorized and non-mechanical. Standard examples of wilderness recreation are hiking, backpacking, horseback riding, mountain climbing, rock climbing, skiing, snowshoeing, canoeing, kayaking, rafting, photography, hunting, fishing, camping, etc. When we include everything non-motorized and non-mechanized we could possibly call "recreation" in a wilderness area, this list may become quite lengthy: reading, talking, listening, building campfires, making snow angels in snow, throwing pebbles, meditating, having sex, etc. Because someone could probably engage in at least *some* form of primitive recreation within any given wilderness area, federal agencies may be hard-pressed to prove that any given area lacks outstanding *opportunities* for primitive recreation.

There are two conclusions to be drawn here. First, the human solitude/recreation requirement of wilderness stipulates that there must be outstanding opportunities for solitude or primitive recreation. It is more difficult to prove that an area could not provide any such opportunities than it is to establish that an official did not feel that she or he was in solitude or was primitively recreating in an outstanding way at a given moment. Federal agencies' ability to prove the latter does not disprove the former. Second, although outstanding wilderness solitude and primitive recreation appear to rest on a mere state of mind and to be highly subjective (Torrey 1992, 493), they both occur *within* wilderness. It makes no sense to say that we can have wilderness solitude in a closet or primitive recreation in Disneyland. Because we must first have an untrammeled and roadless area to establish the naturalness of wilderness, and because it makes no sense to talk about wilderness solitude and primitive recreation without a wilderness setting, the naturalness requirement of wilderness is logically prior to the human solitude/(primitive) recreation requirement of wilderness.

WILDERNESS MANAGEMENT

The above two requirements of wilderness capture what many of us think about when we think about wilderness. Section 2(c) (3) of the Act states that the preservation of wilderness must be made "practicable" in an "unimpaired condition." An "unimpaired condition" of wilderness means a condition of naturalness. What does it mean to say that the preservation of wilderness is to be made "practicable?"

To say that something is practicable is simply to say that it can be done or is possible in practice. Practicable wilderness preservation for a particular tract of land means that the maintenance of the tract as wilderness will accompany the designation of the tract as a wilderness area. Thus, the Act tells us that wilderness preservation must be maintained over time after legal wilderness designation. Federal agencies responsible for wilderness preservation rely on the practice of *wilderness management* to maintain wilderness (Hendee et al. 1990). Much of wilderness management concerns what we as humans can and cannot do in wilderness areas so that these areas can be maintained as wilderness. These uses and disuses of wilderness are legally cashed out and constrained by Section 4 of the Act. In effect, this section of the Act stipulates management requirements federal agencies follow for managing wilderness areas (Rohlf and Honnold 1988, 259–62). The grazing of domestic animals is allowed to continue in wilderness areas managed by the FWS, the USFS, and the BLM when grazing had occurred prior to wilderness designation. Hardrock mining and mineral (energy minerals) leasing are allowed in FWS, USFS, and BLM wilderness areas when a valid claim or valid lease existed prior to wilderness designation. Logging and road building in wilderness areas are allowed in limited circumstances when such activities are necessary for grazing, hardrock mining, mineral leasing, and the management of wilderness (primarily to control fires, insects, and diseases). Water development in wilderness areas is allowed in limited circumstances when it is necessary for grazing, hardrock mining, and mineral leasing; water development is also allowed when authorized by the President of the United States.[22] Primitive recreation, hunting, and fishing are allowed in many wilderness areas (hunting is usually banned in NPS wilderness areas). Finally, allowances are made for the continued use of aircraft and motorboats in wilderness areas where such use had been previously established.

Many legal commentators and wilderness managers claim that the provisions of Section 4 are not requirements that must be met in order for an area to qualify as wilderness but, rather, are prescriptive requirements for the management of areas once they become legal wilderness areas (McCloskey 1966, 309; Hendee et al. 1990, 117). Thus, wilderness is legally defined by naturalness and opportunities for solitude or primitive recreation before we can talk about what we can and cannot do in wilderness areas. But Section 2(c) (3) of the Act states that wilderness preservation must be made practicable. If an area could not be managed in a practicable manner, this would seem to preclude its qualification as wilderness. Thus, how wilderness is managed (along with legal limitations of what we can and cannot do in wilderness areas) seems to be a part of the legal definition of wilderness. I shall call this the management requirements of wilderness.[23]

PARADOXES OF FEDERAL WILDERNESS PRESERVATION

When we map the above three requirements of wilderness onto the rationales and exceptions of the Act, several paradoxes of federal wilderness preservation emerge. The preservation rationale enjoins us to preserve wilderness independent of human uses of it. What in wilderness do we preserve? Naturalness. The recreation rationale enjoins us to preserve wilderness so that we can use it for wilderness recreation and solitude. What in wilderness are we preserving? Outstanding opportunities for primitive recreation and solitude. As indicated, the naturalness requirement of wilderness has logical priority over the human solitude/recreation requirement because wilderness must first exist before we can find wilderness solitude and participate in wilderness recreation. This means that the preservation rationale has logical priority over the recreation rationale.

These rationales conflict, however, when wilderness is managed both for the preservation of its naturalness and for opportunities for solitude and primitive recreation. Primitive recreation today means multitudes of people with high-tech gear backpacking, hiking, horseback riding, skiing, snowshoeing, climbing, canoeing, kayaking, rafting, hunting, fishing, and photographing everything and everywhere possible in wilderness areas. As Roderick Nash (1982) has observed, we are "loving our wilderness to death." Wilderness enthusiasts demand the right to have wilderness solitude and primitive recreation and leave impacted campsites, mazes of trails, chalk (on rock climbing routes), feces, garbage, and so forth all over wilderness areas. In spite of the logical priority the preservation rationale holds over the recreation rationale, the latter supports trammeling the naturalness of wilderness over preserving such naturalness.

We run into more problems when we bring exceptions to strict preservation into the picture. Activities such as limited logging, water development, road-building, grazing, hardrock mining, and mineral leasing can compromise, erode, and destroy the naturalness of wilderness. All of these activities are instances of trammeling nature, and, as per the Wilderness Act of 1964 and many subsequent wilderness acts, they can occur within many wilderness areas. Given the legal language and federal management of wilderness, these exceptions to strict preservation take logical priority over the preservation rationale. Written into the very language of the Act that established federal wilderness preservation is a political compromise that expresses anti-wilderness values and that allows for and legally sanctions the commercial trammeling and economic exploitation of wilderness.

The first paradox of federal wilderness preservation is that its creation can legally sanction its destruction. We may wish to view the recreation rationale and the exceptions to full wilderness preservation (grazing, mining, and such) simply as politically pragmatic compromises that were necessary to get the Act and subsequent wilderness acts passed in Congress. Nonetheless, the creation of federal wilderness areas and the reasons used to justify such creation, whether aimed at preserving or spoiling the untrammeled wilderness character of such areas, inhibit the preservation of wilderness by allowing activities in wilderness areas that denude and destroy naturalness.

The need for wilderness management is, in part, created by the naturalness-eroding, allowable uses of wilderness.[24] Requirements for wilderness management can supersede both requirements for naturalness and for solitude and recreation when human activities allowed in wilderness areas necessitate intervention. Wilderness preservation begins to look like multiple use with the presence of miners, ranchers, backpackers, climbers, and so on competing to use wilderness areas set aside supposedly to preserve an untrammeled condition of naturalness. Simply put, naturalness is trammeled when we legally define wilderness areas (natural, untrammeled areas) in terms of solitude and recreational use and in terms that permit using wilderness for certain kinds of economic exploitation.

In sum, many people quote Section 2(c) of the Act as *the* definition of wilderness, but management requirements, as well as the solitude/recreation requirement, mandate activities that are legally permissible in wilderness areas at the expense of naturalness. This glaring paradox of federal wilderness preservation emerges from the general legal definition of wilderness as untrammeled, natural areas in combination with definitional specifications that mandate how humans may develop and use wilderness in ways that trammel such areas. Wilderness designation does more than merely tolerate "exceptions" to wilderness preservation. The very language used to designate wilderness—how wilderness is legally defined—further stipulates requirements for how wilderness may be developed and trammeled by humans.

A conceptual paradox underlies this legal paradox. Wilderness preservation is regarded within American law as a resource. When we consider federal wilderness preservation, we must remember that we are talking about a particular classification of federal public lands. From a historical and legal perspective, the major issues concerning federal lands have centered around use: which uses and which uses in compatible combinations should be allowed on these lands which comprise 29 percent of the nation's total acreage? Legal analysts describe uses by using the term *resources:* land usage allocation decisions on federal public lands may be understood as resource allocation decisions. Wilderness preservation, described by the Act at Section 2(a) as a "resource," can be characterized as a resource because it describes a limited use or nonuse of an area (Coggins et al. 1993, 967). From a legal standpoint, the question for Congress (the legislating authority for federal public lands) is how to allocate wilderness preservation in conjunction with the six other major resources of federal public lands: water, minerals, timber, range (for grazing), wildlife, and recreation (Coggins et al. 1993, 14–15).

Generally, to label something as a "resource" is to say that it has value. But we associate a resource with a specific kind of value—instrumental value—that can be used to gain something else of value. If wilderness preservation is labeled as a "resource," then the value of wilderness preservation is instrumental because it is used to attain something else of value. What is this further value? It could be the aesthetic, religious, and recreational value that we can attain by finding solitude and/or primitive recreation in wilderness—the recreation rationale. Or it could be economic value we can attain by grazing, mining, or using wilderness for our own economic goals—"exceptions" to full wilderness preservation. But the preservation rationale of the Act strongly suggests that wilderness has more than mere instrumental use value to humans. It suggests that wilderness is intrinsically valuable. For when we define wilderness in terms of untrammeled naturalness, we are setting wilderness apart from instrumental use value. Is wilderness preservation first and foremost a resource for humans or does it legally mandate valuing a condition of naturalness apart from human utility? An examination of federal wilderness preservation within American environmental law yields no clear answer to this question.

NOTES

I wish to express my gratitude to J. Baird Callicott, Dale Jamieson, Holmes Rolston III, and Charles F. Wilkinson for constructive comments and criticisms on several earlier drafts of this paper. I am indebted also to J. Steven Kramer, Paul Veatch-Moriarty, Michael P. Nelson, James W. Nickel, Graham Oddie, and Juliette Strauss who were all instrumental in offering useful comments and criticisms.

1. The Wilderness Act of 1964 mandated the USFS, the NPS, and the predecessor of the FWS to preserve and manage wilderness (the FWS came into existence in 1966 through the National Wildlife Refuge Administration Act). The BLM was mandated by Congress to preserve and manage wilderness by the Federal Land Policy and Management Act (FLPMA) of 1976.

2. I lack the space here to discuss the political history of federal wilderness preservation. For more about the sixty-five versions of wilderness acts from 1956 to 1964 see Allin (1982).

3. For example, see Rolston (1988) and Taylor (1986).

4. The "use and enjoyment rationale" also enshrines other reasons for valuing and preserving wilderness such as the following: wilderness provides a setting for unique aesthetic experiences, wilderness provides for a unique spiritual setting, wilderness gives us unique character-building and/or symbolic experiences, and wilderness serves as a unique place for scientists to study nature. Many argue that this rationale historically rests upon scenic or aesthetic values (e.g., Nash 1982).

5. The BLM also subscribes to a multiple-use, sustained-yield philosophy (via FLPMA of 1976), and the BLM's management of wilderness areas is also "supplemental" to commercial extraction and development interests. Although the FWS is mandated primarily to conserve wildlife in national wildlife refuges and game refuges, the FWS is also mandated to manage these refuges for various development such as mining, mineral leasing, grazing, water development, United States Air Force test bombing, and so on, and wilderness is "supplemental" to all of these uses of the refuges. Finally, the National Park Service Organic Act of 1916 has justified the NPS's opening of national parks (and other areas they manage) via roads, concessionaires, and other forms of "industrial tourism," and wilderness is "supplemental" to this.

6. By "de facto wilderness" I mean a natural area that can meet legal criteria for

inclusion within the NWPS (as opposed to de jure wilderness—as a legally designated wilderness area).

7. Primarily because of abundant precipitation in the eastern United States, human scars in potential and actual eastern wilderness areas tend to disappear under new floral growth more quickly than in wilderness areas found elsewhere.

8. Since 1975, numerous other wilderness acts have been passed that have added more wilderness areas in the eastern United States to the NWPS. The Eastern Wilderness Act was a landmark act that opened the door for the eastern expansion of the NWPS (Browing et al. 1988).

9. The USFS's purity definition of wilderness was rejected by the District Court of Colorado in 1970 in the case of *Parker v. United States.* The court in this case granted an injunction to prevent the USFS from awarding a timber contract to log the East Meadow Creek Area in the White River National Forest in Colorado. This area was adjacent to a roadless area being studied for possible wilderness designation. The court determined that logging could not occur until the USFS adequately studied the wilderness suitability of this area. The USFS argued that because this area contained an access road and a "bug" road (the latter blazed in the 1950s to fight bark beetle infestation; this road had been closed for years), it did not meet naturalness qualifications for possible wilderness designation. The court concluded that the access road was "substantially unnoticeable from approximately 100 yards away" and the bug road had grown over; neither of these roads necessarily prevented the area from wilderness study suitability (*Parker v. United States,* 309 F. Supp. 593, 600 [D.Colo. 1970]).

10. Statement by Senator Buckley of New York, "Hearings on Legislation to Designate Additional Areas to the Wilderness Preservation System, Before the Subcommittee on Public Lands of the Senate Committee on the Interior," 119 *Congressional Record,* Senate 438 (Daily ed., January 1973); quoted in Haight (1974, p. 288).

11. To be sure, many wilderness areas, especially those in the western United States, probably bore less impact simply because many of these wilderness areas are "rock and ice" areas—areas of rock, ice, and tundra high in mountain ranges; historically, there has been less human habitation and development above timberline in mountain ranges than below timberline.

12. To what degree must, of course, be qualified. I lack the space here to discuss this.

13. Many (perhaps most) groups of non-modern humans have built roads. Indeed, numerous wilderness areas in the United States contain vestiges of roads blazed and used by aboriginal inhabitants. Many of these former roads have become trails or other used routes. Insofar as these roads of yesterday have become trails or have been overgrown by natural forces, they are no longer roads. The forward-looking condition of untrammeled wilderness supports the idea that an

area containing a road hundreds of years ago can be considered a roadless area today.

14. House of Representatives Report Number 94-1163, 94th Congress, 2nd Session at 17 (1976). This definition of the term "roadless" was formulated in the legislative history leading up to the FLPMA of 1976 (Harvey 1980, 493–94).

15. *Bureau of Land Management, U.S. Department of the Interior, Wilderness Inventory Handbook (WIH)* 5 (1976).

16. Nash's phenomenological definition of wilderness is that wilderness is whatever one thinks it is: "... wilderness is ultimately a feeling about a place, a state of mind, that varies from person to person ..." (Nash 1982, 384). Obviously, this is much broader than a mere perceptual definition.

17. Some, such as Holmes Rolston III (1988), reject this perceptual condition and claim that naturalness can exist whether or not we can perceive it. On an earlier draft of this paper, Rolston said, "We are objectively detecting the actual degree of human impact." I lack the space here to discuss whether naturalness is objectively in wilderness or subjectively in our minds.

18. More times than not federal agencies recommend against wilderness designation.

19. A strict application of this standard would eliminate opportunities for wilderness solitude in literally hundreds of wilderness areas where outside sights and sounds are visible and audible. For example, commercial airplanes flying to and from the West Coast are constantly visible from the Box-Death Hollow Wilderness Area in Utah, the glow of Denver is visible from many high peaks in the Indian Peaks Wilderness Area in Colorado, and one is hard-pressed to escape the sound of nearby traffic in the Bristol Cliffs Wilderness Area in Vermont.

20. The Kaiparowits Plateau harbors an estimated 5 to 7 billion tons of recoverable coal. Many believe that the BLM is reluctant to thwart the extraction of this coal by wilderness designation of the plateau. Despite the reasons the BLM gives in its Draft Environmental Impact Statements of the Kaiparowits WSAs for recommending against wilderness designation, its support of coal leasing may be the real reason for its lack of wilderness enthusiasm here.

21. On more than one occasion BLM officials conducted a search for wilderness solitude (and for outstanding opportunities for a primitive type of recreation) by flying into a remote area of the Kaiparowits (and into other WSAs elsewhere in Utah) via a helicopter, walking around for ten minutes, and flying out of the WSA (UWC 1990).

The problems I cite here in the BLM wilderness inventory of the 1980s should not be attributed solely to BLM officials looking for wilderness solitude in the field. One BLM wilderness field specialist, Peter Viavant, reported: "Most of the input came from the line management, not the wilderness staff.... There were many times when a decision would be changed after a recommendation was written, and one would just go back and rewrite it again. I mean, you'd write, 'The canyons are 500 feet deep, providing great solitude . . .' And then they'd change their minds and decide it should be out of the wilderness inventory, and you'd come back and write, 'there's only 500 feet of relief and this does not provide very good solitude'" (Wheeler 1990, 36).

22. To be sure, very little hardrock mining, mineral leasing, logging, road building, and water development have occurred in actual wilderness areas (grazing occurs on a much more frequent basis). The point I am trying to make here is that American law prescribes the permissibility of these activities, regardless of whether or not these activities actually occur (and they have occurred in various cases).

23. The legal literature, case law, and judicial reviews concerning legal wilderness preservation are dominated almost exclusively by management issues that revolve around Section 4 of the Act. From a legal standpoint, management requirements for wilderness are central to what wilderness is.

24. Management requirements for wilderness also contain prohibitions against using wilderness in ways that erode the unimpaired condition of wilderness, and requirements for wilderness management guard against the trammeling of wilderness. But management requirements cut both ways and also allow for the trammeling of wilderness (as per Section 4 of the Act).

BIBLIOGRAPHY

Allin, C. W. 1982. *The Politics of Wilderness Preservation.* Westport, Conn.: Greenwood Press.

Browing, J. A., J. C. Hendee, and J. W. Roggenbuck. 1988. *103 Wilderness Laws: Milestones and Management Direction in Wilderness Legislation, 1964–1987.* University of Idaho, College of Forestry, Wildlife and Range Sciences Bulletin 51, Moscow, Id.

Callicott, J. B. 1991. "The Wilderness Idea Revisited: The Sustainable Development Alternative." *Environmental Professional* 13:235–47. (Included in this volume.)

Coggins, G. C., C. F. Wilkinson, and J. D. Leshy. 1993. *Federal Public Land and Resources Law.* 3rd ed. Westbury, N.Y.: Foundation Press.

Denevan, W. M. 1992. "The Pristine Myth: The Landscape of the Americas in 1492." *Annals of the Association of American Geographers* 8:369–85. (Included in this volume.)

Elliot, R. 1982. "Faking Nature." *Inquiry* 25:81–93.

Foote, J. P. 1973. "Wilderness—A Question of Purity." *Environmental Law* 3:255–66.

Foreman, D., and H. Wolke. 1992. *The Big Outside: A Descriptive Inventory of the Big Wilderness Areas of the United States.* Rev. ed. New York: Harmony Books.

Haight, K. 1974. "The Wilderness Act: Ten Years After." *Environmental Affairs* 3:275–326.
Harvey, D. M. 1980. "Exempt from Public Haunt: The Wilderness Study Provisions of the Federal Land Policy and Management Act." *Idaho Law Review* 16:481–510.
Hendee, J. C., G. H. Stankey, and R. C. Lucas. 1990. *Wilderness Management.* 2nd ed. Golden, Col.: Fulcrum Publishing.
McCloskey, M. 1966. "The Wilderness Act of 1964: Its Background and Meaning." *Oregon Law Review* 45:288–321.
Nash, R. 1982. *Wilderness and the American Mind.* 3rd ed. New Haven, Conn.: Yale University Press.
Rickart, T. M. 1980. "Wilderness Land Preservation: The Uneasy Reconciliation of Multiple and Single Use Land Management Policies." *Boston College Environmental Affairs and Law Review* 8:873–917.
Rohlf, D., and D. L. Honnold. 1988. "Managing the Balances of Nature: The Legal Framework of Wilderness Management." *Ecology Law Quarterly* 15: 249–79.
Rolston, H., III. 1988. *Environmental Ethics: Duties to and Values in the Natural World.* Philadelphia: Temple University Press.
Taylor, P. W. 1986. *Respect for Nature: A Theory of Environmental Ethics.* Princeton, N.J.: Princeton University Press.
Torrey, R. S. 1992. "The Wilderness Inventory of the Public Lands: Purity, Pressure, and Procedure." *Journal of Energy, Natural Resources & Environmental Law* 12:453–520.
Utah. 1986. *BLM Statewide Wilderness Draft Environmental Impact Statement.* Volumes I and III. Part B. (Publication date listed as 1985; issued in February 1986).
UWC (Utah Wilderness Coalition). 1990. *Wilderness at the Edge: A Citizen Proposal to Protect Utah's Canyons and Deserts.* Salt Lake City: Foundation for the Utah Wilderness Coalition.
Vest, J. H. C. 1987. "The Philosophical Significance of Wilderness Solitude." *Environmental Ethics* 9:303–30.
Wheeler, R. 1990. "The BLM Wilderness Review." In The Utah Wilderness Coalition. *Wilderness at the Edge: A Citizen Proposal to Protect Utah's Canyons and Deserts.* Salt Lake City: Foundation for the Utah Wilderness Coalition.
Wilderness Act of 1964, Public Law 88-577.
Worf, W. A. 1980. "Two Faces of Wilderness—A Time for Choice." *Idaho Law Review* 16:423–37.

Michael P. Nelson

An Amalgamation of Wilderness Preservation Arguments (1998)

NUMEROUS AND DIVERSE ARGUMENTS have been put forth by people of sundry backgrounds and times on behalf of the preservation of what they took to be wilderness. From backpackers to bureaucrats, Romantics to rednecks, socialists to suburbanites, historians to hunters, philosophers to philanthropists, people have sung the praises of areas which they assumed to exist in their "pristine state." It is safe to think that there will continue to be wilderness defenders regardless of the challenge presented to the very concept of wilderness found in the next two sections of this book. In the present essay I attempt to summarize in one place the many traditional and contemporary arguments proffered on behalf of "wilderness."

To review such arguments for the sake of historical interest and to observe how the received view of wilderness is tellingly manifested in such arguments is worthwhile. But there is another reason for wanting such a review. The rationales we employ on behalf of anything, including wilderness preservation, reflect our attitudes and values. Our attitudes toward and valuation of those places we have thought of as wilderness are revealed in the many traditional defenses of those places. Moreover, our attitudes and values profoundly affect the manner in which we treat something, including the places we call "wilderness." As environmental historian Roderick Nash observes, "So it is that attitudes and values can shape a nation's environment just as do bulldozers and chain saws."[1] Consequently, we might better understand our current environmental policies if we look at the historical rationales for protecting and defending certain wild places. And if we indeed do need to rethink our classical concept of wilderness—and therefore our current policies with regard to those places taken to be wilderness—a review of where we came from can surely aid us in such an undertaking.[2]

Wilderness preservation arguments have been previously catalogued by Roderick Nash, Holmes Rolston III, William Godfrey-Smith (now William Grey), Warwick Fox, George Sessions, and Michael McCloskey. Here I try to integrate and reconcile these disparate compilations. In the process, I rename and recategorize many of the arguments found in these sources. To them, I add hitherto unexplored wilderness preservation rationales. Hence, what follows is more an amalgamation than a taxonomy or typology; although there is a general attempt to move from narrowly instrumental, egocentric, and anthropocentric values to broader social, biocentric, and even intrinsic values attributed to putative wilderness.

Admittedly, an inherent tension exists in such a project. Most of the wilderness preservation arguments contained herein take the existence of wilderness for granted. However, as the next two sections of this anthology document, the usefulness of the concept of wilderness is correctly subject to intense debate: a great new wilderness debate.

Further, all of the following arguments for wilderness preservation are significantly biased in two major ways. First, they assume a terrestrial and not an oceanic or even extraterrestrial sense of wilderness. One might argue that they really ought also to apply to marine wildernesses and to the other, so far untrammeled, planets. Accordingly, I will interject a non-terrestrial perspective into the following arguments when it seems appropriate to do so. Second, the received wilderness idea, and hence many of these arguments for wilderness preservation, has an Australian-American bias. Many Europeans, for example, have no wilderness to worry about preserving. As histories of land settlement and tenure differ so do senses and views of the landscape. Arguably, it takes designated wilderness areas, or at least some recent memory of or belief in a once pristine landscape, to have a received view of wilderness. Actually, then, we are only referring to certain cultures—Protestant Christian, colonial, and postcolonial cultures in particular—when we refer to "a received view of wilderness."

1. THE NATURAL RESOURCES ARGUMENT

In many of those areas we refer to as wilderness, there exist significant quantities of untapped yet precious physical resources. Certain designated wilderness areas are great repositories of a wide variety of natural resources and we humans can render our future more secure by preserving these resource reserves. Clearly this is the most narrowly anthropocentric, instrumental, and simpleminded preservation argument that one could advance. It is, therefore, also the most popular and effective argument for the preservation of purported wilderness.

Some writers even suggest that their value as physical resources for *present* or *immediate* exploitation is a rationale for wilderness *preservation!* "Market Value," is what Rolston calls this liquidatable wilderness value.[3] B. L. Driver, Nash, and Glenn Haas refer to wilderness resources as individual and societal "Commodity-Related Benefits," and cite water, timber, minerals, and forage as examples of what they mean by wilderness-area-produced resources.[4] However, a bit of reflection suggests that this entire argument is actually paradoxical, if not thoroughly self-contradictory. In theory, *designated wilderness areas are places where the extraction of resources is strictly prohibited;* ideally wilderness and resource extraction don't mix.[5] It would seem that if we use an area as a goods resource then we are no longer entitled to call it wilderness. Can we have our wilderness and eat it too? Can we extract resources from an area and still call it wilderness? Maybe we could harvest an area's resources on a very small, sustainable, and non-trammeling scale and the area might still fit the description of wilderness. This argument, however, advanced by Rolston, is conceptually problematic as even small-scale natural resource exploitation arguably runs contrary to the received view of wilderness as untouched by human hands. Moreover, resource use is a matter of degree and "small scale" is a relative term. Such ambiguities leave the present resource extraction argument for wilderness preservation paradoxical and troubling. Less problematic is the argument that we could look at wilderness areas as resource reserves to be exploited only by future generations. Obviously, this is not paradoxical. If we are saving vast resource-rich areas, not exploiting them, we are preserving them. And if future generations did use those presently preserved wilderness areas for resource extraction then they would, at that point, no longer be wilderness areas. Resource reserves, then, are only wilderness areas insofar as their use is potential and not actual. Further, one might argue for preserving as wilderness some areas that may harbor *potential* resources whose existence and instrumental worth has not yet been discovered.[6]

Proponents of the resource reserve argument for wilderness preservation seem to assume that the world without wilderness areas is a world without the many natural resources found (and unfound) in them. That is, they seem to assume that natural resource conservation depends upon wilderness preservation. Wilderness is the untapped pool of natural resources. And this may be true of some resources. Old growth timber and grizzly bear hides come immediately to mind.

Some types of natural resources are so commonly cited as depending upon wilderness preservation that they merit separate discussion—to wit, hunting and medicine.

2. THE HUNTING ARGUMENT

One of the earliest and most popular wilderness preservation arguments asserted that areas of what was taken to be wilderness were worthy of protection because some of them provided terrific venues for hunting or supplied the natural resource of wild game.

Aldo Leopold, one of the earliest and most passionate advocates of wilderness preservation, was keen on what he thought of as wilderness hunting. He traveled to places like the Sierra Madre Occidental in northern Mexico to hunt wild game. He wrote essays with titles like "A Plea for Wilderness Hunting Grounds." And he urged the U.S. Forest Service to set aside certain bits of pristine land dedicated to nothing but wilderness hunting. "The establishment of wilderness areas," he wrote, "would provide an opportunity to produce and hunt certain kinds of game, such as elk, sheep, and bears, which do not always 'mix well' with settlement. . . . Wilderness areas are primarily a proposal to conserve at least a sample of a certain kind of recreational environment, of which game and hunting is an essential part."[7] Hence, in order to hunt these more charismatic and "wilderness" dependent megafauna, their home ranges (purported wilderness) must be maintained.

This special case of wilderness preservation as a sort of big, fierce, almost tribal proving grounds has been especially championed or ridiculed because of its identification with virility, masculinity, and machismo. Leopold once wrote, "Public wilderness areas are essentially a means for allowing the more virile and primitive forms of outdoor recreation to survive."[8] But nowhere was the association of wilderness preservation, hunting, and masculinity more vehemently expressed than in the thoughts and writings of rough-riding U.S. President Theodore Roosevelt. In the twenty-three volumes of his collected written works, Roosevelt often refers to what was in his mind wilderness as a hunting grounds and laments its loss as such. Referring to modern Americans as overcivilized, slothful, and flabby, Roosevelt calls upon Americans to regain and develop those "fundamental frontier values," to lead a "life of strenuous endeavor," and to revel in the "savage virtues."[9] For Roosevelt, what he referred to as wilderness hunting was the means to accomplish this; for only in such wilderness hunting can a person (man) "show the qualities of hardihood, self-reliance, and resolution."[10] Roosevelt declared:

> Every believer in manliness and therefore in manly sport, and every lover of nature, every man who appreciates the majesty and beauty of wilderness and of wild life, should strike hands with the farsighted men who wish to preserve our material resources, in an effort to keep our forests and our game beasts, game-birds, and game-fish—indeed, all the living creatures.[11]

Three comments on this argument. First, we can easily expand on this argument and include fishing as well. In certain terrestrial designated wilderness areas the surface waters harbor the biggest trout that put up the fiercest fight. And deep-sea fishing is reported to be some of the most exciting fishing there is. Second, this argument pertains, however, only to certain putative wilderness areas and not to others. Places inhabited by animals that humans desire to hunt or have a historical predator/prey relationship with are worth preserving as wilderness, but those that are largely devoid of big game are not. Third, a century later the big-game-hunting argument for wilderness preservation is an embarrassment to many wilderness advocates. This is not an argument that contemporary wilderness advocates often employ. It would be like mounting an argument for the establishment of zoos as places where people could go to taunt animals. In fact, many would ban hunting from wilderness areas as the epitome of an intrusive, exploitative, and destructive use of wild places and their denizens.

3. THE PHARMACOPOEIA ARGUMENT

Another special case of ostensible wilderness resource extraction is medicine.[12] The actual and potential pharmaceutical use of what some of us think of as wilderness is perhaps the single most prevalent and persuasive contemporary wilderness preservation argument. The areas of the earth many commonly referred to as wilderness—such as the Amazon Rainforest and the forests of the Pacific Northwest—contain and support the most species on earth.[13] Since around 80 percent of the world's medicines are derived from life forms,[14] these "wilderness" areas therefore contain the greatest source of medicinal natural resources. As these places are "developed," many of the species that live in them become extinct. Thus, we lose forever any medicinal use they may have had. Donella Meadows calls this the "Madagascar periwinkle argument," referring to the celebrated rosy periwinkle (*Catharanthus roseus*) plant of Madagascar from which were derived the drugs vincristine and vinblastine, used in the treatment of leukemia.[15]

This argument seems unpersuasive if constructed in terms of the proven medicinal uses of wild species, since many medically useful species can be cultivated in plantations and laboratories or their active ingredients can be isolated and synthesized. The argument is most forceful in reference to potential, and yet unknown, medicinal uses of wild species. As noted, rainforests, old-growth forests, and the world's oceans house the greatest numbers of species. Most such species have not been described by systematists, let alone assayed for their medicinal potential. Therefore, these same areas arguably also house the greatest source of potential medicines. If these areas are destructively developed, we will lose a significant portion of the species that live in them, and thus we will lose any medicinal use of those species as well. Hence, it is argued, these purported wilderness areas should be saved because they shelter both potential as well as actual medicinal resources.

This argument deserves comment as well. First, many designated wilderness areas in North America and Australia are not species-rich rainforests and old-growth forests. Hence the Madagascar periwinkle argument does little to support their preservation. Second, in conjunction with this argument, it is often noted that the people most knowledgeable about the medicinal uses of rainforest species are the local indigenous inhabitants. But an area inhabited and used by human beings is not, by definition, wilderness.

4. THE SERVICE ARGUMENT

In addition to the natural resource goods provided by certain putative wilderness areas, innumerable and invaluable services are said to be provided by many of these areas as well. Wetlands benefit humans indirectly by serving to protect important river headways. Unbroken forests remove carbon dioxide from the atmosphere and replenish its oxygen, as do the world's oceans. Since we humans depend upon clean air and water, such services are vital to our continued existence.

Thinking critically about this argument for a moment, we realize that wilderness is indeed a sufficient condition for the performance of these services, but it does not seem to be a necessary one. That is, these services are not unique to uninhabited or uncultivated places; they are performed by non-wilderness ecosystems as well. Iowa corn plants purify air and remove atmospheric carbon dioxide just as Douglas firs do. Moreover, recovering forests composed of fast-growing young trees do an even better job of this than do old-growth forests.

Nevertheless, certain ecological services can only be performed in large tracts of relatively untouched land. For example, some designated wilderness areas provide nurseries for species such as salmon. Conservation biologists tell us that certain species, like the grizzly bear, require large tracts of unbroken land to exist. And the earth's oceans help to moderate temperatures. Again, to be sure, certain non-wilderness areas do perform some of these same services. But they do not perform these services as efficiently and thriftily as alleged wilderness areas do. Some wilderness areas are irreplaceable sources of clean air and pure water.

Potentiality comes into play in this argument as well. Since we are not entirely sure exactly what all is in those areas we think of as wilderness, we cannot be entirely confident of all the unique and crucial services provided by them. Hence, for their potential services we ought to preserve them as well.

5. THE LIFE-SUPPORT ARGUMENT

Holmes Rolston explains that there exists "a parallel between the good of the system and that of the individual." Further, we depend upon the healthy functioning of various ecosystems. Ecosystems often have the greatest value for us when they are the most independent of us; or, as Rolston concludes, "So far as [we] are entwined with ecosystems, our choices . . . need to be within the capacities of biological systems, paying some attention to ecosystem value."[16]

George Sessions points out that this prudential argument was made famous in the 1960s and 1970s "ecological revolution" by thinkers such as Rachel Carson, Barry Commoner, and Anne and Paul Ehrlich. The Ehrlichs, for example, liken species eradication to the popping of an airplane's rivets. Rivet-popping will eventually lead to the demise of spaceship Earth.[17]

So, as a mechanism for supporting and ensuring human existence (and the existence of many other species for that matter), so-called wilderness areas not only should but must be preserved, it is argued. Is this argument persuasive? As an argument for wilderness, over and above an argument for species preservation, its proponents must prove that the only way to preserve species is to preserve wilderness. They must explain how species diversity coexisted with people in species-rich areas of the world for hundreds, even thousands of years. Indeed, if this argument is to work, two links must be made. The preservation of wilderness must be linked with the preservation of species, and the preservation of species must be linked with human survival. This latter link is also questionable. Are species rivets? Is Earth a spaceship? One might take "survival" in a literal sense. We might survive, but only in diminished numbers as impoverished creatures in an ecologically impoverished environment. In sum, such an argument opens up a virtual Pandora's box of difficult questions with tough answers.

6. THE PHYSICAL THERAPY ARGUMENT

It is argued that designated wilderness area-related activities are wonderful and essential ways to enhance and even remedy our physical health. Primitivists, for example, claim that the more closely we are associated with nature the more physically healthy we will be. Hence, if putative wilderness is the purest representative form of nature, we would be healthier if we took our physical exercise in such places, some argue. Socialist, wilderness advocate, and cofounder of the Wilderness Society, Robert Marshall once asserted that there exist great physical benefits to "wilderness activities."[18]

This may appear to be quite a weak argument for wilderness preservation, since people in many parts of the world are physically fit despite having no access to designated wilderness areas. Traditionally the world's greatest middle-distance runners have been British. And there are no wilderness areas in Britain. Exercising in what Marshall took to be wilderness is at best only a sufficient, not a necessary, condition for physical well-being. However, wilderness advocates might still argue that these proffered wilderness areas are the *best* source and measure of physical health. Hence to lose them is to lose the *greatest* source and measure of physical therapy.

This wilderness preservation argument is what Godfrey-Smith refers to as the Gymnasium argument. However, it seems that his classification actually has two separate arguments: one is that designated wilderness areas provide us with a source and measure of physical health and the other is that these places serve as a great place to engage in certain sports (which may also aid our physical health), or what I refer to here as the Arena argument.

7. THE ARENA ARGUMENT

Even more elementary than supplying a source and measure of physical fitness, wilderness preservation is sometimes urged on the grounds that many designated wilderness areas provide us with superb and incomparable locales for athletic and recreational pursuits.

In various designated wilderness areas we can engage in a variety of activities: we can go cross-country skiing, hunting, scuba diving, snowmobiling, rock climbing, swimming, kayaking, backpacking, canoeing, horseback riding, hiking, camping, and mountaineering. Those engaged in these pursuits argue that designated wilderness areas allow us an unprecedented place to test our skills, hone our muscles, and experience all of the joys associated with these activities. Aldo Leopold saw this as one of the primary goals of public wilderness areas:

> Wilderness areas are, first of all, a means of perpetuating, in sport form, the . . . primitive skills in pioneering travel and subsistence . . . a series of sanctuaries for the primitive arts of wilderness travel, especially canoeing and packing [he was referring here to mule and horse packing]. . . . Recreation is valuable in proportion to the intensity of its experiences, and to the degree to which it differs from and contrasts with workaday life.[19]

We need places to roam, to use our leg muscles, places to use our hands in the grasping of natural things, so the argument goes. In the civilized world we keep fit and tone our bodies by running on the indoor treadmill, lifting weights, swimming laps, and going to aerobics classes at the local health spa. Wilderness areas provide us with places where we can develop our muscles and realize our strength by hiking, climbing, paddling, and so forth. . . . Obviously, it can contribute to physical well-being and even rehabilitate the disabled.

But we might ask why we need designated wilderness areas to do these things; I can paddle my canoe in a dam-created reservoir, hike in an industrial monoculture pine forest, and climb on a modern climbing-wall. The wilderness advocate may respond by claiming that these "artificial" places are pale substitutes for the "real" thing; that designated wilderness areas provide the *best* locales for these sorts of activities. They are unmatched and unmatchable outdoor gymnasia, places to play. They have all—and more—of the sporting accoutrements that our more "civilized" physical fitness facilities have.

Some wilderness advocates even argue that just as we require certain cultural spaces for certain cultural activities (e.g., football fields to play football on, theaters to see movies in, etc.), so some wilderness athletic activities simply require wilderness areas to do them in. Just as deep-powder skiers require deep powder, mountain climbers mountains, and deep-sea divers deep seas, wilderness backpackers require wilderness to do their thing. It seems that an essential ingredient in these activities is solitude and a pristine natural arena. Without a "wilderness" condition, enthusiasts of wilderness activities cannot pursue those activities.

There are really two different arguments here: first, wilderness is the best locale for certain activities; and, second, wilderness as the only locale for other activities. While one might grant that designated wilderness areas provide the *best* locale for certain activities—say deep-powder skiing and deep-sea diving—they do not provide the *sole* locale. Deep powder may lie a mere saunter from a ski resort. Deep seas exist under shipping lanes. These activities do not depend solely on the existence of designated or even *de facto* wilderness. On the other hand, it seems tautological and circular to claim that wilderness backpackers need wilderness. Still, it is an argument for wilderness preservation: one can hardly deny that wilderness recreationalists need wilderness in order to pursue their sports.

8. THE MENTAL THERAPY ARGUMENT

Perhaps even more prevalent than the above physical benefits of proffered wilderness is the argument for wilderness preservation on behalf of its actual and potential psychological health benefits. Wilderness advocates have often claimed that what are taken to be wilderness experiences can be psychologically therapeutic and can even significantly help treat psychologically disturbed persons.

Reflecting on the ever-increasing human desire to visit America's national parks, John Muir cites psychological dysfunction as a major cause, and refers to city people as "tired, nerve-shaken, overcivilized," "half-insane," "choked with care like clocks full of dust," and bursting with "rust and disease," "sins and cobweb cares."[20] According to Muir, visiting designated wilderness—which, as he realized, both necessitates and threatens its preservation—is the cure to these problems. In Muir's prescription for mental health, "wildness is a necessity," and wild places are "fountains of life."[21] Others, such as Sigurd Olson, Robert Marshall, and even Sigmund Freud have argued that civilization represses, frustrates, and often breeds unhappiness and discontent in humans that can best be alleviated by periodic escape to what they took to be wilderness. More contemporary studies show that drug abusers, juvenile delinquents, and over-stressed executives can and do often benefit from an occasional dose of wilderness experience.[22]

Wilderness helps us to put our "civilized" lives in perspective; it simplifies living; reacquaints us with pain, fear, and solitude; provides us with a necessary sense of challenge; and helps us discover what is really important and essential to our existence. Or so some say. Wilderness experience is also said to be a great form of stress relief and serves as a superb pressure release for those living in metropolitan areas. Primitivists claim that just as there are great physical benefits to close association with nature in wilderness, such association also has the mentally therapeutic benefits of making us happier and more psychologically stable and balanced. On this same note, Rolston claims that "wildlands absorb a kind of urban negative disvalue . . . and provide a 'niche' that meets deep seated psychosomatic needs."[23]

If one believes that the collective mental well-being of a society is but an aggregate of the mental well-being of its individual members, then the individual mental health of those who visit designated wilderness areas arguably contributes to the quality of a society's life and vitality. Some wilderness advocates even go so far as to assert that experience with wild places functions as a gauge or measure for our individual and collective sanity. Wallace Stegner, for example, asserts that his vision of wilderness is a "means of reassuring ourselves of our sanity as creatures."[24]

None of the proponents of the mental health argument for wilderness preservation explain exactly how the existence of designated wilderness areas contributes to sanity. But even if we did grant that the existence of designated wilderness areas is a sufficient condition for instilling mental health, is it a necessary condition? There are seemingly many psychologically fit people who have never visited a designated wilderness area and there are undoubtedly "wilderness junkies" who are not mentally healthy or stable. Surely other methods of ensuring and gauging our mental health and sanity exist.

Nevertheless, some wilderness proponents claim that purported wilderness is necessary for human mental health. In his book *Nature and Madness,* ecologist Paul Shepard argues that we humans *need* wilderness. He claims that in the natural and healthy growth process there needs to be close association, experience, and bonding with the wild things that inhabit wild places.[25] Warwick Fox echoes Shepard's sentiments when he claims that this sort of wilderness preservation argument "emphasizes the importance of the nonhuman world to humans for the development of healthy (sane) minds."[26]

However, even if these areas of alleged wilderness are only sufficient conditions for mental well-being, one might still wish to argue that they are the best means of assuring mental health. Moreover, one might argue that these "wild" places succor our souls in a much more socially and individually cost-effective manner than do therapists, twelve-step programs, support groups, churches, prisons, and mental institutions.

9. THE ART GALLERY ARGUMENT

Many people search out putative wilderness areas for aesthetic experience. Both beauty and sublimity may be found in these places, they say. Therefore, we should preserve them because they are sites of the beautiful and the sublime. "Wild" places, it is argued, are like gigantic art galleries.[27]

In fact, some have argued that aesthetic experience, of the sort so-called wilderness offers, can border on the religious or mystical. Roderick Nash, for example, maintains that the experience of wild things involves "awe in the face of large, unmodified natural forces and places—such as storms, waterfalls, mountains and deserts."[28] And William Wordsworth wrote that experiencing the beauty of what his vision of wilderness was produces "a motion and a spirit, that impels . . . and rolls through all things."[29] Further, some argue that designated wilderness areas are places where the very meaning of aesthetic quality can be ascertained and that, therefore, all beauty is dependent upon such sites. Muir, for example, claims that "None of Nature's landscapes are ugly so long as they are wild."[30]

The destruction of a designated wilderness area—and hence wilderness-dependent species such as wolves and grizzlies—it is claimed, would be as bad as, even worse than, the destruction of a painting by da Vinci or a sculpture by Michelangelo. In principle, works of art can be re-created, but wilderness-dependent species—like any species—cannot. As Aldo Leopold puts it in his essay "Goose Music," "In dire necessity somebody might write another Iliad, or paint another 'Angelus,' but fashion a goose?"[31] Now, geese are admittedly not a wilderness-dependent species and Leopold is not speaking directly of wilderness in this essay, but Leopold's point clearly applies to wolves and grizzlies as well as to geese. What cannot almost unequivocally be recreated (*Jurassic Park*–considerations to the side) is an extinct species. And since species like wolves and grizzlies are dependent upon wilderness for their continued existence, the loss of wilderness is tantamount to the *permanent* loss of those species.

The intensity and type of beauty found in unique land forms, waterfalls, mountains, oceans, outer space, deserts, plants, and animals—all shaped by natural forces—cannot be replicated in urban or even pastoral settings, some wilderness advocates maintain. According to them, these places are both necessary and sufficient conditions for a true sense of beauty. Hence, if the loss of this beauty is to be avoided the preservation of wilderness areas is mandated.

10. THE INSPIRATION ARGUMENT

Many claim that putative wilderness areas are important to maintain because they provide inspiration for the artistically and intellectually inclined. In the process, these designated wilderness areas add to and help shape culture. Numerous artists—such as painters Thomas Cole, Thomas Moran, and Albert Bierstadt; photographers Ansel Adams and Galen Rowell; writers James Fenimore Cooper, Colin Fletcher, and "Cactus" Ed Abbey (not to mention Emerson and Thoreau); musicians John Denver and "Walkin'" Jim Stoltz; and poets Walt Whitman, William Wordsworth, and Robinson Jeffers, to name but a few—find their inspiration in what they take to be wilderness. For them, "wilderness" provides an excellent and unique motif for art.

Some even assert that wilderness serves to inspire those in the intellectual arts as well. Philosophers, for example—especially environmental philosophers—often regard what they take to be wilderness experience to be a contemplative catalyst or cognitive genesis for the really big questions of philosophy: What is the meaning of the universe (a question evoked especially by extraterrestrial wilderness, it seems); where we all came from; what we are all doing here; where we are going; what the character of our existence is, and what our moral place in the world is. Now of course there are other catalysts for artistic and philosophic inspiration: cattle, cities, and factories to name but a few. The point is not that wilderness areas are the only muses for art, but rather that they are excellent and unique ones, and that to lose any such inspirational kindling would be tragic.

11. THE CATHEDRAL ARGUMENT

For some, ostensible wilderness is a site for spiritual, mystical, or religious encounters: places to experience mystery, moral regeneration, spiritual revival, meaning, oneness, unity, wonder, awe, inspiration, or a sense of harmony with the rest of creation—all essentially religious experiences. Wilderness areas are also said to be places where one can come to an understanding of and engage in the celebration of the creation—an essentially religious activity. Hence, for people who think like this, designated wilderness areas can and do serve as a sort of (or in lieu of) a church, mosque, tabernacle, synagogue, or cathedral. We should, then, no more destroy wilderness areas than we should raze Mecca or turn the Sistine Chapel into a giant grain silo. For some, wild places represent and reflect the various spiritual and religious values that they hold dear.

To go one step further, some even claim that since designated wilderness areas are the closest thing we have on earth to the original work of God, to destroy them would be tantamount to the destruction of God's handiwork, forever altering God's original intent.

John Muir believed that the closer one was to nature, the closer one was to God. To Muir, "wilderness" was the highest manifestation of nature and so was a "window opening into heaven, a mirror reflecting the Creator," and all parts of it were seen as "sparks of the Divine Soul."[32] Yosemite's Hetch Hetchy Valley was, for Muir, a place epitomizing wilderness, a shrine to a higher existence, the destruction of which was tantamount to sacrilege. For this reason, Muir vehemently defended Hetch Hetchy and said of its would-be desecrators:

> These temple destroyers, devotees of ravaging commercialism, seem to have a perfect contempt for Nature, and, instead of lifting their eyes to the God of the mountains, lift them to the Almighty Dollar.
>
> Dam Hetch Hetchy! As well dam for water-tanks the people's cathedrals and churches, for no holier temple has ever been consecrated by the heart of man.[33]

Transcendentalists, such as Ralph Waldo Emerson, Thoreau, and William Cullen Bryant, went so far as to claim that one could only genuinely understand moral and aesthetic truths in what they took to be a wilderness setting. For these thinkers, civilization only fragments and taints one's genuine moral and aesthetic understanding.

Actually, this appears to be a quite powerful political argument for legally preserving designated wilderness areas. In the United States, for example, if the sorts of experiences and activities just listed are presumed to be essentially religious experiences and activities, then designated wilderness areas can be said to serve a religious function. Hence, designated American wilderness areas could be defended on the constitutional grounds of freedom to worship as one chooses. Regardless of concerns about the size or type of the area required or the fact that only a minority of people actually "go to church in the woods," designated wilderness areas might still merit protection, both ethically and legally, as places of worship.

12. THE LABORATORY ARGUMENT

Some (mostly wildlife, marine, and conservation biologists) argue that the preservation of designated areas of wilderness is important because it provides scientists with an unprecedented location and the raw materials for certain kinds of scientific inquiry. In order to conduct their scientific queries, these wilderness advocates require many types and varieties of geographical locales which remain in their pristine state. This scientific study is said to be important not only for the sake of knowledge itself but also more instrumentally because a society can use this knowledge to form a better understanding of itself, the world around it, and hence its proper role in that world. Wilderness is viewed by these people as one end of the spectrum of locales for such study.

Admittedly, there is a potential paradox involved with this argument—as there is with all wilderness preservation arguments that entail the use areas of putative wilderness by humans. If we use wilderness areas as laboratories (or gymnasiums, cathedrals, resource pools, etc. . . .) in too dense or intrusive a fashion, they would then cease to be wilderness areas. The use, then, of wilderness areas by humans possibly is or could be a threat to such areas themselves. This tension ought always to be kept in mind when considering these sorts of arguments.

13. THE STANDARD OF LAND HEALTH ARGUMENT

Aldo Leopold's arguments on behalf of wilderness preservation shifted and expanded as his thought progressed. The more mature Leopold believed that what he took to be wilderness was important for scientific as well as for recreational reasons. Leopold proclaimed that "all wild areas . . . have value . . . for land science. . . . Recreation is not their only, or even their principal, utility." For Leopold, the main scientific use of "wilderness" was as a base-datum or measure of land health and as a model of a normal ecologically balanced landscape. According to Leopold, wilderness areas serve as a measure "of what the land was, what it is, and what it ought to be," providing us with both an actual ecological control sample of healthy land and a normative measure of what we ought to strive toward.

Throughout his life, Leopold became increasingly interested in land use and the science of land health. In order to have such a science, Leopold declared that we need "first of all, a base datum of normality, a picture of how healthy land maintains itself as an organism." According to Leopold, the "most perfect norm is wilderness" for "in many cases we literally do not know how good a performance to expect of healthy land unless we have a wild area for comparison with sick ones." Hence, we can easily see how "wilderness," in Leopold's words, "assumes unexpected importance as a laboratory for the study of land-health."[34]

Rolston develops Leopold's original idea: "We want to know what the unmolested system was in order to fit ourselves more intelligently in with its operations when we do alter it."[35] Further, as a measure of healthy land, areas of "wilderness" are said to be of value, according to Warwick Fox, as a sort of "early warning system" whose job it is "to warn us of more general kinds of deterioration in the quality or quantity of the free 'goods and services' that are provided *by* our 'life support system' "; wilderness can function as the proverbial canary in the coal mine.[36]

"But," one might ask, "why do we need *so much* wilderness; *so many* wilderness areas?" "Couldn't we have a base datum and measure of ecological health with but one wilderness area?" The answer is no. In order to serve as a measure of land health, it is argued, designated wilderness areas must be large and varied; for there are many distinct types of biotic communities. Leopold tells us—in a passage remarkably in line with contemporary conservation biology—that "each biotic province needs its own wilderness for comparative studies of used and unused land. . . . In short all available wild areas, large or small, are likely to have value as norms for land science."[37]

Interestingly, this argument avoids the potential paradox of overuse found in all use-oriented arguments because such control areas are *not* places to conduct invasive or manipulative research. Science only takes the pulse of these areas—so to speak.

Although this appears to be a strong argument for preserving designated wilderness areas it also appears, at least in part, to buy into the ontologically impossible notion that there are places totally untouched and unaffected by human actions. In order for "wilderness" to serve as a standard of land health, we must recognize that "wilderness" needs to be conceived of as a relative and tenuous concept.

14. THE STORAGE SILO ARGUMENT

Taking the Laboratory argument one step further, it is often asserted by conservation biologists, among others, that many supposed terrestrial and aquatic wilderness areas are worth saving because they contain vast amounts of biodiversity; especially genetic information or species diversity, or what Harvard biologist E. O. Wilson refers to as the "diversity of life." Beyond the argument that humans have no moral right to muck with the evolutionary and ecological workings of *all* ecosystems of a given type, maintaining these genetic reservoirs intact is instrumentally important because they function as a great safety device; holding a large portion of the world's accumulated evolutionary and ecological wisdom as they do. These proponents of wilderness claim that the whole of this information can only be properly maintained in its original context. Hence, some wilderness areas can serve as places where various forms of biodiversity can be stored for a time when they might be needed for genetic engineering, agricultural rejuvenation, or some other crucial purpose. Biotically rich and untrammeled wilderness areas are better than trammeled areas at providing for the continuation of the crucial processes of evolution and ecology since biotically rich and untrammeled areas have larger gene pools and are places where evolution can work unfettered to bring forth new species. Hence, biotically rich and untrammeled wilderness areas store the information that can help us to better manage and rebuild our natural environment. Certain wilderness areas, some believe, are therefore the key to life on earth. As David Brower puts it, "Wilderness holds the answers to the questions we do not yet know how to ask."[38]

Obviously, it would be foolish to knowingly extirpate natural processes.

According to Aldo Leopold, "To keep every cog and wheel is the first precaution of intelligent tinkering."[39] Hence, to destroy biotically rich wilderness areas is tantamount to the destruction of vast amounts of hitherto untapped and unused, but crucial, information; or, as Holmes Rolston analogizes, "destroying wildlands is like burning unread books."[40]

15. THE CLASSROOM ARGUMENT

Those of us who regularly or only occasionally visit designated wilderness areas are very aware that these locales often function as a sort of classroom where a plethora of valuable lessons can be learned.

Obviously, these experiences can increase our taxonomical environmental education: we can learn to identify Norway maples, magnolia warblers, spotted joe-pye, or timber wolf scat, for example. Granted, we can learn taxonomy elsewhere, but only in would-be wilderness can we encounter and learn of the habits and behaviors of some of these "others"; for only in large tracts of unbroken land do certain animals exist. According to many, "wilderness" experiences are also the *best* way to teach us such tangible skills as navigation and survival, and help us attain a feeling for a particular geographic region and features. However, it is also thought that additional important lessons can be learned through exposure to what many refer to as wilderness.

Nearly all advocates of wilderness preservation note the way that "wilderness experience" can help us put things in perspective or put our priorities in order, how it can teach us proper values and a sense of valuation, how it can instill within us a sense of humility and help build our individual and collective characters. Many also claim that periodic trips to designated wilderness areas force us to recognize our proper place as stewards, not masters, of the land; endow us with long-sighted and ecological vision; train us to make better public-policy decisions; furnish us with a sense of individual responsibility; promote our self-confidence and self-image; teach us how to cooperate successfully with others and how to assess and take wise and appropriate risks; and instill within us a reverence for all life and a proper sense of beauty. Groups such as Outward Bound and the Girl and Boy Scouts, for example, depend upon and utilize designated wilderness areas as classrooms to teach such lessons. In addition, if considered as places where natural processes continue unfettered, designated wilderness areas might also be said to provide a necessary and unique place to glean insights into the precious scientific studies of evolution and ecology.

And finally, for some environmental ethicists at least, the most important pedagogical aspect of "wilderness experience" lies in the fact that the lessons learned through and only through such exposure plays a necessary role in the development and support of a proper and sound environmental ethic. Hence, they argue, one could not develop an ethical relation toward the nonhuman natural environment without the presence of wilderness areas. Given the arguments against the very existence of wilderness found in this book, then, this would seem to imply that if there is no such thing as wilderness then there would be no such thing as an appropriate environmental ethic.

On a critical note: one must ask two questions. First, does this classroom argument for the preservation of designated wilderness areas hold true for all of these areas? That is, are all, or only some, designated wilderness areas sources of these important lessons? And second, while it may be true that the presence of designated wilderness areas is a sufficient condition for providing these valuable lessons, is it a necessary one? Can we not learn these lessons elsewhere? Or, is the argument that these are nonduplicable locales for environmental, life, and ethical wisdom? This argument appears to be weakened either if it does not hold true for all designated wilderness areas or if there are other ways or places to learn these lessons.

16. THE ONTOGENY ARGUMENT

Individually and collectively we human beings, like all living things, have evolved within a specific context. We are what we are, have become what we have become, because of the environment in which we have flourished. Some argue that the context of this historical development includes "wilderness" to a great extent—originally cosmic, then aquatic, and most recently terrestrial.

Homo sapiens and their communities, like all species and their respective communities, are deeply entrenched in nature and nature's processes. In other words, we fit our context. One might go on from this simple premise to argue that since what is thought of as wilderness is the paradigmatic form of nature and its processes, that this wilderness—first oceanic and then terrestrial—is and continues to be the source of our evolution. Hence, putative wilderness in its many forms not only ought to but must be preserved for continued human evolution.

We could, then, view wilderness preservation as a symbolic gesture of love and respect for our evolutionary ontogeny and perhaps as a defense of our evolutionary future. Edward Abbey adds such an argument to his list of wilderness preservation rationales. "The love of wilderness . . . ," he wrote, "is . . . an expression of loyalty to the earth, the earth which bore us and sustains us, the only home we shall ever know, the only paradise we ever need—if only we had the eyes to see." Abbey goes on to claim that the destruction of these areas of "wilderness" would be a sin, then, against our origins, or the "true original sin."[41]

Additionally, there is a nonsymbolic sense of this argument. Walt Whitman believed that those who remain closer and more in touch with their evolutionary context become better people. As he wrote in *Leaves of Grass,* "Now I see the secret of the making of the best persons. / It is to grow in the open air, and to eat and sleep with the earth."[42] Nash points out that a variety of primitivists—from Jean-Jacques Rousseau to Edgar Rice Burroughs—concur with Whitman and likewise declare that "the wild world produces a superior human being."[43]

Our ontogenetic setting, for Whitman and Nash at least, deserves special consideration since our evolutionary past, present, and future is intimately tied to what we now call wilderness. Therefore, according to this argument, these places should be preserved in order to ensure our context, our physio/psycho-genesis and natural individual development. And excursions to these areas, then, appear to function as visitations to our ancestral residence or what Abbey once referred to as the "journey home."

There is a definite irony to this argument. If "wilderness" is a place where humans are but visitors who do not remain or an area beyond that which is cultivated or inhabited continuously by humans, how in the world could it be our literal ancestral home? In fact, "wilderness," designated or not, is often portrayed as an area undefiled by human habitation, a place where people do not live. So how could it be our home? And if we have evicted ourselves from our ancestral paradise home irrevocably, how can wilderness be the locus or future of our human evolution?

17. THE CULTURAL DIVERSITY ARGUMENT

In one of his most familiar works, "Walking," Thoreau writes,

> Our ancestors were savages. The story of Romulus and Remus being suckled by a wolf is not a meaningless fable. The founders of every state which has risen to eminence have drawn their nourishment and vigor from similar wild sources. . . . In such soil grew Homer and Confucius and the rest, and out of such a wilderness comes the Reformer eating locusts and wild honey.[44]

So it is argued that, just as human beings generically derive from a context, specific cultures are derived from and are dependent upon a certain ontogenetic context as well. And the wide variety of cultural variation or diversity stems from the fact that there has been a wide variety of natural ecosystems. As Leopold put it, "The rich diversity of the world's cultures reflects a corresponding diversity in the wilds that gave them birth."[45] The vigor for culture, some deduce, comes from wilderness. Or, as Nash writes, "The wild world is cultural raw material."[46]

A wide variety of designated wilderness areas ought to be preserved, it is thought, because they function as the foundation for the world's myriad cultures. As much as we value individual cultures and the diversity of cultures, we must to the same extent value their respective areas of purported wilderness. "We want some wilderness preserved," Rolston claims, "because it comes to express the values of the culture superimposed on it, entering our sense of belongingness and identity."[47]

There seems to be a paradox implicit in this argument. If the diverse cultures remain in the wilderness that gave them birth, then the places they are in are not, by definition, wilderness. If various areas of the world are *designated* wilderness areas and the cultures they spawned are expelled from them (if these cultures still exist), as the Ik were from the Kidepo in Uganda and the !Kung were from the Etosha National Park in Namibia, then the cultures are exterminated in order to preserve the wildernesses that spawned them.[48]

Furthermore, to argue that areas of wilderness ought to be preserved as museum pieces in honor of a part of the environment that helped shape each unique culture commits the fallacy of appeal to tradition. Slavery helped to form much of the American Deep South; the oppressive Hindu caste system largely shaped modern India; various acts of violence, wars, and systems of patriarchy have helped to mold various world cultures. Should remnant enclaves of these institutions likewise be preserved because they are the roots of various cultures? Should we designate one or two counties in Mississippi as Old South cultural reserves in which the plantation system, including slavery, is preserved? And so on.

18. THE NATIONAL CHARACTER ARGUMENT

A specific example of the above two arguments is national character, and even more specific is colonial American national character.

In the United States, many see designated wilderness areas as monuments; symbolically enshrining national values. Our Euro-American cultural identity, for example, is often said to be deeply entwined with the existence of designated wilderness areas. For many Euro-Americans, the United States is the place where the eagle flies, the buffalo roam, and the deer and the antelope play. Wallace Stegner calls the wilderness idea "something that has helped form our character and that has certainly shaped our history as a people." Designated wilderness areas ought to be preserved then, because they are a "part of the geography of hope."[49] Many seminal American historical thinkers—such as Gertrude Stein and Frederick Jackson Turner—felt that designated wilderness areas in the United States serve as what Roderick Nash refers to as the "crucible of American character."[50]

Hence it is argued that designated wilderness areas ought to be preserved because they and their resident species helped form and continue to enshrine our most fundamental and powerful Euro-American values. Nash notes that American president Theodore Roosevelt was particularly fond of the argument which asserts that since "wilderness shaped our national values and institutions, it follows that one of the most important roles of nature reserves is keeping those values and institutions alive."[51] Environmental philosopher Mark Sagoff even goes so far as to submit that wilderness preservation has constitutional clout:

> If restraints on the exploitation of our environment are to be adequate, then, they must be found in the Constitution itself.... To say that an environmental policy can be based on the Constitution does not require, of

> course, a constitutional passage or article which directly concerns the environment; rather the argument would rest on the concept of nationhood, the structure created by the Constitution as a single instrument functioning in all of its parts. It is reasonable to think that cultural traditions and values constitute a condition—at least a causal one—of our political and legal freedom; and therefore insofar as the Constitution safeguards our nation as a political entity, it must safeguard our cultural integrity as well. Citizenship, then, can be seen to involve not only legal and political but also cultural rights and responsibilities.... The right to ... demand that the mountains be left as a symbol of the sublime, a quality which is extremely important in our cultural history, ... the right to cherish traditional national symbols, the right to preserve in the environment the qualities we associate with our character as a people, belongs to us as Americans. The concept of nationhood implies this right; and for this reason, it is constitutionally based.[52]

Moreover, for all the reasons that we preserve in perpetuity the dwelling places of Americans of European descent (e.g., Puritans or Mormons) as national landmarks denoting their cultural heritage, we ought also to preserve many areas of putative wilderness as well. These designated wilderness areas, we might argue, would function as a similar sort of cultural monument, since they are seen by some to be the historical home of aboriginal American peoples like the Sioux, Hopi, Iroquois, or Inuit.[53] As such, of course, they were not wilderness areas, since these people were people (not wildlife) and did live there (not just visit).

Now there is admittedly yet another inherent paradox involved with the above three ontogeny-type arguments, and especially with the National Character argument. If wilderness preservation is a good thing because it is a representation of Euro-American ontogeny then wilderness destruction seems to be a good thing as well. That part of our Euro-American ontogeny, most recently evolved is the tendency to work to destroy, or at least to severely alter and interrupt, what colonizing Euro-Americans took to be wilderness. Like belief in the existence of a vast North American wilderness, the transformation of that wilderness could be said to be an important aspect of our national character. So, we might argue that Euro-Americans and Euro-American wilderness colonization and destruction is part of Euro-American national character. Hence, destroying designated wilderness areas would be more consistent with the Euro-American national character than preserving it. Wilderness preservation might be seen as both good and bad at the same time then. It is good because the North American "wilderness" is the raw material of Euro-American culture and is valuable as such. It is bad because the destruction of those same places is part of that same national ontogeny and we would be interfering with that ontogeny if we were to artificially restrain or frustrate it. The root problem, it seems, is to make the argument that X is good merely because X is part of our heritage. Slavery, violence, and sexism are parts of our heritage, but that is no reason to keep them around as treasured national institutions. Logicians call this the Genetic Fallacy.

19. THE SELF-REALIZATION ARGUMENT

One of the fundamental tenets of Deep Ecology is the notion of self-realization. Relying heavily upon the works of Muir, Thoreau, Leopold, and the Romantic and American Transcendental traditions, Deep Ecologists assert that—in order truly and appropriately to perceive and understand the world, our place in it, and our duties to it—we must first dismiss the assumed but inaccurate bifurcation between self and nature. We must grasp the depth of the relational reality of all things, including the non-human world. In addition to the general character and self-image building mentioned above, wilderness preservation becomes crucial for Deep Ecologists because designated wilderness areas are, for them, necessary components in this process of self-realization—a sort of asylum of reorientation where this relational self ideal can take form. We must, therefore, maintain areas of "wilderness" in order to achieve a complete and appropriate view of self. Designated wilderness areas are crucial, according to this argument, for individual development and continued existence. No designated wilderness areas means no self-realization; no Deep Ecology; no proper view of self and world; no appropriate treatment of self and world; no continued self-existence.[54]

On a more critical note, this argument appears to fall prey to the recurring problem of confusing necessary and sufficient conditions. Merely because designated wilderness areas provide an arguably sufficient condition for self-realization, it remains to be proven that they are a necessary condition for such self-realization.

20. THE DISEASE SEQUESTRATION ARGUMENT

This is a very new and perhaps very worthy wilderness preservation argument.

As noted, more than half of the earth's extant species live in the tropics; and most of these species are not known to science. Since most all species host viruses, we can assume that there are at least as many viruses as there are living species, if not some multiple of that number.

One thing we do know is that viruses adapt to their hosts. A successful virus either does not, or "learns" not to, kill its host so as to have a residence for its continued existence. However, as human population continues to surge and human invasion into places such as tropical habitats becomes more prevalent, humans cross never-before-traversed ecological and spatial boundaries. Hence, humans increasingly encounter new species; in Star Trek phraseology, we "boldly go where no one has gone before." As humans intrude, put pressure upon, or destroy supposed wilderness, viruses will adapt and jump hosts, taking humans as their new hosts. The effects of this "host jumping" are unknown and potentially dreadful, especially when dealing with lethal viruses for which there is no vaccine nor cure. Known viruses, such as Guanarito and Marburg, and currently "emerging" viruses like Q fever and Monkeypox, are part of a long list of viruses which have already jumped to human hosts with deleterious results.[55] In his insightful—and truly frightening—essay in *The New Yorker,* Richard Preston explains:

> When an ecosystem suffers degradation, many species die out and a few survivor-species have population explosions. Viruses in a damaged ecosystem can come under extreme selective pressure. Viruses are adaptable: they react to change and can mutate fast, and they can jump among species of hosts. As people enter the forest and clear it, viruses come out, carried in their survivor-hosts—rodents, insects, soft ticks—and the viruses meet Homo sapiens.[56]

In fact, human immunodeficiency virus (HIV) is another, more familiar, contemporary example. The reigning theory asserts that HIV is a mutant zoonotic virus that originally resided in the rain forests of Central Africa which jumped to humans when they had some sort of intimate contact (i.e., sexual contact, hunters touching the bloody tissue of the victim, etc.)

with the sooty mangabey, an African monkey. We do not know the source for certain; HIV may have come from chimpanzees or even from other previously isolated humans, or it may just be a viral mutation of some sort. Nevertheless, as Preston points out,

> The emergence of AIDS appears to be a natural consequence of the ruin of the tropical biosphere. Unknown viruses are coming out of the equatorial wildernesses of the earth and discovering the human race. It seems to be happening as a result of the destruction of tropical habitats. You might call AIDS the revenge of the rain forest. AIDS is arguably the worst environmental disaster of the twentieth century, so far.[57]

And HIV is not even as dangerous as some. Yes, it is lethal, but it is relatively noninfectious considering how it might be transmitted. Just imagine HIV as an airborne pathogen, which is not impossible. Ebola Zaire, a lethal airborne filovirus that emerged in fifty-five African villages in 1976 and subjected nine out of ten of its victims to hideous deaths within days of infection, is an example of an encountered virus even more lethal than HIV. And viruses are only part of this gloom-and-doom story. As Preston tells us, "mutant bacteria, such as the strains that cause multidrug-resistant tuberculosis, and protozoans, such as mutant strains of malaria, have become major and growing threats to the [world's] population" as well.[58]

Since these disease-causing agents are for the most part sequestered in many of the earth's remaining wild areas, our intrusion into these areas is a definite issue—one that deserves our utmost attention. So, not only do many of the wild areas of the earth serve as a disaster hedge by acting as a buffer, but much ostensible wilderness left intact is thought to protect us from potential viral and bacterial decimation. For this reason, any sane person would agree, some regions of the Earth—those harboring tropical viruses—deserve special consideration and preservation.

However, this argument appears to be less an argument for preserving designated wilderness areas and more an argument for not intruding into those areas where these viruses thrive; less an argument for preserving the Bob Marshall Wilderness Area and more an argument for staying out of some very specific (i.e., tropical) places. Moreover, with this argument also there are elements of paradox and irony. These lairs of deadly viruses are not exactly those places that we wish to *visit* for recreation, inspiration, religious awe, and so forth.

21. THE SALVATION OF FREEDOM ARGUMENT

In his novel celebrating the arenaceous wilds of southern Utah's Arches National Monument, *Desert Solitaire,* Edward Abbey defends the preservation of what he imagines to be wilderness for what he calls "political reasons." He claims that we need designated and de facto wilderness areas, whether or not we ever set foot in them, to serve as potential sanctuaries from oppressive governmental structures. As Abbey writes,

> We may need it [wilderness] someday not only as a refuge from excessive industrialism but also as a refuge from authoritarian government, from political oppression. Grand Canyon, Big Bend, Yellowstone and the High Sierras may be required to function as bases for guerrilla warfare against tyranny.

Abbey reminds us of some of the supposed findings of modern political science—that our cities might easily be transformed into concentration camps and that one of the key strategies in the imposition of any dictatorial regime in any country is to "Raze the wilderness. Dam the rivers, flood the canyons, drain the swamps, log the forests, strip-mine the hills, bulldoze the mountains, irrigate the deserts and improve the national parks into national parking lots."

In order to attempt to alleviate the obvious charge that this is only a paranoid survivalist-type fantasy, Abbey cites as historical fact that the worst of the world's tyrannies have occurred in those countries with the most industry and the least of what Abbey thinks of as wilderness. Centralized oppressive domination flourished in Germany, Hungary, and the Dominican Republic, according to Abbey, because "an urbanized environment gives the advantage to the power with the technological equipment." On the other hand, more rural insurrections, such as those in Cuba, Algeria, the American colonies, and Vietnam, have favored the revolutionaries since there remained in those countries "mountains, desert, and jungle hinterlands;" or areas of would-be wilderness.[59] Further support for Abbey's wilderness preservation argument might be drawn from the Old Testament story of the Exodus and the New Testament story of John the Baptist. The political dangers of a "wilderness-free" world are fictionally portrayed in Aldous Huxley's *Brave New World* and George Orwell's *1984.*

It should be noted, however, that Abbey's evidence is quite shaky since the counter-examples of the Soviet Union, with its vast expanses of "wilderness," and Cuba, since Castro, throw a monkey wrench into his political theory. Moreover, philosopher Michael Zimmerman has recently chronicled how the Nazi Germans advocated nature preservation more ardently than any other Europeans.[60]

To sum up, according to this argument, areas of "wilderness" in general provide us with a place to escape to and combat a totalitarian police state; as well as providing us with the very standard or meaning of freedom. Hence, in the minds of those such as Ed Abbey, in order to preserve freedom we must preserve alleged wilderness areas.

Now, obviously, not all designated wilderness areas would provide us with base camps for guerrilla warfare against tyrannical governments and one has to allow for the possibility at least that there might be some non-wilderness areas which would. This argument seems, then, to be more an argument for areas from which to oppose tyranny rather than necessarily an argument for the preservation of designated wilderness areas.

Moreover, the fact that something would provide us with the means to oppose tyrannical government is *not* sufficient justification for its preservation. Private atomic-bomb factories in each of our basements would put us in a good position to oppose tyrannical governments too, but clearly (or hopefully) Abbey would not advocate *that.*

22. THE MYTHOPOETIC ARGUMENT

Some contemporary thinkers, such as Deep Ecologists and postmodernists, argue that wilderness preservation is critical for mythopoetic reasons. Those places they view as wilderness serve as the optimum location for the viewing of the history of myth and are absolutely crucial for the building of the myth of the future. In his mythopoetic book *The Idea of Wilderness,* Max Oelschlaeger writes, in reference to a postmodern conception of nature, "the idea of wilderness in postmodern context is . . . a search for meaning—for a new creation story or mythology—that is leading humankind out of a homocentric prison into the cosmic wilderness."[61] Putative wilderness areas function as an essential source of meaning, vital for the future of humanity.

The current "Men's Movement" is an excellent example of a mythopoetic use of putative wilderness. In their search for a new way of understanding or realizing their manhood, those involved in the movement have a strong tendency to gather in "wild places" because they seek to recover the roots of their maleness as hunters and companions of animals originally thought to be found in "wilderness." Adding to this, Robert Bly (the central figure in the "mythopoetic" branch of the movement) promotes the ideal of the "wildman," which has central significance as one of the movement's goals. Men should strive, according to Bly, to be whole, healthy, and energized by realizing the "wildman" deep inside their psyches. He is not here referring to any irrational lunatic sense of wild but rather arguing that men need to get in touch with their more primitive or natural roots. Male mythologizing, accordingly, requires areas of would-be wilderness, and for those in the Men's Movement and others, without this "wilderness" a sense of the history of, standard for, or place for the future of mythbuilding is lost.

This argument appears to rest on some unproved premises: that "wilderness" is the sole or best source of future mythologizing, that our future is impoverished without this source of myth, and that areas other than designated wilderness areas cannot serve as adequate or even more appropriate mythological sources.

23. THE NECESSITY ARGUMENT

As we have seen above, certain wilderness advocates believe in the truism that, historically speaking, no civilized world would have evolved without the prior existence of wilderness. As Leopold claims, "wilderness is the raw material out of which man has hammered the artifact called civilization."[62] But one could, on a more conceptual level, argue that "wilderness" is *necessary* in a more philosophical sense. One might claim that an idea or concept of wilderness is logically and metaphysically necessary (a sense of necessity not historically dependent) for the existence and complete understanding of the concepts of culture and civilization.

Although some may contend that this is more of an explanation of than a justification for wilderness, and that it does not require the preservation of anything but, at the most, a very small bit of wilderness (and at the least only the concept of wilderness and hence no areas of wilderness at all), others might argue that to truly understand concepts such as culture, civilization, freedom, primitive, development, and perhaps others, we need physical wildness (such as that found in designated wilderness areas) to serve as a model or foil of contrast. Holmes Rolston explains:

> Humans can think about ultimates: they can espouse worldviews: indeed, they are not fully human until they do. No one can form a comprehensive worldview without a concept of nature, and no one can form a view of nature without evaluating it in the wild. . . . In that sense, one of the highest cultural values, an examined worldview, is impossible to achieve without wild nature to be evaluated as a foil to and indeed source of culture.[63]

There can be no finish without a start; no good without evil; no yin without yang; and (according to folksinger Arlo Guthrie) "no light without a dark to stick it in." Similarly then, wilderness might be said to be logically (wherever "civilization" has meaning, "wilderness" does also) and metaphysically (where one thing exists, its opposite must also) necessary for a complete and proper understanding of civilization. According to this argument, there can be no proper understanding of civilization when there is no concept of wilderness, and there can be no complete and proper understanding of the concept of wilderness without genuine designated wilderness areas. Thus the move to rid the world of designated wilderness areas is tantamount to an attempt to deny and dismantle a necessary component for a complete understanding of the world.[64]

24. THE DEFENSE OF DEMOCRACY ARGUMENT

Enemies of wilderness preservation and environmentalism in general are fond of charging wilderness advocates and environmentalists with committing the sin of elitism. They claim that wilderness preservation ought *not* be pursued since only a minority of people ever visit designated wilderness areas. Environmentalists, they allege, are selfish people who want to set aside vast stretches of land that only the physically fit and economically able can experience. Wilderness preservation, then, only benefits the elite few and therefore does not serve the general welfare—the greatest good for the greatest number.

However, even though this populist argument is often quite effective, it is also quite easy to "set it on its ear" and turn the charge of elitism into a pro-wilderness preservation argument. Without denying the charge of elitism made by the opponents of wilderness preservation, it could be claimed that precisely because wilderness preservation shows respect for the needs of a minority, wilderness preservation is, therefore, indicative of good democracy. The existence of things like opera houses, softball diamonds, art galleries, public swimming pools and designated wilderness areas—all places used by only a minority—is a display of respect for minority rights or, as Nash says, "the fact that these things can exist is a tribute to nations that cherish and defend minority interests as part of their political ideology."[65] Concerning such an argument, Leopold wrote:

> There are those who decry wilderness sports as "undemocratic" because the recreational carrying capacity of a wilderness is small, as compared with a golf links or a tourist camp. The basic error in such argument is that it applies the philosophy of mass-production to what is intended to counteract mass-production. . . . Mechanized recreation already has seized nine-tenths of the woods and mountains; a decent respect of minorities should dedicate the other tenth to wilderness.[66]

Robert Marshall was also quite fond of pointing to this as a benefit of designated wilderness areas. "As long as we prize individuality and competence," Marshall says, "it is imperative to provide the opportunity for complete self-sufficiency."[67]

In fact, the original charge of elitism may even be more severely wrongheaded. We might argue that designating areas as wilderness does not limit but rather opens up access to more people than would privatization or land development, especially if we consider future generations of humans, since these wilderness areas are public access lands.

Now, there seems to be an irony or paradox involved with this argument as well. If privatization keeps people out and public ownership allows them in, then designating an area as publicly owned wilderness will guarantee that it will be overrun by hordes of backpackers, camera hunters, and the like.

25. THE SOCIAL BONDING ARGUMENT

Expanding on all of those arguments espousing individual benefits, one could argue that many designated wilderness areas serve as valuable mechanisms in the critical process of social bonding. Those who collectively recreate in designated wilderness areas often attest to the benefits that come from social interaction in such a setting. Driver, Nash, and Haas illustrate these benefits when they claim that one of the most pervasive reasons that people choose to spend time in designated wilderness areas

> relates to family cohesiveness and solidarity. Others include strengthening social bonds with small groups of significant others, sharing skills with others, and gaining social recognition or status from demonstration of skills to others and later sharing tales or photographs of enviable experiences.[68]

Because, and to the extent that, we are social animals and our continued survival depends on effective social interaction, we ought to value social bonding. Exposure to "wilderness" is thought to intensify experience and provide a vector for high-level and successful interpersonal cohesion. Therefore, as an important social mechanism, designated wilderness areas are assumed to be valuable and worthy of preservation.

As an expansion of individual-use values of designated wilderness areas, this argument is subject to the same criticisms leveled at many of the other arguments catalogued here. Surely designated wilderness areas are sufficient and effective in facilitating social bonding, but just as surely they are not necessary. Social bonding can and does take place in highly artificial settings like classrooms, concert halls, offices, saloons, and salons. Moreover, we can just as easily imagine the cohesiveness of a group being completely destroyed by the pressures of wilderness travel. And furthermore, the use of designated wilderness areas by groups of people greatly jeopardizes the "wilderness" integrity of a place.

26. THE ANIMAL WELFARE ARGUMENT

By exploring a non-anthropocentric justification for wilderness preservation we might claim that, like us, wild animals too depend upon their respective home environments for their existence. And since they should be considered as members of the community of beings deserving moral consideration, we owe it to those animals not to destroy their homes. Along similar lines, it was popular in the 1970s to argue that the "wild" things on earth had the right to go about their business unmolested and unharmed by Homo sapiens. We humans, therefore, had no right to interfere with their freedom and dignity. And, since designated wilderness areas were seen as the places where this freedom was most fully realized, these areas should be preserved as "reservoirs of ecological freedom."

As intuitively attractive as this sounds, it must be recognized that most of the animals that animal welfare ethicists are concerned with do not depend on designated wilderness areas for their habitat needs: chickens and cows live on farms, squirrels and pigeons live in cities, deer live in rural areas around farms, cats and dogs live in human homes. Indeed, most animals, both wild and domestic, do not require "wilderness" at all. However, some few animals do require large tracts of unbroken land as habitat: wolves and bears are good examples. For the proper respect of and caring for these animals, then, it might be argued that those areas which are their homes, which would include only certain designated wilderness areas, ought to be preserved.

Bringing this argument back to the human realm by reinserting the anthropocentric side of the coin for a moment, one might argue, as does Paul Shepard, that wild animals are a human necessity. Shepard claims that "the human mind needs [animals to exist in their wild habitats] in order to develop and work. Human intelligence is bound to the presence of [these wild] animals."[69] This would apparently include "wilderness" dwelling animals. Therefore, wild animals (including wilderness dwellers) and their habitats (including wilderness areas) deserve moral consideration, albeit indirect, in this sense as well.

The more mature Leopold also believed that designated wilderness areas were valuable for the sake of wildlife such as wolves, mountain sheep, and grizzly bears. Moreover, given the validity of these arguments, such areas would have to be large and unbroken so as to be able to accommodate these "far-ranging species." Arguing for permanent grizzly ranges, for example, Leopold once wrote that,

> saving the grizzly requires a series of large areas from which roads and livestock are excluded. . . . Relegating grizzlies to Alaska is about like relegating happiness to heaven; one may never get there. . . . Only those able to see the pageant of evolution can be expected to value its theater, the wilderness, or its outstanding achievement, the grizzly.[70]

In fact, this argument has now become the principal conservation biology argument for wilderness preservation—big reserves for biodiversity.[71]

27. THE GAIA HYPOTHESIS ARGUMENT

Expanding on the above argument, we might apply non-anthropocentric moral consideration to yet another sort of living organism: namely, the earth or Gaia. Scientists such as James Lovelock posit and defend what they call the Gaia Hypothesis, which postulates that the earth itself, as a self-correcting system, is alive or is tantamount to a living organism. Like any living thing, certain of its parts are imperative to its proper functioning and viability. With regard to the earth, or Gaia, certain wild ecosystems are arguably vital to its prospering. Wild ecosystems might be likened to the internal organs of a multicelled organism—or what we traditionally think of as an organism. With Gaia, designated wilderness areas could be said to perform certain services invaluable to the smooth functioning of the earth organism, just as a human liver provides a service invaluable to the human organism. Without a liver a human cannot live; without wilderness earth perhaps cannot either. So, if planetary homeostasis is to continue, and Gaia is to live, areas of "wilderness" must be preserved.

If we owe moral consideration to living beings, and if the earth itself is alive, as the Gaia Hypothesis maintains, then the earth deserves moral consideration. Disrupting Gaia's vital organs, such as putative wilderness, becomes, then, immoral. So, in order to show proper moral respect to the earth organism, these areas ought to be maintained.

Leopold was inspired by the Russian philosopher and mystic P. D. Ouspensky. In his essay "Some Fundamentals of Conservation in the Southwest," Leopold toys with a Gaia-type defense of earth and wilderness preservation. "Philosophy," he writes, ". . . suggests one reason why we can not destroy the earth with moral impunity; namely, that the 'dead' earth is an organism possessing a certain kind and degree of life, which we intuitively respect as such."[72]

The objections to the above two arguments are too complex to cover here. In sum, these arguments rest on the premise(s) that animals and Gaia deserve moral consideration, which corresponds to actions of preservation on behalf of humans. Clearly, one would have to justify this argument prior to making the larger argument stick. Moreover, with regard to the Gaia argument, one would have to prove that designated wilderness areas really are crucial to the life of Gaia.

28. THE FUTURE GENERATIONS ARGUMENT

Another common defense of wilderness preservation revolves around the supposed moral obligations that existing human beings have with regard to future generations of human beings. One might maintain that, among other debts owed, current humans ought to pass the world on to future generations as we inherited it, with as many designated wilderness areas intact as possible. As the old American Indian saying goes, "We do not inherit the earth from our ancestors; we only borrow it from our children."

Destroying putative wilderness areas, when taking into account these future generations, would then become a matter of injustice and unfairness. When we destroy these areas we can also be said to be depriving future generations of valuable resources and services: we are taking away their heritage and identity; we are not providing them with a place for enjoyment, education, aesthetic experience, and self-discovery; we are subjecting them to the consequences of our irreversible land-use decisions; all those things we commonly think of "wilderness" as providing for us. If we accept that future generations merit at least some degree of moral consideration, many of the "wilderness benefits" mentioned above would also apply to future generations of humans. In short, "wilderness" destruction is wrong, according to this argument, because future generations of humans would mourn its loss.

The truth is that we really do not know for certain what future generations will want or need. And, as hard as it is for some to believe, perhaps they will not want or need designated wilderness areas. However, one might argue that this lack of knowledge is reason enough to save designated wilderness areas, not ravage them. We don't *know* what future generations will want or need. Therefore, we should keep their options open. Wilderness preserved is an option for future generations. They can keep and use their wilderness areas or develop them. But if we develop them now, they won't have the option. Further, our protection of "wilderness" arguably sets a good example for future generations, and displays a good ethic of stewardship, which would encourage them, in turn, to keep land-use options open to future future generations. Interestingly, all future generations might arguably be forced to preserve designated wilderness areas because they would be locked into the same "logic of wilderness preservation for future generations" argument that compelled us to save wilderness for them. This logical compulsion might then propel wilderness preservation indefinitely.

Nevertheless, this argument appears to be contingent upon the validity of the other arguments that attempt to show that a world with designated wilderness areas is better than a world without them. Hence, we might note that this is not an entirely independent argument.

29. THE UNKNOWN AND INDIRECT BENEFITS ARGUMENT

For some, one of the greatest reasons why we ought to err on the side of caution, and preserve and designate more wilderness areas, is that theoretically speaking most of the benefits emanating from these areas are thought to be indirect (what Driver, Nash, and Haas refer to as "spinoff" value and what economists commonly refer to in part as "option value") or unknown. The potential for goods and services may be unlimited, and the possible harms unknown. Many of the indirect social benefits can only be guessed at. We simply do not have all the information in as of yet. E. O. Wilson maintains that "the wildlands are like a magic well: the more that is drawn from them in knowledge and benefits, the more there will be to draw."[73] But if we destroy designated wilderness areas, we would then apparently destroy tremendous amounts of information and potential benefits along with them. Hence, it would seem to be prudent to save designated wilderness areas because of these unknown and indirect benefits, and as many and as large of these areas as possible at that. The downside to this argument is that it makes the preservation of wilderness contingent only on its potential utility.

For certain wilderness proponents the promise of wilderness preservation lies ahead, in the future. Aldo Leopold echoes this sentiment by pronouncing that we "should be aware of the fact that the richest values of wilderness lies not in the days of Daniel Boone, nor even in the present, but rather, in the future."[74] Hence, because we have a responsibility to follow a wise course of action, and since this would include conserving potential benefits to all of humanity, we apparently then have a responsibility to keep designated wilderness areas in existence.

30. THE INTRINSIC VALUE ARGUMENT

Many, many wilderness boosters claim that simply knowing that there exist designated wilderness areas, regardless of whether or not they ever get to experience such areas, is reason enough for them to want to preserve them. For these people, "wilderness" is valuable just because it exists, just because it is. Designated wilderness areas, in this sense, have value in and of themselves; regardless of, or in addition to, their value as a means to some other end—like clean water, recreation, or medicine. "Wilderness," then, is said to posses intrinsic value.

E. O. Wilson asserts that designated wilderness areas have uses not to be ignored: but he quickly points out that the argument for the preservation of these areas does not end there. As he writes, "I do not mean to suggest that every ecosystem now be viewed as a factory of useful products." Utility is not the only measure of "wilderness" value, he declares, "Wilderness has virtue unto itself and needs no extraneous justification."[75] Reiterating Wilson's conviction, and issuing a call to action, Edward Abbey once wrote, "The idea of wilderness needs no defense. It only needs more defenders."[76] Clearly what Wilson, Abbey, and many other wilderness-minded folks are claiming is that, ultimately, "wilderness" defense needs no articulation. Designated wilderness areas just *are* valuable. Such locales, then, join the list of other things, like friends, relatives, children, family heirlooms, and so forth, whose worth is not contingent upon anything other than their mere existence, whose value is intrinsic.

For many environmentalists, to categorize and quantify the benefits of what they take to be wilderness is a fundamentally flawed approach to wilderness preservation. It is, in effect, to play the game of your opponent by trying to bolster your side by including on your team bigger, better, and more players (or, in this case, more human benefit arguments), while the question of whether or not the correct game is being played in the first place goes unanswered and unasked. Many people would challenge the dominant cult of perpetual growth by instead making the more radical claim that areas of "wilderness" are not for humans in the first place, that wild species and ecosystems themselves have a right to exist apart from their uses to humans. "Diffuse but deeply felt," Rolston declares, "such values are difficult to bring into decisions; nevertheless, it does not follow that they ought to be ignored."[77]

If we accept this reasoning, we might get the feeling that designated wilderness areas are important and valuable just because they are there; regardless of whether or not we ever decide to visit, experience, scientifically monitor, or even contemplate them. But this seems to be the real challenge of wilderness preservation. According to William Godfrey-Smith, "the philosophical task is to try to provide adequate justification, or at least clear the way for a scheme of values according to which concern and sympathy for our environment is immediate and natural, and the desirability of protecting and preserving wilderness self-evident."[78]

The initial assumption in this argument is that if purported wilderness areas do indeed possess intrinsic value, their defense and preservation become self-evident; they just *are* of value and, therefore, worthy of preservation. However, there is obviously a lot more to the debate surrounding intrinsic value and the intrinsic value of "wilderness" than I have presented here. In fact, as Godfrey-Smith says, providing for or grounding the intrinsic value of things like "wilderness" is where the real work needs to be done. This might be seen as perhaps one of the central roles of environmental philosophy in this debate; and we might urge philosophers to begin dealing with and answering questions about how we ground the claim that putative wilderness has intrinsic value, how to sort out both instrumental and intrinsic competing value claims, and how, when, and why the intrinsic value of something like wilderness—if such a thing actually exists—trumps these other value claims.

If we can justify the intrinsic value of designated wilderness areas, and if we can locate their level of moral consideration, then wilderness preservation immediately becomes a moral issue. Designated wilderness areas would gain considerable ethical clout. And this changes the argument about wilderness preservation quite a bit; the burden of proof would apparently be shifted. Those who would destroy designated wilderness areas would then have the burden of proof; they would have to demonstrate that something of great social value would be lost if a wilderness area stood in its way; they would have the difficult task of showing that something possessing intrinsic value should be sacrificed for the sake of something of merely instrumental value. "Wilderness is innocent 'til proven guilty," David Brower once quipped, "and they're going to have a tough time proving it guilty."[79]

NOTES

I owe debts of gratitude to Don Fadner, Alan Holland, Kate Rawles, John Vollrath, Dôna Warren, Mark Woods, and especially Baird Callicott for assistance with this essay.

1. Roderick Nash, "The Value of Wilderness," *Environmental Review* 3 (1977): 14–25, p. 25.

2. Nash's wilderness preservation arguments originally appear in *Wilderness and the American Mind,* all editions (New Haven: Yale University Press, 1982), and are later summarized and indexed in "The Value of Wilderness"; see also Nash, "Why Wilderness?," in Vance Morton, ed., *For the Conservation of the Earth* (Golden, Col.: Fulcrum, 1988), pp. 194–201; and Nash with B. L. Driver and Glenn Haas, "Wilderness Benefits: A State-of-Knowledge Review," in Robert Lucus, ed., *Proceedings—National Wilderness Research Conference,* Intermountain Research Station General Technical Report, INT-220 (Ogden, Utah: U.S. Forest Service, 1987), pp. 294–319. Holmes Rolston III presents his wilderness preservation rationales in "Valuing Wildlands," *Environmental Ethics* 7 (1985): 23–48; and in "Values in Nature" and "Values Gone Wild," both in Rolston, *Philosophy Gone Wild: Essays in Environmental Philosophy* (Buffalo, N.Y.: Prometheus, 1986), pp. 74–90 and 118–42, respectively. William Godfrey-Smith's arguments appear in "The Value of Wilderness," *Environmental Ethics* 1 (1979): 309–19. Warwick Fox expands on Godfrey-Smith's work in *Toward a Transpersonal Ecology: Developing New Foundations for Environmentalism* (Boston: Shambhala, 1990), pp. 154–61. George Sessions adds to this lineage in "Ecosystem, Wilderness, and Global Ecosystem Protection," in Max Oelschlaeger, ed., *The Wilderness Condition: Essays on Environment and Civilization* (Washington, D.C.: Island Press, 1992), pp. 90–130. And Michael McCloskey's taxonomy of wilderness values and benefits is in "Evolving Perspectives on Wilderness Values: Putting Wilderness Values in Order," in Patrick C. Reed, ed., *Preparing to Manage Wilderness in the 21st Century: Proceedings of the Conference,* "Southeastern Forest Experiment Station General Technical Report," SE-66 (Athens, Ga.: U.S. Forest Service, 1990), pp. 13–18.

3. Rolston, "Valuing Wildlands," p. 27.

4. Driver, Nash, and Haas, "Wilderness Benefits," p. 304.

5. I say "in theory" and "ideally" because, as Mark Woods's essay in this volume indicates, this is not true in practice. Grazing and other extractive uses *are* permitted in some designated wilderness. It would appear that the resource view is that

there are a suite of commodities that are "wilderness commodities"—timber, fodder, game. The difference between these and other commodities is that they are there for the taking and do not have to be planted and tended.

6. To avoid confusion, two senses of "potential" need to be distinguished. First is the notion of postponing the use of those resources whose uses we are aware of (e.g., turning trees into timber). Second is the sense of the yet unknown uses of something (e.g., undiscovered medicinal uses of known and unknown plants).

7. Aldo Leopold, "A Plea for Wilderness Hunting Grounds," in David E. Brown and Neil B. Carmony, *Aldo Leopold's Wilderness: Selected Early Writings by the Author of A Sand County Almanac* (Harrisburg, Penn.: Stackpole Books, 1990), p. 160; originally published in *Outdoor Life,* November 1925.

8. Leopold, "Wilderness as a Form of Land Use," in Susan Flader and J. Baird Callicott, eds., *The River of the Mother of God and Other Essays by Aldo Leopold* (Madison, Wis.: University of Wisconsin Press, 1991), p. 138; this essay is included in this volume and was originally published in *The Journal of Land and Public Utility Economics,* October 1925.

9. Theodore Roosevelt, "The Pioneer Spirit and American Problems," in *The Works of Theodore Roosevelt* (New York: Charles Scribner's Sons, 1923), 18:23.

10. Roosevelt, "The American Wilderness: Wilderness Hunters and Wilderness Game," in *Works,* 2:19, included in this volume.

11. Roosevelt, "Wilderness Reserves: The Yellowstone Park," in *Works,* 3:267–68.

12. Some medical researchers might discontinue using this argument because of their increased confidence in the ability of computers to generate synthetic models of medicines and the ability to be able to artificially produce such medicines in laboratories. If their optimism were justified, we would no longer need to "mess about in the jungle" for medicine and, as an argument for wilderness preservation, this argument would become contingent only upon the state of our ability to artificially produce medicine.

13. In fact, in her recent book *Coyotes and Town Dogs: Earth First! and the Environmental Movement* (New York: Viking Penguin, 1993), a review of American environmentalism of the past few decades as seen through the eyes, lives, and voices of members of Earth First!, Susan Zakin tells us that ecologist Jerry Franklin discovered upon study "that the Pacific Northwest forest is the most densely green place on earth. One acre of old-growth Douglas fir forest contains more than twice the living matter of an acre of tropical rainforest. Some stands of trees harbor as many as 1,500 species of plants and animals" (p. 240). This finding is substantiated in a recent book on biodiversity by conservation biologists Reed Noss and Allen Cooperrider, *Saving Nature's Legacy: Protecting and Restoring Biodiversity* (Washington, D.C.: Island Press, 1994), p. 101.

14. N. R. Farnsworth, "Screening Plants for New Medicines," in E. O. Wilson, ed., *Biodiversity* (Washington, D.C.: National Academy Press, 1988), pp. 83–97.

15. Donella Meadows, "Biodiversity: The Key to Saving Life on Earth," *Land Stewardship Letter* (Summer 1990).

16. Rolston, "Valuing Wildlands," pp. 26 and 27, respectively. Note that it is quite common for wilderness advocates to equate "wilderness" with "nature" or "environment" and to use these terms interchangeably. Rolston is a good example of someone who quite liberally uses these terms in such a manner. However, these are not the same things and it seems improper to conflate them. But note also that the life-support argument turns not on the equivocation of these terms but rather on the relationship that exists between nature and wilderness.

17. Sessions, "Ecocentrism, Wilderness, and Global Ecosystem Protection," p. 99. Paul and Anne Ehrlich, *Extinction: The Causes and Consequences of the Disappearance of Species* (New York: Random House, 1981), pp. xi–xiv, 77–100; and Paul Ehrlich, "The Loss of Biodiversity: Causes and Consequences," in E. O. Wilson, *Biodiversity,* pp. 21–27, both cited in Sessions.

18. Robert Marshall, "The Problem of the Wilderness," *The Scientific Monthly* 30 (February 1930): 142; included in this volume.

19. Leopold, *A Sand County Almanac: With Essays on Conservation from Round River* (New York: Ballantine, 1966; originally published in 1949), pp. 269–72, his emphasis. See also Leopold, "Wilderness as a Form of Land Use," in this volume.

20. John Muir, *Our National Parks* (Boston: Houghton Mifflin, 1901; San Francisco: Sierra Club Books, 1991), chap.1, portions of which are included in this volume.

21. Ibid., p. 1.

22. See the studies cited in Driver, Nash, and Haas, "Wilderness Benefits," p. 301.

23. Rolston, "Valuing Wildlands," p. 30.

24. Wallace Stegner, "The Wilderness Idea," in David Brower, ed., *Wilderness: America's Living Heritage* (San Francisco: Sierra Club Books, 1961), p. 102.

25. Paul Shepard, *Nature and Madness* (San Francisco: Sierra Club Books, 1982).

26. Fox, *Toward a Transpersonal Ecology*, p. 160.

27. See, for example, Robert Marshall's "The Problem of the Wilderness" in this volume.

28. Nash, "Why Wilderness?," p. 198.

29. William Wordsworth, "Lines Composed a Few Miles Above Tintern Abbey" (1798).

30. Muir, *Our National Parks*, p. 4. This point has also been made by Allen Carlson in "Nature and Positive Aesthetics," *Environmental Ethics* 6 (1984): 5–34.

31. Aldo Leopold, "Goose Music," in Luna B. Leopold, ed., *Round River: From the Journals of Aldo Leopold* (Minocqua, Wis.: NorthWord Press, 1991), p. 245.

32. Muir, quoted in Nash, "The Value of Wilderness," p. 23.

33. Muir, *The Yosemite* (San Francisco: Sierra Club Books, 1988), pp. 196–97.

34. These three quotations, respectively, are from Leopold, *A Sand County Almanac,* p. 276; Leopold, "The Arboretum and the University," *Parks and Recreation* 78 (1934): 60 (interestingly, this quotation is used by Nash in *Wilderness and the American Mind* and "The Value of Wilderness," as well as by others, to refer to wilderness; however, what Leopold is actually referring to is the Arboretum at the University of Wisconsin, which is nothing remotely like a wilderness area. It seems safe to assume, as Nash and I do, however, that Leopold would say the same of putative wilderness); and *A Sand County Almanac,* pp. 274–75.

35. Rolston, "Valuing Wildlands," p. 27.

36. Fox, *Toward a Transpersonal Ecology,* p. 158 (his emphasis).

37. Leopold, *A Sand County Almanac,* pp. 274 and 276, respectively (my emphasis).

38. David Brower, quoted in Nash, "Why Wilderness?," p. 198.

39. Leopold, *A Sand County Almanac,* p. 190.

40. Rolston, "Valuing Wildlands," p. 28.

41. Edward Abbey, *Desert Solitaire: A Season in the Wilderness* (New York: Avon Books, 1968), p. 190.

42. Walt Whitman, *Leaves of Grass* (Ithaca, N.Y.: Cornell University Press, 1961; originally published in 1860), p. 319.

43. Nash, "Why Wilderness?," p. 199.

44. Henry David Thoreau, "Walking," in *Excursions* (Boston: Ticknor and Fields, 1863), pp. 185 and 191, respectively.

45. Leopold, *A Sand County Almanac,* p. 264.

46. Nash, "Why Wilderness?," p. 199.

47. Rolston, "Valuing Wildlands," p. 29.

48. A good example of this type of wilderness preservation rationale is found in Aldo Leopold's essay "Conserving the Covered Wagon," in Flader and Callicott, eds., *The River of the Mother of God,* pp. 128–32.

49. Stegner, "The Wilderness Idea," pp. 97 and 102, respectively.

50. Nash quoted in McCloskey, "Evolving Perspectives on Wilderness Values," p. 15.

51. Nash, "The Value of Wilderness," p. 22.

52. Mark Sagoff, "On Preserving the Natural Environment," *The Yale Law Review* 84 (1974): 266–67.

53. To take this line of thought even further, if that which is unspoiled is of value, and if completely pristine wilderness no longer really does exist, then our current wilderness areas (no matter how tainted by human interference), stand as a memorial to and a symbol of hope for the lost unsullied wildness of the earth and should be preserved as such.

54. Michael Zimmerman makes this point in *Contesting Earth's Future: Radical Ecology and Postmodernism* (Berkeley: University of California Press, 1994), p. 120.

55. The eradication of a virus's familiar host species exacerbates this problem by forcing the virus to jump hosts in order to survive.

56. Richard Preston, "A Reporter at Large: Crisis in the Hot Zone," *The New Yorker,* October 26, 1994: 62.

57. Ibid., p. 62.

58. Ibid., p. 80. See also Preston's book-length treatment of this issue, *The Hot Zone* (New York: Random House, 1994). This is also the topic of the 1995 Hollywood movie *Outbreak.*

59. Edward Abbey, *Desert Solitaire,* pp. 148–51.

60. Michael Zimmerman, "The Threat of Ecofascism," *Social Theory and Practice* 21 (1995): 207–238.

61. Max Oelschlaeger, *The Idea of Wilderness* (New York and London: Yale Univ. Press, 1991), p. 231.

62. Leopold, *A Sand County Almanac,* p. 264.

63. Rolston, *Conserving Natural Value* (New York: Columbia University Press, 1994), p. 15.

64. Note that both Nash and Oelschlaeger see wilderness as a contrasting value. For the ideas and insights in this section I owe much credit to the thought and work of Arthur Herman in *The Problem of Evil and Indian Thought,* 2nd ed. (New Delhi: Motilal Bansaridas, 1993), in which he discusses the logical and metaphysical necessity of evil as a possible solution to the philosophical and theological problem of evil.

65. Nash, "The Value of Wilderness," p. 24.

66. Leopold, *A Sand County Almanac,* pp. 271–72.

67. Marshall, "The Problem of the Wilderness," p. 143.

68. Driver, Nash, and Haas, "Wilderness Benefits," p. 302.

69. Paul Shepard, *Thinking Animals: Animals and the Development of Human Intelligence* (New York: Viking Press, 1978), pp. 246–52. See also Shepard's new book, *The Others: How Animals Made Us Human* (Washington, D.C.: Island Press, 1996) on this topic.

70. Leopold, *A Sand County Almanac,* pp. 276–78.

71. For an explanation of modern conservation biology and the plight of the grizzly bear, see R. Edward Grumbine, *Ghost Bears: Exploring the Biodiversity Crisis* (Washington, D.C.: Island Press, 1992). See also Noss and Cooperrider, *Saving Nature's Legacy.*

72. Leopold, "Some Fundamentals of Conservation in the Southwest," *Environmental Ethics* 1 (1979): 140; originally written in the early 1920s.

73. E. O. Wilson, *The Diversity of Life* (Cambridge: Harvard University Press, 1992), p. 282.

74. Leopold, "Wilderness Values," *1941 Yearbook of the Parks and Recreational Services* (Washington, D.C.: National Park Service, 1941), p. 28.

75. Wilson, *The Diversity of Life,* p. 303.

76. Edward Abbey, *The Journey Home* (New York: Dutton, 1977), p. 223.

77. Rolston, "Valuing Wildlands," p. 30.

78. Godfrey-Smith, "The Value of Wilderness," p. 319.

79. David Brower, quoted in *Wild by Law,* a film by Lawrence Hott and Dianne Garey, Florentine Films, 1990.

PART TWO

Third and Fourth World Views of the Wilderness Idea

Chief Luther Standing Bear

Indian Wisdom (1933)

THE 'GREAT OUT-DOORS' WAS REALITY and not something to be talked about in dim consciousness. And for them there was perfect safety. There were not the dangers that seem to surround childhood of today. I can recall days—entire days—when we roamed over the plains, hills, and up and down streams without fear of anything. I do not remember ever hearing of an Indian child being hurt or eaten by a wild animal.

Every now and then the whole village moved ten or fifteen miles to a grassier spot, but this was not considered much of a job. It was less trouble than moving a house from the front to the back of a city lot. Miles were to us as they were to the bird. The land was ours to roam in as the sky was for them to fly in. We did not think of the great open plains, the beautiful rolling hills, and winding streams with tangled growth, as 'wild.' Only to the white man was nature a 'wilderness' and only to him was the land 'infested' with 'wild' animals and 'savage' people. To us it was tame. Earth was bountiful and we were surrounded with the blessings of the Great Mystery. Not until the hairy man from the east came and with brutal frenzy heaped injustices upon us and the families we loved was it 'wild' for us. When the very animals of the forest began fleeing from his approach, then it was that for us the 'Wild West' began.

NATURE

The Lakota was a true naturist—a lover of Nature. He loved the earth and all things of the earth, the attachment growing with age. The old people came literally to love the soil and they sat or reclined on the ground with a feeling of being close to a mothering power. It was good for the skin to touch the earth and the old people liked to remove their moccasins and walk with bare feet on the sacred earth. Their tipis were built upon the earth and their altars were made of earth. The birds that flew in the air came to rest upon the earth and it was the final abiding place of all things that lived and grew. The soil was soothing, strengthening, cleansing, and healing.

This is why the old Indian still sits upon the earth instead of propping himself up and away from its life-giving forces. For him, to sit or lie upon the ground is to be able to think more deeply and to feel more keenly; he can see more clearly into the mysteries of life and come closer in kinship to other lives about him.

The earth was full of sounds which the old-time Indian could hear, sometimes putting his ear to it so as to hear more clearly. The forefathers of the Lakotas had done this for long ages until there had come to them real understanding of earth ways. It was almost as if the man were still a part of the earth as he was in the beginning, according to the legend of the tribe. This beautiful story of the genesis of the Lakota people furnished the foundation for the love they bore for earth and all things of the earth. Wherever the Lakota went, he was with Mother Earth. No matter where he roamed by day or slept by night, he was safe with her. This thought comforted and sustained the Lakota and he was eternally filled with gratitude.

From Wakan Tanka there came a great unifying life force that flowed in and through all things—the flowers of the plains, blowing winds, rocks, trees, birds, animals—and was the same force that had been breathed into the first man. Thus all things were kindred and brought together by the same Great Mystery.

Kinship with all creatures of the earth, sky, and water was a real and active principle. For the animal and bird world there existed a brotherly feeling that kept the Lakota safe among them. And so close did some of the Lakotas come to their feathered and furred friends that in true brotherhood they spoke a common tongue.

The animal had rights—the right of man's protection, the right to live, the right to multiply, the right to freedom, and the right to man's indebtedness—and in recognition of these rights the Lakota never enslaved the animal, and spared all life that was not needed for food and clothing.

This concept of life and its relations was humanizing and gave to the Lakota an abiding love. It filled his being with the joy and mystery of living; it gave him reverence for all life; it made a place for all things in the scheme of existence with equal importance to all. The Lakota could despise no creature, for all were of one blood, made by the same hand, and filled with the essence of the Great Mystery. In spirit the Lakota was humble and meek. 'Blessed are the meek: for they shall inherit the earth,' was true for the Lakota, and from the earth he inherited secrets long since forgotten. His religion was sane, normal, and human.

Reflection upon life and its meaning, consideration of its wonders, and observation of the world of creatures, began with childhood. The earth, which was called *Maka,* and the sun, called *Anpetuwi,* represented two functions somewhat analogous to those of male and female. The earth brought forth life, but the warming, enticing rays of the sun coaxed it into being. The earth yielded, the sun engendered.

In talking to children, the old Lakota would place a hand on the ground and explain: 'We sit in the lap of our Mother. From her we, and all other living things, come. We shall soon pass, but the place where we now rest will last forever.' So we, too, learned to sit or lie on the ground and become conscious of life about us in its multitude of forms. Sometimes we boys would sit motionless and watch the swallow, the tiny ants, or perhaps some small animal at its work and ponder on its industry and ingenuity; or we lay on our backs and looked long at the sky and when the stars came out made shapes from the various groups. The morning and evening star always attracted attention, and the Milky Way was a path which was traveled by the ghosts. The old people told us to heed *wa maka skan,* which were the 'moving things of earth.' This meant, of course, the animals that lived and moved about, and the stories they told of *wa maka skan* increased our interest and delight. The wolf, duck, eagle, hawk, spider, bear, and other creatures, had marvelous powers, and each one was useful and helpful to us. Then there were the warriors who lived in the sky and dashed about on their spirited horses during a thunder storm, their lances clashing with the thunder and glittering with the lightning. There was *wiwila,* the living spirit of the spring, and the stones that flew like a bird and talked like a man. Everything was possessed of personality, only differing with us in form. Knowledge was inherent in all things. The world was a library and its books were the stones, leaves, grass, brooks, and the birds and animals that shared, alike with us, the storms and blessings of earth. We learned to do what only the student of nature ever learns, and that was to feel beauty. We never railed at the storms, the furious winds, and the biting frosts and snows. To do so intensified human futility, so whatever came we adjusted ourselves, by more effort and energy if necessary, but without complaint. Even the lightning did us no harm, for whenever it came too close, mothers and grandmothers in every tipi put cedar leaves on the coals and their magic kept danger away. Bright days and dark days were both expressions of the Great Mystery, and the Indian reveled in being close to the Big Holy. His worship was unalloyed, free from the fears of civilization.

I have come to know that the white mind does not feel toward nature as does the Indian mind, and it is because, I believe, of the difference in childhood instruction. I have often noticed white boys gathered in a city by-street or alley jostling and pushing one another in a foolish manner. They spend much time in this aimless fashion, their natural faculties neither seeing, hearing, nor feeling the varied life that surrounds them. There is about them no awareness, no acuteness, and it is this dullness that gives ugly mannerisms full play; it takes from them natural poise and stimulation. In contrast, Indian boys, who are naturally reared, are alert to their surroundings; their senses are not narrowed to observing only one another, and they cannot spend hours seeing nothing, hearing nothing, and thinking nothing in particular. Observation was certain in its rewards; interest, wonder, admiration grew, and the fact was appreciated that life was more than mere human manifestation; that it was expressed in a multitude of forms. This appreciation enriched Lakota existence. Life was vivid and pulsing; nothing was casual and commonplace. The Indian lived—lived in every sense of the word—from his first to his last breath.

The character of the Indian's emotion left little room in his heart for antagonism toward his fellow creatures, this attitude giving him what is sometimes referred to as 'the Indian point of view.' Every true student, every lover of nature has 'the Indian point of view,' but there are few such students, for few white men approach nature in the Indian manner. The Indian and the white man sense things differently because the white man has put distance between himself and nature; and assuming a lofty place in the scheme of order of things has lost for him both reverence and understanding. Consequently the white man finds Indian philosophy obscure—wrapped, as he says, in a maze of ideas and symbols which he does not understand. A writer friend, a white man whose knowledge of 'Injuns' is far more profound and sympathetic than the average, once said that he had been privileged, on two occasions, to see the contents of an Indian medicine-man's bag in which were bits of earth, feathers, stones, and various other articles of symbolic nature; that a 'collector' showed him one and laughed, but a great and world-famous archeologist showed him the other with admiration and wonder. Many times the Indian is embarrassed and baffled by the white man's allusions to nature in such terms as crude, primitive, wild, rude, untamed, and savage. For the Lakota, mountains, lakes, rivers, springs, valleys, and woods were all finished beauty; winds, rain, snow, sunshine, day, night, and change of seasons brought interest; birds, insects, and animals filled the world with knowledge that defied the discernment of man.

But nothing the Great Mystery placed in the land of the Indian pleased the white man, and nothing escaped his transforming hand. Wherever forests have not been mowed down; wherever the animal is recessed in their quiet protection; wherever the earth is not bereft of fourfooted life—that to him is an 'unbroken wilderness.' But since for the Lakota there was no wilderness; since nature was not dangerous but hospitable; not forbidding but friendly, Lakota philosophy was healthy—free from fear and dogmatism. And here I find the great distinction between the faith of the Indian and the white man. Indian faith sought the harmony of man with his surroundings; the other sought the dominance of surroundings. In sharing, in loving all and everything, one people naturally found a measure of the thing they sought; while, in fearing, the other found need of conquest. For one man the world was full of beauty; for the other it was a place of sin and ugliness to be endured until he went to another world, there to become a creature of wings, half-man and half-bird. Forever one man directed his Mystery to change the world He had made; forever this man pleaded with Him to chastise His wicked ones; and forever he implored his Wakan Tanka to send His light to earth. Small wonder this man could not understand the other.

But the old Lakota was wise. He knew that man's heart, away from nature, becomes hard; he knew that lack of respect for growing, living things soon led to lack of respect for humans too. So he kept his youth close to its softening influence.

RELIGION

The Lakota loved the sun and earth, but he worshiped only Wakan Tanka, or Big Holy, who was the Maker of all things of earth, sky, and water. Wakan Tanka breathed life and motion into all things, both visible and invisible. He was over all, through all, and in all, and great as was the sun, and good as was the earth, the greatness and goodness of the Big Holy were not surpassed. The Lakota could look at nothing without at the same time looking at Wakan Tanka, and he could not, if he wished, evade His presence, for it pervaded all things and filled all space. All the mysteries of birth, life, and death; all the wonders of lightning, thunder, wind, and rain were but the evidence of His everlasting and encompassing power.

Wakan Tanka prepared the earth and put upon it both man and animal. He dispensed earthly blessings, and when life on earth was finished, provided a home, *Wanagi yata,* the place where the souls gather. To this home all souls went after death, for there were no wicked to be excluded.

Roderick Nash

The International Perspective (1982)

I personally am not very interested in animals. I do not want to spend my holidays watching crocodiles. Nevertheless, I am entirely in favor of their survival. I believe that after diamonds and sisal, wild animals will provide Tanganyika with its greatest source of income. Thousands of Americans and Europeans have the strange urge to see these animals. Julius Nyerere, ca. 1961

From 1854 to 1857 Sir St. George Gore, a British nobleman, vacationed in the wilderness of the upper Missouri River. Gore traveled through what later became Wyoming and Montana with 40 assistants, 112 horses, 24 mules, six yoke of oxen, a large pack of staghounds and greyhounds, and three milk cows. He shot 2,000 buffalo, 1,600 deer and elk and 105 bears.[1] From April 1909 to March 1910 ex-President Theodore Roosevelt vacationed in the wilderness of British East Africa. Roosevelt traveled through what later became the nations of Kenya and Uganda with 200 trackers, skinners, porters, gun bearers, and tent "boys." Roosevelt and his son shot, preserved, and shipped to Washington, D.C., over 3,000 specimens of African wildlife.[2]

In the half century between Gore's safari and Roosevelt's the United States changed from an exporter to an importer of wild nature. The changeover might be thought of as occurring in the 1890s when the American frontier officially ended and the cult of wilderness began.[3] Previously, foreign tourists seeking wildness found a mecca in the trans-Missouri West. St. George Gore's trip exemplifies the efforts of wealthy and socially prominent Europeans to experience wild America while it lasted. Contemporary Americans competed too closely with the wild to hear its call. Their relationship to it was that of transformers, not tourists, and they did their work well. By the time a later generation of Americans, represented by Theodore Roosevelt, became civilized enough to appreciate wildness, it had largely vanished from the American West. Africa became the new mecca for nature tourists like Roosevelt who were wealthy enough to import from abroad what had become scarce at home.

Thinking of wild nature as an actively traded commodity in an international market clarifies appreciation and largely explains the world nature protection movement. The export-import relationship underscores the irony inherent in the fact that the civilizing process which imperils wild nature is precisely that which creates the need for it. As a rule the nations that have wilderness do not want it, and those that want it do not have it. Nature appreciation is a "full stomach" phenomenon, that is confined to the rich, urban, and sophisticated. A society must become technological, urban, and crowded before a need for wild nature makes economic and intellectual sense. A Marxist formulation is tempting. There seems to be a social and economic class of nature lovers whose national affiliations are not as strong as their common interest in enjoying and saving wilderness wherever it exists. These people organize, confer, correspond, and raise money for nature preservation. A social profile of their ranks would reveal an inordinately high proportion of scientists, writers, artists—people of quality and the affluence to pay for it.

More than a metaphor is involved in nature importing; it has an economic value.[4] Wildness is actually bought and sold and not for trifling amounts. Except in the case of trophies and the live capture of animals for zoos, nature does not physically leave the exporting country. The traded commodity is experience. The importers consume it on the premises. In addition, there are many armchair nature enthusiasts. Their eagerness to consume motion pictures, television specials, magazines and books about wildlife, and to support nature philanthropy is an important form of nature importing. But wealthy tourists, following in the footsteps of Gore and Roosevelt, have been the mainstay of the nature business. Their willingness to pay heavily to see wild nature is a major factor in the economies of the nations where it still exists.

To extend the export-import metaphor, national parks and wilderness systems might be thought of as the institutional "containers" that developed nations send to underdeveloped ones for the purpose of "packaging" a fragile resource. Personnel sent to run the parks or to train native managers have a key role in the transfer of wildness for money.

Although less utilitarian arguments certainly do exist, in actual fact money is the most important reason for preserving nature in most cultures. As the scope of the Gore and Roosevelt trips suggests, nature exporting can be lucrative. It subsidizes nature preservation. Less developed countries can afford to maintain wildness, while necessarily restraining development, if the exportation of nature pays sufficient dividends. A poster intended for natives in Africa makes the point explicitly: "OUR NATIONAL PARKS BRING GOOD MONEY INTO TANZANIA—PRESERVE THEM." Local people are reminded, for instance, that an adult male lion in Amboseli National Park in Kenya generates $515,000 in tourist revenue over the course of its lifetime. For a poacher, the meat and skin might bring as much as $1,150.[5] On the basis of the revenue they generate by attracting tourists, lions or elephants may be the most valuable animals in the world, race horses included.

The tension between the nature exporters and the nature importers is historic and continuing. It should be clear that the exporters do not as a rule recognize the marketability of their product. Africans, for example, have lived with wild animals as long as they can remember. You cannot interest a Masai in seeing and photographing a giraffe any more than you can interest a New Yorker in a taxicab. Similarly, the restrictions on grazing and farming in an African park or preserve are as perplexing to the natives as a law that prevents a New Yorker from living in and using ten square blocks of midtown Manhattan would be. Not sharing the developed world's conception of the value of wild nature, the less developed world sees no reason not to continue to exploit resources in the accustomed manner. But as Tanzania's president Julius Nyerere's remark opening this chapter suggests, if incomprehensible foreign tourists want to travel thousands of miles just to *look* at wild animals, and especially if they spend money in the process, exporters will not protest.[6]

Exporting and importing nature also has a regional or *intra-national* significance. The urban segment of a population may support preservation of wilderness in hinterlands, the inhabitants of which are indifferent or actively hostile. In the United States, the East and civilized islands in the West, like San Francisco, reached the nature importing stage several generations before the still-wild West. The first nature tourists came from the these areas.[7] So did the first stirrings of the nature preservation movement. Henry David Thoreau and Theodore Roosevelt were Harvard men. John Muir, like Sir St. George Gore, came from Great Britain, and when he organized the Sierra Club in 1892 it was dominated by an elite from Berkeley and San Francisco.[8]

Parallels exist throughout the world. The concern in Tokyo protects what wildness remains on Japan's northernmost island, Hokkaido. Australia's outback is of primary interest to residents of Sydney and Melbourne. The Malaysian Nature Society has little support outside the nation's metropolis, Kuala Lumpur. The national parks of Norway and Sweden are the concerns of urban people in the southern portions of those countries. And, to return to the American experience, nature preservation efforts in Alaska have been led by outsiders from the rest of the United States. Robert Marshall, for example, was a classic nature importer, amply endowed with the money and free time to indulge his passion for wilderness during the depths of the Great Depression. . . .

Until very recently traveling for pleasure almost always entailed movement from less civilized to more civilized areas. The trapper or farmer came out of the woods for a few days of fun in the biggest city available. If people traveled in the other direction their purpose was invariably to transform wilderness into civilization in the manner of the Pilgrims and the Mormons. No one went to New England or Utah in the early years for recreation. The intellectual revolution that made unmodified nature *per se* a mecca for travelers is the principal subject of this book [*Wilderness and the American Mind*]. It depended upon the emergence of a group of affluent and cultured persons who resided in urban environments. For such persons wilderness could become an intriguing novelty and even a deep spiritual and psychological need. But the civilized conditions that cause interest in wilderness also destroy it. Travel was the solution. . . .

Jet travel and the expansion of the clientele for wildness beyond a few aristocrats like Sir St. George Gore and Theodore Roosevelt have recently increased the leverage of the nature importers. Purveyors of nature tours, such as Questers of New York and Mountain Travel, based in California, have simplified the task of importing nature. The cost is still considerable, but apparently enough well-to-do nature lovers exist to fill dozens of trips. Hotels specializing in bringing civilized people and wildness together under comfortable circumstances—Treetops in Kenya and Tiger Tops in Nepal are famous examples—are solidly booked months in advance. Leaders of the nature travel industry are aware that the business they generate underwrites nature preservation and thereby ensures their own future. "Since tourism is an increasingly important source of revenue to countries around the world," declares Questers' president, Michael L. Parkin, "we believe that international . . . travel to wildlife sanctuaries and nature reserves will help encourage further conservation efforts."[9]

Yet even as nature tourism increases, doubts remain about its effectiveness in preserving wildness. Even nonhunting tourism is not always compatible with preservation. Those who can afford to import nature are, in general, older people unprepared to rough it in the wilds. They demand hotels, restaurants, roads, motorized vehicles, and small towns of supporting servants, all located, more often than not, inside the park or reserve.[10] This luxury tourism yields the biggest economic rewards for the nature exporters. Backpackers are notoriously low spenders, preferring self-sufficiency to service. Roadless wilderness does not generate much revenue, yet this kind of environment is precisely what some importers and most scientists accord the highest value. If the only reason for a park is to make money, restrictions on revenue-producing tourism, even in the cause of protecting nature, are unlikely.

A second argument against tourism as the mainstay of global nature protection centers is on the final distribution of the importers' money. Although local people are supposed to be compensated for what they forgo in income by *not* developing wilderness, the bulk of the tourism revenue goes to entrepreneurs from the developed world. The extreme examples are the cruise ships such as those Lindblad Travel sends to wild places throughout the world. The modest sums spent on shore by the passengers for souvenirs and postcards do not constitute a strong argument for protecting nature. The economics of land-based tours work out better for native people, but it is still foreign-owned airline companies, hotel chains, and travel agents who chiefly benefit. On a forty-four day, $3,600 trip on a wild river in Peru,

Amazon Expeditions, headquartered in Erie, Pennsylvania, does not even buy its food locally. Each member of the trip is sent a bag of foodstuff before leaving the United States. The consumption of local goods and services is limited to a few nights' lodging and the services of a local truck and driver before and after the river trip. Hydropower developers wishing to dam the river can plausibly claim that nature importing does not constitute a viable economic alternative.[11] In addition, political disturbances, such as those of the late 1970s which closed all of Uganda and the Kenya-Tanzania border to tourists, could completely remove the protection to nature afforded by nature tourism.

The alternative, and the ultimate extension of nature importing, is outright ownership, or at least control, of important natural environments. Though it is unfair to think of international collaboration to this end as neocolonial, its purpose is much the same as that of the earlier European park promoters in South Africa, the Belgian Congo, and Kenya. The central concept of what might be thought of as an international park is found as early as 1834 in the statement of Andrew Reed and James Natheson regarding one of North America's scenic wonders. "Niagara," these tourists wrote, "does not belong to Canada or America. Such spots should be deemed the property of civilized mankind."[12]

Since Niagara Falls certainly did in the traditional sense "belong" to Canada and the United States, the logic behind this and many similar statements needs clarification. Apparently Reed and Natheson looked on Niagara as a scenic resource of value to all mankind. By geographical accident the Falls happened to be in Canada and the United States, but this did not give these nations the right to destroy them. Every human being, now and for all time, had a stake in such treasures and individual nations must not be allowed to act unilaterally in regard to their future. Pressed to the logical conclusion, this line of reasoning would authorize the world community to intervene, forcibly if necessary, to halt the destructive activities of a country within its own borders. Institutionalizing this idea was extremely difficult since it touched upon the uniquely sensitive issue of sovereignty.

Moreton Frewen, the English aristocrat, anticipated the international park in the 1870s when he bought land in Wyoming to protect it from American frontiersmen. Recent world control of natural treasures began in 1956: Masai incursions upon Serengeti National Park, and the prospect of African independence, led nature conservationists like Bernhard Grzimek to propose to buy or otherwise arrange for the Serengeti to be made international property under the United Nations. Nothing came of the idea, but in the following decade the Sierra Club discussed the prospects for an Earth International Park, while Friends of the Earth preferred Earth National Park. In 1971 Wildlife Conservation International arranged with the government of Zambia to manage, under a twenty-five-year lease, the Zambia International Wildlife Park. The Nature Conservancy, based in Arlington, Virginia, preferred to work by purchasing or receiving actual title to endangered natural environments. After two decades of concentration on the United States, the Nature Conservancy began in the mid-1970s to look into the acquisition of title to lands in foreign countries. In 1975 the Conservancy received a gift of 950 acres of pristine tropical rainforest on the Caribbean island of Dominica, leased to the Dominican government for management as a park.[13] Much broader in scope is the Man and the Biosphere Program (MAB) set in motion under the auspices of UNESCO in 1970. Participating nations link appropriate sites into a worldwide network of ecologically significant environments. Science, not recreation, is served by MAB. Participation is voluntary, but the program generates pressure, if only that of world opinion, to keep reserved sites protected. At Stockholm in 1972 the United Nations Conference on the Human Environment gave full support to MAB.[14]

The other international gathering of nature protectors in 1972 occurred in northwestern Wyoming in honor of the centennial of Yellowstone National Park.[15] It was a pleasant experience for Americans, scarred as a society by the Vietnam War, political assassinations and the developing Watergate scandals. Delegate after delegate from around the world rose to credit the United States with inventing the national park.[16] A few speakers added that, in frankness, the American experience in preserving wilderness served as warning as well as inspiration. The 1913 removal of Hetch Hetchy Valley from Yosemite National Park was an example of what not to do. So was the overcrowded condition of Yosemite Valley. But in general the gathering celebrated Cornelius Hedges, John Muir, and Stephen T. Mather as visionaries who anticipated human needs for nature and worked to institutionalize wilderness preservation. Thinking about the future, the four hundred delegates from eighty nations resolved that in the second century of national parks "the concept of world parks should be promoted." As a start the conference further resolved that "the nations party to the Antarctic Treaty should negotiate to establish the Antarctic Continent and surrounding seas as the first world park."[17] The United Nations would provide management of this unprecedented international wilderness.

Russell Train took the platform to offer another idea: the world heritage trust. Train explained it as an international extension of the national park concept. Certain natural features had such outstanding value "that they belong to the heritage of the entire world."[18] As examples Train cited Mount Everest, the Galapagos Islands, the Serengeti Plain, Angel Falls in Venezuela, the Grand Canyon of Arizona, and certain animal species like the mountain gorilla. Train's idea was to marshal the world's financial, technical, and managerial resources on behalf of these places and life forms.

Moving toward institutionalization of the world heritage idea, UNESCO drafted a convention in November 1972, and the United States became the first nation to ratify it on December 7, 1973.[19] Not until September 1978, however, did eleven nations actually place areas on the "World Heritage List." Yellowstone, Grand Canyon, Everglades, and Redwoods national parks are among the American listings. Forty-two countries have ratified the convention to date, and more areas will be added to the list, but the degree of protection obtained thereby is not great. The chronic problem is that national sovereignty is left unchallenged. Participating nations may delete areas listed at will or, for that matter, denounce the entire convention. There are no reprisals. What a nation does to nature within its own borders remains its own business as Principle 21 of the Declaration of the United Nations Conference on the Human Environment made clear in Stockholm in 1972. Yet the world heritage concept does give more recognition than ever before to the international significance of natural environments and the international responsibility for their protection.

The economics inherent in trading wild nature are not, to be sure, the whole story. Other motives—some would say better or higher ones—exist for protecting the natural world, and the less developed nations may eventually evolve economically and intellectually to the point where nature protection is more than a business. In the meantime preservation of the world's remaining wilderness will depend as it has depended since its beginning a century ago, on the exporting and importing of an increasingly rare commodity.

NOTES

1. Francis Haines, *The Buffalo* (New York, 1970), pp. 146–47; Wayne Gard, *The Great Buffalo Hunt* (New York, 1959), pp. 62–64; F. George Heldt, "Narrative of Sir St. George Gore's Expedition, 1854–1856," *Contributions of the Historical Society of Montana,* 1 (1876): 128ff.

2. Theodore Roosevelt, *African Game Trails: An Account of the African Wanderings of an American Hunter-Naturalist* (New York, 1910); R. L. Wilson, *Theodore Roosevelt: Outdoorsman* (New York, 1971), pp. 172–202; Paul Russell Cutright, *Theodore Roosevelt: The Naturalist* (New York, 1956), pp. 186–224.

3. See Nash, *Wilderness and the American Mind,* Third edition (New Haven: Yale University Press), Chapter 9.

4. Norman Myers, "Wildlife of Savannahs and Grasslands: A Common Heritage of the Global Community" in *EARTHCARE: Global Protection of Natural Areas,* ed. Edmund A. Schofield (Boulder, Col., 1978), pp. 396ff., and Boyce Rensberger, *The Cult of the Wild* (Garden City, 1978), pp. 217–51, are important recent recognitions of the economics of world nature protection.

5. Philip Thresher, "The Present Value of an Amboseli Lion" (unpublished manuscript, April 1977), p. 1.

6. Quoted in Wolfgang Engelhardt, *Survival of the Free,* trans. John Coombs (New York, 1962), p. 112.

7. Earl Pomeroy, *In Search of the Golden West: The Tourist in Western America* (New York, 1957).

8. Holway R. Jones, *John Muir and the Sierra Club: The Battle for Yosemite* (San Francisco, 1965).

9. Michael L. Parkin in the introduction to *Questers Directory of Worldwide Nature Tours, 1975* (New York, 1974), p. 2.

10. See, for example, Norman Myers, *The Long African Day* (New York, 1972). Another critical look at African nature tourism is Colin Turnbull, "East African Safari," *Natural History* 90 (1981): 26–34.

11. Henry Pelham Burn, "Packaging Paradise: The Environmental Costs of International Tourism," *Sierra Club Bulletin* 60 (May 1975), 25; Myers, "Wildlife of Savannahs and Grasslands."

12. As quoted in Charles M. Dow, ed., *Anthology and Bibliography of Niagara Falls,* 2 vols. (Albany, 1921), 2:1070–71.

13. R. Michael Wright, "Private Action for the Global Protection of Natural Areas," in *EARTHCARE,* pp. 715–39; R. Michael Wright, "After God the

Earth—Rainforest Preserved in Dominica, West Indies," *Nature Conservancy News* 25 (Spring 1975): 8–11; R. Michael Wright, Director, International Program, Nature Conservancy, to Roderick Nash, January 23, 1976.

14. International Co-ordinating Council of the Programme on Man and the Biosphere, *UNESCO MAB: Final Report* (Paris, 1971); Michel Batisse, "The Beginning of MAB," in *World National Parks: Progress and Opportunities,* ed. Richard van Osten (Brussels, 1972), pp. 178–79.

15. The proceedings are available as Hugh Elliott, ed., *Second World Conference on National Parks* (Morges, Switzerland, 1974). Related publications are Freeman Tilden, *National Parks Centennial, 1872–1972—Yellowstone, the Flowering of an Idea* (Washington, D.C., 1972) and *A Gathering of Nations: A Time of Purpose—In Commemoration of the Centennial Celebration of Yellowstone and the Second World Conference on National Parks* (Washington, D.C., 1973).

16. On this point see Roderick Nash, "The American Invention of National Parks," *American Quarterly* 22 (1970): 726–35. Another essay evaluates the evidence that Australia may have been the first nation to establish a national park: Roderick Nash, "The Confusing Birth of National Parks," *Michigan Quarterly Review* 19 (1980): 216–26.

17. Elliot, *Second World Conference on National Parks,* pp. 443–44.

18. Russell E. Train, "An Idea Whose Time Has Come: The World Heritage Trust, A World Need and a World Opportunity," in Elliott, *Second World Conference,* pp. 378–79. Train first raised the idea of a world heritage trust in 1965: Committee on Natural Resources Conservation and Development, *Report to the White House Conference on International Cooperation* (Washington, D.C., 1965), pp. 17–19.

19. UNESCO, *Convention Concerning the Protection of the World Cultural and Natural Heritage* (Paris, 1972).

David Harmon

Cultural Diversity, Human Subsistence, and the National Park Ideal (1987)

INTRODUCTION

One of the main goals of the international conservation movement has been the creation of protected-area systems in all countries. Protected areas—land or water held in trust (often on the national level) and brought under a systematic managerial regime primarily to benefit the public—exist in a variety of forms, but by far the best known is the "national park." One of the first and most famous protected areas, Yellowstone, received this designation when it was created in 1872. Afterward the number of protected areas worldwide rose continually except during World War II. There were remarkable increases in the 1970s, a decade during which about twice as many protected areas were created as existed altogether in 1969.[1]

In October 1982, representatives from seventy countries met in Denpasar, Bali, Indonesia, to discuss this rapid expansion and assess the state of protected-area systems around the world. The meeting was the third "World Congress on National Parks" and was organized by the International Union for Conservation of Nature and Natural Resources (IUCN). Despite the meeting's title, its focus was as much on alternatives to national parks as anything else. The IUCN has vigorously promoted the diversification of protected-area categories beyond national parks.[2] In his introduction to the collected papers of the Congress, Jeffrey A. McNeely sums up the IUCN's perspective:

> National parks are generally considered to be areas of outstanding natural significance where the influences of humans are minimal. But in a period of increasing human populations, economic uncertainty and social instability, many governments are finding that the traditional national park approach is no longer sufficient to meet their needs for recreation, education, genetic resource management, watershed protection, and the many other goals and services produced by protected area conservation. . . . [P]eople and governments around the world still appreciate the values of national parks, but they also realize that the stringent protection required for such areas is not necessarily appropriate for all areas which should be kept in a natural or seminatural state.[3]

"National park," nevertheless, is still the most used protected area designation in the world[4]—a fact due in large measure to the preponderant influence the conventional national park idea has had on international protected-area conservation. The conventional national park conserves a predominantly natural area, typically of great scenic beauty, by putting it into government ownership and banning human residency and resource extraction within. There is an inherent tension between this idea and the needs associated with the physical subsistence of human beings, needs which can only be satisfied through resource production and extraction. The success of national parks—to the exclusion of other, perhaps more suitable, protected-area categories—raises the question of whether the conventional national park idea is being misapplied when it becomes the centerpiece of protected-area systems in modern nations whose governments are not wealthy and whose land is at a premium for the rural population.

NATIONAL PARKS AS A CULTURAL PHENOMENON

No nation in the world has contributed more to the intellectual and social framework of the national park idea than the United States. Many authors have come straight out and called national parks an American idea. Historians Roderick Nash and Alfred Runte, for example, find the idea's origins to be squarely in the United States of the nineteenth century. Nash attributes it to the confluence of three factors that were found only in America: "the nation's unique experience with nature in general and wilderness in particular," the presence of a democratic ideology, and "the existence of a sizable amount of undeveloped land at the moment when the first two influences combined to produce a desire for its protection."[5] Runte sees national parks as having arisen out of a pervasive atmosphere of cultural insecurity in which Americans, feeling obliged to hold themselves up to the European mirror, put forth monumental examples of wild nature as a way of compensating for their own culture's comparative lack of human-made achievements.[6] In either interpretation, the parks are a phenomenon of affluent culture.[7]

Since then the phrase "national park idea" has evolved into a shorthand for the belief that there are certain places inherently worth being preserved, places apart from run-of-the-mill locales, places special enough to have the capacity to fix themselves on the public consciousness to the degree that they become part of an amorphous "national heritage." The conventional national park idea is, at bottom, an ideal.

CONTEXT, CONSERVATION, AND ETHICS

National parks, then, are products of an affluent culture. They grew up in a context of boundless wealth, under the expectation that the natural resources left outside them were inexhaustible. Are there any ethical implications attendant to these facts?

The conservation of natural resources by means of protected areas cannot go forward without reference to the culture in which it is taking place. Since it is a kind of social choice, conservation is not to one side of culture, but rather is a part of it. Conservation is not the same in the United States as it is in Canada or Europe or Africa or anywhere else. Not only are the choices not the same (this is no great revelation), but so are the approaches to the act of choice. The way to conservation is everywhere different.

Others make similar points. According to Keith Garratt, an IUCN official in New Zealand, "A basic principle which is fundamental to this subject is that, although protected areas are primarily for the protection of *natural* species and systems, the whole process of defining differing types of land tenure, controlling land use by legislation and deciding the positions of boundaries is entirely a human *cultural* concept." Bernard Nietschmann of the University of California goes so far as to say that natural resources themselves are "cultural appraisals," as much cultural concepts as biological facts, because "those things that are differentiated as 'resources' by one group of people may not be recognized by another group of people—even in the same environment":

> For a specific society and place, culture is a resource in itself because through culture, environments are conceptually constituted, the means and controls of exploitation are organized, and cumulative resource knowledge stored, taught, and used.[8]

If one assumes that cultural diversity is good and desirable (because contrast enlivens the individual and collective human spirit), it requires but a small step from the observations of Garratt and Nietschmann to the conclusion that there are serious problems with the wholesale transfer of a protected-area concept from one culture to another. The problems are not just ones of management, but of ethics: if conservation is part and parcel with culture, can it be right to take a conservation concept which is itself closely allied with a particular type of culture and promote (or acquiesce in) its dissemination, in unmodified form, around the world? Nevertheless, that is what has happened with the national park ideal.[9]

NATIONAL PARKS AS A "DEMONSTRATION EFFECT"

But is it fair to say the problem is that the rich countries "export" the conventional national park idea to poor ones? No. It would be simplistic to attribute the problem solely to proselytizing (although a missionary element is there)[10] or to some sort of "conservation colonialism" (although it is not inconceivable that it might be in the economic interest of wealthy countries to have the Third World adopt conventional national parks). The national park ideal has been as much imported as exported. Its spread, particularly to the Third World, can be attributed to the *demonstration effect.*

According to theorists of culture, the demonstration effect is produced by a desire for "superior" technologies and consumer goods from somewhere else that develops within a society when it is exposed to those technologies and goods. The term is usually used in the context of culture clash between "traditional" or "tribal" or "native" or "indigenous" people and "modern" society: the traditionals become less isolated from the moderns because of advances in transportation or communication, and as a result are increasingly exposed to new ways of living, ways which they sense are easier than their own, and which they soon come to want. The *effect* comes after the *demonstration* has succeeded: the traditionals obtain new technologies, and are either by that means assimilated into modern society at the loss of their traditional culture, or find themselves unable to adjust completely to the change, becoming stuck somewhere in between, their lives a disjointed limbo between past and present.

Anthropologists who accept the demonstration effect as a basis for culture change have been criticized on the grounds that they thereby grossly underestimate the importance of instances of forced acculturation.[11] The term also carries the unpleasant implication that the collective resolve of traditional cultures is somehow inherently "weak," with their people unable to resist "temptation" and perhaps even susceptible to being duped by opportunistic outsiders. I propose to sidestep these controversies by using the term in a slightly different way: for my purposes, the demonstration effect of the conventional national park idea can be said to operate, not between modern and traditional societies, but between modern society and the modern governments of countries containing traditional societies. Exposure to the national park ideal has proven irresistible to the governments of many developing countries. The embrace of these governments, however, is not necessarily shared by the peoples (traditional or not) who will be affected by a national park's establishment. In other words, here the demonstration effect operates not between the First World and the Third World, as it is usually supposed to do, but between the First World and the First Worlders in Third World countries.

The demonstration effect is fueled by the high status of the appellation "national park" both here and abroad. Take a look at almost any coffee-table book on the United States national park system: it will preponderantly, and often exclusively, concern itself with the approximately fifty units (out of more than 330 systemwide) which are actually designated as national parks. For many governments which have just recently developed protected-area systems, national parks are an inexpensive status symbol.

National parks are chic, and "are sometimes more successful in attracting financial support than other less-publicized categories of protected areas."[12] Every country, it seems, wants to have its Yellowstone, and the consequences be damned.

PROBLEMS WITH NATIONAL PARK HEGEMONY

But the consequences can be terrible. In northeast Uganda there is a remote valley called Kidepo at the foothills of Mount Morungole. The Kidepo was home to the Ik, a tribe of nomadic hunter-gatherers who had until recently remained virtually uncontacted. By all accounts the Ik were stewards of their environment: they considered the overhunting of the valley's game, upon which they depended heavily, to be the greatest sin a person could commit. They also maintained a strong spiritual identification with the landscape, especially the mountains around Morungole.

But in 1962 the government decided to establish Kidepo National Park. Since people are not allowed to live inside national parks, the Ik were relocated to packed villages on the periphery of the new park boundary. Since in national parks hunting is banned, the Ik were forbidden to hunt in the valley and were instead encouraged to farm. Taken from their mobile existence, the Ik were force-fit into a sedentary, close-quarters life style. Deprived of their social economy and separated from the physical locale that was the basis for their spiritual beliefs, the culture of the Ik literally disintegrated into a travesty of humanity.[13]

The destruction of the Ik was caused by the government's blind adherence to the national park ideal. Although both the Ik and the cause of conservation would have been better served if some other protected-area designation had been chosen for Kidepo, such is the power of national park hegemony that none was considered. If the Ik had been allowed to remain as residents of some other kind of protected area, their wise use of resources could have been made an object lesson to visitors.

A major drawback of using conventional national parks in countries which still have active traditional societies is that such parks tend to ride roughshod over cultures with a highly localized sense of what the geographer Yi-Fu Tuan calls "geopiety": a complex social and emotional (and often manifestly physical) attachment to place.[14] The Ik are a case in point: their culture was locale specific and could not survive even a short move. For many societies one place is not as good as another when it comes to "home." Research on the Maasai of east Africa has found that they require the presence of certain very precise environmental preconditions before they choose a settlement site. Their sites are decided by a number of factors not readily apparent to the outside observer, such as subtle variations in slope.[15] To displace such people from a newly created national park, even if the move is only a short distance to a place with close resemblance, can have devastating effects on their culture.

Another problem related to human populations is that conventional national parks have been established in countries where land and monetary affluence do not exist.[16] In some cases, park area residents then become convinced that resources essential to their well-being have been "locked up." This is just what happened when Nepal designated its first national park in 1973. Local people had traditionally used the area included in Royal Chitwan National Park to hunt wildlife for food and gather elephant grass to thatch their roofs. Upon designation these activities became illegal in accordance with American models. Since then the wildlife protected by the park, including the increasingly rare tiger and Nepal's only population of the great one-horned Asiatic rhinoceros, has flourished, to the point where some animals have wandered out of the park into adjacent areas, killing villagers. This, along with the withdrawal of lands traditionally used for livestock grazing, has understandably created local resentment of the park.

The making of Chitwan brought with it the making of a boundary—one which has often since been thought of by local people as a boundary of injustice. The following story is a typical example:

> Ram Preshad was proud of a buffalo he bought with savings from nearly four years of hard labor. Unfortunately the animal could not resist the lush green grass across the river inside the park and one morning he wandered about a kilometer inside the boundary, where he was killed and devoured by a big male tiger. Since it is illegal to let domestic cattle into the park, Ram could have been fined. He has no legal recourse but he will hold a grudge against the park for as long as he lives.[17]

Belatedly, Nepal's park authority has made some moves at Chitwan away from the national park ideal. While grazing and hunting are still banned, since 1978 the government has allowed a fifteen-day controlled harvest of elephant grass each year. Because the grass had been denuded from areas outside Chitwan, the existence of the national park is now seen by locals as providing at least one tangible benefit.[18] Area residents have also been given regular, formal access to the workings of the park authority. Lay co-management and controlled resource extraction, both alien to the conventional national park idea, are indispensable in the Nepalese context.

The national park ideal is especially intransigent when it comes to people living permanently inside park borders. This is an issue which extends beyond the Third World to industrialized countries. Great Britain, for one, has gotten around it by departing completely from American criteria: not only are whole villages allowed in the national parks of England and Wales, their presence is reckoned essential to the success of these protected areas. Even in such countries, nevertheless, there is often regret at not being able to have unpopulated national parks as land-rich nations do. For example, a paper presented at Bali by the British delegation linked their country's dense population with "limited choices" in the selection of protected-area categories. Private land tenure is also sometimes regarded as an inconsistency requiring special explanation. Since it is allowed within Japanese park boundaries, the Japanese delegate to the Bali conference felt compelled to defend the practice: "This method may not be necessarily ideal for conservation of a national park. But in Japan, where so many people live in a small limited space, it is hardly possible to allow any vast tract of land for the purpose of setting up a national park." The defensive tone of such comments is part of the pall cast by the success of the conventional national park idea.[19]

From the international conservationist's point of view, the biggest problem with the national park ideal is that its success has made it synonymous in the public mind with the entire field of protected-area conservation. According to an Indian delegate to the Bali conference, it is this perception which most impedes the establishment and effective management of protected areas in his country and throughout southeast Asia, for "even in the minds of the educated and of the decision-makers, protected areas are still identified as an elitist concept, as non-utilized tracts, and hence a waste of resources" in a region of intense demographic pressures and dire economic development needs.[20]

PARKS AND PEOPLE: REFOCUSING AND MANAGEMENT DEBATE

What is the responsibility of a government with respect to the needs of the people who will be directly affected by the creation of a national park? This is the question which nags at the cross-cultural transfer of the conventional national park idea. Rather than attempt a survey of the many innovative management ideas that are being offered around the world, I would like to reexamine one of the premises behind the basic question.[21]

Throughout the literature on the issue of human residents and the management of protected areas there is a lack of precise language. Almost everywhere, emphasis is placed on the needs of "traditional societies," "native peoples," and "indigenous peoples." These three phrases and their variants are used euphemistically, much in the same way "primitive" was once used to describe people and cultures not readily understandable to Western affluents. These more recent euphemisms reflect a newfound desire to be more sensitive to the wishes of unassimilated societies. However well-intentioned, the terms remain euphemisms. Going through the literature, one finds them used carelessly, even interchangeably. They are phrases that are easily politicized, and the supercharged atmosphere enveloping the debate shows that most already have been.

The World Bank is one of the few organizations working in this area to have published definitions for some of these vague terms:

Aboriginal. Implies having no known race preceding in the occupation of the region. Hence, the term includes national peasants with a traditional way of life.

Native. Implies birth or origin in the region in question, and so includes peasants and all others born in an area. It also implies a collection of different ethnic or racial groups.

Indigenous. Adds to "native" the implication of not having been introduced from another region of the country. It includes people who are surviving descendants of the earliest populations of the area.[22]

The World Bank did not define *traditional societies,* but the term conveys the notion of groups (usually native, aboriginal, etc.) who have never embraced modern technology and so stand apart from the dominant "modern" society.

It will be noticed that the three terms dominating the debate—*traditional, native,* and *indigenous*—are all temporal; they make sense only in comparison with some unnamed (and unspecified) "mainstream" culture. "Traditional societies" are only traditional relative to "modern society." Yet how old does a society (or a technology, or whatever) have to be before it becomes "traditional"? How long do colonists have to live in a place before they cease to be colonists and become "natives" or "indigenous people"?

When people involved in the management of protected areas use these terms, the distinction they seem to want to make is between those who live harmoniously (in a strictly ecological sense) with their immediate environment, and those who don't. In the current management literature it is too often presumed that traditional societies, native peoples, and indigenous peoples do; modern societies, colonists, and other non-indigenous peoples don't.

The presumption is illustrated by the ninth resolution of the Bali conference. Entitled "Protected Areas and Traditional Societies," it declares that "traditional societies *which have survived to the present in harmony with their environment*" [emphasis added] deserve our respect for their wise stewardship of natural resources, stewardship worthy of our emulation.[23] The delegates apparently were not recognizing all traditional societies, only those who live in harmony with their environment. But the implication is that *only* traditional societies live in harmony with their environment, or are entitled to the benefit of the doubt. The possibility that modern cultures could just as appropriately live in harmony with the environment—and therefore could just as appropriately remain as residents of protected areas—is ignored.

The conferees, who wished to promote both cultural diversity and protected-area conservation, unnecessarily and detrimentally limited themselves by focusing so much on the needs of traditional societies. Cultural diversity among modern social groups should be afforded as much recognition as that which is given to the juxtaposition of traditional with modern societies.

The adjective *traditional* has its uses, as I have already suggested in my discussion of the demonstration effect and the Ik. So, perhaps, do the terms *native, indigenous, aboriginal,* and so on. But the debate over the nuts-and-bolts *management* of protected areas should be taken out of this value-laden, temporal, political context. What is needed is a valueless, spatial term that cannot be staked out by politicians, and *resident population* is just such a term. *Resident population* can be defined as any group of people—traditional or modern—who occupy, reside, or otherwise use a specific physical territory within an established or proposed protected area on a regular or repeated basis.

The term embraces both continuous and continual residencies, and therefore covers recurring seasonal or other periodic use—but only if such nomadic residency is site-specific to the protected area. For example, when a protected area is created encompassing a specific locale that is essential to the subsistence (either physical, cultural, or spiritual) of a particular nomadic group, the group would qualify as a resident population if it can be demonstrated that it returns to that locale on a regular periodic basis. However, if that same group had a history of returning to the locale only on an irregular basis, and did not identify itself with the locale, and in fact indiscriminately used similar locales outside the protected area, then it would not qualify as a resident population.

Extreme caution must be taken by those who would make such judgments of mobile groups. Often it is difficult to tell whether a group is nomadic: some, such as the Nuer of Africa, may make residential use of a particular site only in cycles of long periodicity. It can be generally stated that the protected-area category "national park" does not tolerate well the presence of mobile societies; in countries where nomadism or transhumance is prevalent, they are a poor choice of category for protected areas. Again, this judgment assumes that one of the goals of conservation should be the protection of cultural diversity: it is, therefore, bad to create a national park if its creation threatens the integrity of a culture that diverges significantly from the nation's mainstream.

The real value of the term *resident population* is that it carries no minority-group connotations. For purposes of definition, it exists in space, not time: either the people make residential use of the protected area or they don't. And because it has no inherent cultural associations, the term *resident population* is hard to politicize.

Once the common language of the debate has been demystified and depoliticized, then everyone concerned can move on to the crux of the issue: whether a given human residency is ecologically and managerially compatible with the purposes of a given protected area. For instance, the resolution that was made by the 1982 world national parks congress could have profitably stricken all references to "traditional societies" and replaced them with "resident populations." All human populations in existing or proposed protected areas should be evaluated on equally objective terms before any managerial decision substantially affecting them is carried out.

THE MORAL IMPERATIVE OF A GOOD IDEA

Many, and perhaps most, protected-area planners working outside the United States appear ready to go beyond the national park ideal to a broader use of protected-area categories. "There has been a tendency," says D. A. Johnstone, the director of the National Parks and Wildlife Service of New South Wales, Australia, "to apply criteria developed in one or more countries to other countries, or similarly to apply criteria developed in the past to the problems of today. It is my contention that this is unecological. It is accepted that physical and ecological differences should be taken into account when formulating plans of management for different protected areas around the world; similarly, cultural variations also should be taken into account if ecologically sustainable results are to be achieved." Another Bali delegate, Arthur Lyon Dahl of New Caledonia, declares that "much more flexibility is needed in the approaches to the protection of species, ecosystems, and habitats. Most legislation in the [Oceanian] region was copied from often inappropriate examples elsewhere. . . . "[24]

Norman Myers, the noted conservation theorist, put the whole matter most cogently when he wrote that the whole notion of "setting aside" has in fact done great damage to the conservation movement around the world and that "far from being considered as 'set aside,' a park should be viewed as *brought into* the main arena of human affairs."[25]

The conventional national park idea is a good idea. It is, however, also a potentially dangerous idea because it is context-sensitive: even if the motives behind it were universal, its method would not be universally applicable. As merely a good idea, it is incumbent upon affluent nations to refrain from encouraging its wholesale adoption elsewhere, and to make sure that authorities (and not just in Third World countries) are aware of other, possibly more suitable, protected-area categories.

This prescription will be difficult to accept, for it seems almost unthinkable to play down success. But if we value cultural diversity in the world it is what we must do. The old saying holds that "Nothing succeeds like success." Sometimes, however, nothing fails like it.

NOTES

1. Jeremy Harrison, Jeffrey McNeely, and Kenton Miller, "The World Coverage of Protected Areas: Development Goals and Environmental Needs," in *National Parks, Conservation, and Development: The Role of Protected Areas in Sustaining Society, Proceedings of the World Congress on National Parks, Bali, Indonesia, 11–22 October 1982,* ed. Jeffrey A. McNeely and Kenton R. Miller (Washington, D.C.: Smithsonian Institution Press, 1984), p. 24. Hereafter cited as *NPCD.*

2. IUCN Commission on National Parks and Protected Areas, *1982 United Nations List of National Parks and Protected Areas* (Gland, Switzerland: IUCN, 1982), pp. 15–27.

3. Jeffrey A. McNeely, "Protected Areas are Adapting to New Realities," in *NPCD,* p. 1.

4. *1982 United Nations List,* pp. 35 ff.

5. Roderick Nash, "The American Invention of National Parks," *American Quarterly* 22, no. 3 (Fall 1970): 726.

6. Alfred Runte, *National Parks: The American Experience* (Lincoln: University of Nebraska Press, 1979), pp. 1–47.

7. More generally, the contention that preservationism has no basis in Western culture has been rebutted in Eugene C. Hargrove, "The Historical Foundations of American Environmental Attitudes," *Environmental Ethics* 1 (1979): 209–40.

8. Keith Garratt, "The Relationship between Adjacent Lands and Protected Areas: Issues of Concern for the Protected Area Manager," in *NPCD,* p. 66; Bernard Nietschmann, "Indigenous Island Peoples, Living Resources, and Protected Areas," in *NPCD,* p. 334.

9. Countries with long-established protected-area systems, such as Japan and New Zealand, openly acknowledge Yellowstone as the benchmark against which their first efforts were measured. Nash, "American Invention," p. 735.

10. The redemptive, missionary strain continues in the thinking of American park theorists today. Indeed, moral righteousness is a central theme of the best philosophical analysis of this country's national parks to have appeared in recent years, Joseph L. Sax's *Mountains Without Handrails: Reflections on the National Parks* (Ann Arbor: University of Michigan Press, 1980). See also A. Dan Turlock, "For Whom the National Parks?" *Stanford Law Review* 34, no. 255 (November 1981): 259–61.

11. John H. Bodley, *Victims of Progress,* 2d ed. (Menlo Park, Calif. Benjamin/Cummings, 1982), p. 44.

12. John Blower, "National Parks for Developing Countries," in *NPCD,* p. 723.

13. John B. Calhoun, "Plight of the Ik and Kaiadilt Is Seen as a Chilling Possible End for Man," *Smithsonian* 3, no. 8 (November 1972): 27–29. A full-length study of the Ik is Colin M. Turnbull, *The Mountain People* (New York: Simon and Schuster, 1972); see esp. p. 24. For the creation of Kidepo National Park, see F. Kayanaja and Iain Douglas-Hamilton, "The Impact of the Unexpected on the Uganda National Parks," in *NPCD,* p. 88.

14. Yi-Fu Tuan, "Geopiety: A Theme in Man's Attachment to Nature and to Place," in *Geographies of the Mind: Essays in Historical Geosophy in Honor of John Kirtland Wright,* ed. David Lowenthal and Martyn J. Bowden (New York: Oxford University Press, 1976), pp. 11–39.

15. David Western and Thomas Dunne, "Environmental Aspects of Settlement Site Decisions Among Pastoral Maasai," *Human Ecology* 7, no. 1 (1979): 75–98.

16. Cf. Nash's fourth reason, Nash, "American Invention," p. 726.

17. Hemanta R. Mishra, "A Delicate Balance: Tigers, Rhinoceros, Tourists and Park Management *vs.* The Needs of the Local People in Royal Chitwan National Park, Nepal," in *NPCD,* p. 200.

18. Ibid., p. 202.

19. John Foster, Adrian Phillips, and Richard Steele, "Protected Areas in the United Kingdom: An Approach to the Selection, Establishment, and Management of Natural and Scenic Protected Areas in a Densely Populated Country with Limited Choices," in *NPCD,* pp. 426–37; Masaaki Sakurai, "Adjustment Between Nature and Human Activity in National Parks in Japan," in *NPCD,* p. 480.

20. M. K. Ranjitsinh, "The Indomalayan Realm," in *NPCD,* p. 150.

21. The ideas in this section were developed in association with Steven R. Brechin of the University of Michigan School of Natural Resources. Much of this discussion is taken from "Resident Populations and Protected Areas: Management Concepts and Issues," a chapter we co-authored for *Resident Populations and National Parks in Developing Nations: Social Impacts and Alternative Management Concepts* (Ann Arbor: University of Michigan Natural Resource Sociology Research Laboratory Monograph Series No. 5, in press). However, I have made numerous changes and the conclusions here are mine alone.

22. Robert Goodland, *Tribal Peoples and Economic Development* (Washington, D.C.: International Bank for Reconstruction and Development, 1982), p. vii.

23. World National Parks Congress, *Declaration,* Bali, Indonesia, 1982.

24. D. A. Johnstone, "Future Directions for the Australian Realm," in *NPCD,* p. 302; Arthur Lyon Dahl, "Future Directions for the Oceanian Realm," in *NPCD,* pp. 360–61.

25. Norman Myers, "Eternal Values of the Parks Movement and the Monday Morning World," in *NPCD,* p. 657.

Ramachandra Guha

Radical American Environmentalism and Wilderness Preservation (1989)

A Third World Critique

Even God dare not appear to the poor man
except in the form of bread.
Mahatma Gandhi

I. INTRODUCTION

THE RESPECTED RADICAL JOURNALIST Kirkpatrick Sale recently celebrated "the passion of a new and growing movement that has become disenchanted with the environmental establishment and has in recent years mounted a serious and sweeping attack on it—style, substance, systems, sensibilities and all."[1] The vision of those whom Sale calls the "New Ecologists"—and what I refer to in this article as deep ecology—is a compelling one. Decrying the narrowly economic goals of mainstream environmentalism, this new movement aims at nothing less than a philosophical and cultural revolution in human attitudes toward nature. In contrast to the conventional lobbying efforts of environmental professionals based in Washington, it proposes a militant defence of "Mother Earth," an unflinching opposition to human attacks on undisturbed wilderness. With their goals ranging from the spiritual to the political, the adherents of deep ecology span a wide spectrum of the American environmental movement. As Sale correctly notes, this emerging strand has in a matter of a few years made its presence felt in a number of fields: from academic philosophy (as in the journal *Environmental Ethics*) to popular environmentalism (for example, the group Earth First!).

In this article I develop a critique of deep ecology from the perspective of a sympathetic outsider. I critique deep ecology not as a general (or even a foot soldier) in the continuing struggle between the ghosts of Gifford Pinchot and John Muir over control of the U.S. environmental movement, but as an outsider to these battles. I speak admittedly as a partisan, but of the environmental movement in India, a country with an ecological diversity comparable to the U.S., but with a radically dissimilar cultural and social history.

My treatment of deep ecology is primarily historical and sociological, rather than philosophical, in nature. Specifically, I examine the cultural rootedness of a philosophy that likes to present itself in universalistic terms. I make two main arguments: first, that deep ecology is uniquely American, and despite superficial similarities in rhetorical style, the social and political goals of radical environmentalism in other cultural contexts (e.g., West Germany and India) are quite different; second, that the social consequences of putting deep ecology into practice on a worldwide basis (what its practitioners are aiming for) are very grave indeed.

II. THE TENETS OF DEEP ECOLOGY

While I am aware that the term *deep ecology* was coined by the Norwegian philosopher Arne Naess, this article refers specifically to the American variant.[2] Adherents of the deep ecological perspective in this country, while arguing intensely among themselves over its political and philosophical implications, share some fundamental premises about human-nature interactions. As I see it, the defining characteristics of deep ecology are fourfold:

First, deep ecology argues, that the environmental movement must shift from an "anthropocentric" to a "biocentric" perspective. In many respects, an acceptance of the primacy of this distinction constitutes the litmus test of deep ecology. A considerable effort is expended by deep ecologists in showing that the dominant motif in Western philosophy has been anthropocentric—i.e., the belief that man and his works are the center of the universe—and conversely, in identifying those lonely thinkers (Leopold, Thoreau, Muir, Aldous Huxley, Santayana, etc.) who, in assigning man a more humble place in the natural order, anticipated deep ecological thinking. In the political realm, meanwhile, establishment environmentalism (shallow ecology) is chided for casting its arguments in human-centered terms. Preserving nature, the deep ecologists say, has an intrinsic worth quite apart from any benefits preservation may convey to future human generations. The anthropocentric-biocentric distinction is accepted as axiomatic by deep ecologists, it structures their discourse, and much of the present discussion remains mired within it.

The second characteristic of deep ecology is its focus on the preservation of unspoilt wilderness—and the restoration of degraded areas to a more pristine condition—to the relative (and sometimes absolute) neglect of other issues on the environmental agenda. I later identify the cultural roots and portentous consequences of this obsession with wilderness. For the moment, let me indicate three distinct sources from which it springs. Historically, it represents a playing out of the preservationist (read *radical*) and utilitarian (read *reformist*) dichotomy that has plagued American environmentalism since the turn of the century. Morally, it is an imperative that follows from the biocentric perspective; other species of plants and animals, and nature itself, have an intrinsic right to exist. And finally, the preservation of wilderness also turns on a scientific argument—viz., the value of biological diversity in stabilizing ecological regimes and in retaining a gene pool for future generations. Truly radical policy proposals have been put forward by deep ecologists on the basis of these arguments. The influential poet Gary Snyder, for example, would like to see a 90 percent reduction in human populations to allow a restoration of pristine environments, while others have argued forcefully that a large portion of the globe must be immediately cordoned off from human beings.[3]

Third, there is a widespread invocation of Eastern spiritual traditions as forerunners of deep ecology. Deep ecology, it is suggested, was practiced both by major religious traditions and at a more popular level by "primal" peoples in non-Western settings. This complements the search for an authentic lineage in Western thought. At one level, the task is to recover those dissenting voices within the Judeo-Christian tradition; at another, to suggest that religious traditions in other cultures are, in contrast, dominantly if not exclusively "biocentric" in their orientation. This coupling of (ancient) Eastern and (modern) ecological wisdom seemingly helps consolidate the claim that deep ecology is a philosophy of universal significance.

Fourth, deep ecologists, whatever their internal differences, share the belief that they are the "leading edge" of the environmental movement. As the polarity of the shallow/deep and anthropocentric/biocentric distinctions makes clear, they see themselves as the spiritual, philosophical, and political vanguard of American and world environmentalism.

III. TOWARD A CRITIQUE

Although I analyze each of these tenets independently, it is important to recognize, as deep ecologists are fond of remarking in reference to nature, the interconnectedness and unity of these individual themes.

1. Insofar as it has begun to act as a check on man's arrogance and ecological hubris, the transition from an anthropocentric (human-centered) to a biocentric (humans as only one element in the ecosystem) view in both religious and scientific traditions is only to be welcomed.[4] What is unacceptable are the radical conclusions drawn by deep ecology, in particular, that intervention in nature should be guided primarily by the need to preserve biotic integrity rather than by the needs of humans. The latter for deep ecologists is anthropocentric, the former biocentric. This dichotomy is, however, of very little use in understanding the dynamics of environmental degradation. The two fundamental ecological problems facing the globe are (i) overconsumption by the industrialized world and by urban elites in the Third World and (ii) growing militarization, both in a short-term sense (i.e., ongoing regional wars) and in a long-term sense (i.e., the arms race and the prospect of nuclear annihilation). Neither of these problems has any tangible connection to the anthropocentric-biocentric distinction. Indeed, the agents of these processes would barely comprehend this philosophical dichotomy. The proximate causes of the ecologically wasteful characteristics of industrial society and of militarization are far more mundane: at an aggregate level, the dialectic of economic and political structures, and at a micro-level, the life style choices of individuals. These causes cannot be reduced, whatever the level of analysis, to a deeper anthropocentric attitude toward nature; on the contrary, by constituting a

grave threat to human survival, the ecological degradation they cause does not even serve the best interests of human beings! If my identification of the major dangers to the integrity of the natural world is correct, invoking the bogy of anthropocentricism is at best irrelevant and at worst a dangerous obfuscation.

2. If the above dichotomy is irrelevant, the emphasis on wilderness is positively harmful when applied to the Third World. If in the U.S. the preservationist/utilitarian division is seen as mirroring the conflict between "people" and "interests," in countries such as India the situation is very nearly the reverse. Because India is a long settled and densely populated country in which agrarian populations have a finely balanced relationship with nature, the setting aside of wilderness areas has resulted in a direct transfer of resources from the poor to the rich. Thus, Project Tiger, a network of parks hailed by the international conservation community as an outstanding success, sharply posits the interests of the tiger against those of poor peasants living in and around the reserve. The designation of tiger reserves was made possible only by the physical displacement of existing villages and their inhabitants; their management requires the continuing exclusion of peasants and livestock. The initial impetus for setting up parks for the tiger and other large mammals such as the rhinoceros and elephant came from two social groups, first, a class of ex-hunters turned conservationists belonging mostly to the declining Indian feudal elite and second, representatives of international agencies, such as the World Wildlife Fund (WWF) and the International Union for the Conservation of Nature and Natural Resources (IUCN), seeking to transplant the American system of national parks onto Indian soil. In no case have the needs of the local population been taken into account, and as in many parts of Africa, the designated wildlands are managed primarily for the benefit of rich tourists. Until very recently, wildlands preservation has been identified with environmentalism by the state and the conservation elite; in consequence, environmental problems that impinge far more directly on the lives of the poor—e.g., fuel, fodder, water shortages, soil erosion, and air and water pollution—have not been adequately addressed.[5]

Deep ecology provides, perhaps unwittingly, a justification for the continuation of such narrow and inequitable conservation practices under a newly acquired radical guise. Increasingly, the international conservation elite is using the philosophical, moral, and scientific arguments used by deep ecologists in advancing their wilderness crusade. A striking but by no means atypical example is the recent plea by a prominent American biologist for the takeover of large portions of the globe by the author and his scientific colleagues. Writing in a prestigious scientific forum, the *Annual Review of Ecology and Systematics,* Daniel Janzen argues that only biologists have the competence to decide how the tropical landscape should be used. As "the representatives of the natural world," biologists are "in charge of the future of tropical ecology," and only they have the expertise and mandate to "determine whether the tropical agroscape is to be populated only by humans, their mutualists, commensals, and parasites, or whether it will also contain some islands of the greater nature—the nature that spawned humans, yet has been vanquished by them." Janzen exhorts his colleagues to advance their territorial claims on the tropical world more forcefully, warning that the very existence of these areas is at stake: "if biologists want a tropics in which to biologize, they are going to have to buy it with care, energy, effort, strategy, tactics, time, and cash."[6]

This frankly imperialist manifesto highlights the multiple dangers of the preoccupation with wilderness preservation that is characteristic of deep ecology. As I have suggested, it seriously compounds the neglect by the American movement of far more pressing environmental problems within the Third World. But perhaps more importantly, and in a more insidious fashion, it also provides an impetus to the imperialist yearning of Western biologists and their financial sponsors, organizations such as the WWF and IUCN. The wholesale transfer of a movement culturally rooted in American conservation history can only result in the social uprooting of human populations in other parts of the globe.

3. I come now to the persistent invocation of Eastern philosophies as antecedent in point of time but convergent in their structure with deep ecology. Complex and internally differentiated religious traditions—Hinduism, Buddhism, and Taoism—are lumped together as holding a view of nature believed to be quintessentially biocentric. Individual philosophers such as the Taoist Lao Tzu are identified as being forerunners of deep ecology. Even an intensely political, pragmatic, and Christian influenced thinker such as Gandhi has been accorded a wholly undeserved place in the deep ecological pantheon. Thus the Zen teacher Robert Aitken Roshi makes the strange claim that Gandhi's thought was not human-centered and that he practiced an embryonic form of deep ecology which is "tradi-

tionally Eastern and is found with differing emphasis in Hinduism, Taoism and in Theravada and Mahayana Buddhism."[7] Moving away from the realm of high philosophy and scriptural religion, deep ecologists make the further claim that at the level of material and spiritual practice "primal" peoples subordinated themselves to the integrity of the biotic universe they inhabited.

I have indicated that this appropriation of Eastern traditions is in part dictated by the need to construct an authentic lineage and in part a desire to present deep ecology as a universalistic philosophy. Indeed, in his substantial and quixotic biography of John Muir, Michael Cohen goes so far as to suggest that Muir was the "Taoist of the [American] West."[8] This reading of Eastern traditions is selective and does not bother to differentiate between alternate (and changing) religious and cultural traditions; as it stands, it does considerable violence to the historical record. Throughout most recorded history the characteristic form of human activity in the "East" has been a finely tuned but nonetheless conscious and dynamic manipulation of nature. Although mystics such as Lao Tzu did reflect on the spiritual essence of human relations with nature, it must be recognized that such ascetics and their reflections were supported by a society of cultivators whose relationship with nature was a far more *active* one. Many agricultural communities do have a sophisticated knowledge of the natural environment that may equal (and sometimes surpass) codified "scientific" knowledge; yet, the elaboration of such traditional ecological knowledge (in both material and spiritual contexts) can hardly be said to rest on a mystical affinity with nature of a deep ecological kind. Nor is such knowledge infallible; as the archaeological record powerfully suggests, modern Western man has no monopoly on ecological disasters.

In a brilliant article, the Chicago historian Ronald Inden points out that this romantic and essentially positive view of the East is a mirror image of the scientific and essentially pejorative view normally upheld by Western scholars of the Orient. In either case, the East constitutes the Other, a body wholly separate and alien from the West; it is defined by a uniquely spiritual and nonrational "essence," even if this essence is valorized quite differently by the two schools. Eastern man exhibits a spiritual dependence with respect to nature—on the one hand, this is symptomatic of his prescientific and backward self, on the other, of his ecological wisdom and deep ecological consciousness. Both views are monolithic, simplistic, and have the characteristic effect—intended in one case, perhaps unintended in the other—of denying agency and reason to the East and making it the privileged orbit of Western thinkers.

The two apparently opposed perspectives have then a common underlying structure of discourse in which the East merely serves as a vehicle for Western projections. Varying images of the East are raw material for political and cultural battles being played out in the West; they tell us far more about the Western commentator and his desires than about the "East." Inden's remarks apply not merely to Western scholarship on India, but to Orientalist constructions of China and Japan as well.

> Although these two views appear to be strongly opposed, they often combine together. Both have a similar interest in sustaining the Otherness of India. The holders of the dominant view, best exemplified in the past in imperial administrative discourse (and today probably by that of "development economics"), would place a traditional, superstition-ridden India in a position of perpetual tutelage to a modern, rational West. The adherents of the romantic view, best exemplified academically in the discourses of Christian liberalism and analytic psychology, concede the realm of the public and impersonal to the positivist. Taking their succour not from governments and big business, but from a plethora of religious foundations and self-help institutes, and from allies in the "consciousness industry," not to mention the important industry of tourism, the romantics insist that India embodies a private realm of the imagination and the religious which modern, western man lacks but needs. They, therefore, like the positivists, but for just the opposite reason, have a vested interest in seeing that the Orientalist view of India as "spiritual," "mysterious," and "exotic" is perpetuated.[9]

4. How radical, finally, are the deep ecologists? Notwithstanding their self-image and strident rhetoric (in which the label "shallow ecology" has an opprobrium similar to that reserved for "social democratic" by Marxist-Leninists), even within the American context their radicalism is limited and it manifests itself quite differently elsewhere.

To my mind, deep ecology is best viewed as a radical trend within the wilderness preservation movement. Although advancing philosophical rather than aesthetic arguments and encouraging political militancy rather than negotiation, its practical emphasis—viz., preservation of unspoilt nature—is virtually identical. For the mainstream movement, the function of wilderness is to provide a temporary antidote to modern civilization. As a special institution within an industrialized society, the national park "provides an opportunity for respite, contrast, contemplation, and affirmation of values for those who live most of their lives in the workaday world."[10] Indeed, the rapid increase in visitations to the national parks in postwar America is a direct consequence of economic expansion. The emergence of a popular interest in wilderness sites, the historian Samuel Hays points out, was "not a throwback to the primitive, but an integral part of the modern standard of living as people sought to add new 'amenity' and 'aesthetic' goals and desires to their earlier preoccupation with necessities and convenience."[11]

Here, the enjoyment of nature is an integral part of the consumer society. The private automobile (and the life style it has spawned) is in many respects the ultimate ecological villain, and an untouched wilderness the prototype of ecological harmony; yet, for most Americans it is perfectly consistent to drive a thousand miles to spend a holiday in a national park. They possess a vast, beautiful, and sparsely populated continent and are also able to draw upon the natural resources of large portions of the globe by virtue of their economic and political dominance. In consequence, America can simultaneously enjoy the material benefits of an expanding economy and the aesthetic benefits of unspoilt nature. The two poles of "wilderness" and "civilization" mutually coexist in an internally coherent whole, and philosophers of both poles are assigned a prominent place in this culture. Paradoxical as it may seem, it is no accident that Star Wars technology and deep ecology both find their fullest expression in that leading sector of Western civilization, California.

Deep ecology runs parallel to the consumer society without seriously questioning its ecological and socio-political basis. In its celebration of American wilderness, it also displays an uncomfortable convergence with the prevailing climate of nationalism in the American wilderness movement. For spokesmen such as the historian Roderick Nash, the national park system is America's distinctive cultural contribution to the world, reflective not merely of its economic but of its philosophical and ecological maturity as well. In what Walter Lippman called the American century, the "American invention of national parks" must be exported worldwide. Betraying an economic determinism that would make even a Marxist shudder, Nash believes that environmental preservation is a "full stomach" phenomenon that is confined to the rich, urban, and sophisticated. Nonetheless, he hopes that "the less developed nations may eventually evolve economically and intellectually to the point where nature preservation is more than a business."[12]

The error which Nash makes (and which deep ecology in some respects encourages) is to equate environmental protection with the protection of wilderness. This is a distinctively American notion, born out of a unique social and environmental history. The archetypal concerns of radical environmentalists in other cultural contexts are in fact quite different. The German Greens, for example, have elaborated a devastating critique of industrial society which turns on the acceptance of environmental limits to growth. Pointing to the intimate links between industrialization, militarization, and conquest, the Greens argue that economic growth in the West has historically rested on the economic and ecological exploitation of the Third World. Rudolf Bahro is characteristically blunt:

> The working class here [in the West] is the richest lower class in the world. And if I look at the problem from the point of view of the whole of humanity, not just from that of Europe, then I must say that the metropolitan working class is the worst exploiting class in history. . . . What made poverty bearable in eighteenth or nineteenth-century Europe was the prospect of escaping it through exploitation of the periphery. But this is no longer a possibility, and continued industrialism in the Third World will mean poverty for whole generations and hunger for millions.[13]

Here the roots of global ecological problems lie in the disproportionate share of resources consumed by the industrialized countries as a whole *and* the urban elite within the Third World. Since it is impossible to reproduce an industrial monoculture worldwide, the ecological movement in the West must begin by cleaning up its own act. The Greens advocate the creation of a "no growth" economy, to be achieved by scaling down current (and clearly unsustainable) consumption levels.[14] This radical shift in consumption and production patterns requires the creation of alternate economic and political structures—smaller in scale and more amenable to social participation—but it rests equally on a shift in cultural values. The expansionist character of modern Western man will have to give way to an ethic of renunciation and self-limitation, in which spiritual and communal values play an increasing role in sustaining social life. This revolution in cultural values, however, has as its point of departure an understanding of environmental processes quite different from deep ecology.

Many elements of the Green program find a strong resonance in countries such as India, where a history of Western colonialism and industrial development has benefited only a tiny elite while exacting tremendous social and environmental costs. The ecological battles presently being fought in India have as their epicenter the conflict over nature between the subsistence and largely rural sector and the vastly more powerful commercial-industrial sector. Perhaps the most celebrated of these battles concerns the Chipko (Hug the Tree) movement, a peasant movement against deforestation in the Himalayan foothills. Chipko is only one of several movements that have sharply questioned the nonsustainable demand being placed on the land and vegetative base by urban centers and industry. These include opposition to large dams by displaced peasants, the conflict between small artisan fishing and large-scale trawler fishing for export, the countrywide movements against commercial forest operations, and opposition to industrial pollution among downstream agricultural and fishing communities.[15]

Two features distinguish these environmental movements from their Western counterparts. First, for the sections of society most critically affected by environmental degradation—poor and landless peasants, women, and tribals—it is a question of sheer survival, not of enhancing the quality of life. Second, and as a consequence, the environmental solutions they articulate deeply involve questions of equity as well as economic and political redistribution. Highlighting these differences, a leading Indian environmentalist stresses that "environmental protection per se is of least concern to most of these groups. Their main concern is about the use of the environment and who should benefit from it."[16] They seek to wrest control of nature away from the state and the industrial sector and place it in the hands of rural communities who live within that environment but are increasingly denied access to it. These communities have far more basic needs, their demands on the environment are far less intense, and they can draw upon a reservoir of cooperative social institutions and local ecological knowledge in managing the "commons"—forests, grasslands, and the waters—on a sustainable basis. If colonial and capitalist expansion has both accentuated social inequalities and signaled a precipitous fall in ecological wisdom, an alternate ecology must rest on an alternate society and polity as well.

This brief overview of German and Indian environmentalism has some major implications for deep ecology. Both German and Indian environmental traditions allow for a greater integration of ecological concerns with livelihood and work. They also place a greater emphasis on equity and social justice (both within individual countries and on a global scale) on the grounds that in the absence of social regeneration environmental regeneration has very little chance of succeeding. Finally, and perhaps most significantly, they have escaped the preoccupation with wilderness preservation so characteristic of American cultural and environmental history.[17]

IV. A HOMILY

In 1958, the economist J. K. Galbraith referred to overconsumption as the unasked question of the American conservation movement. There is a marked selectivity, he wrote, "in the conservationist's approach to materials consumption. If we are concerned about our great appetite for materials, it is plausible to seek to increase the supply, to decrease waste, to make better use of the stocks available, and to develop substitutes. But what of the appetite itself? Surely this is the ultimate source of the problem. If it continues its geometric course, will it not one day have to be restrained? Yet in the literature of the resource problem this is the forbidden question. Over it hangs a nearly total silence."[18]

The consumer economy and society have expanded tremendously in the three decades since Galbraith penned these words; yet his criticisms are nearly as valid today. I have said "nearly," for there are some hopeful signs. Within the environmental movement several dispersed groups are working to develop ecologically benign technologies and to encourage less wasteful life styles. Moreover, outside the self-defined boundaries of American environmentalism, opposition to the permanent war economy is being carried on by a peace movement that has a distinguished history and impeccable moral and political credentials.

It is precisely these (to my mind, most hopeful) components of the American social scene that are missing from deep ecology. In their widely noticed book, Bill Devall and George Sessions make no mention of militarization or the movements for peace, while activists whose practical focus is on developing ecologically responsible life styles (e.g., Wendell Berry) are derided as "falling short of deep ecological awareness."[19] A truly radical ecology in the American context ought to work toward a synthesis of the appropriate technology, alternate life style, and peace movements.[20] By making the (largely spurious) anthropocentric-biocentric distinction central to the debate, deep ecologists may have appropriated the moral high ground, but they are at the same time doing a serious disservice to American and global environmentalism.[21]

NOTES

The author is grateful to Mike Bell, Tom Birch, Bill Burch, Bill Cronon, Diane Meyerfeld, David Rothenberg, Kirkpatrick Sale, Joel Seton, Tim Weiskel, and Don Worster for helpful comments.

1. Kirkpatrick Sale, "The Forest for the Trees: Can Today's Environmentalists Tell the Difference," *Mother Jones* 11, no. 8 (November 1986): 26.

2. One of the major criticisms I make in this essay concerns deep ecology's lack of concern with inequalities *within* human society. In the article in which he coined the term *deep ecology,* Naess himself expresses concerns about inequalities between and within nations. However, his concern with social cleavages and their impact on resource utilization patterns and ecological destruction is not very visible in the later writings of deep ecologists. See Arne Naess, "The Shallow and the Deep, Long-Range Ecology Movement: A Summary," *Inquiry* 16 (1973): 96.

3. Gary Snyder, quoted in Sale, "The Forest for the Trees," p. 32. See also Dave Foreman, "A Modest Proposal for a Wilderness System," *Whole Earth Review,* no. 53 (Winter 1986–87): 42–45.

4. See, for example, Donald Worster, *Nature's Economy: The Roots of Ecology* (San Francisco: Sierra Club Books, 1977).

5. See Centre for Science and Environment, *India: The State of the Environment 1982: A Citizens Report* (New Delhi: Centre for Science and Environment, 1982); R. Sukumar, "Elephant-Man Conflict in Karnataka," in Cecil Saldanha, ed., *The State of Karnataka's Environment* (Bangalore: Centre for Taxonomic Studies, 1985). For Africa, see the brilliant analysis by Helge Kjekshus, *Ecology Control and Economic Development in East African History* (Berkeley: University of California Press, 1977).

6. Daniel Janzen, "The Future of Tropical Ecology," *Annual Review of Ecology and Systematics* 17 (1986): 305–6; emphasis added.

7. Robert Aitken Roshi, "Gandhi, Dogen, and Deep Ecology," reprinted as appendix C in Bill Devall and George Sessions, *Deep Ecology: Living as if Nature Mattered* (Salt Lake City: Peregrine Smith Books, 1985). For Gandhi's own views on social reconstruction, see the excellent three volume collection edited by Raghavan Iyer, *The Moral and Political Writings of Mahatma Gandhi* (Oxford: Clarendon Press, 1986–87).

8. Michael Cohen, *The Pathless Way* (Madison: University of Wisconsin Press, 1984), p. 120.

9. Ronald Inden, "Orientalist Constructions of India," *Modern Asian Studies* 20 (1986): 442. Inden draws inspiration from Edward Said's forceful polemic, *Orien-*

talism (New York: Basic Books, 1980). It must be noted, however, that there is a salient difference between Western perceptions of Middle Eastern and Far Eastern cultures respectively. Due perhaps to the long history of Christian conflict with Islam, Middle Eastern cultures (as Said documents) are consistently presented in pejorative terms. The juxtaposition of hostile and worshipping attitudes that Inden talks of applies only to Western attitudes toward Buddhist and Hindu societies.

10. Joseph Sax, *Mountains Without Handrails: Reflections on the National Parks* (Ann Arbor: University of Michigan Press, 1980), p. 42. Cf. also Peter Schmitt, *Back to Nature: The Arcadian Myth in Urban America* (New York: Oxford University Press, 1969), and Alfred Runte, *National Parks: The American Experience* (Lincoln: University of Nebraska Press, 1979).

11. Samuel Hays, "From Conservation to Environment: Environmental Politics in the United States Since World War Two," *Environmental Review* 6 (1982): 21. See also the same author's book entitled *Beauty, Health and Permanence: Environmental Politics in the United States, 1955–85* (New York: Cambridge University Press, 1987).

12. Roderick Nash, *Wilderness and the American Mind,* 3rd ed. (New Haven: Yale University Press, 1982).

13. Rudolf Bahro, *From Red to Green* (London: Verso Books, 1984).

14. From time to time, American scholars have themselves criticized these imbalances in consumption patterns. In the 1950s, William Vogt made the charge that the United States, with one-sixteenth of the world's population, was utilizing one-third of the globe's resources. (Vogt, cited in E. F. Murphy, *Nature, Bureaucracy and the Rule of Property* [Amsterdam: North Holland, 1977, p. 29]). More recently, Zero Population Growth has estimated that each American consumes thirty-nine times as many resources as an Indian. See *Christian Science Monitor,* 2 March 1987.

15. For an excellent review, see Anil Agarwal and Sunita Narain, eds., *India: The State of the Environment 1984–85: A Citizens Report* (New Delhi: Centre for Science and Environment, 1985). Cf. also Ramachandra Guha, *The Unquiet Woods: Ecological Change and Peasant Resistance in the Indian Himalaya* (Berkeley: University of California Press, forthcoming).

16. Anil Agarwal, "Human-Nature Interactions in a Third World Country," *The Environmentalist* 6, no. 3 (1986): 167.

17. One strand in radical American environmentalism, the bioregional movement, by emphasizing a greater involvement with the bioregion people inhabit, does indirectly challenge consumerism. However, as yet bioregionalism has hardly raised the questions of equity and social justice (international, intranational, and intergenerational) which I argue must be a central plank of radical environmentalism. Moreover, its stress on (individual) *experience* as the key to involvement with nature is also somewhat at odds with the integration of nature with livelihood

and work that I talk of in this paper. Cf. Kirkpatrick Sale, *Dwellers in the Land: The Bioregional Vision* (San Francisco: Sierra Club Books, 1985).

18. John Kenneth Galbraith, "How Much Should a Country Consume?" in Henry Jarrett, ed., *Perspectives on Conservation* (Baltimore: Johns Hopkins University Press, 1958), pp. 91–92.

19. Devall and Sessions, *Deep Ecology,* p. 122. For Wendell Berry's own assessment of deep ecology, see his "Amplifications: Preserving Wildness," *Wilderness* 50 (Spring 1987): 39–40, 50–54.

20. See the interesting recent contribution by one of the most influential spokesmen of appropriate technology—Commoner, "A Reporter at Large: The Environment," *New Yorker,* 15 June 1987. While Commoner makes a forceful plea for the convergence of the environmental movement (viewed by him primarily as the opposition to air and water pollution and to the institutions that generate such pollution) and the peace movement, he significantly does not mention consumption patterns, implying that "limits to growth" do not exist.

21. In this sense, my critique of deep ecology, although that of an outsider, may facilitate the reassertion of those elements in the American environmental tradition for which there is a profound sympathy in other parts of the globe. A global perspective may also lead to a critical reassessment of figures such as Aldo Leopold and John Muir, the two patron saints of deep ecology. As Donald Worster has pointed out, the message of Muir (and, I would argue, of Leopold as well) makes sense only in an American context: he has very little to say to other cultures. See Worster's review of Stephen Fox's *John Muir and His Legacy,* in *Environmental Ethics* 5 (1983): 277–81.

David M. Johns

The Relevance of Deep Ecology to the Third World (1990)

Some Preliminary Comments

INTRODUCTION

The appearance in *Environmental Ethics* of work by Third World environmentalists is to be welcomed, and should be greatly encouraged. If the movements for environmental protection anywhere in the world are to be relevant, they must address issues within the global framework. This can only be done in conjunction with and by engaging other movements around the globe. Only through the genuine amalgamation of the various and specific historical experiences can we move toward a new direction(s) for human society. Ramachandra Guha's "sympathetic critique" of deep ecology is an important step and a good example of the necessity of such exchanges, for he raises several issues concerning the tenets of deep ecology that are most easily visible from outside the Western industrial world.[1]

In this paper, I comment on two of these issues: wilderness preservation and the usefulness of the anthropocentric/biocentric distinction.

WILDERNESS: ORIGINS AND VALUES

Guha criticizes deep ecology for equating environmental protection with wilderness preservation. This flaw, he argues, is due to deep ecology's lack of awareness of its historical roots and limitations. Adherence to this position, he goes on to argue, actually obscures the real sources of environmental degradation and thus helps to perpetuate the existing order. Moreover, deep ecology fails to recognize the impact of its commitment to wilderness in the Third World. I discuss each of these in turn.

Deep ecology is obviously rooted in the culture of those who espouse it; this is the case of every movement. The very process of transcendence or dialectical working out, moreover, assumes a history; nevertheless, simply pointing out the origins of a particular historical experience does not invalidate it. There is no question that the circumstances of development in the United States—including the pattern of settlement over the huge geographical area available—have helped to shape the response to environmental degradation, e.g., an emphasis on wilderness preservation. There is also little question that in many respects the existence of wilderness may "fit in" with the cultural categories of a consumer society, as a retreat from the insanity of (sub)urban-industrial life—an alternative that only a country like the U.S. can afford, inheriting as it did a virtually unexploited continent still underpopulated compared to the rest of the world, and living off wealth extracted throughout the world.

To the contrary, however, while there may be some cultural fit between wilderness and the existing order that results from the particular experience of material development in the U.S., in most respects it does not "fit." From the very beginning and increasingly so the wilderness system, wildlife refuges, and old-growth forests are under fierce attacks by those who say we cannot afford them because they undermine the viability of an economy based on endless growth.

The real issues are whether Guha's claim that environmental protection means protection of wilderness is an accurate description of deep ecology and whether such a position is wrongheaded in substance. Related to these issues is a larger question: how should humans interact with the rest of the biosphere, and must wilderness preservation stand in opposition to an approach that integrates human livelihood and environmental protection?

I believe that Guha is partly wrong in stating that deep ecology equates environmental protection with wilderness protection and simply wrong in calling wilderness protection untenable or incorrect as a global strategy for environmental protection. The deep ecological support for wilderness is predicated upon two important fact/values: (1) that the Earth can support a limited amount of biomass and the more of it that is composed of humans or turned to human use, the less is available for other life, and (2) that humans do not have the right to so alter the composition of the biomass that there is a resulting destruction, in Leopold's words, of "the integrity, stability and beauty" of the ecosystem. The basis for these values may lie in the experience of Self-realization or through identification with nature as the real community of which one is a part. Whether it is called a transcendence of alienation in its various forms or the healing of a crippled heart, it is supposed to support the claim that human life is no more valuable than any other form of life, life being broadly construed to include plants, animals, ecosystems, rivers, mountains, and the Earth.

Associated with this understanding of the human/Earth relationship is a recognition that in much of the world almost any human impact is destructive of the biosphere. In many ecosystems human livelihood—beyond very minimal numbers and very limited technology—is simply not compatible with maintaining the integrity of the biosphere. Such situations are most obvious when one looks at the fate of other large mammals. Ecosystems must normally be healthy to support them. Their disappearance is a good indication of degradation. Grizzly bears, orangutans, tigers, elephants, and many other species cannot easily coexist with humans in any numbers or with very exploitative technologies. Many ecosystems, moreover, cannot easily accommodate significant human presence without serious deterioration in diversity and balance. Recognition of other species, of ecosystems, and the Earth as valuable in and of themselves, individually or collectively, apart from their usefulness to humans, means that in practice much of the Earth cannot be used for permanent human settlement. Existing devastation, the ever increasing spread of humans into new areas, and the nature of those human incursions, makes the task of protection of areas still in their natural state ever more urgent. Returning large areas to wilderness is only slightly less urgent.

LIVELIHOOD AND WILDERNESS

While preservation of wilderness may seem to be the overriding focus of deep ecology given the ever accelerating destruction of species, ecosystems, and possibly the planet itself, there is a profound recognition that humans have their place *in* nature as well. With regard to places where it is appropriate for humans to settle, how to combine livelihood with environmental integrity is a major emphasis and how to move toward the reestablishment of a real community, embedded in the local ecosystem is a priority of the deep ecology movement. While it may be a valid criticism that much of the thinking in this area is fuzzy, naive, or falls victim to mystification, it is not true that wilderness is the single goal of deep ecologists. Given the human-nature relationship that deep ecology espouses—that to be effective in allowing nature to heal itself, one must also heal one's own self and community—it seems odd to suggest that deep ecology is unconcerned with human communities and their place in nature.[2]

SOURCES OF ENVIRONMENTAL DEGRADATION

Another criticism that Guha makes of deep ecology is that the focus on humans in general as the problem obscures the real causes of environmental degradation, namely overconsumption and militarization. Although his criticism has much merit, I believe he overstates the case. There can be no doubt that in explaining the particular developments that have resulted in so much destruction over the past two to three hundred years, industrialization, imperialism, overconsumption in the developed world, and the huge commitment of resources to armaments, are paramount. Guha is correct when he says that many in the movement see the problem as simply too many people behaving stupidly, without any regard for the nature of the system in which they live, its dynamics, and the fact that it victimizes most people as well as nature. Because it is probably true that most people who are victimizing nature are themselves victims of the social order, he is right in suggesting that the obstacles to significant changes in the relationship to nature are structural, not simply a matter of altering one's world view.

Yet, for every bit of evidence that this criticism is valid, there is evidence that it is only partially valid. Deep ecology and the German Greens do not see things as differently as Guha suggests. Indeed, I believe the Green movement in Germany and in particular Bahro have informed the deep ecology movement in the U.S.[3] There is a widespread recognition within deep ecology of the great inequality that exists in the world with regard to consumption, great differences in the existing power of various groups to shape a society's relationship with nature, and a recognition that the solution to ecological problems must address the issues of class, gender, and ethnicity. In addition, there is a recognition that all forms of domination are linked, as evidenced by the ongoing debates between deep ecology and social ecology, between deep ecology and ecofeminism, and between deep ecology and Marxism and other socialisms.

The nature of these linkages is certainly not settled; nonetheless, deep ecology may be distinctive in believing that the resolution of equity issues among humans will not automatically result in an end to the human destruction of the biosphere. One can envision, depending upon the theoretical version chosen, a society without class distinctions, without patriarchy, and with cultural autonomy that still attempts to control or manage the rest of nature in a utilitarian fashion that results in the deterioration of the biosphere. Such social changes would probably lessen the destructiveness of the relationship for two reasons: much of the technology of the last three hundred years is incompatible with a truly egalitarian society and much of the alienation that distorts the expression of human energy into schemes of control would not exist. Thus, it can be argued that although a significant change in the way humans relate to each other is a necessary component in a changed relationship to nature, it is not sufficient to bring about, by itself, the recognition and inclusion of nature as part of the moral community.

Deep ecologists point out, correctly I believe, that in terms of the integrity of an ecosystem, it makes little difference if an old-growth forest and its inhabitants are destroyed to build one house for a North American or fifty simple structures in the Third World. From a strictly human standpoint the latter is much more justifiable than the former. Nevertheless, even if North Americans were to sharply restrict their consumption, the fact remains that it is human numbers as well as levels of consumption that count.[4] There is, I believe, widespread agreement among Greens and deep ecologists that fewer humans (and especially less extensive occupation of the globe) as well as equitable and drastically curtailed consumption are essential to restoring the balance of the planet.

Guha does have much to tell us about the situation in the Third World and the problems of wilderness preservation there. While those of us engaged in political activity in North America are used to confronting the issue of jobs versus environment, it is important to understand that in the Third World "jobs" often equates with actual survival. While sparing old growth in the U.S. within the existing economic structure may cause hardship, sparing tropical forests within the existing economic structure may mean immediate hunger. (However, clearing tropical forests may mean eventual hunger as well, depending on the quality of the land cleared.) What Guha is telling us is that efforts to protect the environment by establishing wilderness areas in the Third World hurt the poorest of the poor—they are just more examples of imperialism, the same imperialism that pushes the poor and others into the wilderness in the first place.

The alternative, Guha suggests, is to recognize that wilderness is not appropriate; instead, one must integrate livelihood with environmental protection. Certainly this is the preferred path, when one is discussing how humans should interact with the rest of nature in areas appropriate for human settlement, but it does not address the needs of other species (such as elephants and tigers)—those that cannot coexist in the same area with all but the fewest humans living very simply—or the fact that the integrity of many ecosystems is negatively impacted by the settlement of any humans, even those living at subsistence levels. It is the sheer extensiveness of human settlement in much of Asia (and Europe, and parts of Africa and North America) that is a problem. Humans compete for habitat with other species, threaten their destruction, and otherwise degrade the environment, even diminishing its human carrying capacity.

Wilderness is needed in the Third World as much as it is in Europe and other long settled parts of the globe; nevertheless, it is also important to realize that the structure of imperialism makes the manner in which wilderness is created/protected in the Third World often unjust from a human standpoint. This fact desperately needs to be taken into account by environmentalists. How? First, by understanding how imperialism has created and continues to feed much of the dynamic that threatens ecosystems and species in the Third World from the Amazon to Malaysia; by understanding how countries that have broken or are attempting to break with their historical place in the existing structure in an effort to survive find themselves adopting economic strategies that are environmentally destructive; and by understanding how the wealth extracted from the Third World makes possible the culture of consumption in the First World.[5]

Second, based upon the understanding just set out, we must come to terms with the severe limits of what can be achieved to protect the environment within the framework of a system based on endless material growth and extreme socioeconomic inequality. We need to grasp the necessity of moving beyond the choices offered by the powers that be; only by pushing beyond the limits of what is acceptable to the existing political-economic order can constraints on ecological-political choices be transcended.

Finally, we must recognize that we cannot alter the existing biocidal order without broad-based support. Only if our thinking and feeling is informed by an understanding of human social relations can we develop successful strategies for protecting the Earth and its diversity. If we are to move beyond the existing order we need to understand who our potential allies are, as well as what the obstacles are. If we treat the poor—who go to the rain forest to farm because they have been driven off the land they formerly cultivated by the wealthy who can make higher profits producing cash crops for the international market—as the problem, rather than the system which constrains their choices, we will fail. If we do not forge alliances with those who oppose the existing order—albeit on the basis of its injury to the poor, to women, to oppressed ethnic groups—we will also fail. The work of EPOCA in Nicaraguan reforestation efforts and in Central America generally and the Greenpeace campaign directed at the IMF and the World Bank are both examples of environmental action that reflect at least some elements of what is necessary. Making common cause where possible in pushing for reform and ultimately transformation is essential.

In the short term—given the continued existence of an international political economic system committed to growth and great inequality, and given an international state system in which those who would resist such domination adapt to it to survive—how do we resolve conflicts between particular groups of humans, often the most oppressed, and other species? Even if deep ecologists and other advocates of wilderness creation and protection do attempt to ensure that such activities are not taken at the expense of the oppressed, they will not always be able to achieve both ends: protecting the environment and the poor. By what method do we choose what is to be done? There is no getting around these uncomfortable questions, and previous attempts to address them, including Naess's notion of "near and vital," are not adequately developed.[6]

ANTHROPOCENTRIC/BIOCENTRIC DISTINCTION

In attempting to sort out the questions raised in the previous section the biocentric/anthropocentric distinction is critical. If nature and nonhuman life are held by humans to be as valuable as human life, different answers and courses of action certainly follow than if one only regards human life and needs as valuable. Nonetheless, Guha argues that this distinction has little meaning when it comes to addressing two of the major problems undermining the health of the planet: overconsumption and militarization. I argue to the contrary that significantly different practical consequences follow from adherence to a biocentric rather than an anthropocentric system of values: a biocentric world view constrains human activity much more significantly and distinctly than even the most self-enlightened anthropocentric view. Moreover, a biocentric critique of overconsumption and militarization fully takes account of the underlying dynamic which produces these problems. Initially, a few words need to be said concerning the meaning of anthropocentrism.

Guha suggests that the existing social order in the West is not truly anthropocentric in the sense that it does not value and benefit humans per se, but only some humans, e.g., political and economic elites. Humans who do not fit into these categories are often treated much the way other species are—they are valuable only insofar as they are useful to those at the top of the social hierarchy. Such an analysis is certainly accurate; nevertheless, in contrast to the various forms of domination that are salient features of society, the stated values of the elites are in content anthropocentric: all humans have equal dignity and value. Thus, a social system which benefits the few is justified by arguing that it benefits most or all. Certainly no other justification would do in this age. The contradiction between myth and behavior, however, is most apparent when we look at those who have become superfluous as either producers or consumers: they are "other," objects to be managed by welfare or simply repressed. Widespread toleration of this gap between values and behavior exists for a variety of reasons that we cannot go into here.

Are such systems *anthropocentric,* as deep ecologists argue, or are they better characterized by another term such as *patriarchy* or *class society?* Calling such societies *anthropocentric* seems to miss the fact that only some humans have value (in practice, dominant myths notwithstanding), that control over the human relationship with nature is not shared equally by all, and that the fruits of the exploitation of nature are not shared equally. Yet terms such as *patriarchy* or *class society* seem to ignore the degradation of nonhuman life and the Earth that is so fundamental in almost all existing human societies—and the way in which most humans, regardless of social position, participate in that degradation.[7] Even where critics of the dominant social order recognize the fundamental importance of the human relationship with nature, they continue to share many of the assumptions of the order they criticize: for example, faith in human reason and its ability to solve all problems and the centrality of humans in the universe. Even the phrase "relationship with nature" itself suggests something very different than "place within nature."

My purpose here is not to resolve the issue, but to point out the need for clarification when using these terms as well as the importance of the distinction between values as a world view and as actual behavior. In the discussion that follows, anthropocentrism is used to denote any ideology or system of values which is human centered.[8] Such a definition for the purposes of this paper is not meant to slight the reality of the social systems in which these values are embedded—systems which in practice have yet to come anywhere close to living up to their avowed norms. Nor does such a definition imply that the very real and significant differences among human-centered systems of values are unimportant.

OVERCONSUMPTION

In what ways then is a biocentric system of values meaningful in dealing with overconsumption and militarization? Let's consider overconsumption first. The very meaning of overconsumption differs depending upon whether one takes a biocentric or anthropocentric view. A biocentric view, by giving moral considerability to other species and ecosystems, much more sharply limits human consumption—not only as individuals or groups, but as a species, i.e., it implies a limit on human numbers—than an anthropocentric view which sees value in nature only insofar as it is useful to humans.

If nonhuman nature is valued for itself, if the integrity of the biosphere as a community is valued for itself, then human consumption which disrupts it is wrong: it would constitute overconsumption. Most modern forms of agriculture, forestry, mining, energy extraction and use, housing, transportation, and the like are part of a system of biocidal carnage, and therefore can clearly be called overconsumption.

In a human-centered system of values, overconsumption is primarily seen as a social relationship, a problem of distribution between wealthy and poor, a problem of economic ownership.[9] Overconsumption occurs when some consume more than they need at the expense of those who do not have what they need. Generally speaking, material growth and rising levels of consumption are equated with quality of life; the poor can become better off through economic growth and/or more egalitarian distribution. To this end technology and social organization need to be applied. Such a view does not admit to any finite limit on consumption; nor does it recognize injury to the biosphere as a factor except insofar as it may affect the continued use of the biosphere for human benefit.

Even with most forms of weak anthropocentrism—a view that is sensitive to long-range sustainability—we are still left with a system of values which can and does justify monoculture, high use of energy, massive reclamation projects, conversion of self-regulating ecosystems into cities, and suburbs and agricultural land managed for human use. Such a system continues to view nature as primarily a resource (or a nuisance) and only places limits on consumption which do not affect the sustainability of exploitation. The conversion for human benefit of vast portions of the biosphere is not viewed as wrong even though countless other species are reduced to minimal numbers or to extinction, and ecosystems impoverished or destroyed. Moreover, faith in the centrality of human abilities, particularly as expressed through technology, makes the constraints imposed by concerns for sustainability so vague as to be ineffective as limits on consumption. In contrast, constraints imposed by regarding the ecosystem and other species as valuable in and of themselves narrow the range of appropriate human behavior very sharply: if it injures the biosphere, don't do it.

The distinction between the two views goes much deeper when we examine the roots and social function of high levels of consumption. On a psychological level much consumption is a result of alienation, not just from nature, but also from self (nature within). Endless accumulation and the distractions it offers are essential features of developed societies and of the elites and middle classes elsewhere in the world. Such pathetic attempts to substitute possession of things in lieu of empowerment, sense of place, and authentic relationships is never satisfactory. A hunger for more always remains.[10] On a social level consumption is used by elites to manage large segments of the population. Give people enough stuff and they will forget their pain and powerlessness. The poor make do with the promise of some adequate level of consumption in the distant future and in the meantime they turn to other forms of distraction, often drugs.

Dominant Western or liberal capitalist views tend to deny that there is such a thing as overconsumption. To liberalism, high levels of consumption are viewed as a true measure of the success of our civilization and the individuals within it. It represents the triumph of control and technique, of humans over nature, the fullest flowering of our human faculties. It embraces dualism, hierarchy, atomism, all the machinery of control; nature is fodder, the "other," something to be mastered and managed. Man (intentionally masculine) is the centerpiece of the universe.

There are many human-centered theories which do recognize the pathological roots and role high levels of consumption play in many societies. The Marxisms of Reich, Marcuse, Gorz, and others are concerned with how factors such as consumption are both the result of and further feed alienation. Nevertheless, most Marxism remains wedded to some kind of control over nature, and thus embraces dualism as well as open-ended material growth through progress in technology and social organization.[11] It espouses an unlimited faith in human intelligence and rationality, the ability to understand, control, shape, and improve upon not just human social organization, but all of nature. It is assumed that the evolution of human consciousness will keep pace with any problems, and that we will learn to guide cultural evolution. On the plus side, however, Marxism does reject the view of the world as essentially atomized. As Ollman has so ably demonstrated, Marx saw things as constituted by their relationship and the field of relationships.[12] One can neither change nature without changing oneself nor change an element in a system without changing the system. Despite the limitations of Marxism, a profoundly ecological sort of truth is recognized in such a perspective.

Much radical feminist theory rejects all institutionalized hierarchy and control as being inimical to authenticity.[13] The social problem is not so much who has power, but the power or domination itself. Relationships and community are essential values in this understanding. Both feminists and those concerned with domination based on ethnic differences have been central in pointing out how the category of "the other" runs throughout civilization, justifying oppression, exploitation, and cruelty toward anything that falls within it.[14]

Although there are several anthropocentric world views which object to Cartesian dualism, liberal atomism, and so on, nature and other species remain excluded from the community either explicitly or by silence. One is left with the gulf between humanity and nature, spirit and body, and with an ungrounded faith in the human mission to make over the planet in its own image.

Some anarchist, Marxist, and feminist theory does suggest that part of realizing one's fullest humanity, i.e., part of the process of transcending alienation, involves embracing one's place in nature. In other words, non-alienated being requires a biocentric view of some sort. In such cases being a "citizen rather than conqueror" is biocentric if the natural as well as the human community is recognized as valuable; however, if one simply values the human interest in non-alienation, then the dualism—and the anthropocentrism—remains and serves as a theoretical foundation for views which rationalize structures of control.

A biocentric view rejects both the enlightenment values which justify the social order that thrives on overconsumption as well as the various forms of faith and dualism which underlie world views critical of much of the existing order. This is not to say that much or even most of feminist, anarchist, or other critical social theory is fundamentally incompatible with biocentrism; nevertheless, where such theory continues to accept the human species as the centerpiece of evolution, with the rest of nature existing solely for humanity's use, it fails to address a central form of domination. As such, even under such critical value systems the biosphere is open to suffer the consequences of the arrogance implied in human attempts to manage nature. If species hierarchy is justified, then hierarchy is justified. Much of what such critiques abhor follows from a human-centered view.

Biocentrism draws a clear line. To reject the human/nature dualism is to reject the "triumph" of the enlightenment attempt to control nature. It is to reject the triumph of knowledge and technique and analysis over Earth wisdom, understanding, and connectedness. It is to reject the focus on things rather than relationships. By rejecting these and valuing nature in and of itself, a biocentric view limits human consumption more fundamentally than any anthropocentric view can; it does so by thoroughly rejecting the roots of such consumption. In its place biocentrism values the web of life, as well as its parts, of which we are one.

MILITARIZATION

As with overconsumption we should ask which system of values will constrain militarism more: the human- or the biosphere-centered? By recognizing the valuableness of nature and other species apart from their usefulness to humans, a significant constraint is imposed on human activity with regard to both the conduct of war and more importantly the economic activity that is essential to preparation for war. Indeed, more than war itself, it is the consumption of "resources" to create and maintain the industrial capacity geared to arms production—for whatever purpose—that is so destructive of the biosphere. All human centered value systems necessarily fall prey to the easy rationalization of militarism.

If one is concerned only with humans, with the perpetuation and protection of particular social systems against internal or external threats, the constraints placed upon the consumption of nature are weak indeed. Even when limits on resources may temper overconsumption generally, there is a real tendency in this sphere of "national security" to literally let the future take care of itself and commit all to the current struggle. Certainly aesthetic regard for nature falls by the wayside. If the machine needs oil, then drill. The Soviet Union, as an example, has some of the strictest environmental legislation in the world. These laws also provide a giant loophole for any endeavor related to the security of the state, virtually negating restrictions.[15] Most countries start with weaker laws to begin with before embracing the exceptions.

There are many human-centered value systems, religious and secular, critical of militarization—and all are largely ineffective. The failure comes in part from the wedding of values to structures of power—be they church or state—that depend upon force for their survival. Insofar as these pacifistic values are taken up by those "outside" these structures they provide some check. But because they are human-centered—the point of opposing militarization is to end human waste and suffering—it is easy to neutralize them by appeal to other human values and to other forms of suffering even worse than war or the costs of deterrence. The other great weakness is that much pacifistic thinking does not address adequately the roots of militarism, something I attempt to do below.

If one values nature in and for itself, then human goals and needs are placed within the context of a larger community. The value placed on the integrity of that community militates heavily against any human-centered rationalization for exploitation. A biocentric view quite simply limits the conversion of ecosystems and biomass to human use to any extensive degree. Although such a view may seem utopian, because it poses a threat to the survival of particular social systems or the system of historical social systems, it does not pose a threat to the survival of the species as some would argue. Quite the opposite, the threat to both us and the planet comes from this system of systems. It is here that biocentrism provides understanding which human-centered approaches cannot, for the latter accept fundamental values which justify the very structures that give rise to the outcomes they criticize.

Consider the roots of militarism. Because modern militarism is particularly virulent, attempts to understand and criticize this blight are often limited to the modern period. Certainly the combination of enlightenment arrogance, science, and technology, embedded in the international political economy resulting from the European expansion, has produced a very dangerous world.[16] It is, however, necessary to look more deeply into human history to grasp the underlying dynamic of militarism. While it may have reached new proportions, it is not new, but rather an essential feature of something very old: civilization.[17] It is inseparable from social systems based upon hierarchy (class, gender, and ethnic), control of nature, the denial of self, and the emotions and bonds which constitute the self. It is an essential feature of those societies in which the state exists, the process by which the state attempts to substitute itself for authentic human community is well underway, and conflict between communities has been replaced by the institutionalized conflict of center and periphery and between competing centers.[18] Civilization, and the process of its formation and emergence in the neolithic, is the story of the human attempt to adapt through various strategies of control—control of nature and of people through technology and social organization. It is this attempt to control nature that separates us from it, that constitutes the core of our alienation from life, and that becomes the foundation for social development that includes patriarchy, class domination, statism, and militarism.

While most, but by no means all human-centered *value* systems eschew militarism, civilization is held as a crowning achievement. Some value systems praise the military spirit, while the majority that condemn it usually do so as a necessary evil, i.e., they simultaneously justify it to one degree or another. The point to be made here is that civilization is based upon and is constituted by relationships of domination that invariably and necessarily produce the conflict and inequality which make militarism inevitable. Certainly some human-centered theory recognizes aspects of the roots of militarism, and it recognizes the terrible price humans have paid, even if ignoring the price nature has paid. Nevertheless, critics maintain a fervent faith in the human mission to manage, in the human ability to disentangle what is inextricably linked. They speak from within the perspective of civilization and cannot see that they must transcend the precarious ground on which they (we) teeter.[19]

Critical theory shares much in common with liberal theory in this area. Some Marxist analysis of the genesis of modern militarism is sound. The notion that many human ills would be solved with the end of class society is also appealing. But the end of class is not the end of the state or of domination, and hence not the end of social systems which produce militarism. (Nor is the end of capitalism the end of class.) The control of nature and the human control of social and cultural evolution are values deeply embedded in most Marxism. Although it has developed useful models for understanding social transformation, the assumptions, perspective, and the content of the transformative vision are very much within the human-centered tradition that is part of the problem.[20]

Some feminism gets much closer to the source of the problem in its critique of hierarchy generally and in particular in its understanding of the central role of patriarchy to militarism and to producing humans amenable to domination. At times, however, feminist theory falls into a kind of intraspecific dualism, i.e., human males are the problem (while at the same time claiming credit for the fact that females created agriculture, which became the economic foundation for the emergence of hierarchy), ignoring that systems adapt to and alter the environment, and individuals adapt to (even while they resist) the roles created by the system's division of labor.[21] Even where this dualism is not at issue, most feminism, like Marxism, remains human-centered. Values such as community, spontaneity, and integration of emotion and intellect militate against the worst features of mainstream human-centered values, but still fail to take account of the relationship with nature as fundamental to all hierarchical systems. Or they remain anthropocentric and fail to address the separation from nature which not only makes possible the superexploitation of the biosphere for the maintenance of the military apparatus, but also underlies the social structures which produce militarism.

While Marxism, feminism, and other critical social theory have contributed much to understanding the dynamic of our civilization, they tend to miss the point that if nonhuman life is not valued for itself, then life is not valued for itself. Any system of values that does not transcend nature-as-other cannot limit destruction of the biosphere as effectively as one that embraces nonhuman life as intrinsically valuable. Nor can such a value system help to heal the fundamental split in the human psyche which makes possible civilization and militarism.

Biocentrism is not alone in grasping that the dynamic of human evolution over the last six or seven thousand years may be at a dead end. Certainly the huge growth in human numbers, the displacement of "simpler" societies by more "complex" ones, ones with greater capacity to exploit nature, capture and use energy, and so on suggests that the underlying dynamic is highly adaptive, at least at first glance. What is increasingly clear, however, is that if this dynamic continues we stand a very good chance of killing ourselves along with a good portion of the rest of the planet. The latter is well under way—it's business as usual.

Biocentrism offers a direction for human society based upon a thoroughly fundamental transformation which stresses the centrality of finding our place *in* nature. Such a transformation is as fundamental as the neolithic or industrial revolutions.

A life-centered or planet-centered value system requires that we move toward transcending the split with nature both within our own psyches and in our material relationships: how we consume and alter the biosphere. Far fewer humans, far lower levels of consumption for many, much improved levels for others, the recreation of authentic communities that reintegrate the human into the natural, and the abandonment of the instrumentalities of control—these are a few of the implications of such an ethic.

In contrast, a human-centered approach focuses on wiser if not greater human control. In its more progressive forms we hear words like stewardship rather than ownership; nevertheless, underlying both is the notion that we can replace nature with our intellect, that we can manage our way out of any problems, that we as a species are not only unique (as every species and ecosystem is), but that our uniqueness means we are godlike, better than the others. In short, it is the same arrogance, the same split that has brought us to the current crisis.

VALUES AND CULTURE

All systems of values are part of a broader cultural framework that mediates human behavior by shaping personality and thought. Culture organizes human experience and gives it meaning. Biocentric values are no exception—they are part of a larger cultural framework, albeit an emergent one. Part of that culture includes an understanding of the role of culture generally as well as the critique of particular cultures.

A biocentric approach presents human-centered cultures as both rooted in and perpetuating a split between our species and the rest of nature. This split, which is manifest as both a chronic and debilitating inner tension and as a stressful warlike antagonism toward the environment, is the source of our experience of estrangement. Disembodied, our cortex is a shadow of life, ever busy trying to rationalize the irrational while telling us it is all. Biocentrism offers us back our body by recognizing that the Earth is our real community—that by healing our split from it, by healing the split between cortex and heart, and by healing nature within, we can begin to heal all of nature. It also helps us to understand how that split is possible.

To point to the neolithic as the origin of the culture of control is not enough. A biocentric view places these events in context. It helps us understand how the capacity for culture and the resulting plasticity in human behavior, thought and emotion, and our ability to learn and pass on learning (attitudes and world views as well as technical or social information) enables us to divide ourselves. This capacity for culture allowed humans threatened with localized overpopulation in neolithic times to increase the human carrying capacity by altering both their behavior and the environment in substantial ways. With the new dynamic set in motion the fundamental injury to other needs, the split itself, was probably not very obvious. First, it was cumulative over a long time. Second, the very capacity for culture allowed them (and us) to deny the estrangement, even required such denial for both psychological and social reasons. Third, the emerging social dynamic of hierarchy distributed the costs and benefits of the new adaptive strategies unequally, favoring the decision makers and shapers of a society's values.

Culture has allowed us to trade our place in nature for larger human numbers spread over the entire planet, converting large amounts of the biosphere to our purposes so long as we are willing to pay the price of the various forms of domination and the accompanying anguish with which we are so familiar—and so long as we are willing to deny the value of other life and allow nature to pay the price. The plasticity with which evolution has endowed us allows us to create alienating and biocidal sociocultural systems, but it does not require it; it is not natural in the sense of being necessary or in the sense of being in tune with our deepest nature. (We should not forget that while cancer is a part of nature, it kills its host.) There are other cultural possibilities, including biocentric ones. Indeed, for most of the time humans have been around we have lived in communities which included the rest of nature. We can do so again, this time with full knowledge of what the alternatives are and their price. To limit our biocidal possibilities is not unnatural, as Callicott quite rightly argues, because cultural systems always limit behavior.[22] Culture is always prescriptive.

The roots of biocentrism are deep and its emergence in modern form is a result of both the resilience of Earth wisdom and the current crisis—just as surely as human-centered values and cultural systems are a result of the crisis of the neolithic. Both Marxist and feminist anthropology have traced the roots of class and patriarchal domination and have contributed much to understanding the dynamic which emerged. Marx and some Marxists have rightly regarded the split with nature as a decisive milestone in human cultural evolution, albeit a positive one. Some feminist writing has addressed itself to the split with nature, both in the modern epoch and in the distant past, but usually as something ancillary to the development of human forms of domination. A biocentric understanding suggests that the culture of control over nature is part of an adaptation to scarcity and lays the groundwork for other forms of domination.[23]

By accepting biocentric limits upon our behavior we directly undermine the split from nature and the resulting culture of domination which arises from it. In doing so we accept constraints on overconsumption and militarism that no human-centered system of values could impose. Domination and hierarchy, the attempt to control that gives rise to high levels of consumption and militarism, will be unshakable problems until we recognize we cannot substitute our intellect for nature. The degree to which we attempt to do so is a function of our own estrangement.

ALLIANCES

Guha gives us much to think about and we ignore his voice at our peril. Although the Earth and other species need wilderness, we will lose the battle for the planet if we do not realize, as Guha suggests, that imperialism and militarism are our enemies as well as anthropocentrism. We cannot dismiss the struggles over human social structure and realize a deep ecological vision. The land ethic is not compatible with most of the existing order of things. In struggling to alter that order it is necessary to understand how it works, for if we do not, the vision in the hearts of a few will not be enough.

NOTES

1. Ramachandra Guha, "Radical American Environmentalism and Wilderness Preservation: A Third World Critique," *Environmental Ethics* 11 (1989): 71–83. [Included in this volume.]

2. Some critics, though not Guha, have accused deep ecology of being fundamentally misanthropic. No doubt there are genuine misanthropes about, but in my reading of the deep ecological literature, both scholarly and popular, I find criticism aimed at human behavior resulting from alienation and disease, not the species per se. Even the angriest statements of those struggling on the front lines against biocide can best be understood in this context. Humans have a place *in* nature; it is when they try to separate themselves that their behavior becomes destructive.

3. Bahro, one of the leading German Green theorists, comes from a Marxist background, as do many other German Greens. *The Alternative* (London: New Left, 1978), his Marxist critique of "actually existing socialism," is a major contribution to understanding human society. It earned him a lengthy jail sentence in the GDR before he was allowed to emigrate. His later, Green writings are more widely read in the U.S. than his first book, but his historical perspective and concern with human society runs through all of his work, which is read by environmentalists. It is unfortunate, however, that U.S. Marxists are more familiar with his later work than deep ecologists are with his earlier work.

4. Human population pressure combined with human capacity for culture are probably the two most important factors in explaining the dynamic of social evolution and the subsequent alteration of the biosphere. This is discussed more fully below in the section "Values and Culture." In conjunction with this thesis, see note 17 and Mark N. Cohen, *Food Crisis in Prehistory* (New Haven: Yale University Press, 1977), and Marshall Sahlins, *Stone Age Economics* (Chicago: Aldine, 1972).

5. We live in a world shaped by the European expansion. Most of Africa, Asia, and Latin America are still bound tightly into an international political economy and state system which keeps them subservient. The wealth continues to flow north. Most of the environmental degradation in the Third World can only be understood in this context. This is also the case with regard to many efforts at conservation, which are often at the expense of the poor. Attempts to break with this international system have sometimes been successful, as with the Russian and later revolutions; nevertheless, the results of such revolutions for the environment have not been impressive. Much more than Marx's nineteenth-century notion of progress (and in direct contradiction with many of his revolutionary goals), it has been the international political environment of hostility that has greeted these revolutionaries, combined, of course, with the inertia of hierarchy inherent in civilized societies, that has resulted in their systems following a very similar road of super-exploitation of the environment. Both capitalist and statist societies eat and despoil nature, notwithstanding different internal dynamics. There are similarities as well: increasing competition in weapons and higher levels of consumption. More recently there has been a recognition among revolutionaries and even some non-revolutionary governments that protection of the environment is an important value in and of itself and that industrialization and its products are not desirable goals. However, the costs of defending themselves from a hostile world (and the costs of giving in) lead to environmental degradation. Clearly solutions must be global and systemic.

6. Arne Naess has suggested in "Identification as a Source of Deep Ecological Attitudes" (Michael Tobias, ed., *Deep Ecology,* 2d ed. [San Marcos, Calif.: Avant Books, 1980], p. 270) that conflicts between humans and other species can be resolved by balancing the competing interests based upon how "near and vital" the interests are to the species involved. Given the large numbers of Homo sapiens and

their extensive settlement, it is difficult to see how this approach could lead to a redress of the current imbalance unless one takes a global perspective. There can be little question, for example, that humans need to give way to tigers, chimps, elephants, grizzlies, and other species. With five billion people and only a few thousand members of certain other species, restoring a balance can only mean a movement in one direction: more room for other species. Of course, the impact on humans of making room for other creatures will not affect all humans equally. Specific humans will have to make way. How are the costs to be spread? If one takes a strictly localized perspective, trying to balance the interests of a local human population only with the interests of a local nonhuman population, an assessment of competing interests gives a result less favorable to nonhuman life. Once one takes the extensive human presence as a given, human interests in their existing livelihood must be weighed without taking into account significant human numbers elsewhere. In this way, the pressure on already diminished populations of other species would continue.

7. Under what circumstances and in what situations humans can be treated as a species rather than as classes, genders, and ethnic groups or as individuals is a contentious matter among theoreticians and activists concerned with the environment and human justice. It is an important issue, and is best resolved in regard to specific questions. For some questions a species approach is appropriate; for others another level of aggregation is required. See note 21 below for a related discussion.

8. After submission of this manuscript Carolyn Marchant published a discussion paper, "Environmental Ethics and Political Conflict: A View from California," *Environmental Ethics* 12 (1990): 45–68, in which she distinguishes egocentric values from anthropocentric values. The dominant values in the U.S. and most other developed societies are most properly termed egocentric rather than anthropocentric.

9. Critical theory distinguishes between natural poverty, which is due to a lack of development of productive forces, and social poverty, which is the result of exploitation and inequality. Both are regarded as oppressive. It is fair to say that Marx saw the human/nature schism and subsequent struggle through eyes sensitized to the class struggle. Thus, there can be no essential harmony until the struggle is over, i.e., until the human species brings the rest of nature under its control.

10. In *Ecology as Politics* (Boston: South End, 1980), pp. 28–42, Andre Gorz has a very useful discussion on the social definition of poverty and its role in social management. On a psychological level, the existence of narcissistic tendencies (if not clinical narcissism) is important in explaining how individuals become susceptible to the social dynamic of consumption. When people are not aware of their real needs, they often seek socially defined and approved substitutes which necessarily cannot satiate them. Toys, drugs, and even television cannot effectively fill the emptiness left by the inability to experience intimacy, for example. The lack of a developed self leaves individuals particularly vulnerable to manipulation. Christopher

Lasch examines this phenomenon in *The Culture of Narcissism* (New York: Norton, 1978). On the psychological aspects of the genesis of narcissism, see the various works of Alice Miller, James F. Masterson, Heinz Kohut, and Otto Kernberg. Alexander Lowen has provided a more popular treatment. The function of the family generally in shaping the malleable young for the roles society requires has long been the subject of some of the best psychological writing, including that of Wilhelm Reich, Erik Erikson, Dorothy Dinnerstein, Nancy Chodorow, and others. The phenomenon is not new. Ruskin noted in the last century that the two objects of civilized life are: "Whatever we have—to get more; and wherever we are—to go somewhere else." The only thing we might add today is: to get there faster.

11. Bertell Ollman, *Alienation,* 2d ed. (Cambridge: Cambridge University Press, 1976), pp. 14–32.

12. For Marx the labor process, by which humans transform nature to make their living, defines the human relationship to nature and is central to human biological and sociocultural evolution. This is linked not only to the conquest or appropriation of nature as central to our "species-being," but to an open-ended notion of human *material* needs. It is in this area that more modern anthropological critiques offer a deeper understanding of the circumstances of human evolution than nineteenth-century theorists had available. One notable point of difference involves the notion of communal ownership which Marx and Engels, following Morgan, viewed as a central feature of egalitarian society. This idea was important because it informed their vision of a future egalitarian society: communism. They believed that the failure of primitive egalitarian society was based upon scarcity; communism would be wealthy, taking advantage of previously developed forces of production. Modern anthropology suggests, nevertheless, that while primitive societies were egalitarian they did not have a sense of ownership, either private or communal. Animals, plants, and nature were part of a community. Much anthropological research also calls into question the notion that there is a deep-seated need in humans to fundamentally transform the environment.

13. Feminists have offered some of the most cogent criticism of power and hierarchy as such. The work of Susan Griffin, Dorothy Dinnerstein, Kathy E. Ferguson, Mary Daly, and others offers enormously valuable insights into the difference between power (over the other) and empowerment, the nonalienating experience of being a full member in a real community. If one extends the notion of community to include other species and the biosphere, one has, I believe, a version of biocentrism. See Judith Plant, ed., *Healing the Wounds* (Philadelphia: New Society, 1989), for example.

14. Marjorie Spiegel, in *The Dreaded Comparison* (Santa Cruz: New Society Publishers, 1988), cogently compares the similarities in the arguments that attempt to justify exploitation and domination of and cruelty to humans, animals, and nature. The manner in which some humans separate themselves from other people and from the natural community invariably involves a process in which differences

(both real and imagined) are translated into value distinctions simply on the basis of difference. Thus, because Africans or women or wolves are different from Europeans, men, or humans, they are less valuable or of no value.

15. See, for example, Boris Kamarov, *Destruction of Nature in the Soviet Union* (White Plains: M. E. Sharpe, 1980).

16. The literature on imperialism and the world that it has produced is enormous. A good general introduction is Charles K. Wilber, ed., *The Political Economy of Development and Underdevelopment,* 4th ed. (New York: Random House, 1988) or L. S. Stavrianos, *Global Rift* (New York: Morrow, 1981). From there one may pursue more specialized works in economics, politics, conflict, and so on. For those concerned with the environment explicitly, the work of Alfred Crosby stands out; both *The Columbian Exchange* (New York: Greenwood Press, 1973) and *Ecological Imperialism* (Cambridge: Cambridge University Press, 1986) are truly major contributions. Carolyn Merchant's excellent *The Death of Nature* (San Francisco: Harper & Row, 1980) describes the emergence of the scientific world view that formed an essential part of the cultural framework justifying the domination of other people, nature, and women.

17. The institutions and processes that constitute civilization are not matters for serious debate. The state, urbanization, extensive division of labor, class structure, patriarchy, militarism, and monumental architecture are among those elements identified by Elman Service, Robert M. Adams, K. C. Chang, Morton Fried, William Sanders, Barbara Price, and others. On the origins and evolution of the problems of agriculture see Wes Jackson, *Altars of Unhewn Stone* (San Francisco: North Point, 1987).

18. The earliest human communities include nature, but do distinguish the "other," namely, humans belonging to other societies or communities, especially those not involved in some reciprocal relationship. Conflict between communities certainly predates civilization and the neolithic period. Much of this conflict, however, was symbolic and sharply limited in its destructiveness to both persons and the environment. As some communities followed a path toward hierarchical social organization, the ability of the more hierarchical groups to displace, conquer, or otherwise make nonhierarchical groups dependent on them developed. Institutionalized relationships of domination became the order of the day. To escape domination other groups were forced to move, submit, or resist by adopting similar social strategies. The origins of center and periphery and the attendant exploitation, brutalization and conflict between center and periphery and competing centers are thus very old indeed. How egalitarian communities might relate noncompetitively will be a critical issue for humans and others on this planet. Certainly understanding both the psychological dynamic that generates the notion of the "other" and the social/ecological dynamic that creates pressure toward conflict and hierarchical solutions goes a long way toward solving the problem, although it is not sufficient in itself.

19. To say that civilization must somehow be fundamentally transcended is to say that the dynamic founded on and constituted by various relationships of domination must be overcome. It may represent a kind of return to the past, but in the service of the future. For the last six thousand years our species has behaved much like one might expect adolescents from a severely dysfunctional family to act. We must go back to where things went wrong—to the origins of our estrangement—and pick up from there. In doing so we make use of all that has occurred in the interim. We have already paid dearly for the lessons.

20. There are attempts within the Marxist tradition to break out of this approach, but they are limited in degree. For instance, Gorz has argued that socialist industrialization is as bad as capitalist industrialization if it results in the same kind of environmental damage. Nevertheless, in his attempt to integrate environmental constraints into a Marxist framework, he remains concerned with the fate of Homo sapiens, not with the biosphere as such. In the West it is painfully obvious that the Marxist concern with the environment is largely a response to the environmental movement, rather than something generated internally. It remains to be seen how both dialogue and praxis will develop. Recently a journal was founded in this area, *Capitalism, Nature, Socialism.* The analysis of environmental policy in statist or state socialist countries demonstrates that their path reflects a dynamic based on maintaining a bureaucratic oligarchy at home and defending themselves from a hostile world capitalist order. This dynamic, while differing from what moves capitalist societies, offers little in the way of hope for a significant alternative relationship with the environment. Genuinely socialist or communitarian experiments in the Third and First Worlds that reject massive industrialization are another matter.

21. The degree of choice that individual humans and human collectives exercise is a matter of serious debate. Factors such as consciousness, social position of particular individuals and groups, and the limits of historical possibility certainly all play a role. In the course of social struggle people necessarily make certain assumptions: that it is possible to realize their goals, that they understand the operation of the social (or natural) world enough to bring about the desired results, and that those opposed to them must be held accountable for their resistance as if they understood what they were doing and recognized the choices available to them. The rather consistent failure in the social realm to realize expectations calls these assumptions into question to varying degrees. Some things, nevertheless, do seem clear: when people make decisions, it is the unintended consequences that tend to outweigh in importance the intended one—a point which is often never realized. No one set out to invent the state, but a series of decisions over time had that result. There have been "moments" in human history (and prehistory) when real choices about basic human social structure and interaction with nature were possible. At such moments it was possible to switch a train from one track to another. When these moments occur in the future, there must be an awareness that the tracks are

available and that awareness must be shared by enough people to make a difference. Most of the time choices are more constrained, with humans squabbling over who gets what seat on the train, rather than where its going. This is not to say, of course, that who sits where is not important to both humans and the rest of nature, for who sits where has much to do with the choices eventually taken when switching tracks is possible. On an individual level, both the powerful and the weak are socialized to roles that neither created, and they are limited in their options by structure and consciousness. Certainly those who benefit more seek to perpetuate and strengthen their position and this contributes to the perpetuation of the system. Even though those who benefit less are more inclined to resist their roles and seek change, they also cooperate in their subjugation because they have been socialized to do so and because they fear repression. The degree to which socialization limits possibilities varies with time and the conjunction of a number of factors. On the other hand, it is important to keep in mind the often overlooked distinction between explaining behavior and excusing it. The former by no means implies the latter. Although people may not choose freely, their actions have consequences and accountability is fundamental to constraining both individuals and collectives. Without it there is no learning.

22. See J. Baird Callicott's discussion on the biosocial role of culture in *In Defense of the Land Ethic* (Albany: SUNY Press, 1989), pp. 63–73. See also Dorothy Lee's *Freedom and Culture* (Englewood Cliffs: Prentice-Hall, 1959).

23. See, for example, Peggy Reeves Sanday, *Female Power & Male Dominance* (Cambridge: Cambridge University Press, 1981).

Ramachandra Guha

Deep Ecology Revisited (1998)

I.

"RADICAL ENVIRONMENTALISM and Wilderness Preservation: A Third World Critique" was written at the end of an extended period of residence in the United States, which followed directly upon several years of research on the origins of Indian environmentalism. That background might explain the puzzlement and anger which, in hindsight, appear to mark the essay. To my surprise, the article evoked a variety of responses, both pro and con. The veteran Vermont radical, Murray Bookchin, himself engaged in a polemic with American deep ecologists, offered a short (three-line) letter of congratulation. A longer (thirty-page) response came from the Norwegian philosopher Arne Naess, the originator of the term "deep ecology." Naess felt bound to assume responsibility for the ideas I had challenged, even though I had distinguished between his emphases and those of his American interpreters. Other correspondents, lesser known but no less engaged, wrote in to praise and to condemn.[1] Over the years, the essay has appeared in some half-a-dozen anthologies, as a voice of the "Third World," the token and disloyal opposition to the reigning orthodoxies of environmental ethics.[2]

The essay has acquired a life of its own. This postscript allows me to look at the issues anew, to expand and strengthen my case with the aid of a few freshly arrived examples.

II.

Woodrow Wilson once remarked that the United States was the only idealistic nation in the world. It is indeed this idealism which explains the zest, the zeal, the almost unstoppable force with which Americans have sought to impose their vision of the good life on the rest of the world. American economists urge on other nations their brand of energy-intensive, capital-intensive, market-oriented development. American spiritualists, saving souls, guide pagans to one or another of their eccentrically fanatical cults, from Southern Baptism to Moral Rearmament. American advertisers export the ethic of disposable containers—of all sizes, from coffee cups to automobiles—and Santa Barbara.

Of course, other people have had to pay for the fruits of this idealism. The consequences of the forward march of American missionaries include the undermining of political independence, the erosion of cultures, and the growth of an ethic of sheer greed. In a dozen parts of the world, those fighting for political, economic, or cultural autonomy have collectively raised the question whether the American way of life is not, in fact, the Indian (or Brazilian, or Somalian) way of death.

One kind of U.S. missionary, however, has attracted virtually no critical attention. This is the man who is worried that the rest of the world thinks his country has a dollar sign for a heart. The dress he wears is also colored green, but it is the green of the virgin forest. A deeply committed lover of the wild, in his country he has helped put in place a magnificent system of national parks. But he also has money, and will travel. He now wishes to convert other cultures to his gospel, to export the American invention of national parks worldwide.

The essay to which these paragraphs are a coda was one of the first attacks on an imperialism previously reckoned to be largely benign. After all, we are not talking here of the Marines, with their awesome firepower, or even of the World Bank, with its money power and the ability to manipulate developing country governments. These are men (and more rarely, women) who come preaching the equality of all species, who worship all that is good and beautiful in Nature. What could be wrong with them?

I had suggested in my essay that the noble, apparently disinterested motives of conservation biologists and deep ecologists fueled a territorial ambition—the physical control of wilderness in parts of the world other than their own—that led inevitably to the displacement and harsh treatment of the human communities who dwelt in these forests. Consider in this context a recent assessment of global conservation by Michael Soulé, which complains that the language of policy documents has "become more humanistic in values and more economic in substance, and correspondingly less naturalistic and ecocentric." Soulé seems worried that in theory (though certainly not in practice!) some national governments and international conservation organizations (or ICOs) now pay more attention to the rights of human communities. Proof of this shift is the fact that "the top and middle management of most ICOs are economists, lawyers, and development specialists, not biologists." This is a sectarian plaint, a trade union approach to the problem spurred by an alleged takeover of the international conservation movement by social scientists, particularly economists.[3]

Soulé's essay, with its talk of conspiracies and takeover bids, manifests the paranoia of a community of scientists which has a *huge* influence on conservation policy but yet wants to be the sole dictator of it. A scholar acclaimed by his peers as the "dean of tropical ecologists" has expressed this ambition more nakedly than most. Daniel Janzen, in a paper in the *Annual Review of Ecology and Systematics,* urges upon his fellow biologists the cultivation of the ability to raise cash so as to buy space and species to study. Let me now quote from a report he wrote on a new national park in Costa Rica, whose tone and thrust perfectly complement the other, ostensibly "scientific" essay. "We have the seed and the biological expertise: we lack control of the terrain," wrote Janzen in 1986. This situation he was able to remedy for himself by raising enough money to purchase the forest area needed to create Guanacaste National Park. One can only marvel at Janzen's conviction that he and his fellow biologists know all, and that the inhabitants of the forest know nothing. He justifies the taking over of the forest and the dispossession of the forest farmer by claiming that "today virtually all of the present-day occupants of the western Mesoamerican pastures, fields and degraded forests are deaf, blind, and mute to the fragments of the rich biological and cultural heritage that still occupies the shelves of the unused and unappreciated library in which they reside."[4]

This is an ecologically updated version of the White Man's Burden, where the biologist (rather than the civil servant or military official) knows that it is in the native's true interest to abandon his home and hearth and leave the field and forest clear for the new rulers of his domain. In Costa Rica we only have Janzen's word for it, but elsewhere we are better placed to challenge the conservationist's point of view. A remarkable recent book on African conservation has laid bare the imperialism, unconscious and explicit, of Western wilderness lovers and biologists working on that luckless continent. I cannot here summarize the massive documentation of Raymond Bonner's *At the Hand of Man,* so let me simply quote some of his conclusions:

> Above all, Africans [have been] ignored, overwhelmed, manipulated and outmaneuvered by a conservation crusade led, orchestrated and dominated by white Westerners.
>
> Livingstone, Stanley and other explorers and missionaries had come to Africa in the nineteenth century to promote the three C's—Christianity, commerce and civilization. Now a fourth was added: conservation. These modern secular missionaries were convinced that without the white man's guidance, the Africans would go astray.
>
> [The criticisms] of egocentricity and neocolonialism ... could be leveled fairly at most conservation organizations working in the Third World.
>
> As many Africans see it, white people are making rules to protect animals that white people want to see in parks that white people visit. Why should Africans support these programs? ... The World Wildlife Fund professed to care about what the Africans wanted, but then tried to manipulate them into doing what the Westerners wanted: and those Africans who couldn't be brought into line were ignored.
>
> Africans do not use the parks and they do not receive any significant benefits from them. Yet they are paying the costs. There are indirect economic costs—government revenues that go to parks instead of schools. And there are direct personal costs [i.e., of the ban on hunting and fuel collecting, or of displacement].[5]

Bonner's book focuses on the elephant, one of the half-a-dozen or so animals that have come to acquire "totemic" status among Western wilderness lovers. Animal totems existed in most pre-modern societies, but as the Norwegian scholar Arne Kalland points out, in the past the injunction not to kill the totemic species applied only to members of the group. Hindus do not ask others to worship the cow, but those who love and cherish the elephant, seal, whale, or tiger try and impose a worldwide prohibition on its killing. No one, they say, anywhere, anytime, shall be allowed to touch the animal they hold sacred even if (as with the elephant and several species of whale) scientific evidence has established that small-scale hunting will not endanger its viable populations and will, in fact, save human lives put at risk by the expansion, after total protection, of the *lebensraum* of the totemic animal. The new totemists also insist that their species is the "true, rightful inhabitant" of the ocean or forest, and ask that human beings who have lived in the same terrain (and with the animals) for millennia be taken out and sent elsewhere.[6]

I turn, last of all, to an ongoing controversy in my own bailiwick. The Nagarhole National Park in southern Karnataka has an estimated forty tigers, the species toward whose protection enormous amounts of Indian and foreign money and attention have been directed. Now Nagarhole is also home to about 6,000 tribals, who have been in the area longer than anyone can remember, perhaps as long as the tigers themselves. The state Forest Department wants the tribals out, claiming they destroy the forest and kill wild game. The tribals answer that their demands are modest, consisting in the main of fuel wood, fruit, honey, and the odd quail or partridge. They do not own guns, although coffee planters living on the edge of the forest do. Maybe it is the planters who poach big game? In any case, they ask the officials, if the forest is only for tigers, why have you invited India's biggest hotel chain to build a hotel inside it while you plan to throw us out?

Into this controversy jumps a green missionary, passing through Karnataka. Dr. John G. Robinson works for the Wildlife Conservation Society in New York, for whom he oversees 160 projects in 44 countries. He conducts a whistle-stop tour of Nagarhole, and before he flies off to the next project on his list, hurriedly calls a press conference in the state capital, Bangalore. Throwing the tribals out of the park, he says, is the only means to save the wilderness. This is not a one-off case but a sacred principle, for in Robinson's opinion "relocating tribal or traditional people who live in these protected areas is the single most important step towards conservation." Tribals, he explains, "compulsively hunt for food," and compete with tigers for prey. Deprived of food, tigers cannot survive, and "their extinction means that the balance of the ecosystem is upset and this has a snowballing effect."[7]

One does not know how many tribals Robinson met (none, is the likely answer). Yet the Nagarhole case is hardly atypical. All over India, the management of parks has sharply pitted the interests of poor tribals who have traditionally lived in them against those of wilderness lovers and urban pleasure seekers who wish to keep parks "free of human interference"—that is, free of other humans. These conflicts are being played out in the Rajaji sanctuary in Uttar Pradesh, in Simlipal in Orissa, in Kanha in Madhya Pradesh, and in Melghat in Maharashtra.[8] Everywhere, Indian wildlifers have ganged up behind the Forest Department to evict the tribals and relocate them far outside the forests. In this they have drawn sustenance from American biologists and conservation organizations, who have thrown the prestige of science and the power of the dollar behind the crusade to kick the original owners of the forest out of their home.

Specious nonsense about the equal rights of all species cannot hide the plain fact that green imperialists are possibly as dangerous and certainly more hypocritical than their economic or religious counterparts. For the American advertiser and banker hopes for a world in which everyone, regardless of color, will be in an economic sense an American—driving a car, drinking Pepsi, owning a fridge and a washing machine. The missionary, having discovered Jesus Christ, wants pagans also to share in the discovery. The conservationist wants to "protect the tiger (or whale) for posterity," yet expects *other* people to make the sacrifice.

Moreover, the processes unleashed by green imperialism are well-nigh irreversible. For the consumer titillated into eating Kentucky Fried Chicken can always say, "once is enough." The Hindu converted to Christianity can decide later on to revert to his original faith. But the poor tribal, thrown out of his home by the propaganda of the conservationist, is condemned to the life of an ecological refugee in a slum, a fate, for these forest people, which is next only to death.

III.

The illustrations offered above throw serious doubt on Arne Naess's claim that the deep ecology movement is "from the point of view of many people all over the world, the most precious gift from the North American continent in our time."[9] For deep ecology's signal contribution has been to privilege, above all other varieties and concerns of environmentalism, the protection of wild species and wild habitats, and to provide high-sounding, self-congratulatory but nonetheless dubious moral claims for doing so. Treating "biocentric equality" as a moral absolute, tigers, elephants, whales, and so on will need more space to grow, flourish and reproduce, while humans—poor humans—will be expected to make way for them.

By no means do I wish to see a world completely dominated by "human beings, their mutualists, commensals and parasites." I have time for the tiger and the rainforest, and wish also to try and protect those islands of nature not yet fully conquered by us. My plea rather is to put wilderness protection (and its radical edge, deep ecology) in its place, to recognize it as a distinctively North Atlantic brand of environmentalism, whose export and expansion must be done with caution, care, and above all, with humility. For in the poor and heavily populated countries of the South, protected areas cannot be managed with guns and guards but must, rather, be managed with full cognizance of the rights of the people who lived in (and oftentimes cared for) the forest before it became a national park or a world heritage site.[10]

Putting deep ecology in its place is to recognize that trends it derides as "shallow" ecology might in fact be varieties of environmentalism that are more apposite, more representative, and more popular in the countries of the South. When Arne Naess says that "conservation biology is the spearhead of scientifically based environmentalism"[11] one wonders why "agroecology," "pollution abatement technology" or "renewable energy studies" cannot become the "spearhead of scientifically based environmentalism." For to the Costa Rican peasant, the Ecuadorian fisherman, the Indonesian tribal, or the slum dweller in Bombay, wilderness preservation can hardly be more "deep" than pollution control, energy conservation, ecological urban planning, or sustainable agriculture.

NOTES

1. I speak here of private communications: published responses to my essay include David M. Johns, "The Relevance of Deep Ecology to the Third World: Some Preliminary Comments," *Environmental Ethics* 12 (Fall 1990), included in this volume; and J. Baird Callicott, "The Wilderness Idea Revisited: The Sustainable Development Alternative," *The Environmental Professional* 13 (1991), also included herein.

2. These anthologies include Thomas Mappes and Jane Zembaty, eds., *Social Ethics: Morality and Public Policy,* 4th ed. (New York: McGraw-Hill, 1992); Carolyn Merchant, ed., *Key Concepts in Critical Theory: Ecology* (Atlantic Highlands, N.J.: Humanities Press, 1994); Louis Pojman, ed., *Environmental Ethics: Readings in Theory and Application* (Boston: Jones and Bartlett, 1994); Lori Gruen and Dale Jamieson, eds., *Reflecting on Nature: Readings in Environmental Philosophy* (New York: Oxford University Press, 1994); Larry May and Shari Collins Sharriat, eds., *Applied Ethics: A Multicultural Approach* (Englewood Cliffs, N.J.: Prentice-Hall, 1994); Andrew Brennan, ed., *The Ethics of the Environment* (Brookfield, Vt.: Dartmouth Publishers, 1995).

3. Michael Soulé, *The Tigress and the Little Girl* (manuscript of forthcoming book), Chapter 6, "International Conservation Politics and Programs."

4. Daniel H. Janzen, *Guanacaste National Park: Tropical Ecological and Cultural Restoration* (San José, Costa Rica: Editorial Universidad Estatal a Distancia, 1986). Also David Rains Wallace, "Communing in Costa Rica," *Wilderness* 181 (Summer 1988), which quotes Janzen as wishing to plan "protected areas in a way that will permanently accommodate solitude seeking humans as well as jaguars, tapirs, and sea turtles." These solitude-seeking humans might include biologists, backpackers, deep ecologists, but not, one supposes, indigenous farmers, hunters, or fishermen.

5. Raymond Bonner, *At the Hand of Man: Peril and Hope for Africa's Wildlife* (New York: Alfred A. Knopf, 1993), pp. 36, 65, 70, 85, 221.

6. Arne Kalland, "Seals, Whales and Elephants: Totem Animals and Anti-Use Campaigns," in *Proceedings of the Conference on Responsible Wildlife Management* (Brussels: European Bureau for Conservation and Development, 1994). Also, Arne Kalland, "Management by Totemization: Whale Symbolism and the Anti-Whaling Campaign," *Arctic* 46 (1993).

7. Quoted in *The Deccan Herald,* Bangalore, 5 November 1995.

8. A useful countrywide overview is provided in Ashish Kothari, Saloni Suri, and Neena Singh, "Conservation in India: A New Direction," *Economic and Political Weekly,* 28 October 1995.

9. Arne Naess, "Comments on the Article 'Radical American Environmentalism and Wilderness Preservation: A Third World Critique' by Ramachandra Guha," typescript (1989), p. 23.

10. Recent writings by Indian scholars strongly dispute that conservation can succeed through the punitive guns-and-guards approach favored by most wildlife conservationists, domestic or foreign. For thoughtful suggestions as to how the interests of wild species and the interests of poor humans might be made more compatible, see Kothari et al., "Conservation in India"; M. Gadgil and P. R. S. Rao, "A System of Positive Incentives to Conserve Biodiversity," *Economic and Political Weekly,* 6 August 1994; and R. Sukumar, "Wildlife-Human Conflict in India: An Ecological and Social Perspective," in R. Guha, ed., *Social Ecology* (New Delhi: Oxford University Press, 1994).

11. Arne Naess, *Ecology, Community, and Lifestyle,* translated by David Rothenberg (Cambridge: Cambridge University Press, 1990), p. 45.

Arne Naess

The Third World, Wilderness, and Deep Ecology (1995)

I

THIS ARTICLE IS MOTIVATED by listening to some people from the Third World who express a suspicion that Deep Ecology is a new variant of Western domination and "neocolonialism": they fear that people of the Third World will be pushed out of their homes to make more room for spectacular animals. Some authors have expressed the opinion that Deep Ecology is for the rich nations that can afford the luxury of vast wilderness as habitat for wild species. In my opinion, however, it would indeed be tragic if such ideas were going to spoil the much-needed cooperation between supporters of the Deep Ecology movement throughout the various regions of the globe, including the Third World.

Throughout most of human history, all humans have lived in what we now call wilderness. As Gary Snyder points out:

> Just a few centuries ago, when virtually *all* was wild in North America, wilderness was not something exceptionally severe. Pronghorn and bison trailed through the grasslands, creeks ran full of salmon, there were acres of clams, and grizzlies, cougar, and bighorn sheep were common in the lowlands. There were human beings, too: North America was *all populated*. . . . The fact is, people were everywhere. . . . All of the hills and lakes of Alaska have

been named in one or another of the dozen or so languages spoken by the native people.[1]

Prior to agriculture, our ancestors left few traces. Ecosystems were not appreciably changed for the most part, except by large fires, and probably through the extermination of some large animal species. But for the most part, landscapes and ecosystems were not irreversibly reduced in richness and diversity, and the basic ecological conditions of life were maintained. There is not today, nor was there ever, any essential conflict between humans in moderate numbers and a state of wilderness or wildness. There are reasons today, however, for some areas to be left entirely devoid of human settlement, and for limiting even short carefully arranged visits by scientists to a minimum, but this should be looked upon as an exceptional situation.

At present, there are old growth forests in Australia, for example, which are inhabited by ecologically conscious and careful people. This situation illustrates the essential compatibility of people living in wilderness with a presumably high quality of life—a "rich life with simple means." They use plants for food and other purposes, but they do not, of course, engage in subsistence agriculture.

What is considered a normal lifestyle in industrial countries is clearly incompatible with living in wilderness. Industrial people interfere so severely with natural processes that even a very small number of them can significantly alter the landscape. For example, it is widely recognized that people doing research in the Antarctic should use extreme care not to damage the ecosystems, but it is also clear that the rules are widely disobeyed.[2] Bad habits are difficult, but not impossible, to change!

It is unavoidable that some people concerned with the protection of wildlife and natural ecosystems tend to see a direct and global antagonism between human settlement and wilderness. But supporters of the Deep Ecology movement, like many others, know that wilderness, or wildness, need not be destroyed by people living in these areas (or nearby) and that they may enjoy a high quality of life.

It is not possible for people living in the United States to interfere as little with the wilderness as did the traditional American Indians, and Gary Snyder (and other articulate American supporters of the Deep Ecology movement) insist that there should be no further destruction of wilderness in America. Even what is now set aside in the United States as designated wilderness is interfered with too much. The traditional point of view of the U.S. Forest Service still has a lot of influence: "Wilderness is for people.... The preservation goals established ... are designed to provide values and benefits to society.... Wilderness is not set aside for the sake of its flora or fauna, but for people."[3] It is not only the "*but* for people" that makes all the difference, from Gary Snyder's point of view, but also the term "society." People who live in wilderness, or who have their roots in wilderness, form communities rather than "societies." There is a vast difference between the slogan of the World Wildlife Fund ("wilderness for people") and the meaning of the U.S. Forest Service phrase: "wilderness for American society."

Those people in the United States who are actively trying to stop the destruction of wilderness do not tend to publish *general proposals on how to treat apparently similar problems in the Third World.* At least this is true of theoreticians of the Deep Ecology movement. Nevertheless, there are writers who look upon "radical environmentalism," including Deep Ecology, as a threat to the poverty stricken people of the Third World. The opinion is not uncommon that people in the rich Western world tend to support wild animals and wilderness rather than poor people. However, the real question is: *How* can the poor be helped in a way that is sustainable in the long run?

Close cooperation between supporters of the Deep Ecology movement and ecologically concerned people in the poor countries requires that the latter trust the former's concern for the economic progress of the poor. But what is progress in this case? Is consumerism progress?

The principle formulated by Gary Snyder is applicable in Third World countries: that is, there is no inherent antagonism between human settlement and free nature, for it all depends on the *kind* of culture humans have. It should be a universal goal for mankind to avoid all kinds of consumerism and concentrate, instead, on raising the basic quality of life for humans, including the satisfaction of their economic needs.

The number of poor people in Third World countries is too large for all of them to dwell non-destructively in the tropical forests; more and more subsistence agriculture in these forests serves neither the best interests of the poor, nor does it protect the forests from destruction. Millions of people now live in the tropical forests in a broadly sustainable way; that is, without reducing the richness and diversity of life forms found there. But what is now happening is an *invasion* of these areas resulting in major disruption of the people and the communities who have been living there in harmony. The forests are clear-cut and burned, and subsistence agriculture is introduced. These practices cannot help the poor reach the goals of long-term economic progress. This is true as well of the large industrial operations in the forests and along the rivers.

The present ecological world situation requires a focusing of attention upon *urban* settlements; changing them in ways so that they will be appropriate and habitable places for the thousands of millions of people who now, and in the next century, will need a place to live. This gigantic effort will require mutual help between rich and poor countries. Significant economic progress for the poor is not possible through the extensive use of less fertile lands for agriculture. *There is no way out except through urbanization,* together with the willingness of the rich to buy products from the poor.

It has been pointed out that, from an ecological long-range perspective, the economies of some traditional North American native cultures were superbly sustainable in a broad sense. It has been noted that the philosophical, religious, and mythological basis for these economies, and for their social relations in general, was expressed through sayings which are eminently consistent with the fundamental attitudes found in the Deep Ecology movement. Similar sayings found in Eastern cultures have had an even greater impact. As the Indian social ecologist Ramachandra Guha (who has published what he sees as a Third World critique of Deep Ecology) claims, "This coupling of (ancient) Eastern and (modern) ecological wisdom seemingly helps consolidate the claim that deep ecology is a philosophy of universal significance."[4] The total views suggested among supporters of the Deep Ecology movement do, in a sense, couple "(ancient) Eastern and (modern) ecological wisdom." But there are reasons to be cautious here.

To cherish some of the ecosophic attitudes convincingly demonstrated by people from the East does not imply the doctrinal acceptance of any past definite philosophy or religion conventionally classified as Eastern. Heavy influence does not imply conformity with any beliefs: the history of ideas and contemporary philosophizing are different subjects. At any rate, there is ample reason for supporters of the Deep Ecology movement to refrain from questioning each other's ultimate beliefs. Deep cultural differences are more or less cognitively unbridgeable and will remain so, I hope.

Desperate people (including desperately poor, hungry people) will naturally have a narrow utilitarian attitude towards their environment. But overall, the people of the Third World, apart from the desperate minority, manifest a positive concern for the protection of free nature, and a respect for nonhuman living beings. At least this has been my experience while living among poor people in India, Pakistan and Nepal (and others in the Third World agree with me on this point). Without these experiences, I would not have talked about the international basis of a Deep Ecology movement.

Temporarily pressing problems of material need might monopolize their attention, but this is also true of people in similar circumstances in the West, despite their affluence. In short, there is a sound basis for *global* cooperation between supporters of the Deep Ecology movement and ecologically concerned people in the Third World, and also with people who try to understand and lessen the poverty in those regions. These people cooperate in movements against poverty which do not entail further large-scale deforestation. And there is no tendency to support animals at the expense of humans within the framework of this cooperation.

II

To Social Ecologists in countries which are less affluent than the United States, it may look threatening when environmental activists in the United States declare "an unflinching opposition to human attacks on undisturbed wilderness."[5] Some activists even engage in un-Gandhian ecotage; for instance, destroying vehicles and other machinery while making sure that no one gets hurt in the process. So far, there have been very few authenticated cases of anyone being seriously hurt. Considering the vehemence of these struggles, and the passions involved, this should be considered a great victory.

Clearly, these intense personally involved activists are speaking about wilderness primarily in the United States, not necessarily about the situation on other continents. At least, this is true of supporters of the Deep Ecology movement, but this point can be easily overlooked by observers in the Third World. Unflinching opposition to the cutting down of *any* trees, or to the establishment of *any* new human settlements in any wilderness *what so ever* is a preposterous idea presumably held by no one. The real issue here for the Third World is: How much wilderness and wildlife habitat is it acceptable to continue to modify and destroy, and for what purposes?

In the richest nations of the world, the destruction of old growth forests still goes on. There is ample justification for activists in the United States to focus on these destructive, mindless, irreversible activities. The term "ecocriminality" is a suitable word to use for this forest destruction, and a question of great importance arises here: given their own unecological practices, do the rich nations deserve any *credibility* when preaching ecological responsibility to the poor countries?

One has to distinguish between three things: (1) the present dismal situation concerning the lack of protection of wilderness; (2) the estimates published by conservation biologists concerning the size of wilderness areas needed for continued speciation; and (3) the more-or-less realistic plans put forth by established environmental organizations (e.g. the World Wildlife Fund and the International Union for the Conservation of Nature) concerning how to improve the present state of affairs for protecting wildlife and wild ecosystems. (It should be pointed out that, given the estimates of Frankel and Soulé that an area on the order of six hundred square kilometers is necessary for the speciation of birds and mammals, of course nothing *specific* follows concerning how to achieve what is deemed necessary for this purpose.)

Is the idea that "the biosphere as a whole should be zoned" considered threatening to some in the Third World? Actually, it should be considered more of a threat directed against First World practices than toward any other nations. For, according to Odum and Phillips, establishment of protection zones "may be the only way to limit the destructive impact of our technological-industrial-agribusiness complex upon the earth."[6] This is clearly a warning directed more toward the destructive practices of the First World. Of course, if Third World elites try to copy First World excesses, then the situation would change.

III

The movements supporting the establishment of "green" societies, and for a global Green movement, have their origin among people in the rich countries. It is understandable that they have not had much impact so far among people in the Third World, and that they are met with suspicion. The priorities among First and Third World countries are and, to some extent, must be, different. Furthermore, "green utopias," and even the everyday conceptions of what constitutes "greenness," tend to be rather uniform, as if green societies, in spite of the deeply different cultures and traditions of the world, would look very much alike.

It is to be hoped that there would be no standard green societies, no *Gleichschaltung* of human institutions and behavioral patterns. Hopefully, economically sound societies of Africa, South America, and Southeast Asia would not resemble present rich countries except in certain rather superficial ways.

Some people think that "ecological sustainability" will be attained when policies have been adopted which will protect us from great ecological catastrophes. But it is beneath human dignity to have this as a supreme ecological goal! Ecological sustainability, in a more proper sense, will be achieved only when policies on a global scale protect the full richness and diversity of life forms on the planet. The former goal may be called "narrow"; the latter "wide" ecological sustainability. In short, it is my opinion that *a necessary, but not sufficient, criterion of the fully attained greenness of a society is that it is ecologically sustainable in the wide sense.* (The Bruntland Report admits of various interpretations, but it does envisage a sustainable "developed" country to be one that satisfies the wide sense of ecological sustainability.)

(A small digression: When I do not go into complex argumentation, but just announce that "it is beneath human dignity to aspire to less than *wide* ecological sustainability," I intend to express a personal view (and as with other assertions) thought to be *compatible* with supporters of the Deep Ecology movement. My assertions *that* supporters of the Deep Ecology movement have such and such attitudes or opinions are, of course, more or less certain, and should not be taken to assert that strictly everybody has those attitudes or opinions.)

IV

It should be clear that the realization of wide ecological sustainability will require deep changes in the rich societies of the world having to do, in part, with policies of growth and overconsumptive lifestyles. If we accept that the realization of the goals of the Deep Ecology movement imply wide sustainability, two questions immediately arise: (1) does the realization of wide sustainability presuppose or require acceptance of the views of the Deep Ecology movement? and (2) does the realization of wide sustainability require significant changes in Third World societies?

If we answer "yes" to the first question, this might be interpreted as asserting that the realization of wide sustainability would require that most members of the relevant societies must accept the views of the Deep Ecology movement. As I see it, this is not necessary (and it would imply a change of heart of an extremely unlikely kind!). But a "yes" to the first question might be interpreted as the assertion that a sufficiently strong minority would be needed to bring about wide sustainability. This situation may well arise. (I don't mean to claim here that a definite answer to the first question is conceptually implied. A decisive "no" to this question is thinkable. It does seem clear, however, that the more people who explicitly or implicitly accept the views of the Deep Ecological movement, the better.)

As to question (2), a "yes" answer seems warranted as far as I can judge. In Third World countries at present there is a general tendency to attempt to follow an "economic growth and development" path which emulates the rich countries. This must be avoided, and to avoid it requires significant changes in the orientations of these societies.

What kinds of changes are necessary? A discussion of the nature of these changes has intentionally been left abstract and general in the Deep Ecology "platform." Point 6 of the "platform" states that "Policies must therefore be changed. These policies affect basic economic, technological, and ideological structures. The resulting state of affairs will be deeply different from the present."

It is obviously pertinent to ask: "Exactly which changes need to be made?" But times change. A short answer to this question seems much more difficult to provide in 1993 than it was in 1970. Practically every major concrete change envisaged in 1970 today seems either more difficult to realize, or not unreservedly desirable in the form it was proposed in 1970.

As a preliminary to serious practical discussion, one must specify which country, state, region, society, and community one has in mind. The distinction First World/Second World/Third World/Fourth World is still relevant. But practically all Deep Ecology literature has focused on rich countries even though there are many supporters in other kinds of countries. The "sustainability" literature is fortunately more diverse.

As an example of social and political change that was highly recommended in 1970, but not in the nineties, one may mention various forms of decentralization. Today the global nature of all the major ecological problems is widely recognized, along with the stubborn resistance of most local, regional, and national groups to give global concerns priority over the less-than-global, even when this is obviously necessary in order to attain wide global sustainability. To the slogan "Think globally, act locally" should be added a new one: "Think globally, act globally." Even if we take it for granted that your body is geographically at a definite place, nevertheless every action influences the Earth, and many of these may be roughly positive or negative. Actions are global in whatever locality you act. Many fierce local or regional conflicts have a global character, crossing every border and level of standard of living.

The moderately poor people in the Second and Third Worlds may seem more helpless, for example, than the coastal people of rich Arctic Norway, but the ecological conflicts are, to a remarkable degree, of the same kind. Communities who live largely by fishing within a day's distance from land in Arctic Norway are in extreme difficulties, because the resources of the Norway Sea, and even the vast uniquely rich Barent Sea, have been badly depleted. For the coastal people it is "a question of sheer survival" but, because Norway is a rich welfare state of sorts, there is no chance that they will go to bed hungry. If the policy makers had seen the intrinsic value, the inherent greatness of the ocean with its fullness of life (and not *only* its narrow usefulness as the source of big profits; e.g., trawling, ocean-factories), then the coastal people could have retained their way of life. They would not have lost their self-esteem by having to migrate to the cities. The supporters of the Deep Ecology movement in the rich countries are not in conflict with Deep Ecology supporters in the Third and Fourth worlds. Such behavior would be strange indeed, because the global perspective reveals the basic similarity of the situation among poor and rich.

The Sami people (wrongly referred to as Lapplanders), a Fourth World nomadic people living in the Arctic Soviet Union, Finland, Sweden, and Norway, have resisted being completely dominated by these four powerful states for the last four hundred years. When a big dam was proposed in their lands (as part of an unnecessary hydroelectric development), thousands of First World people joined them in protest. When a Sami was arrested for standing "unlawfully" on the shore of the river, the police asked him, "Why do you stay here?" He answered, "This place is part of myself." I know of no major ecological conflict anywhere which has not manifested the power and initiative of people who are not alienated from "free nature," but who protect it for its own sake as something which has meaning in itself, independently of its narrow human utility. This kind of motivation for protection of "free nature" adds substantially to the strong, but narrow, utilitarian motivation.

Sometimes the "environmental concern" of poor Third World communities seems to Westerners to relate to the "environment per se." As an example, the people of the Buddhist community of Beding (Peding) in the Rolwaling Himalaya live with the majestic holy mountain Gauri Shankar (Tseringma) straight above their heads. It has long been the object of religious respect. Some of us (mountaineers and Deep Ecology supporters) asked the people whether they wished to enjoy the profits they would get from expeditions by Westerners and Japanese trying to "conquer the mountain," or whether they preferred to *protect the Mountain itself* from being trodden upon by humans with no respect for its cultural status. The families of the community came together and unanimously voted for protection. I had the honor of walking for a week with the chief of the community, Gonden, to deliver a document addressed to the King of Nepal in Kathmandu, asking him to prohibit the climbing of Gauri Shankar. There was no reply. The rich Hindu government of Nepal is economically interested in big expeditions, and the opinion of the faraway Buddhist communities of poor people carry little weight.

The work of Vandana Shiva and others shows how women in rural India continue to try to protect an economy that is largely ecologically sustainable. But do they have the power to resist Western inspired unecological development?

Consider an example from Africa. Large areas where the Masai live may be classified as areas of "free nature," if not wilderness. The Masai are not disturbed by the vast populations of spectacular animals on their lands, such as lions and leopards, together with hundreds of others, nor are these animals severely disturbed by the Masai. For a long time, there has been a remarkable compatibility between people and wild animals. As more or less nomadic herders, the Masai do not need land set apart for agricultural purposes.

What holds true for the Masai holds as well for a great number of other peoples and cultures in the Third World. Ecologically sustainable development may proceed in direct continuity with their traditional culture as long as population pressures remain moderate.

Lately, the Masai have been using more and more money for motor vehicles and other products they don't make themselves. This makes it tempting to sell parts of their territory to farmers looking for land for their many children. From the point of view of economic development, such sales are unfortunate because the relevant kind of subsistence agriculture does not lead to economic progress. The Masai can get sufficient cash through very carefully managed tourism and still have the traditional use of the land and preserve their cultural continuity. Some supporters of the Deep Ecology movement are working with the Masai to help them keep what is left of their land intact. An increase in subsistence farming, in this situation, is a blind alley. But the alternatives are all problematic and there are no easy answers to be found anywhere.

Individual arguments can be singled out and used and misused to defend a variety of mutually incompatible conclusions. In his paper, Ramachandra Guha warns that such is the case with arguments used by supporters of the Deep Ecology movement. And this does not happen only to Deep Ecology supporters in the United States.

After a speech I gave in Norway in favor of considering the Barent Sea seriously as a whole complex ecosystem (together with treating the living beings, including the tiny flagellates, as having intrinsic value) the politician considered to be the most powerful proponent of big fishing interests is said to have remarked, "Naess is of course more concerned about flagellates than about people." My point was that the present tragic situation for fishermen could have been avoided if policy makers had shown a little more respect for all life, not less respect for people. In every such case, one has reason to say that communication on the part of the supporters of Deep Ecology was imperfect. In this case, I certainly should have talked more about people than I did, but not to the exclusion of flagellates, radiolarians and all the other life forms which attract the interest of only a minority of people, and certainly not to the exclusion of ecosystems as a whole.

In 1985, at the international conservation biology conference in Michigan, a representative of a Third World country stood up and asked, "What about *our* problems?" Of course, it was strange for this person, and other representatives of the tropical countries, to hear discussions, day after day, on the future of biological processes in their countries without touching the main social and economic problems facing them. If the conference had been organized by the Green movement, the agenda would have been somewhat different. The discussions concerning how to deal with the ecological crisis would have taken up, let us say, only one third of the time. The other two-thirds would have concerned mainly social problems ("social justice," I would say) and peace. The representatives of the Third World could have introduced the latter two areas of concern and could have stressed that efforts to protect what is left of the richness and diversity of Life on Earth must not interfere with efforts to solve the main problems they have today.

Supporters of the Deep Ecology movement, however, might have raised the following question for discussion. "How can the increasing global interest in protecting all Life on Earth be used to further the cause of genuine economic progress and social justice in the Third World?"

Such questions will inevitably bring forth different and, in part, incompatible proposals. But as we explore these incompatible proposals, we must never lose sight of the importance of all humans everywhere of preserving the richness and diversity of Life on Earth.

NOTES

1. Gary Snyder, *The Practice of the Wild* (San Francisco: North Point Press, 1990), pp. 6–7.

2. See the Greenpeace report: *Greenpeace Antarctic Expedition 1989/90,* p. 53. Waste disposal procedures have improved but "many more changes are still needed if stations are to comply with the new waste disposal guidelines contained in ATCM Recommendation XV-3. Indeed, most stations have not even met the minimal guidelines agreed to by the treaty States in 1975."

3. Quoted from a valuable survey of the wilderness issues: George Sessions, "Ecocentrism, Wilderness, and Global Ecosystem Protection," in *The Wilderness*

Condition: Essays on Environment and Civilization, edited by Max Oelschlaeger (San Francisco: Sierra Club Books, 1992).

4. Ramachandra Guha, "Radical American Environmentalism and Wilderness Preservation: A Third World Critique," in *Environmental Ethics* 11 (1989). [Included in this volume.]

5. Ibid., p. 72. "In contrast to the conventional lobbying efforts of environmental professionals based in Washington, [Earth First!] proposes a militant defense of 'Mother Earth,' and unflinching opposition to human attacks on undisturbed wilderness."

6. Sessions, "Ecocentrism," p. 112.

Arturo Gómez-Pompa and Andrea Kaus[1]

Taming the Wilderness Myth (1992)

Despite nearly a century of propaganda, conservation still proceeds at a snail's pace; progress still consists largely of letterhead pieties and convention oratory. On the back forty we still slip two steps backward for each forward stride. The usual answer to this dilemma is "more conservation education." No one will debate this, but is it certain that only the volume of education needs stepping up? Is something lacking in the content as well?

Aldo Leopold

NEVER BEFORE HAS THE WESTERN WORLD been so concerned with issues relating to humankind's relationship with the environment. As concerned members of this industrialized civilization, we have recognized that humanity is an integral part of the biosphere, at once the transformer and the self-appointed protector of the world. We assume that we have the answers. We assume that our perceptions of environmental problems and their solutions are the correct ones, based as they are on Western rational thought and scientific analysis. And we often present the preservation of wilderness as part of the solution toward a better planet under the presumption that we know what is to be preserved and how it is to be managed.

However, we need to evaluate carefully our own views of the environment and our own self-interests for its future use. Until now, a key component of the environmental solution has been left out of our conservation policies and education. The perspectives of the rural populations are missing in our concept of conservation. Many environmental education programs are strongly biased by elitist urban perceptions of the environment and issues of the urban world. This approach is incomplete and insufficient to deal with the complex context of conservation efforts at home and abroad. It neglects the perceptions and experience of the rural population, the people most closely linked to the land, who have a firsthand understanding of their surrounding natural environment as a teacher and provider. It neglects those who are most directly affected by the current policy decisions that are made in urban settings regarding natural resource use. It neglects those who feed us.

Environmental policies and education reflect a collective perception of nature, the consolidation of what is held to be true about the natural world and what is necessary to pass on to future generations. And this perception underlies and shapes the visions of alternative actions and appropriate actions formed by individuals and conservation groups. How accurate and sound is this vision? Our perceptions and knowledge of the environment are based on common beliefs, basic experience, and scientific research. Through time and generations, certain patterns of thought and behavior have been accepted and developed into what can be termed a Western tradition of environmental thought and conservation.

WESTERN CONCEPTS OF WILDERNESS

Traditional conservationist beliefs have generally held that there is an inverse relationship between human actions and the well-being of the natural environment. The natural environment and the urban world are viewed as a dichotomy and the concern is usually focused on those human actions that negatively affect the quality of life by urban standards. Mountains, deserts, forests, and wildlife all make up that which is conceived as "wilderness," an area enhanced and maintained in the absence of people. According to the 1964 U.S. Wilderness Act, wilderness is a place "where man himself is a visitor who does not remain." These areas are seen as pristine environments similar to those that existed before human interference, delicately balanced ecosystems that need to be preserved for our enjoyment and use and that of future generations. The wilderness is valued for its intrinsic worth, as places of reverence for nature, as sacred places for the preservation of the wilderness image.[2]

Wild lands are also seen as areas useful to modern civilization. They are presented to the public as natural-resource banks of biodiversity that merit protection from human actions and as outdoor laboratories that deserve unhindered exploration by the scientific community. And they are seen as a vital part of the environmental machinery that must be maintained to provide an acceptable quality of life in developed areas, as exemplified by current concerns about air pollution, global climate change, and deforestation. All of these concepts fall under the general term *conservation,* yet they represent mostly urban beliefs and aspirations. All too often they do not correspond with scientific findings or first-hand experience of how the world works.

In addition, the validity of widely accepted environmental truths needs to be challenged, from our belief in the virgin nature of the tropical forests to our newly developing thoughts on global warming. Scientific findings are often accepted as if they are the gospel word. But a scientific truth is really a conclusion drawn from a limited data set. It is an explanation of what scientists know to date about a topic, based on their training and interpretation of the information available. It may be replaced by another truth in light of new information that does not fit the old paradigm.

Concepts of climax communities and ecological equilibrium, for example, have been used for most of this century as a basis for scientific research, resource management, and conservation teaching. As long-term studies are analyzed and their findings tested against the old truths, the previous paradigms are being challenged.[3] Today, few ecologists defend the equilibrium and climax concepts. Nonequilibrium models now influence ecological theory, and nature is increasingly perceived as being in a state of continuous change. Some changes are in part random and independent of each other, whereas others are human-induced.

Other accepted truths relating to the environment are myths about nature that come from nonscientific sources. For instance, the concept of wilderness as an area without people has influenced thought and policy throughout the development of the western world.[4] People see in the wilderness a window to the past, to the remote beginnings of humankind long before the comforts of modern life. We wish to set aside and preserve that which both reminds us of our place in evolutionary time and contrasts with our beliefs of human nature. Yet, recent research indicates that much wilderness has long been influenced by human activities.[5]

> The ongoing public conversation about the environment is grounded in the ancient dichotomy of man versus nature. So far we have sought to resolve the argument through a series of truces—either sequestering large tracts of wilderness in a state of imagined innocence, say, or limiting the ways in which man can domesticate nature's imagined savagery.[6]

The Western world has also seen the wilderness as a challenge, a frontier to be tamed and managed. Agricultural landscapes are often admired for their intrinsic beauty, as living masterpieces created by human hands from the wild. They are confirmation of an underlying belief in human technological superiority over primal forces. It confirms our faith in our ability to manage the environment, a legacy from the Industrial Revolution rooted in the concept of progress and a biblical notion of human dominion over nature. In Genesis (1:28), God says to Adam and Eve, "Be fruitful, and multiply, and fill the earth, and subdue it."

The danger is that this theoretical delineation between the realms of civilized and wild, of the intrinsic value of each realm separately, and of human mastery over natural forces, has only too tangible consequences. Emerging from Western history and experience in temperate zones, a belief in an untouched and untouchable wilderness has permeated global policies and politics in resource management from the tropics to the deserts, causing serious environmental problems.

We must begin to challenge some of our most fundamental and contradictory beliefs regarding the natural environment: the scientific capacity and knowledge available to harness and manage nature the way we see fit, and the perceived pristine state of uninhabited areas. Both beliefs, combined with the concept of the balance of nature, have led to unrealistic and contradictory tenets in our natural-resource management policies. On the utilitarian side, these policies are permeated with an acceptance of destructive practices, generated from a belief that mitigating measures can halt or reverse environmental depletion and degradation. Yet, on the preservationist side, conventional resource-management policy also includes practices based on the belief that setting aside so-called pristine tracts of land will automatically preserve their biological integrity. Neither belief takes into consideration the possibilities for natural-resource management that might arise from the integration of alternative environmental perceptions and current scientific information.

ALTERNATIVE PERCEPTIONS AND CONSERVATION PRACTICES

The concept of wilderness as the untouched or untamed land is mostly an urban perception, the view of people who are far removed from the natural environment they depend on for raw resources. The inhabitants of rural areas have different views of the areas that urbanites designate as wilderness, and they base their land-use and resource management practices on these alternative visions. Indigenous groups in the tropics, for example, do not consider the tropical forest environment to be wild; it is their home. To them, the urban setting might be perceived as a wilderness.

> As a city dweller never looks at bricks, so the Indian never looks at a tree. There are saplings for making bows, and jatoba for making canoes, and certain branches where animals like to sit, but there are never trees noticeable for self-conscious reasons—beauty, terror, wonder.[7]

Many agriculturalists enter into a personal relationship with the environment. Nature is no longer an object, an *it,* but a world of complexity whose living components are often personified and deified in local myths. Some of these myths are based on generations of experience, and their representations of ecological relationships may be closer to reality than scientific knowledge. Conservation may not be part of their vocabulary, but it is part of their lifeway and their perceptions of the human relationship to the natural world.

Throughout the world, communally held resources have been managed and conserved by diverse human societies via cultural mechanisms that attach symbolic and social significance to land and resources beyond their immediate extractive value.[8] In the Brazilian Amazon, the Kayapó belief system and ecological management, as described by Posey,[9] revolves around maintaining an energy balance between the natural and spiritual world by regulating animal and plant use via ritual and custom. Indigenous fishermen of northern California used to place a ritual moratorium on fishing during the first few days of the salmon runs, thereby protecting the perpetuation of the salmon resource and maintaining intergroup relations along the river.[10]

External economic and political demands for natural resources have placed conflicting demands on land and resources maintained by local inhabitants. Often backed by powerful government or corporate business interests, conflicting perceptions of how the land and resources should be used have led to the replacement or collapse of previous resource management systems and subsequent unrestricted or uneducated use of the region. In Chiapas, Mexico, for example, the Lacandon Maya perceived the forest as the provider of subsistence. Forests were converted temporarily to agricultural lands for corn, beans, and squash within a shifting cultivation system, or the forest fallows managed to attract wildlife.[11] Before the entry of outsider groups with other objectives and other interests, Maya people lived within the tropical ecosystem of southern Mexico and Guatemala for centuries in ways that allowed continuous forest regeneration. Yet, the majority of Maya groups that inhabited the Lacandon forest were never consulted in the government policy decisions of land use that ultimately led to its destruction.

These same lands have been, and still are, viewed from the outside as lands to be conquered, colonized, grazed, or preserved. The forests contain hardwoods valuable on the international market. The cleared forest then provides land for the landless and pastures for the cattle industry. Deforestation is not perceived as a problem by the representatives of these interests but rather as a mechanism by which to gain land tenure rights. Traditional conservationists, on the other hand, see the aesthetic, biological, and ecological value of the same land but do not necessarily see the people. They often fail to see the effects of past or current human actions, to differentiate among types of human use, or to recognize the economic value of sustainable use.

The well-known cycle of initial lumber or mineral extraction followed by colonization, land acquisition, and subsequent conversion to pasture lands has been a common denominator in most of the American tropics.[12] Though we tend to focus on the actions of the local people, on what is immediately observable, such actions are often the result of higher-level policies, such as government concessions to extractive industries.[13]

Even with the documentation of this cycle, even with the evidence that it is our own outside interests that are ultimately responsible for the greater part of tropical deforestation, we continue to place the blame on poverty and on the land practices of the rural sector when they are only the visible symptoms of much deeper underlying problems. More important, our beliefs and assumptions blind us to the fact that, in many cases, the traditional land-use practices of the rural sector are responsible for maintaining and protecting the biodiversity of our wilderness and have often provided the genetic diversity that strengthens the world's major food crop varieties.[14]

FOOTPRINTS IN THE WILDERNESS

Scientific findings indicate that virtually every part of the globe, from the boreal forests to the humid tropics, has been inhabited, modified, or managed throughout our human past.[15] Although they may appear untouched, many of the last refuges of wilderness our society wishes to protect are inhabited and have been so for millennia. In any current dialogue regarding tropical forests, for instance, the Amazon Basin is usually mentioned as a vital area to be left untouched and protected. Yet, archaeological, historical, and ecological evidence increasingly shows not only a high density of human populations in the past and sites of continuous human occupation over many centuries but an intensively managed and constantly changing environment as well.[16]

The Amazon is still the homeland for many indigenous groups who have inhabited the area since long before the arrival of Europeans, and it contains the resources on which they and other nonindigenous people depend for their livelihood. The Kayapó of Central Brazil currently occupy a two-million hectare Indian reserve, but at one time they practiced their nomadic agriculture in an area approximately the size of France.[17] In addition, new evidence from the Maya region suggests that the seemingly natural forests we are trying to protect from our version of civilization supported high densities of human populations and were managed by past civilizations.

Present-day parks, reserves, and refuges in the region are filled with archaeological sites. According to Turner[18] the Maya population of southeastern Mexico may have ranged from 150 to 500 people per km^2 in the Late Classic Period, contrasting sharply with current population densities of 4.5 to 28.1 people per km^2 in the same region.[19] These past civilizations apparently managed the forests for food, fiber, wood, fuel, resins, and medicines.[20] Many of the tree species now dominant in the mature vegetation of tropical areas were and still are the same species protected, spared, or planted in the land cleared for crops as part of the practice of shifting agriculture.[21]

It is only relatively recently that research on shifting agriculture and other tropical production systems has started to change from its previous focus on the initially cleared field to an examination of the management of the fallow after abandonment of the area for annual crops. The current composition of mature vegetation may well be the legacy of past civilizations, the heritage of cultivated fields and managed forests abandoned hundreds of years ago. Our late realization of this possibility stems from the long-held belief that only cleared and planted areas are managed, as in the ploughed fields of our experience, and that mature vegetation represents a climax community, a stable endpoint reflecting the order of nature given no human interference. Until we understand and teach that the tropical forests are "both artifact and habitat,"[22] we will be advocating policies for a mythical pristine environment that exists only in our imagination.

As our knowledge and understanding of the anthropogenic influence on the composition of mature vegetation increases, it is necessary to redefine and qualify what is meant by undisturbed habitat. The issue is not simply the presence or density of humans, but the tools, technologies, techniques, knowledge, and experience that accompany a given society's production system. The ancient societies discussed previously, for example, were more closely bound to the local environment and more dependent on regional resources for basic subsistence. Increased productivity would have come from principally internal modifications and increased human labor for more intensive ecosystem management. Production systems that were viable remained; those that failed disappeared.

In contrast, modern production systems have advanced technologies, from chemical fertilizers to hydroelectric dams, that are external to the local environment. These technologies have the potential to impose irreversible transformations on the environment that cannot be predicted by traditional knowledge (i.e., cumulative knowledge specific to the local environment). What is recognized by the environmental and conservation movements is an ability to destroy the environment at a much greater scale than ever before in human history. When we speak of protecting undisturbed habitat or wilderness, then, it is important to clarify that the word *undisturbed* refers to the absence of disturbance by modern technologies.

However, not all modern societies use destructive technologies, and the benefits of human interference in ecological processes are not restricted to tropical zones or past times. Present-day farmers in remote areas all over the world have managed, conserved, and even created some of the biodiversity we value so highly.[23] In the Sonoran Desert, a study of two oases on either side of the Mexico/United States border indicates that the customary land-use practices of Papago farmers on the Mexican side of the border contributed to the biodiversity of the oasis. In turn, the protection from land use of an oasis 54 km to the northwest, within the U.S. Organ Pipe Cactus National Monument, resulted in a decline in the species diversity over a 25-year period.[24]

In addition, many rare varieties and species related to our major food crops can be found maintained within or bordering agricultural fields in cultivated regions. In the Sierra de Manantlan (Jalisco, Mexico), the discovery of a new perennial corn, *Zea diploperennis,* led to the establishment of a biosphere reserve to protect this species and the ecosystem in which it survives.[25] (Biosphere reserves are part of an international reserve system established by the UNESCO Man and the Biosphere Programme, which contain zones of human use within the overall management agenda; theoretically, a biosphere reserve integrates the goals and strategies of conservation, development, research, and education.) The difficulty is that *Z. diploperennis* is a secondary species that grows in abandoned cornfields. To protect the species, the slash-and-burn techniques of this form of traditional agriculture have to be continued to provide the habitat that it requires. Without all the human cultural practices that go with the habitat, the species will be lost forever. Yet, this dimension of conservation has been neglected in our own tradition of natural-resource management.

ANTHROPOGENIC FIRES IN NATURAL-RESOURCE MANAGEMENT

It is of utmost importance to understand both the beneficial and destructive ecological consequences of anthropogenic perturbations and to incorporate this knowledge in research and education programs. Future scientists, leaders, farmers, fishermen, and ranchers need to be exposed to alternative views and taught to see natural environmental issues within their historical, social, and cultural contexts. The view of the white ashes of forest trees that have been felled and burned for an agricultural plot may appear to an urbanite outsider to be a desecration of the wilderness, but a farmer may see it as an essential stage of renewal. One could argue that the trees felled are the representatives of rare and endangered species, and, in selected sites, this argument might be reasonable. However, most often, many of the cut or burned trunks resprout, providing the bulk of the new forest.

Slash-and-burn agriculture has been an integral part of the tropical forests ecosystems for millennia. This ancient form of agriculture is not to be confused with the widespread destructive fires set by recent colonists or squatters who have little local experience or land-tenure security. Fire is today used to open new forest land, often on the edge of new timber or mining roads, or, even worse, used as a mechanism to vent anger at the impotence of poverty and inadequate government programs. Though such rapid clearing of the forest by landless peasants is also improperly referred to as slash-and-burn or shifting agriculture, in reality the planted areas are not fallowed, but successively replanted and eventually abandoned. This sequence is very different from the continual process of clearing, planting, and fallowing that is typical of more long-standing forms of shifting agriculture, which creates a mosaic of different ages of forest growth, including large patches of mature vegetation.

To give a concrete example, when a major forest fire in 1989 burned 120,000 hectares near Cancún, Mexico, the news media conveyed an image of ecocide, covering the fire's daily progress with statements about the extinction of species and the loss of invaluable forests. Environmentalists, conservationists, and most nongovernmental organizations connected to environmental issues protested the lack of any fire management plan to prevent, stop, or control forest fires. Yet, no attempt was made to understand why a fire of this magnitude had occurred in the first place.

The Cancún fire began in several different places at the same time, and its cause remains unclear. It is unlikely that it was the result of an agricultural fire that escaped from an area cleared for crops. In all of the tropical Maya region, no official form of fire control has ever existed beyond that of the individual farmers. Yet, fires have seldom been as large or extensive as this one. Agricultural fires are carefully controlled by the farmers. One of the most critical decisions they must make is when to burn the slash: when the conditions are finally dry enough, but before the first seasonal rains. The farmers know the winds, the annual climatic shifts, and past fire histories. They know how to control the size and intensity of their fires to protect the neighboring forests from burning.

The patchy mosaic of forests, forest fallows, and agricultural plots is an ideal landscape for controlling forest fires. The view from a helicopter flying over the burned area around Cancún revealed that the line of fire had stopped in areas of slash-and-burn agriculture. Local residents and forest authorities say that the forest burned most dramatically in areas where the valuable woods had been mined out and been subsequently devastated by Hurricane Gilberto.[26] The actual commercial and biological value of the forest was low. Biological surveys indicate that the burned zone was in fact not rich in endemic organisms.[27]

Although Mexico contains a multitude of unique sites in which rare and endangered species are truly threatened, these sites have not received the same visibility in the public consciousness as did the Cancún fire. But neither are they situated so close to a major international tourist location. The concern over the Cancún fire was due to a desire to have an attractive forest landscape for the increasing tourist trade in the area. This concern is not necessarily invalid, but the entire fire coverage was presented out of context and based on unfounded arguments.[28] The underlying problem was the general lack of understanding of the ecological processes that shape forests and landscapes. We confuse all too easily the great need to protect rare and endangered species with the protection of wilderness, and we confuse our admiration of forests with the conservation of nature.

Cancún is not an isolated example. Ongoing research on the chaparral environments on both sides of the Mexico/United States California border has shown the role of fire in combating fire.[29] These studies indicate that the mosaic vegetation pattern in Baja California, the result of repeated small burns, has prevented the large catastrophic burns so characteristic of the equivalent ecological zone in Southern California. The composition and structure of so-called virgin forests and wilderness areas are in part artifacts of previous burns, both natural and anthropogenic.[30] A policy of fire suppression in the United States has eliminated natural barriers to fires.

Fire control in wilderness areas, from the big trees of California to the Northern and Northeastern forests, has also led to undesirable changes in their environments.[31]

Due to our limited knowledge of the role and experience of local human populations in managing fire, fire suppression remains the dominant policy in our management of natural resources and many national parks. We fear, and are trying to prevent, a repeat of the 1988 fire in Yellowstone National Park without fully understanding the underlying causes for its great extent, intensity, and damage. In addition, without the knowledge of the role of fire in a given ecosystem, we have developed areas so that they can no longer be subjected to prescribed burns without great risk. Yet, in so doing, these areas are also at risk from fires that cannot be controlled once they start.

THE INTEGRATION OF ALTERNATIVE VIEWS OF THE ENVIRONMENT

The differences between the perceived and actual environmental effects of forest fires, fire suppression, shifting agriculture, or preservationist policies are only a few examples of the contradictions and confusion that exist in connection with environmental issues and conservation. In the city or in the rural areas, inaccurate information is passed from our own education system to the rest of society and the next generations of environmental users, managers, and abusers. Research and education programs need to be redesigned to inform urban as well as rural populations (from children to adults) about appropriate and alternative resource management practices and policies. Most policy agendas and education curricula neglect rural perceptions of the environment or traditional systems of food production and resource management. They do not address the current difficulties confronted by these systems and lifeways or their contributions to conservation and our own survival. Beyond opening our eyes to the realities of the areas we call wilderness, we must learn how to listen to their caretakers (both good and bad ones) to include local needs, experience, and aspirations within our perspectives.[32]

To adjust our recommendations for better use of the environment to reflect reality rather than myths, we must learn how local inhabitants in rural areas understand their environment and must bring this vision into both the urban and rural classrooms. The first step is to recognize that conservation traditions exist in other cultural practices and beliefs that are separate from Western traditional conservation. The rural sector is not a homogeneous group, however, and research and education efforts also need to be directed toward the social or economic constraints and incentives that lead to destructive practices or conflict with institutional conservation policies.

Several priorities for research and education programs can be mentioned to improve the information and alternatives available for natural resource management programs and future resource managers:

- Research on the influence of human activities on past and present environments to understand the influence of all forms of management, whether modern or traditional, intensive or extensive, on the shape and content of the environment.
- Long-term monitoring of environmental change that includes the social and economic variables affecting such change.
- Documentation of the views and perceptions of nature and conservation found in rural populations and integration of these beliefs and corresponding empirical realities in the general pool of collective knowledge. Knowledge of the beliefs, constraints, and aspirations of local residents in ecologically fragile lands will aid in coordinating conservation and rural development policy and practice.
- Continued emphasis on the coordination of research efforts in different scientific disciplines to present conservation and management alternatives with balanced representations of the different interests in conservation and rural development.
- Collaboration with interested individuals in the rural sector to establish demonstration and experimental sites for resource management alternatives and environmental restoration techniques.
- Development of environmental educational programs that integrate the knowledge and experience of scientists, educators, and local practitioners. This development should include programs that not only take scientists and educators out to rural communities, but also

encourage rural residents with successful land-use techniques to teach, whether in their own communities, other rural areas, or the urban setting.

- Development of graduate programs in conservation and natural-resource management that train a new generation of professionals, scientists, and decision makers with a view of conservation issues that includes the role of humans in both environmental deterioration and enrichment. These programs need to instill a sense of the tremendous responsibility current and future generations have to shape our own environment and of the risks of failing.

We have reached a time where the lines are not drawn between the known and unknown but between belief systems. This situation leads to an unfortunate set of circumstances in which we divide ourselves on issues where our opponents are not villains. They are often others who, like us, are working toward the protection of the environment. Yet, we line up behind the banners of preservation, conservation, development, or restoration and then subdivide on points of human involvement, responsibility, and equity in resource management. The only realities left between these polemics are the resources and the people who use them. This middle ground is where future research and education needs to be concentrated if we are to emerge from this seeming morass of controversial arguments held at a level far removed from the field.

As scientists or conservationists, we need to enter the field, literally. We speak of local participation and of developing a dialogue among the rural, research, and education communities. However, the presence of local rural residents in a classroom or conference hall does not necessarily engender participation. These locations and procedures are a standard part of our own traditional education process; they are unlikely to be familiar to the majority of indigenous or remote communities and unlikely to be conducive to an exchange of information among researchers and local people.

We sometimes forget that experience is often the best teacher and place great emphasis instead on the letters that proceed or follow a person's name, on the way that person talks and on the material he or she writes. In so doing, we have created a barrier of formally structured education and language that is imposing to rural populations. One rancher from North

Mexico once commented on the researchers with whom he had worked, "We tell them what it is like here, but they write about it differently."[33]

We know, in fact, very little about how environmental knowledge held by farmers, ranchers, fishermen, hunters, and gatherers from the deserts to the tropics is passed from one generation or society to another. This understanding requires learning the setting and language that people use to describe their environment and their relationship with the land. It implies understanding the underlying concepts of their words and the corresponding actions that are considered appropriate. Such environmental perceptions may not exactly match concepts of sustainable use or restricted access to limited or fragile resources, but overlaps of conservation concepts and practices do exist.

In an informal survey,[34] 15 people of a remote region of Durango, Mexico, were asked what the word *conservación* meant. None of them knew. "No," they replied, shaking their heads. "¿Qué será? (What would it mean?)" Earlier, one man from this group had pointed out ways in which he and his family were trying to protect the rangeland from the effects of drought and overgrazing and to protect the wildlife from poachers. When asked why, he turned in his saddle, viewing the range stretching away from him into the distance, and said, "Hay que cuidar, ¿verdad? [You've got to care for it, don't you?]"

A shared perception of caring for the land can be emphasized in conservation policy and education. However, integrating this perception requires acknowledging the presence of humans in wilderness areas. Part of the problem in working with local people stems from our perception of wilderness as uninhabited. The attention automatically falls on the land first and the people second. We think of local people living in the buffer zone surrounding an uninhabited area and do not stop to consider that perhaps the buffer zone should be the principal area of conservation.

Botkin[35] describes how resource management policies to both protect and control elephant populations in the Tsavo National Park of East Africa led to a severe deterioration of the land within the park boundaries. The inhabited area surrounding the park remained forested. The clear demarcation of the boundaries in the LANDSAT images and aerial photos appeared "as a photographic negative of one's expectation of a park. Rather than an island of green in a wasted landscape, Tsavo appeared as a wasted island amid a green land."[36]

Perceptions of wilderness and protected areas as uninhabited means that local-level collaboration is often neglected or considered only as an afterthought and in terms of our own priorities. We talk easily of the role of local people in our conservation programs but often do not stop to think of the role we play in their lives. Local cooperation, participation, or collaboration are not free commodities. They influence lives and futures and both deserve and require negotiation. In the Chihuahuan Desert, for example, inhabitants in the region of the Mapimí Biosphere Reserve have included a policy of wildlife conservation and an ecological research program within their own lifeway. Their willingness to stop eating the endangered Bolsón tortoise, *Gopherus flavomarginatus,* and protect it from poaching has resulted in an increase in population of this endemic species within the reserve. In turn, the researchers have opened a window to another world outside that arid basin by providing a vision of the national importance and value of local resources and efforts.

However, the local-level efforts up to now have not been equal. Some of the local people say they have benefited the reserve more than the reserve has benefited them.[37] Why, then, did the local people accept the researchers in the first place? They say it was for *la convivencia,* the willingness of the initial researchers to live and work side by side with them, to accept their help and advice, and to include their concerns in the decision-making process. It was a matter of trust. The local people trust that their perceptions, their world, will be part of what is taught to others who have never set foot in the Bolsón de Mapimí, part of what is taken into consideration by those who wish to alter either local land use or the reserve management.

ENVIRONMENTAL CONSERVATION RESPONSIBILITIES

Cooperative relationships with local residents in ecologically fragile areas are of utmost importance to our understanding of the natural environment and the effects of resource use. Yet, we cannot neglect our responsibilities in such relationships or underestimate the effect (positive or negative) that we can have on a rural community. We need to contribute in turn and impart the information to which we have access. In this way, local people can come to understand their situation in a larger context and make informed decisions about their lives and land. But it also means orienting some of the research toward local benefits and including local-level perspectives in research design and dissemination. More important, it means including the local people in the same education process that we, ourselves, are undergoing to understand the natural environment and society's effects on it.

The benefits of local-level collaboration to our urban communities are perhaps even more than we can realistically offer in return. Perceptions, knowledge, and experience of the rural sector incorporated into the urban classroom can guide our global civilization to more informed decisions about what is termed *wilderness* and about what is meant by *conservation.* The wilderness we have envisioned up to this point is not the same when viewed from the field. In reality, the frontier does not exist between people and the wild, but between the known and unknown.

The point here is not to create a new myth or fall into the trap of the "ecologically noble savage."[38] Not all farmers or ranchers are sages, folk scientists, or unrecognized conservationists. Yet, within the rural sector can be found individuals who directly depend on the land for their physical and cultural subsistence. And within that group of individuals exists a set of knowledge about that terrain, a knowledge of successes and failures that should be taken into account in our environmental assessments. Currently, we are discussing and designing policies for something about which we still know little. And those who do know more have rarely been included in the discussion. The fundamental challenge is not to conserve the wilderness, but to tame the myth with an understanding that humans are not apart from nature.

NOTES

1. The authors would like to thank David Bainbridge and Denise Brown for their timely comments and suggestions and the external reviewers of *BioScience* for their excellent critical assessments and arguments. This article is based on a presentation given at the 19th Annual Conference of the North American Association for Environmental Education, 2–7 November 1990, in San Antonio, Texas.

2. R. F. Nash, "Why Wilderness?" in V. Martin, ed., *For the Conservation of Earth* (Golden, Col.: Fulcrum, 1988), pp. 194–201.

3. See D. B. Botkin, *Discordant Harmonies: A New Ecology for the Twenty-First Century* (New York: Oxford University Press, 1990).

4. R. E. Manning, "The Nature of America: Visions and Revisions of Wilder-

ness," *Natural Resources Journal* 29 (1989), pp. 25–40; R. F. Nash, *Wilderness and the American Mind* (New Haven, Conn.: Yale University Press, 1967); G. H. Stankey, "Beyond the Campfire's Light: Historical Roots of the Wilderness Concept," *Natural Resources Journal* 29 (1989), pp. 9–24; L. White, "The Historical Roots of Our Ecological Crisis," *Science* 155 (1967), pp. 1203–1207.

5. See A. Gómez-Pompa and A. Kaus, "Traditional Management of Tropical Forests in Mexico," in A. B. Anderson, ed., *Alternatives to Deforestation: Steps Toward Sustainable Use of the Amazon Rain Forest* (New York: Columbia University Press, 1990), pp. 45–64.

6. M. Pollan, "Only Man's Presence Can Save Nature," *Journal of Forestry* 88 (7) (1990), pp. 24–33.

7. A. Cowell, *The Decade of Destruction: The Crusade to Save the Amazon Rain Forest* (New York: Henry Holt and Co., 1990).

8. See D. Feeny, B. J. Berkes, B. J. McCay, and J. M. Acheson, "The Tragedy of the Commons: Twenty-two Years Later," *Human Ecology* 18 (1990), pp. 1–19; B. J. McCay and J. M. Acheson, eds., *The Question of the Commons: The Culture and Ecology of Communal Resources* (Tucson: University of Arizona Press, 1990).

9. D. A. Posey, "Indigenous Knowledge and Development: An Ideological Bridge to the Future," *Ciencia e Cultura* 35 (1983), pp. 877–894.

10. S. L. Swezey and R. F. Heizer, "Ritual Management of Salmonid Fish Resources in California," *Journal of California Anthropology* 4 (1982), pp. 7–29.

11. J. D. Nations and R. B. Nigh, "The Evolutionary Potential of Lacandon Maya Sustained-Yield Tropical Forest Agriculture," *Journal of Anthropology Resources* 36 (1980), pp. 1–30.

12. N. Myers, "Deforestation in the Tropics: Who Gains, Who Loses?" in V. H. Sudive, N. Altshuler, and M. Zamora, eds., *Where Have All the Flowers Gone? Deforestation in the Third World* (Williamsburg, Va.: Studies in Third World Societies 13, Dept. of Anthropology, College of William and Mary, 1981), pp. 1–21.

13. J. J. Parsons, "Forest to Pasture: Development or Destruction?" *Rev. Biol. Trop.* 24 (Suppl. 1) (1976), pp. 121–138; W. L. Partridge, "The Humid Tropics Cattle Ranching Complex: Cases from Panama Reviewed," *Hum. Org.* 43 (1984), pp. 76–80; and R. Repetto, "Deforestation in the Tropics," *Sci. Am.* 262 (4) (1990), pp. 36–42.

14. M. A. Altieri and L. C. Merrick, "In Situ Conservation of Crop Genetic Resources Through Maintenance of Traditional Farming Systems," *Econ. Bot.* 41 (1987), pp. 86–96; S. B. Brush, "Genetic Diversity and Conservation in Traditional Farming Systems," *Journal of Ethnobiology* 6 (1986), pp. 151–167; G. P. Nabhan, "Native Crop Diversity in Aridoamerica: Conservation of Regional Gene Pools," *Econ. Bot.* 39 (1985), pp. 387–399; M. L. Oldfield and J. B. Alcorn, "Conservation of Traditional Agroecosystems," *BioScience* 37 (1987), pp. 199–208; and J. P. Reganold, R. I. Papendick, and J. F. Parr, "Sustainable Agriculture," *Sci. Am.* 262 (6) (1990), pp. 112–120.

15. A. Gómez-Pompa, "On Maya Silviculture," *Mexican Studies* 3 (1987), pp. 1–17; P. Kundstadter, "Ecological Modification and Adaptation: An Ethnobotanical View of Lua' Swiddeners in Northwestern Thailand," *Anthropological Papers* 67 (1978), pp. 169–200; C. L. Lundell, *The Vegetation of Peten* (Washington, D.C.: Carnegie Institute, 1937); J. J. Parsons, "The Changing Nature of New World Tropical Forests Since European Colonization," *Proceedings of the International Meeting on the Use of Ecological Guidelines for Development in the American Humid Tropics* (Morges, Switzerland: IUCN Publications New Series 31, 1975), pp. 28–38; and C. Sauer, "Man in the Ecology of Tropical America," *Proceedings of the Ninth Pacific Science Congress* 20 (1958), pp. 105–110.

16. A. B. Anderson and D. A. Posey, "Management of a Tropical Scrub Savanna by the Gorotire Kayapo of Brazil," in D. A. Posey and W. Balée, eds., *Resource Management in Amazonia: Indigenous and Folk Strategies* (Bronx: New York Botanical Garden, 1989), pp. 159–173; W. C. Balée, "The Culture of Amazon Forests," in D. A. Posey and W. Balée, eds., *Resource Management in Amazonia: Indigenous and Folk Strategies* (Bronx: New York Botanical Garden, 1989), pp. 1–21; W. M. Denevan, *The Native Population of the Americas in 1492* (Madison: University of Wisconsin Press, 1976); G. S. Hartshorn, "Neotropical Forest Dynamics," *Biotropica* 12 (Suppl.) (1980), pp. 23–30; S. Hecht and A. Cockburn, *The Fate of the Forest: Developers, Destroyers and Defenders of the Amazon* (New York: Harper Perennial, 1990); and A. Roosevelt, "Resource Management in Amazonia Before the Conquest: Beyond Ethnographic Projection," in D. A. Posey and W. Balée, eds., *Resource Management in Amazonia: Indigenous and Folk Strategies* (Bronx: New York Botanical Garden, 1989), pp. 30–62.

17. S. Hecht and A. Cockburn, *The Fate of the Forest: Developers, Destroyers and Defenders of the Amazon* (New York: Harper Perennial, 1990); and D. A. Posey, "Indigenous Knowledge and Development: An Ideological Bridge to the Future," *Ciencia e Cultura* 35 (1983), pp. 877–894.

18. B. L. Turner II, "Population Density in the Classic Maya Lowlands: New Evidence for Old Approaches," *Geographical Review* 66 (1976), pp. 73–82.

19. J. B. Pick, E. W. Butler, and E. L. Lanzer, *Atlas of Mexico* (Boulder, Col.: Westview Press, 1989).

20. A. Gómez-Pompa, "On Maya Silviculture," *Mexican Studies* 3 (1987), pp. 1–17.

21. A. Gómez-Pompa and A. Kaus, "Traditional Management of Tropical Forests in Mexico," in A. B. Anderson, ed., *Alternatives to Deforestation: Steps Toward Sustainable Use of the Amazon Rain Forest* (New York: Columbia University Press, 1990), pp. 45–64.

22. S. Hecht, "Tropical Deforestation in Latin America: Myths, Dilemmas, and Reality," paper presented at the Systemwide Workshop on Environment and Development Issues in Latin America (Berkeley: University of California, October 16, 1990).

23. J. B. Alcorn, "Indigenous Agroforestry Systems in the Latin American Tropics," in M. A. Altieri and S. Hecht, eds., *Agroecology and Small Farm Development* (Boca Raton, Fla.: CRC Press, 1990), pp. 203–213; R. S. Felger and G. P. Nabhan, "Agroecosystem Diversity: A Model from the Sonoran Desert," in N. L. Gonzalez, ed., *Social and Technological Management in Dry Lands: Past and Present, Indigenous and Imposed* (Boulder, Col.: Westview Press, 1978), pp. 129–149; and S. R. Gliessman et al., "The Ecological Basis for the Application of Traditional Agricultural Technology in the Management of Tropical Agro-ecosystems," *Agro-Ecosystems* 7 (1981), pp. 173–185.

24. G. P. Nabhan et al., "Papago Influences on Habitat and Biotic Diversity: Quitovac Oasis Ethnoecology," *Journal of Ethnobiology* 2 (1982), pp. 124–143.

25. H. H. Iltis, "Serendipity in the Exploration of Biodiversity: What Good Are Weedy Tomatoes?," in E. O. Wilson, ed., *Biodiversity* (Washington, D.C.: National Academy Press, 1988), pp. 98–105.

26. A. Gómez-Pompa, 1989, interviews.

27. J. López-Portillo, et al., "Los Incendios de Quintana Roo: ¿Catástrofe Ecológica o Evento Periódico?," *Ciencia y Desarrollo* 16 (91) (1990), pp. 43–57.

28. Ibid., pp. 43–57.

29. R. A. Minnich, "Fire Mosaics in Southern California and Northern Baja California," *Science* 219 (1983), pp. 1287–1294, and "Chaparral Fire History in San Diego County and Adjacent Northern Baja California: An Evaluation of Natural Fire Regimes and the Effects of Suppression Management," *Nat. Hist. Mus. Los Angel. Cty. Contrib. Sci.* 34 (1989), pp. 37–47.

30. E. V. Komarek, "Ancient Fires," in *Proceedings of the Annual Tall Timbers Fire Ecology Conference No. 12* (Tallahassee, Fla.: Tall Timbers Research Station, 1973), pp. 219–240; C. Savonen, "Ashes in the Amazon," *Journal of Forestry* 88 (9) (1990), pp. 20–25; and D. Q. Thompson and R. H. Smith, "The Forest Primeval in the Northeast—A Great Myth?," in *Proceedings of the Annual Tall Timbers Fire Ecology Conference No. 10* (Tallahassee, Fla.: Tall Timbers Research Station, 1971), pp. 255–265.

31. D. B. Botkin, *Discordant Harmonies: A New Ecology for the Twenty-First Century* (New York: Oxford University Press, 1990); M. L. Heinselman, "Restoring Fire to the Ecosystems of Boundary Waters Canoe Area, Minnesota, and to Similar Wilderness Areas," in *Proceedings of the Annual Tall Timbers Fire Ecology Conference No. 10* (Tallahassee, Fla.: Tall Timbers Research Station, 1971), pp. 9–21; B. M. Kilgore, "The Impact of Prescribed Burning on a Sequoia–Mixed Conifer Forest," in *Proceedings of the Annual Tall Timbers Fire Ecology Conference No. 12* (Tallahassee, Fla.: Tall Timbers Research Station, 1973), pp. 345–375; G. S. Wells, *Garden in the West* (New York: Mead and Co., 1969); and H. A. Wright and A. W. Bailey, *Fire Ecology: United States and Southern Canada* (New York: John Wiley & Sons, 1982).

32. A. Gómez-Pompa and D. A. Bainbridge, "Tropical Forestry as if People

Mattered," in A. E. Lugo and C. Lowe, eds., *A Half Century of Tropical Forest Research* (New York: Springer-Verlag, in press).

33. A. Kaus, interviews, 1990.

34. A. Kaus, interviews, 1989–1990.

35. D. B. Botkin, *Discordant Harmonies: A New Ecology for the Twenty-First Century* (New York: Oxford University Press, 1990).

36. Ibid., p. 16.

37. A. Kaus, interviews, 1990.

38. K. H. Redford, "The Ecologically Noble Savage," *Orion* 9 (3) (1990), pp. 25–29.

Fabienne Bayet

Overturning the Doctrine (1994)

Indigenous People and Wilderness—Being Aboriginal in the Environmental Movement

I AM ABORIGINAL, my mother is Aboriginal, as is my mother's mother. I also consider myself as an environmentalist, conservationist, greenie, whatever. I care about this planet, this landscape we live in. So, am I black or green? Black on the outside and green on the inside, or the other way around? Perhaps there is no clear definition, I am both green and black, swirling colours which combine to make my identity, me. Not a pretty sight.

Seriously, as part of this identity, and in the face of many other greenies, I cannot remove humans from the landscape. Aboriginal people are an integral part of the Australian landscape. We are the land, the land is us. As one Aboriginal woman put it recently at a rally against the Hindmarsh Island Bridge, in South Australia, 'I am the Coorong, I am Goolwa.'

How then do I deal with a common green ideology often advocated, and used as a major selling concept, by some wilderness groups? 'Wilderness,' in this perspective, denotes land which is wild, uninhabited, or inhabited only by wild animals. Such conceptions of wilderness and conservation are yet another form of paternalism and dispossession if they continue to conceptually remove Aboriginal people from the Australian landscape.

How then do I also deal with an issue that has rocked Aboriginal and non-Aboriginal relations for the past eighteen months—Native Title and the overturning of the doctrine of terra nullius? In essence one of my issues is about the overturning of the doctrine of terra nullius through native title and whether this influences concepts of wilderness within the green movement. This then links back to my belief that human beings cannot really be removed from the landscape, regardless of whether it's sustainable or not.

ABORIGINAL PERCEPTIONS OF NATIVE TITLE

Despite constant stereotyping as 'miserable nomads' or 'noble savages' Aboriginal and Torres Strait Islander people lived very complex, diverse lifestyles throughout the continent and its islands before European invasion, although the ways communities acquired food and resources for living differed according to which ecological area they lived in. All Aboriginal and Torres Strait Islander people held spiritual beliefs relating to why they existed and why their world was like it was. Aboriginal people traditionally have a strong physical and spiritual bond with the Australian landscape through the Dreaming. This is the time of creation when the mystical and powerful ancestors of the Aboriginal people moved over the featureless earth and sea, and formed the environmental features found today.

> The descendants of these people have the responsibility of caring for the sacred site and preserving sacred objects by ceremonies that in turn ensure the well-being of the land and the plants and animals upon which their livelihood depends. These [responsibilities], together with the sacred sites and associated sacred objects, constitute, in Aboriginal terms, inalienable title to the land.[1]

NON-ABORIGINAL, COMMON LAW DEFINITION

In the High Court's *Mabo v Queensland* judgement the majority held that, under the introduced English Common Law of 1788 and thereafter, Australia was not legally—or in fact—a vacant territory, but was occupied and possessed by indigenous communities with traditions and customs of their own.[2]

The term native title conveniently describes the interests and rights of indigenous inhabitants in land, whether communal, group or individual, possessed under the traditional laws and the 'traditional' customs acknowledged and observed by the indigenous inhabitants.[3]

The High Court has ruled that native title exists where Aboriginal people have a continuing connection with land which goes back to their ancestors who occupied the land in 1788 (when Europeans came to Australia.)

Legislation defines the expression 'native title' or 'native title rights and interest' as: 'the communal, group or individual rights and interests of Aboriginal or Torres Strait Islanders in relation to land or waters.'[4]

The rights and interests of such a group are to be possessed under the traditional laws acknowledged, and the traditional customs observed, by the Aboriginal or Torres Strait Islanders. The Aboriginal peoples or Torres Strait Islanders, by those laws and customs, must have a connection with the land or waters; and such rights and interests are to be recognized by the common [non-Aboriginal] law of Australia.

THE DREAMING

To understand the cultural differences and to understand the key conceptual basis of such terms as terra nullius, native title and wilderness one must examine the history and culture deriving from and influencing the Aboriginal and the environmental movements.

The Dreaming lays down the laws concerning the accessing of resources from the environment. The environment relates directly to social organization, kinship and social obligations, sacred law, offenses against property and persons, marriage, and an individual's relationship with the land.

Aboriginal land and the meaning behind it passes on information about the environment to each generation, depending on where each person was conceived, when she/he quickened within the womb, and the totemic associations of the father and mother. On this basis certain resources could be gained in different areas by accessing kinship rights. Conversely, restriction and taboos would apply to other persons in other areas. As a result of these checks and balances, sanctuaries occurred throughout the environment where certain species could reproduce without threat of destruction. Aboriginal people perceive the Australian landscape as their cultural domain. It was their traditional duty to be custodians of the land.

The main difference between Aboriginal and non-Aboriginal cultures lies in attitudes to the land and the changes made to the environment through the accessing of resources.

THE SEPARATION OF PEOPLE FROM LAND

Another dimension of the division between culture and the environment held by Western society, relates to the Eurocentric work ethic and the human domination over land.

The Eurocentric work ethic held by the settlers dictated that the 'land must be toiled with your own sweat' and the environment possessed, changed and exploited to its fullest extent. 'To the settlers, . . . natural Australia was an uncouth waste. They were programmed to change, improve, dominate, exploit.'[5]

In this process the original inhabitants were dispossessed of their land and cultural roots. The Western land ethic dictated that 'progress' and development was the ultimate justification for the dispossession of the original custodians.

The Aboriginal population were perceived as savages who did not exploit the land to its potential (according to European agricultural expectations), thus Aboriginal people were considered to have no right of property ownership. The entire continent was perceived to be empty of meaningful human occupation, bringing about the term 'Terra-Nullius': unoccupied wasteland.

Non-Aboriginal society has exploited the land beyond sustainability. This has led to an environmental awakening. The notion of valuing wilderness for its own sake motivates the environmental movement.

Those areas that were deemed to be economically non-viable for development, and some which were aesthetically pleasing according to non-Aboriginal perspectives have been put aside, deemed to be beyond human society and culture.

Wilderness has been idealized as a way to counteract the undesirable element of non-Aboriginal civilization. This has been linked to the idea that civilization, urbanization and technology have made the human species weak in its 'unnatural' state. Wilderness has been perceived as restoring humans to strength, an avenue to 'recharge the batteries.' As a result there has been a movement towards forming reserves and national parks, in order to protect the land from human interference, urbanization and development.

INDIGENOUS PEOPLES VERSUS WILDERNESS

The concept of wilderness as nature without any trace of human interaction dehumanizes the indigenous peoples living within that landscape.

To indigenous peoples the land is no abstract wilderness. The whole of Australia is an Aboriginal artifact. The whole of the continent has been affected by Aboriginal people living out their Dreaming obligations.

For outsiders, Aboriginal people were seen to be so much in tune with nature that they were treated as part of the wilderness, almost like animals. Those concerned about the destruction of the environment have often promoted Aboriginal people as superconservationists. Although this is an acknowledgment that Aboriginal people have sustained the Australian landscape for thousands of years, it stresses the relationship of the noble savage with an idealized 'garden of Eden,' once again distorting the reality of the landscape we live in. In effect the doctrine of terra nullius lives on under the conceptual banner of wilderness: a land without human interaction or impact.

This perception of Aboriginal people denied them an independent dimension from 'wilderness' and made them, in effect, invisible and mute as a society. The potential for self determination was denied until the stark realities of environmental degradation forced environmentalists and others to look at Aboriginal land and resource management in a new light.

Even so, when National Parks were created in order to preserve the wilderness, as written into Australian legislation, Aboriginal people were no longer able to access resources since wilderness was legally defined as land devoid of any human interaction. Consequently, Aboriginal people now perceive National Parks and wilderness legislation as the second wave of dispossession which denies their customary inherited right to use land for hunting, gathering, building, rituals and birthing rites. The concession that allows traditional hunting practices in some national parks is simply another form of colonial thinking. This is regarded by Aboriginal people as an unfair and unrealistic imposition forced upon them by the Western, dominant society.

Land ownership, access and use is considered priority number one for Aboriginal people. Without land there is no base for the structure of Aboriginal culture. Aboriginal people, however, are no longer predominantly hunters and gatherers due to the dispossession and decimation of their societies. Aboriginal people are looking to alternatives for land management, which will meet their traditional obligations, as well as the constraints of contemporary economic survival. Aboriginal communities see no option but to compete within the Australian economy (often on a non-sustainable basis). The Central Land Council writes:

> Our land is our life. We look at it in a different way to non-Aboriginal people. For us, land isn't simply a resource to be exploited. It provides us with food and materials for life, but it also provides our identity and it must be looked after, both physically and spiritually. If we abuse our land, or allow someone else to abuse it, we too suffer. [However] the many resource development projects and commercial enterprises now operating on Aboriginal land show that respecting our land rights can be compatible with national economic development . . . economic activity on Aboriginal land contributes significantly to the Australian economy through a number of avenues.[6]

This is an indication of the cultural clash between Aboriginal interests and the conservation perspective.

Some environmentalists fear that allowing Aboriginal communities to claim National Parks and to hunt in them would threaten the ecological diversity of the area and the survival of endangered species. As reported in *Habitat Australia:*

> No longer will they be National Parks but Government-sanctioned killing fields of our once protected ecology, a long-time member of the Daintree Wilderness Action Group claimed. Almost 100 years of protection ripped up, shot up, burnt up, stuffed.[7]

This is one area where the environmental and Aboriginal movements start parting company. The major difference between the Aboriginal (black) and Environmental (green) movements is that each group has an ideology constructed from differing cultural backgrounds. The environmental movement's wilderness perspective is seen by Aboriginal people as a continuation of paternalism and colonialism. Thus, Toyne and Johnston comment on:

> . . . a new wave of dispossession—the denial of Aboriginal peoples' rights to land in the name of nature conservation. The creation of national parks, wil-

> derness areas or wild life sanctuaries could be every bit as threatening and destructive to Aboriginal people as were the pastoral stations or the farms of past generations.[8]

It is not surprising therefore, that Aboriginal people remain ambivalent about green issues since in their purest form green values deny black rights.

In most states, conservation areas and National Parks have been created without the involvement or consent of Aboriginal people. When the Winychanam people sought to purchase a cattle station in their country on Cape York, the 1976 Bjelke-Petersen National Party declared the property and land around the area as the Archer River National Park. The Winychanam felt they had been denied the chance for economic self determination and saw the creation of the Park as the government's implementation of its racist policies.

Another example of this perceived 'green dispossession' is the World Heritage nomination of north Queensland's Wet Tropic Rainforests in 1987. This brought opposition from the local Yarrabah Aboriginal community as they had hoped to create a saw milling operation from their local lease.

The West Australian Liberal Government of 1977 declared a large area of the Martu's expansive country, in the Gibson Desert, as the Ruddal River National Park in reaction to criticism that Western Australia did not have enough conservation areas. The Martu people are outraged, not just because of the physical disruption to their lives from mining but because of further limitations placed on them by the creation of a National Park. The Martu people were not consulted in its planning. They do not want a National Park imposed on their land, especially as they cannot receive Federal funding for infrastructure because they don't have title to land.

Despite such examples of Aboriginal dispossession through conservation, the green movement and Aboriginal communities are coming together for the benefit of the land. The majority of the green movement is trying to acknowledge Aboriginal ownership of land, their cultural power and their relationships with the land, and their skills in land management.

Successful alliances are beginning to form. For example, the Uluru-Kata Tjuta and Kakadu National Parks are located on Aboriginal land which has been leased to the Australian National Parks and Wildlife Service. Local traditional landowners form a majority on the Park's Boards of Management and are employed as rangers and cultural interpreters. Their knowledge of the country is a vital ingredient in successful environmental management.

Aboriginal people are also using wilderness arguments to protect the Cape York area from the proposed Space Station development.[9]

Increasingly, and successfully, Aboriginal people are becoming more involved in National Park and conservation management as rangers, through representation on park boards of management and most importantly through ownership. However, as yet, Aboriginal ownership proposals for National Parks in South Australia, Western Australia and Tasmania have not yet been made, although it is something that should and will occur in the future, if social and ecological justice is to remain on the social agenda.

In part, these moves represent a recognition on behalf of conservationists that Aboriginal people possess traditional land management skills which will be of great benefit in managing National Parks and conservation areas. It seems that non-Aboriginal environmentalists are putting Aboriginal people back into the landscape; Aboriginal people consider that they never left it.

NATIVE TITLE AND WILDERNESS

Many environmentalists, while supporting Aboriginal lobby groups against non-sustainable industries in the Native Title debate, are now wondering how the overturning of terra nullius will affect areas particularly considered reserves from human interaction and impacts. One concept which is particularly emotive is that of wilderness. The concepts of wilderness could be perceived as conceptually outside the Native Title legislation and Aboriginal land ownership.

What some environmentalists may fail to realize is that Aboriginal people still have to participate in the policy of assimilation by following Eurocentric legal processes in order to justify their claim, regardless of whether the land in question is 'wilderness' or not. Aboriginal people still have to participate in perpetuating the dominant stereotype of 'traditional cultural practices' and 'communal ownership' if they are to retain any remnants of ownership.

Once again Aboriginal people are still expected to be 'civilized' while retaining static stereotypes imposed on them since invasion. The conservation movement will have to address Native Title and the dispossessing doctrine of wilderness if they are to find common ground between cultural self-determination and ecological sustainability.

IS ECOTOURISM THE ANSWER?

A potential avenue currently being pursued by both black and green movements is the new rise of ecotourism. However, ecotourism, whether on a sustainable or non-sustainable basis, raises a number of issues for Indigenous peoples and environmental interest groups, relating to the blending of the concepts nature and culture.

It could be argued that non-Aboriginal environmentalists are once again putting the onus on the victim of the invasion, through the conceptualization of ecotourism. Aboriginal people remain on the lowest socioeconomic rung of Australia's society. The pressures on them to succeed in building a new industry is extraordinary. However, Aboriginal people need to seek alternatives away from non-sustainable development. Many questions need to be debated amongst environmentalists, ecotourism operators and Indigenous peoples. Some of these include:

- Is ecotourism sustainable on a cultural basis?
- Is ecotourism yet another form of pressure on Aboriginal society and communities—this time on those least affected by the invasion and urbanization of Australia?
- Will ecotourism quite simply be perpetuating the stereotype of the non-urban Aboriginal?
- Will ecotourism quite simply be perpetuating the stereotype of the wandering noble savage, the hunter-gatherer, remnant of the stone age or is ecotourism an appropriate avenue of reconciliation?
- Will ecotourism open avenues through which non-Aboriginal people can explore and understand Aboriginal people, eventually venturing into relating with urban Aboriginals?
- Will ecotourism be yet another barrier between urban Aboriginals being ignored for the sake of the 'traditional' Aboriginal stereotype that will sell?

CONCLUSION

While acknowledging that there are major differences between Aboriginal and non-Aboriginal approaches to land management and conservation, it must be emphasized that there is no single overarching 'Aboriginal viewpoint,' any more than there is any single 'green ideology.' Indeed it would be true to say that within the Aboriginal context of Native Title and self-determination, there is a wide variety of approaches to development, use of resources, land management and sustainability. By the same token there are many shades of green, from 'dark' greenies who will not countenance any human involvement within the landscape, up to the pragmatic 'lighter' greenies' who can identify the realities of economic needs.

The environmental movement must stop stereotyping Aboriginal people as noble savages or arch-conservationists and must recognize the rights of Aboriginal people to choose self-determination on a contemporary, realistic basis. Brown writes:

> Only if we negotiate management of significant areas with indigenous communities will we be able to protect both the ecological and cultural values of the land. Management plans which are imposed as preconditions on Aboriginal ownership will never operate effectively.[10]

Nevertheless development cannot continue as though there are no environmental limits. Cultural survival relates very much to ecologically sustainable land management, as it is no use retaining culture now if people cannot sustain themselves on the land for generations to come. In this sense Aboriginal communities should not use the argument of cultural survival, through economics, to initiate or continue non-sustainable land management. By the same token, if conservation groups are going to insist on the dispossession of Aboriginal communities through Wilderness legislation then they must make more of an effort to find alternative means of accessing resources for Aboriginal communities, and may have to back up their argument by providing financial support. Conservationists cannot expect to be received well if they continue to impose environmental responsibility on Aboriginal communities. This could be perceived as a continuation of the 'noble savage' myth. In the same way non-Aboriginal people have to confront and overcome their inherited view of Aboriginal people, and allow themselves to seek an equality in joint land access and management.

Through the environmental movement the Australian community is beginning to recognize that the relationship to the environment must be re-evaluated, and, as the Native Title debate has demonstrated, an integral part in that evaluation lies in the fact that Australia was, and is, Aboriginal land.

NOTES

1. Mildred Kirk, *A Change of Ownership: Aboriginal Land Rights,* (Queensland, Australia: Jacaranda Press, 1986).

2. *The Mabo Judgement* (Canberra: Aboriginal and Torres Strait Islander Commission, 1993).

3. Richard Bartlett, et al., *The Mabo Decision* (Melbourne, Australia: Butterworths, 1993).

4. Native Title Act, Legislation, AGPS, 1993.

5. Derek Whitelock, *Conquest to Conservation* (South Australia: Wakefield Press, 1985).

6. *Our Land, Our Life* (Alice Springs, Northern Territory, Australia: Central and Northern Land Councils, 1991).

7. M. Horstman, "Cape York Peninsula: Forging a Black-Green Alliance," in *Habitat Australia: Caring for Country. Aboriginal Perspectives on Conservation* (Victoria: The Australian Conservation Foundation, vol. 19, no. 3, June 1991), pp. 19–22.

8. P. Toyne and R. Johnston, "Reconciliation, or the New Dispossession," in ibid., pp. 8–10.

9. M. Horstman, "Cape York Peninsula: Forging a Black-Green Alliance," in ibid., pp. 19–22.

10. A. J. Brown, *Keeping the Land Alive: Aboriginal People and Wilderness Protection in Australia* (Sydney, Australia: The Environmental Defender's Office Ltd./ The Wilderness Society, 1993).

Carl Talbot

The Wilderness Narrative and the Cultural Logic of Capitalism (1998)

> *Thousands of tired, nerve-shaken, over-civilized people are beginning to find out that going to the mountains is going home; that wildness is a necessity; and that mountain parks and reservations are useful not only as fountains of timber and irrigating rivers, but as fountains of life. Awakening from the stupefying effects of the vice of over-industry and the deadly apathy of luxury, they are trying as best they can to mix and enrich their own little ongoings with those of Nature, and to get rid of rust and disease. . . . some are washing off sins and cobweb cares of the devil's spinning in all-day storms on mountains.* John Muir[1]

BY FAR THE MOST DOMINANT ROLE of "wilderness" in modern society is as a leisure resource; it serves the role, which Muir identified in 1898, of a refuge for the sufferers of over-exposure to capitalist society, and from the outset has been stage-managed to meet these requirements. As Chris Rojek argues:

> Bourgeois culture equated progress with the cultural domination of Nature. One of the strongest appeals of the wilderness as a leisure resource is that it appears to reverse this order. In the wilderness nature appears to dominate culture. The mountains, the forests, the lakes, the moors, the deserts and the

> national parks seem to have resisted the juggernaut of modernity. Free from the wreckage of the metropolis these places seem to be oases of serenity. Certainly they are widely celebrated as the quintessential escape areas in contemporary society.[2]

While wilderness enjoys this "mythical status" as a sanctuary of untransformed nature, human reorientation, and spiritual fulfillment, in Western ideology the reality is very different. As Steven Vogel remarks with regard to areas of "preserved wilderness," such areas "present us not with 'pure' nature but rather with something highly artificial: a piece of nature that has been withdrawn from the natural order in which human transformative activity plays such a crucial part."[3] Such wilderness is the result of social decision-making and social action insofar as areas are chosen to be preserved. Rojek goes still further with his assessment of the artificiality of wilderness preserves:

> Far from offering us experience of pre-social nature, unscarred by history, class or politics, the parks are, in fact, social constructs, man-made environments in which Nature is required to conform to certain social ideals. For example, it must radiate cleanliness, vastness, emptiness, silence, and peace. In other words it must be the exact opposite of the metropolis. The parks are stage representations of nature.[4]

D. MacCannell also identifies the artificiality of representations of wilderness when he refers to national parks as "museumized nature,"[5] a designation that echoes the sentiments of the initial advocates of the creation of national parks in nineteenth-century North America, who commended Yellowstone National Park as a "museum," a place where it would be possible to observe the "freaks and phenomena of Nature" and "wonderful natural curiosities."[6]

Within the Western tradition the idea of wilderness is closely linked to its function as a salve for a spiritually battered workforce. Hand-in-hand with the physical creation of ludic space in the form of parks and nature preserves goes a philosophy of preservationism that gives this project ideological legitimation and articulates the category of wilderness, which thus conceals the bourgeois pragmatism that informs it. As capitalism reordered human geography and social relations, and revolutionized productive and economic relations, so it reordered space for its purpose. Thus nature was organized so as to meet the spatial, economic, and psychological needs of capitalism.

An understanding of the degrading, alienating, and "unnatural" character of the work process under capitalism was apparent not only to the workforce but also to the bourgeoisie. As Raymond Williams notes, even before the industrial revolution, landed capital was aware of the alienation from nature that capitalism produces:

> The clearing of the parks as "Arcadian" prospects depended on the completed system of exploitation of the agricultural and genuinely pastoral lands beyond the park boundaries.... [These] are related parts of the same process—superficially opposed in taste, but only because in the one case the land is being organized for production, where tenants and labourers will work, while in the other case it is being organized for consumption: the view, the ordered proprietary repose, the prospect. Indeed it can be said of these eighteenth century arranged landscapes not only, as is just, that this was the high point of agrarian bourgeois art, but that they succeeded in creating in the land below their windows and terraces ... a rural landscape ... *from which the facts of production had been banished.*[7]

As industrial capitalism developed the bourgeoisie's eagerness to resist in the sphere of consumption, what was the order of the day for the working class in the sphere of production (that is, alienation from nature) became increasingly significant. The Romantics and American Transcendentalists, such as Byron and Thoreau respectively, were the ideologues of this resistance. However, as the economics of preservationism show, this reaction of the bourgeoisie did not remain solely in the realm of ideology; it took on a concrete, physical form in terms of the production of natural space—parks, nature preserves, rural estates, and so on. But, as David Harvey points out,

> The response rested on a mystification, of course, for it reduced "nature" to a leisure-time concept, as something to be "consumed" in the course of restful recuperation from what was in fact a degrading relation to nature in the most fundamental of all human activities—work. But the mystification had bitten deep into the consciousness of all elements of society. To talk now of the relation to nature is to conjure up images of mountains and streams and seas and lakes and trees and soft grass, far from the coal-face, the assembly line, and the factory, where the real transformation of nature is continuously being wrought.[8]

The production of natural space as an antidote to the debilitating effects of industrial capitalism was not restricted to the consumer needs of the bourgeoisie; reformist sections of the bourgeoisie recognized the economic sense in providing for the psychological needs of their workforce by the creation of parks and sylvan "model" towns and villages[9] where the myth of a romanticized nature could be played out. As Harvey summarizes: "Capital . . . seeks to draw labour into a Faustian bargain: accept a packaged relation to nature in the living place as just and adequate compensation for an alienating and degrading relation to nature in the workplace."[10] The "facts of production" that entail alienation from nature are concealed by a romantic mystification of nature, which banishes any imaginary opportunity for an unalienating relation to nature to the realm of leisure, so that the alienating relation to nature implicit in capitalist production remains unchallenged.[11] As Harvey argues: "As labour seeks relief from a degrading relation to nature in the workplace, so capital seeks to parlay that into a mystified relation to nature in the consumption sphere."[12]

These efforts of capital to manage a relation to nature in the sphere of consumption result in nature emerging as a "stylized spectacle" packaged for easy consumption. The nineteenth century presented this spectacle of nature in the form of zoos, circuses, parks and gardens, exhibits of stuffed animals, natural museums, and so on, thus converting nature into an object to be displayed and consumed. "One no longer lived in nature, one visited it as a tourist from the real world of the metropolis and the money economy."[13] The commodification and presentation of this spectacle has followed advances in technology and production. The simple cages of the past have, in public zoos, largely given way to "artificial habitats" intended to mimic the natural environment of the exhibits, while safari parks offer the possibility of "drive-through" nature.

In capitalist societies exposure to the spectacle of nature is most commonly experienced secondhand through such managed media as television, where wildlife programs enjoy a widespread popularity—our escape to nature can now take place in the convenience of our own room. People's estrangement from nature, even in their pilgrimages to wilderness preserves, is dramatically illustrated by the latest advances in the methods of wilderness presentation: Rojek offers the example of the Zion National Park in Utah, North America, where the park's wilderness is displayed to visitors on a drive-in cinema screen so that they need never leave their vehicles. A spokesperson for the park optimistically noted:

> There's a market for this. The one and a half million people who visit Zion each year won't have to sweat or get their heart rate above wheel-chair level. Whole busloads can come to Springdale, have the Zion experience and be in Las Vegas that night.[14]

"Wilderness," or rather the ideal of it, is a valuable commodity which is traded on the international market.[15] For the most part—as Nash, in the selection from *Wilderness and the American Mind* in this section, notes—the commodity is not removed from the country of origin (except in the form of trophies and exhibits); rather the commodity is the *experience* of wilderness. With the opportunities of mass tourism, a particular experience may be chosen at home from travel brochures, bought in the high-street, and, after being transported to the "wilderness" of one's choice, "consumed on the premises." The economists of those countries where wild nature still exists, recognizing areas of wild nature as a capital asset, in what are often weak economies, have been quick to capitalize on this export opportunity.

While the powerful economic elites of these Southern countries may benefit from the export of "wilderness experience," as Ramachandra Guha notes, "the [international] implementation of the wilderness agenda is causing serious deprivation in the Third World."[16] Calls by such environmentalists as Holmes Rolston, Garrett Hardin, and Dave Foreman for international wilderness designation, or zoning,[17] represent the global export of the North American national park ideal,[18] but, as Guha argues, this North American model of nature preservation is inappropriate for countries of the South. Not only is the implementation of policies of wilderness designation inappropriate, it has also resulted in gross injustice: as Ashish Kothari, the founding member of the Indian environmental group Kalpvriksh, comments with regard to the establishment of national parks in India:

> The approach was essentially one of saving wildlife *from* the people, rather than *with* them. In other words, local communities in wildlife-rich areas were seen to be enemies to the conservation cause, and were either physically displaced or denied access to these areas. While in the short run this approach has led to habitat and species protection in several parts of India, it has also had two detrimental impacts: first severe human rights abuses as peoples' basic survival resources have been denied them; and second, the alienation of local people from the very areas which they could have been instrumental in

> conserving. The approach is therefore not only anti-human, it is also short-sighted from the wildlife conservation point of view. Increasingly, as a sharp backlash, communities are demanding the denotification of national parks and sanctuaries in order to regain their rights over resources.[19]

Wilderness designation, to the exclusion of humans, is an expression of the cultural imperialism of the North. The peoples of the poor countries of the South cannot afford the luxury of a "museumized nature," nor does such an imposition usefully serve the long-term interests of either humans or nonhumans.

The concept of "wilderness" as employed by the narrative of some sections of the environmentalist movement is an invention. As Roderick Nash concludes, "wilderness is a matter of perception—part of the geography of the mind,"[20] but this is only part of the story; the wilderness narrative not only reflects a subjective ordering of nature, but also the objective production of space by capitalism. The process of civilization gave rise to a particular representation of nonhuman nature as "wilderness," as yet untransformed by human agency. The normative connotations ascribed to this conceptualization have, in the twentieth century, been revolutionized: the wilderness is no longer to be feared and vanquished but to be cherished as humanity's spiritual homeland. The cult of wilderness, which emerged from nineteenth-century Romanticism, in the twentieth century has found a home in what purport to be the radical factions of modern environmentalism.[21] But the idea of wilderness is more than a "matter of perception"; it plays a vital role in the cultural logic of capitalism. As a sanctuary away from the degrading consequences of capitalist work relations, the ideal of wilderness reduces nature to a "leisure-time concept." The category of wilderness forms a part in a myth of nature that conceals the facts of production by offering false compensation in the sphere of leisure consumption. Though it is true that under the conditions of capitalist production any possibility for a relation to nature in the arena of work that does not alienate the worker is very small, supposedly our relief comes when we make the pilgrimage (under the management of leisure industries!) to wilderness sanctuaries.[22]

Environmentalism that incorporates this ideal of wilderness unwittingly accepts capital's "Faustian bargain," arguing that humanity's (and subsequently all of nature's) redemption rests with this myth of a salvationist wilderness, conceptualized by capital's ideologues to seal the deal.

This wilderness experience is given priority in the deep ecology doctrine of "Self-realization" (understood as identification with a unified, harmonious natural order), whereby human psychological rehabilitation and development is given an ecological significance. In fact the psychological reorientation of humanity is considered to be a cosmological/ontological imperative by deep ecologists; for example the deep ecologist Warwick Fox has stated that,

> the appropriate framework of discourse for describing and presenting deep ecology is not one that is fundamentally to do with the value of the world . . . but rather one that is fundamentally to do with nature and possibilities of the Self, or we might say, the question of who we are, can become, and should become considered in the larger scheme of things.[23]

The emphasis is on human psychological development, and the stage on which this development is to take place is the wilderness; Edward Abbey's "journey home"[24] is not merely a metaphorical passage but a physical journey to the supposed sanctuary of wilderness. The geography of this journey is itself revealing: the journey leads away from the urban setting, where this environmentalism has little, or nothing, to say about humanity's relation to nature in the sphere of production, to the "wilderness," which is managed as a leisure resource in the capitalist sphere of consumption.

These types of environmentalism, by confining their discourse to the relation humanity should have in the sphere of spiritual or psychological consumption of nature, implicitly accept the parameters imposed by the cultural logic of capitalism. While the sensibility of this environmentalism may be offended by the vulgarity of some of the modern forms of wilderness consumption leading to calls for a return to a more ascetic appreciation,[25] the structure of the myth remains unchallenged.

NOTES

1. John Muir, "The Wild Parks and Forest Reservations of the West," *Atlantic Monthly* 81 (1898): 483; reprinted in *Our National Parks* (Boston: Houghton Mifflin, 1901; San Francisco: Sierra Club Books, 1991). [Included in this volume.]

2. C. Rojek, *Ways of Escape* (London: Macmillian Press, 1993), p. 198.

3. Steven Vogel, "Alienation from Nature," *Social Theory and Practice* 14 (1988): 377.

4. Rojek, *Ways of Escape,* p. 198.

5. D. MacCannell, "Nature, Inc.," *Quarterly Journal of Environmental Design* 6(3) (1990): 25.

6. See Roderick Nash, *Wilderness and the American Mind,* 3rd ed. (New Haven, Conn.: Yale University Press, 1982), p. 113.

7. Raymond Williams, *The Country and the City* (London: Chatto and Windus Limited, 1973), p. 124 (emphasis added).

8. David Harvey, *Consciousness and the Urban Experience* (Oxford: Basil Blackwell, 1985), p. 54.

9. Port Sunlight, built by Lord Lever, is an example of such a town in Britain.

10. Harvey, *Consciousness and the Urban Experience,* p. 56. This is the understanding of alienation developed by Marx in the *1844 Manuscripts* whereby the object that labor produces—labor's product—confronts it as *something alien,* as a *power independent* of the producer. The worker puts his life into the object; but now his life no longer belongs to him but to the object. The alienation of the worker in his product means not only that his labor becomes an object, an *external existence,* but that it exists *outside him,* independently, as something alien to him, and that it becomes a power on its own confronting him. It means that the life that he has conferred on the object confronts him as something hostile and alien. See Karl Marx, *Economic and Philosophic Manuscripts of 1844* (Moscow: Progress Publishers, 1977), pp. 63–64.

11. As Harvey notes, in some sense this mystification is necessary under capitalism: Without it, life would scarcely be bearable (see *Consciousness and the Urban Experience,* p. 55).

12. Ibid., p. 57.

13. Rojek, *Ways of Escape,* p. 119.

14. Ibid., p. 204.

15. Nash, *Wilderness and the American Mind.*

16. Ramachandra Guha, "Radical American Environmentalism and Wilderness Preservation: A Third World Critique," *Environmental Ethics* 11 (1989), p. 71. [Included in this volume.]

17. See for example: Holmes Rolston III, "The Wilderness Idea Reaffirmed," *The Environmental Professional* 13 (1991): 370–77 (included in this volume); Garrett Hardin, "The Economics of Wilderness," *Natural History* 78 (1969): 20–27; Dave Foreman, "A Modest Proposal for a Wilderness System," *Whole Earth Review* 53 (1986–87): 42–45.

18. Nash believes the development of national parks is North America's specific cultural contribution to the cause of nature preservation.

19. Ashish Kothari in an interview with J. M. Alier, "Ecological Struggles in India: Interview with Ashish Kothari," *Capitalism, Nature, Socialism* 4(3) (1993): 111. Environmentalism frequently prescribes such a withdrawal of human enterprise from nature, often summed up in the Heideggerian slogan of "letting Being (Nature) be": Rolston ("The Wilderness Idea Reaffirmed," p. 371), for example,

when discussing wilderness designation argues that "humans ought to draw back and let nature be." Other environmentalists who ascribe to this position include Laura Westra ("Let It Be: Heidegger and Future Generations," *Environmental Ethics* 7 [1985]: 341–50), and Michael Zimmerman ("Toward a Heideggerian Ethos for Radical Environmentalism," *Modern Schoolman* 64 [1986]: 19–43).

20. Nash, *Wilderness and the American Mind,* p. 333.

21. For example, Rolston ("The Wilderness Idea Reaffirmed," p. 370) claims that "the wilderness ideal is critical today"; deep ecologists such as Bill Devall and Max Oelschlaeger, who, in an openly and enthusiastically mythopoeic text (*The Idea of Wilderness* [New Haven and London: Yale University Press, 1991], p. 321) asserts that "the idea of wilderness in the postmodern context is . . . a search for meaning—for a new creation story or mythology."

22. There is an ironic side to this prescription: areas of remaining "wild" nature are threatened by the increasing popularity of the ethic of wilderness appreciation. The problem is such that the seemingly contradictory notion of wilderness management is now a practical necessity in most wilderness preserves. Wilderness recreation is now a serious threat to the future of remaining areas of "wild" nature.

23. Warwick Fox, "Approaching Deep Ecology," University of Tasmania Environmental Studies Occasional Papers 20 (1986): 85.

24. This is a popular metaphor in deep ecological writings and has also been called "the search for our unique spiritual/biological personhood"; see Fox, "Approaching Deep Ecology," pp. 22f. It originates from Edward Abbey's *The Journey Home: Some Words in Defense of the American West* (New York: Plume/Penguin, 1977).

25. It might be argued that such "natural aristocrats" are projecting onto nature the asceticism of middle-class culture in their calls for an "authentic" experience of nature that incorporates such purist values as privacy, cleanliness, and a degree of hardship and austerity in its achievement.

PART THREE

The Wilderness Idea Roundly Criticized and Defended

J. Baird Callicott

The Wilderness Idea Revisited (1991)

The Sustainable Development Alternative

It seems to me that sanctuaries are akin to monasticism in the dark ages. The world was so wicked it was better to have islands of decency than none at all. Hence decent citizens retired to monasteries and convents. Once established these islands became an alibi for lack of private reform. People said: "We pay the bills for all this virtue. Let goodness stay where it belongs and not pester practical folks who have to run the world." . . . The more monasteries or sanctuaries the grimmer the incongruity between inside and outside.

Aldo Leopold (1942, p. 263)

INTRODUCTION

In making selections for the new collection of Aldo Leopold's essays that we prepared for the University of Wisconsin Press, Susan L. Flader and I sometimes disagreed. Because of his close association with the wilderness movement in America, I was for including virtually all of Leopold's papers on wilderness. Flader thought that those from the 1920s were too similar to one another, albeit rhetorically varied, to include more than a few. And she thought, more generally, that the importance of wilderness to Leopold had been exaggerated and that we should not unwittingly abet such a misperception. In Flader's very well informed opinion, Leopold was more interested, or at least more fully engaged, in managing humanly inhabited and used land—not only or even principally for enhanced resource production, but for a flourishing wild flora and fauna—than in campaigning for wilderness set asides. His work, in other words, had a different focus than that of his contemporaries, Robert Marshall, Howard Zahnizer, and Olaus Murie.

Flader, however, was anxious to include Leopold's earliest discussion of the relationship of wilderness to conservation, "The Popular Wilderness Fallacy," first published in 1918 in a long defunct periodical called *Outer's Book—Recreation.* I argued for suppressing it, since it seemed so anomalous. The fallacy that Leopold explores in this essay is that we need wilderness to have game. On the contrary, Leopold argues, game penury is not an inevitable result of the march of empire any more than timber famine is. The eradication of "varmints," he insists, following the reduction of wilderness, will make more game available to increased numbers of hunters; fire suppression will save game and its food from destruction; flowages behind dams will replace drained and seasonally dried up marshes for waterfowl; and so on. In short, everything will be coming up roses (or, rather deer and ducks, in this case) after the complete conquest of the continent by the forces of civilization. Indeed, in "The Popular Wilderness Fallacy" Leopold (1991a, p. 50) even writes, "Nature was actually improved upon by civilization"—at least as far as game goes. Leopold's early persecution of predators is well known and was personally and apologetically acknowledged in *Sand County*'s famous essay, "Thinking Like a Mountain." But that Leopold was also at first a Philistine on the wilderness issue—a cause, by 1918, well publicized and vigorously championed by the late John Muir—seemed to me a sleeping dog that we might just as well let lie.

I now no longer think that "The Popular Wilderness Fallacy" is as aberrant as at first it seemed. Rather, I think that Leopold envisioned throughout his career an ideal of human unity and harmony with nature, rather than a trade-off between human economic activities and environmental quality. One might even go so far as to say that he did not altogether abandon his 1918 proposition respecting the improvement of nature by civilization, though by no stretch of the imagination did he retain it in its original sanguine and laissez faire form. One can imagine the ecologically chastened Leopold suggesting not that nature *was* actually improved upon by civilization, but that a mature and humane civilization *might* actually improve upon it.

On the other hand, Leopold eventually came to see and expressly to formulate the important and absolutely vital role that wild refugia had to play in biological conservation. And I certainly hope that my remarks here will not be construed to deny or undermine the importance and necessity of wild lands in that regard. But the dialectical history of American conservation philosophy has fostered a more recent popular wilderness fallacy. That fallacy has two closely connected formulations. The first is that the New World was, when Christopher Columbus stumbled upon it, in a totally "wilderness condition"—as Roderick Nash (1967) characterizes it. The second is that any human alteration of pristine nature degrades it and therefore that biological conservation is best served by wilderness preservation, that the best way to conserve nature is to protect it from human inhabitation and utilization.

Here I briefly review the history of American conservation philosophy and the role that the wilderness idea has played in it. And then I argue that wilderness preservation, as a conservation stratagem, needs to be . . . not replaced, certainly, since nature reserves now fill and will continue to fill a vital niche in a more broadly conceived struggle to conserve biodiversity, but . . . refined, and augmented by a complementary approach.

Having said here at the beginning what I am critically revisiting, let me set out as clearly and explicitly as I can a few caveats and qualifications.

First, I am as ardent an advocate of those patches of the planet called "wilderness areas" as any other environmentalist. My discomfort is with an idea, the received concept of wilderness, not with the ecosystems so called.

Second, to suggest that something is amiss with the concept of wilderness is to suggest, at the same time, that something is amiss with its antithesis, the concept of civilization. That is the point of the epigram inaugurating this essay. Implicit in the most passionate pleas for wilderness preservation is a complacency about what passes for civilization. If all that we can feature is the present adolescent state of civilization and its mechanical motif continuing indefinitely into the future, then naturally the only way we can conceive of conserving nature is to protect bits of it from destructive development. A harmony-of-man-and-nature conservation philosophy such as Leopold espoused implies re-envisioning civilization as well as critically revisiting the wilderness idea.

Therefore, third, I do not advocate what is euphemistically called "multiple use" for all landscapes. I do not suggest we attempt to mix strip-mining, clearcutting, stock-grazing, four-wheeling, downhill-skiing, motor-camping, etc. with biological conservation. By suggesting that we try to shift the burden of conservation from wilderness preservation to sustainable development, I mean to suggest that we try to think up economic strategies that are compatible with ecosystem health and that are strictly limited by ecological exigencies. There is precious little designated wilderness as things stand. Such areas serve the cause of biological conservation most importantly as refugia for species not tolerant of and/or not tolerated by people. Personally, I would like to see more wild lands designated as wilderness with this purpose in view. But, given a global human population approaching six billion persons, the greatest part of the best land will be put to economic use whether we conservationists like it or not. We conservationists, however, may realistically hope that in the future ecological, as well as technological, feasibility may be taken into account in designing new and in redesigning old ways of human living with the land.

THE WILDERNESS IDEA IN HISTORIC AMERICAN CONSERVATION PHILOSOPHY

Ralph Waldo Emerson and Henry David Thoreau were the first notable American thinkers to insist, a century and a half ago, that wild nature might serve "higher" human spiritual values as well as supply raw materials for meeting our more mundane physical needs. Nature can be a temple, Emerson (1989) enthused, in which to draw near and to commune with God. Too much civilized refinement, Thoreau argued, can over-ripen the human spirit; just as too little can coarsen it. "In wildness," he wrote, "is the preservation of the world" (Thoreau, 1962, p. 185).

Building on the nature philosophy of Emerson and Thoreau, John Muir (1901) spearheaded a national, morally charged campaign for public appreciation and preservation of wilderness. People going to forest groves, mountain scenery, and meandering streams for religious transcendence, aesthetic contemplation, and healing rest and relaxation put these resources to a higher and better use, in Muir's opinion, than did the lumber jacks, miners, shepherds, and cowboys who went to the same places in pursuit of the Almighty Dollar and who were inspired only by the Main Chance.

Critics today, as formerly, may find an undemocratic and un-American presumption lurking in the Romantic-Transcendental conservation philosophy of Emerson, Thoreau, and Muir. To suggest that some of the human satisfactions that nature affords are morally superior to others may only reflect aristocratic biases and class privileges. Let me hasten to say that personally I agree with Muir et al. Birdwatching, for example, is, in my opinion, morally superior to dirtbiking. But there is a contingent of powerful and influential professionals who do not agree. An axiom of neoclassical economics is that all human preferences concerning "resource" use are morally equal and should be weighed one against the other in the marketplace (Randall 1988).

At the turn of the century, Gifford Pinchot, a younger contemporary of John Muir, formulated a novel conservation philosophy that reflected the general tenets of the Progressive era in American history. Notoriously, the country's vast biological capital had been plundered and squandered for the benefit, not of all its citizens, but for the profit of a few. Pinchot (1947, pp. 325–326) crystallized a populist, democratic conservation ethic in a credo—"the greatest good of the greatest number for the longest time"—that echoed John Stuart Mill's well known Utilitarian maxim, "the greatest happiness for the greatest number." He bluntly reduced Emerson's "Nature" (with a capital "N") to "natural resources." Indeed, Pinchot (1947, p. 326) insisted that "there are just two things on this material earth—people and natural resources." He even equated conservation with the systematic exploitation of natural resources. "The first great fact about conservation," Pinchot (1947, p. xix) noted, "is that it stands for *development*"—with the proviso that resource development be scientific and thus efficient. For those who might take the term "conservation" at face value and suppose that it meant saving natural resources for future use, Pinchot (1947, p. xix) was quick to point out their error: "There has been a fundamental misconception," he wrote, "that conservation means nothing but the husbanding of resources for future generations. There could be no more serious mistake." And it was none other than Gifford Pinchot (1947,

p. 263) who first characterized the Muirian contingent of nature lovers as aiming to "lock up" resources in the national parks and other wilderness reserves.

The notorious Schism in the traditional American conservation movement thus was rent. Muir and Pinchot, once friends and allies, quarreled and each followed his separate path (Nash, 1967). Pinchot appropriated the term "conservation" for his utilitarian philosophy of scientific resource development. And Muir and his exponents came to be called "preservationists."

The third giant in twentieth-century American conservation philosophy is, of course, Aldo Leopold. At the Yale Forest School, founded with the help of the Pinchot family fortune, Leopold was steeped in what historian Samuel Hays (1959) called "the gospel of efficiency"—the scientific exploitation of natural resources, for the satisfaction of the broadest possible spectrum of human interests, over the longest time. And for fifteen years Leopold worked for the Forest Service, whose first Chief was Pinchot himself. Leopold's ultimately successful struggle for a system of wilderness reserves in the national forests was consciously molded to the doctrine of highest use, and his new science of game management essentially amounted to the direct transference of the principles of forestry from a standing crop of large plants to a standing crop of large animals (Leopold, 1918; 1921). But Leopold gradually came to the conclusion that Pinchot's utilitarian conservation philosophy was inadequate because it was not well informed by the new kid on the scientific block, ecology (Flader, 1974; Meine, 1988).

As Leopold (1939a, p. 727) put it:

> Ecology is a new fusion point for all the sciences. . . . The emergence of ecology has placed the economic biologist in a peculiar dilemma: with one hand he points out the accumulated findings of his search for utility, or lack of utility, in this or that species; with the other he lifts the veil from a biota so complex, so conditioned by interwoven cooperations and competitions, that no man can say where utility begins or ends.

Conservation, Leopold came to realize, must aim at something larger and more comprehensive than a maximum sustained flow of desirable products (like lumber and game) and experiences (like sport hunting and fishing, wilderness travel, and solitude) garnered from an impassive nature. It must take care to ensure the continued function of ecological processes and the integrity of ecosystems. For it is upon them, ultimately, that human resources and human well-being depend, for the present generation as well as for those to come. Indeed, Leopold quietly transformed the concept of conservation from its pre-ecological to its present deep ecological sense—from conservation understood as the wise use of natural resources to conservation understood as the maintenance of biological diversity and ecological health.

The word "preserve" in the summary moral maxim of Leopold's (1949, pp. 224–225) oft-quoted land ethic—"A thing is right when it tends to preserve the integrity, stability, and beauty of the biotic community. It is wrong when it tends otherwise."—is unfortunate because it seems to ally Leopold with the Preservationists in the familiar Preservation vs. Conservation feud. We tend to think of Leopold as having begun his career in the Conservationist camp and then gradually to have come over, armed with new ecological arguments, to the Preservationist camp. Leopold appears to be a mid-twentieth century conservation prophet emerging from the woods wearing the hat of Gifford Pinchot and speaking with the voice of John Muir. His historical association with the wilderness movement cements this impression. While still with the Forest Service, Leopold had campaigned hard to preserve a few relics of the American frontier in which he and like-minded sportsmen might play at being pioneers (Leopold, 1925a; 1925b; 1925c; 1925d; Allin, 1987). After becoming a professor of wildlife management, he suggested that the designated wilderness areas he had helped to create might serve threatened species as biotic refugia (Leopold, 1936). In the last decade of his life, he suggested that representative undeveloped biomes might serve science as a "base datum of land health" (Leopold, 1941).

While I would be the first to agree that designated and de facto wilderness areas are important as "land laboratories" and vitally important as biotic refugia, Leopold's unfortunate—and unintended—legacy for the American conservation policy debate has been to intensify the familiar alternative: either efficiently exploit the remaining and dwindling wild lands or lock them up and preserve them forever as wildernesses. But a review of his unpublished papers and published but long-forgotten articles (now conveniently collected in the new book of his essays) confirms Flader's opinion that Leopold was primarily concerned, in theory as well as on the ground, with integrating an optimal mix of wildness with human habitation and economic utilization of land.

As *Sand County*'s upshot essay, "Wilderness," shows, Leopold was committed to wilderness preservation no less fully at the end of his career than at the beginning. But his mature vision went beyond the either develop and necessarily destroy or lock up and preserve dilemma of American conservation as he inherited it from the generation of Pinchot and Muir. Wild sanctuaries, for Leopold, were a component of a much broader and subtler conservation philosophy, a conservation philosophy that he regarded as more developed and mature than simply saving wild remnants. As he put it, "This impulse to save wild remnants is always, I think, the forerunner of the more important and complex task of mixing a degree of wildness with utility" (Leopold, 1991b, p. 227).

In a typescript composed shortly after a four-month trip to Germany in 1935 Leopold (1991b, pp. 226–227) wrote,

> To an American conservationist, one of the most insistent impressions received from travel in Germany is the lack of wildness in the German landscape. Forests are there. . . . Game is there. . . . Streams and lakes are there. . . . But yet, to the critical eye there is something lacking. . . . I did not hope to find in Germany anything resembling the great "wilderness areas" which we dream about and talk about, and sometimes briefly set aside, in our National Forests and Parks. . . . I speak rather of a certain quality [—wildness—] which should be, but is not found in the ordinary landscape of producing forests and inhabited farms.

In a more fully developed essay entitled "The Farmer as a Conservationist" Leopold (1939b) regales his reader with a rustic idyll in which the wild and domesticated floral and faunal denizens of a Wisconsin farmscape are feathered into one another to create a harmonious whole. In addition to cash and the usual supply of vegetables, chicken, beef, and pork, lumber, and fuel wood, Leopold's envisioned farmstead affords its farm family venison, quail and other small game, and a variety of fruit and nuts from its woodlot, wetlands, and fallow fields; and its pond and stream yield panfish and trout. It also affords intangibles—songbirds, wild flowers, the hoot of owls, the bugle of cranes, and intellectual adventures aplenty in natural history. To obtain this bounty, the farm family must do more than permanently set aside acreage, fence woodlots, and leave wetlands undrained. They must sow food and cover patches, plant trees, stock the stream and pond, and generally thoughtfully conceive and skillfully execute scores of other modifications, large and small, of the biota that they inhabit.

Further, Leopold (1939b, p. 294) explicitly states the preservationist heresy that human economic activity may not only co-exist with healthy ecosystems, but that it may actually enhance them: "When land does well for its owner, and the owner does well by his land; when both end up better by reason of their partnership, we have conservation. When one or the other grows poorer, we do not."

Like Pinchot, Leopold (1949, p. 207) attempted to distill his philosophy of conservation into a quotable definition. And indeed it is often quoted, but little analyzed or appreciated: "Conservation is a state of harmony between men and land." This definition represents a genuine third alternative to Pinchot's brazenly anthropocentric, utilitarian definition of conservation as efficient exploitation of "resources" and Muir's anti-anthropocentric definition of conservation as saving innocent "Nature" from inherently destructive human economic development.

Can we generalize Leopold's vision of an ecologically well integrated family farm to an ecologically well integrated technological society? Can we reconcile and integrate human economic activities with biological conservation? Can we achieve "win-win" rather than "zero-sum" solutions to development-environment conflicts? Can we design "sustainable economies" rather than zone the planet into ever-expanding sectors of conventional, destructive development and ever-shrinking wilderness sanctuaries? Can we succeed as a global technological society in enriching the environment as we enrich ourselves?

A THIRD WORLD CRITIQUE OF CONSERVATION VIA WILDERNESS PRESERVATION

I think we can. More to the point, I think we have to. The pressure of growing human numbers and rapid development, especially in the Third World, bodes ill for a global conservation strategy focused primarily on "wilderness" preservation and the establishment of nature sanctuaries. Such a strategy represents a holding action at best and a losing proposition at last.

A sobering "Third World critique" of conservation on a global scale via wilderness preservation has been recently advanced by Ramachandra Guha in the pages of *Environmental Ethics* the journal. Guha (1989a, p. 79) points out that we Americans

> possess a vast, beautiful, and sparsely populated continent. . . . America[ns thus] can simultaneously enjoy the material benefits of an expanding economy and the aesthetic benefits of unspoilt nature. The two poles of "wilderness" and "civilization" mutually coexist in an internally coherent whole. . . .

Exporting the American pattern of conventional industrial development counter-balanced by conservation of nature via wilderness preservation to a "long settled, densely populated country" like India, Guha (1989a, p. 75) explains, results in a "direct transfer of resources from the poor to the rich"—because the poor are evicted from their homelands to create nature reserves and only wealthy foreigners and the in-country elite that benefit from industrial development can afford to enjoy the wilderness experience or just the knowledge that a bit of pristine nature is being protected.

Guha's critique notwithstanding, I am pleased to note that the United Nations biosphere reserve concept genuinely de-Americanizes the wilderness idea in that it specifically requires planners to take account of and to integrate local peoples culturally and economically into reserve designs (von Droste, 1988). But faced with the harsh realities of the coming century, the wilderness idea—even become the biosphere reserve concept—is, by itself, too little too late. And it is too defensive to save the planet and all of us, its people, from ecological collapse. We need to integrate wildlife sanctuaries into a broader philosophy of conservation that generalizes Leopold's vision of a mutually beneficial and mutually enhancing integration of the human economy with the economy of nature.

Here let me be both clear and emphatic: I am not suggesting that we open the remaining wild remnants to development, but that we begin to reconceive economic development in the light of ecology. Human economic activities should at least be compatible with the ecological health of the environments in which they occur. And ideally they should enhance it. In Leopold's agrarian idyll of conservation, not only does the ecological farm family actively manage their wild lands, they also reform their farming practices on the fields so that the two, the wild culture and the domestic culture, might better coexist. "Clean farming" was a frequent target of Leopold's pointed criticisms and wry wit. And he may have been the first environmentalist clearly to recognize and lament the "industrialization" of agriculture during the twentieth century and the conversion of the classic, relatively benign and sustainable farm, into an environmentally destructive and unsustainable "food-factory" (Leopold, 1945).

Echoing the Leopold epigraph of this essay, Guha further charges that conservation primarily via wilderness preservation is an American subterfuge. We who enjoy the benefits of modern industrial civilization with all its environmental costs can salve our consciences by pointing to the few odds and ends of arid, rough, scenic, or remote country that we have set aside, first for our own recreational, aesthetic, and spiritual needs and second as ecologic refugia—little places where our fellow denizens of Planet Earth can live and blossom. Thus we can avoid facing up to the fact that the ways and means of industrial civilization lie at the root of the current global environmental crisis. To Guha (1989a, p. 79), the new voice for Third World environmental ethics, conservation via wilderness preservation "runs parallel to the consumer society without seriously questioning its ecological and socio-political basis."

Without such questioning, the call of wilderness advocates for the recrudescence of big wilderness is quixotic. Wilderness is one pole of a dualism. To want more of it is to oppose the forces of conventional industrial development, certainly, but not to challenge the conventional conception of economic development. And since it pits—as a one-or-the-other-but-not-both choice—human economic interests against the interests of nature, it has little political appeal and thus little chance of success.

In the Third World, on the other hand, industrial development has often resulted in human tragedy as well as environmental disaster. The green revolution, for example, has dispossessed small land holders and actually increased chronic hunger as prices fall, costs increase, and crops are exported to First World consumers (Wright, 1984). Industrial forestry has similarly disrupted traditional patterns of sustainable forest use as well as played havoc with forest ecosystems (Guha, 1989b). By contrast, according to Guha (1989a, p. 81), Third World peasants

> have far more basic needs, their demands on the environment are far less intense, and they can draw upon a reservoir of cooperative social institutions and local ecological knowledge in managing the "commons"—forests, grasslands, and the waters—on a sustainable basis.

Concerned to articulate a distinctive Third World environmental philosophy, Guha (1989a) may have too readily conceded that conservation principally via wilderness preservation is even practicable in the United States. The violent confrontations of the Redwood Summer of 1990 in California and the on-going, increasingly acrimonious Old Growth/Spotted Owl impasse in the Pacific Northwest make me wonder if Guha (1989a, p. 79) were not uncritically perpetuating a Third World myth when he describes North America as a "vast," superabundant, and "sparsely populated continent."

The late twentieth-century crescendo of cut/graze/plow/pave *or* preserve conflicts here and abroad leads me to undertake a generalization of Ramachandra Guha's "Third World critique" of the wilderness idea. I wish to strike deeper than he and suggest that the popular wilderness idea is as inherently flawed as its counterpart, the conventional development idea. Most conservation biologists today recognize the paradoxical necessity of managing (and hence artificializing?) wilderness areas in order for them to continue to play their vital part in biological conservation (Reed, 1990). And just as paradoxically refugia for species that require lots of living space may be compromised by wilderness recreation for which purpose such areas were originally set aside pursuant to the popular concept of wildernesses as areas where man is a visitor who does not remain (Reed and Merigliano, 1990).

A THREE-POINT CRITIQUE OF THE RECEIVED CONCEPT OF WILDERNESS

Upon close scrutiny the simple, popular wilderness idea dissolves before one's gaze.

First, the concept perpetuates the pre-Darwinian Western metaphysical dichotomy between "man" and nature, albeit with an opposite spin.[1] So much so, indeed, that one of the principal psycho-spiritual benefits of wilderness experience is said to be contact with the radical "other" and wilderness preservation the letting be of the nonhuman other in its full otherness (Birch, 1990).

Second, the wilderness idea is woefully ethnocentric. It ignores the historic presence and effects on practically all the world's ecosystems of aboriginal peoples.

And third, it ignores the fourth dimension of nature, time. In a recent discussion H. Ken Cordell and Patrick C. Reed (1990, p. 31) say flatly, "Preservation implies cessation of change." But in ecosystems, change is as natural as it is inevitable (Botkin, 1990). Hence trying to preserve in perpetuity—trying to "freeze-frame"—the ecological status quo ante is as unnatural as it is impossible. A more sophisticated and refined concept of wilderness preservation among contemporary conservationists aims rather to perpetuate the integrity of evolutionary and ecological processes, instead of existing "natural" structures (Parsons et al., 1986). Cordell and Reed (1990, pp. 30–31) in fact understand wilderness preservation not as an effort to halt change, but to slow "accelerating rates of change" and to preserve the "dynamic operation of natural processes . . . fire, drought, disease, predation, and geological change." But even such a dynamic, process-sensitive notion of wilderness preservation is incomplete if it ignores the role that Homo sapiens has played historically practically everywhere and if it would deny Homo sapiens the opportunity to reestablish a positive symbiotic relationship with other species and a positive role in the unfolding of evolutionary processes.

The Wilderness Act of 1964 beautifully reflects the conventional understanding of wilderness. It reads: "A wilderness, in contrast with those areas where man and his works dominate the landscape, is hereby recognized as an area where the earth and its community of life are untrammeled by man, where man himself is a visitor who does not remain" (Nash 1967, p. 5).

This definition assumes, indeed it enshrines, a bifurcation of man and nature. That the man-nature dichotomy insidiously infects even our well intentioned and noble efforts to limit our own grasp should not be surprising. A major theme both in Western philosophy, going back to the ancient Greeks, and in Western religion, going back to the ancient Hebrews, is how man is unique and set apart from the rest of nature.

In the Judeo-Christian religious tradition, man alone among all the other creatures is created in the image of God. In the Greco-Roman philosophical tradition, among all the other animals, man is uniquely rational. Subsequently, philosophers as different from one another as Thomas Aquinas, the perennial philosopher of the Catholic Church, and René Descartes, the father of modern philosophy, variously synthesized these two strands of Western thought, Aquinas in the Middle Ages and Descartes at the dawn of the Scientific Revolution. The classical Western segregation of man from nature thus became ever more ingrained, a veritable cachet of Western ideology both religious and scientific. And all the wonderful works of man—from the pyramids of Egypt to the Gothic cathedrals of France, to say nothing of all the marvels of modern technology—seemed to confirm the radical metaphysical rift between us and the brute creation.

Now that man's technological dominion over the earth is virtually absolute, and its community of life will survive only if we permit, a rising chorus of voices is crying, if not in the wilderness, at least for it—for man to show a little mercy, to allow a little untrammeled, unconquered nature to exist here and there (Oelschlaeger, 1991). Until recently, man seemed the up-and-coming hero armed with Promethean science in the struggle with Titanic nature. Now, as the twentieth century winds to a close, victorious man seems to be a tyrant, his conquest a spoils, and nature the victim. For many ardent wilderness advocates, the roles of hero and villain are reversed (Birch, 1990). But the underlying dichotomy goes unchallenged.

Since Darwin's *Origin of Species* and *Descent of Man,* however, we have known that man is a part of nature. We are only a species among species, one among twenty or thirty million natural kinds. The natural works of other species, everyone seems to agree, can help as well as harm the biotic communities of which they are a part. Pursuing their own economic interests, bees assist the reproduction of flowering plants and thus perform an invaluable community service. Hundreds of other similar examples—from nitrogen-fixing bacteria to scavenging turkey vultures—could be cited. Elephants, on the other hand, pursuing their own economic interests, can be very destructive members of their biotic communities. So can deer. And again, hundreds of similar examples—from cow birds to kudzu—could be cited.

If man is a natural, a wild, an evolving species, not essentially different in this respect from all the others, as Gary Snyder (1990) reminds us, then the works of man, however precocious, are as natural as those of beavers, or termites, or any of the other species that dramatically modify their habitats. And if entirely natural, then the works of man, like those of bees and beavers, in principle *may be,* even if now they are usually not, beneficial—judged by the same objective ecological norms—to the biotic communities which we inhabit.

In one important (and relevant) respect we are different from other species. The pollinating services performed by bees and the decomposition of dead wood and soil treatment provided by termites are instinctive behaviors. The migration routes of birds, the hunting techniques of predators, and many other animal behaviors that are learned might be regarded as "cultural" or at least "protocultural" rather than strictly hereditary or "instinctive." But the cultural component in human behavior is so greatly developed as to have become more a difference of kind than of degree. To suggest that the works of man are not natural is not to suggest that they are supernatural or preternatural, but that they are products of culture not instinct. Still, the cultural works of man are evolutionary phenomena no less than are other massive structures created by living things like, say, coral reefs. They are, one and all, natural in that sense of the word (Lemons, 1987). And therefore it is logically possible that they may be well attuned and symbiotically integrated with other contemporaneous evolutionary phenomena, with coral reefs and tropical forests, as well as the opposite.

Precisely because the works of man are largely cultural they are capable of being rapidly reformed. Other animals cannot change what they do in and to their biotic communities, at least not very rapidly, and perhaps not ever consciously and deliberately. We can, since our economic behaviors are determined more by our cultures than by our genes. To totally trammel nature—to "develop" every acre—or, here and there, to forbear, is not the only question man should put to himself. How to work our works in ways that are at once humanly, socially, *and ecologically* benign rather than malignant—that's the more important, and problematic, question.

Now on to my second point, that the popular wilderness idea is ethnocentric.

More than anyone else, Roderick Nash (1967) has molded the popular idea of wilderness in the contemporary American mind. Nash acknowledges but skates rapidly over American Indian complaints that the very concept of wilderness is a racist idea and he expresses no doubt that the first European settlers of North America encountered a "wilderness condition" (Nash, 1967). In the recent (and excellent) Wilderness Idea film by Lawrence Hott and Diane Garey (1989), Nash is even more emphatic. He says that the pilgrims literally stepped off the Mayflower into a wilderness of continental dimensions.

Upon the eve of European landfall most of temperate North America was not, *pace* Nash, in a wilderness condition—not undominated by the works of man—unless one is prepared to ignore the existence of its aboriginal inhabitants and their works or to insinuate that they were not "man," i.e., not fully human human beings. In 1492, Antarctica was the only true wilderness land mass on the planet. (And by now, even a good bit of it has come under the iron heel of industrial man.)

Until rather recently it was possible for environmental historians to minimize the ecological importance of the original human inhabitants of the New World because the decimating effects of Old World diseases had not been taken into account in estimating their Pre-Columbian numbers. The distinguished anthropologist Alfred L. Kroeber (1939) calculated a hemispheric total of eight and a half million souls at contact and placed the population of the lands now comprised of the forty-eight (geographically) United States, Canada, Greenland, and Alaska at fewer than one million. Henry F. Dobyns (1966), adding the impact of Old World diseases on the immunologically innocent New World aborigines to his demographic equations, proposed to increase Kroeber's estimates by, roughly, a factor of ten. Tragically, only one Indian in ten seems to have survived the epidemics of small- and chicken-pox, diphtheria, measles, scarlet fever, and other infections that swept through North America from East to West during the fifteenth and sixteenth centuries—often passed from Indian to Indian before the leading edge of the pale-face tide arrived. As John Witthoft (1965, p. 28) described this inverse decimation, "Great epidemics and pandemics of these diseases are believed to have destroyed whole communities, depopulated whole regions, and vastly decreased the native population everywhere in the yet unexplored interior of the continent."

In the spring of 1492, North America (not including Mexico) may not have been densely populated but it was largely inhabited by some ten million people. Nor were the Indians passive denizens of the continent's forests, prairies, swamps, deserts, and tundra, simply taking—like foraging Mandrill baboons—what usable plants and animals they happened to stumble upon. Most of temperate North America was actively managed by its aboriginal human inhabitants. In addition to domesticating and cultivating an extraordinarily wide range of food and medicine plants, native North Americans managed the continent's forest and savannah communities, principally with fire (Heizer, 1955; Pyne, 1982). Their pyrotechnology helped to determine the mix of species and reset succession in the various plant associations in which they lived (Day, 1953; Martin, 1973; Lewis,

1973). The European immigrants in fact found a man-made landscape, but they thought it was a wilderness because it didn't look like the man-made landscape that they had left behind (Cronon, 1983; Merchant, 1989).

It is important to note, however, that the same kind of country that is now designated or de facto wilderness in the United States was also less frequented and utilized by the American aborigines (Barrett, 1980; Devall, n.d.). The two percent of the forty-eight (geographically) United States presently devoted to wilderness is mostly in high, rough, or arid lands—and often all three. A good argument, therefore, could be made that an expansion of the present system of wilderness reserves to mirror ancient human land use patterns on this continent is consistent with long established Nearctic evolutionary and ecological regimes.

The incredible abundance of wildlife encountered in the western hemisphere by the first European intruders was not, however, a concomitant of a *universal* wilderness condition, that is, not due to the absence of inherently destructive Homo sapiens in significant numbers everywhere. Rather the biological wealth of North America on the eve of European landfall is more attributable to the bioregional management programs of the indigenous population than to low numbers (Hughes, 1983). Further, the ubiquity of grizzly bears throughout the west and big cats and wolves throughout the continent indicates a mutual tolerance of these species with Homo sapiens americana that was, apparently, disrupted when Homo sapiens europi began to persecute them as varmints (Martin, 1978).

I now take up my third point: Wilderness *preservation,* as the popular conservation alternative to destructive land use and development, suggests that, untrammeled by man, a wilderness will remain "stable," in a steady state. But nature is inherently dynamic; it is constantly changing and ultimately evolving. Today, most of the pitifully small fenced-off patches of designated wilderness areas of temperate North America lack major components of their Holocene ecological complement—notably their large predators. Not only that, a fence or the policy equivalent thereof, will not exclude all the exotic species that have accompanied or followed the migration of Homo sapiens europi to North America. Designated wilderness areas, paradoxically, must be actively restored and managed if they are to remain fit habitat for native species. But the necessity of means raises a question of ends, of values. Is maintaining "vignettes of primitive America" the most important and defensible goal of biological conservation?

(Gordon et al., 1989). (Here, again, let me be clear. I am excluding from consideration biologically destructive economic desiderata such as hydroelectric impoundment.) Since we must actively and invasively manage nature to preserve the ecological status quo ante, the possibility of managing nature for more direct, less incidental conservation goals arises (Westman, 1991).

The whole notion of *preserved* wilderness areas—even were we to restore the wolf and the grizzly to representative spots in their original ranges, simulate the effects, where they existed, of Homo sapiens americana's hunting and burning, and maintain a constant vigil against invasion by exotics—defies the fourth dimension of nature, time. In the course of time, ecological succession is continually reset by one or another natural disturbance. Paleo-ecological studies reveal, moreover, that species composition within successional seres—the structure, in other words, of biotic communities—has changed over time (Thompson, 1988). Fluctuations in climate drive migrations of glaciers, forests, deserts, and grasslands (Botkin, 1990). Exotic species, with or without human help, invade new environments and in the course of time become naturalized citizens (Westman, 1990). Indeed, the concepts of exotic/native species are "relative, scale dependent (temporally and spatially), and about as ambiguous as any in our conservation lexicon" (Noss, 1990a, p. 242). Are the feral mustangs roaming the American west, for example, natives or exotics? Michael Soulé (1990, pp. 234–235) speculatively envisions lions, cheetahs, camels, elephants, saiga antelope, yaks, and spectacled bears joining horses in the name of "the restoration to the Nearctic of the great paleomammalian megafauna" which disappeared from this continent "only moments ago in evolutionary time."

The wilderness idea has not only made conservation convenient—if Guha is correct to argue that it has served American conservationists as a subterfuge, allowing us to enjoy the benefits of industrial development and overconsumption, while salving our consciences by setting aside a few undeveloped remnants for nature—it has made conservation philosophy simple and easy. If we conceive of wilderness as a static benchmark of pristine nature in reference to which all human modifications may be judged to be more or less degradations then we can duck the hard intellectual job of specifying criteria for land health in four-dimensional, inherently dynamic landscapes long inhabited by Homo sapiens as well as by other species. The idea of healthy land maintaining itself is more sensitive to the dynamic quality of ecosystems, than is the conventional idea of preserving vignettes of primitive America. And if the concept of land health replaces the popular, conventional idea of wilderness as a standard of conservation, then we might begin to envision ways of creatively reintegrating man and nature.

Conservation biologists are just now coming to grips with the problem of setting out objective criteria of ecological health in dynamic, long-humanized landscapes. Recent efforts to do so have been made by Robert Costanza (1991), Walter E. Westman (1991), and Robert Ulanowicz (1991). While insisting upon the naturalness of change, the importance of rate and scale of change for land health cannot be overemphasized. Cordell and Reed (1990, pp. 30–31) suggest that "'bad' change" is the result of accelerating rates of environmental change and Botkin (1990) again and again warns that, while change per se has always characterized the living earth, the current changes imposed upon nature by global industrial civilization are unprecedentedly rapid and radical and therefore, albeit natural, not normal.

CONSERVATION VIA SUSTAINABLE DEVELOPMENT

The new idea in conservation today is called "sustainable development" (Brundtland et al., 1987). But that term can mean different things to different people. Under the essentially economic interpretation of the Brundtland Report (1987), it means little more than what it says, initiation of human economic activity that can be sustained indefinitely, quite irrespective of whether or not such development is ecologically salubrious. Worse still, some economists would denominate a development path "sustainable" even if it leaves subsequent generations a depauperate natural environment, but sufficient technological know-how and investment capital to invent and manufacture an ersatz world (Passell, 1990). Following Raymond F. Dasmann (1988), I would like to mean by "sustainable development" initiation of human economic activity that is limited by ecological exigencies; economic activity that does not seriously compromise ecological integrity; and, ideally, economic activity that positively enhances ecosystem health.

But is sustainable development so understood possible? The surest proof of possibility is actuality. Here are some actual examples of mutually sustaining and enhancing human-nature symbioses.

The Desert Smells Like Rain by ethnobotanist Gary Nabhan is about present-day Papago dry farmers in the desert Southwest. From time immemorial two oases some thirty miles apart, A'al Waipia and Ki:towak, had been inhabited by Papago. The former lies in the United States, in the Organ Pipe Cactus National Monument, and the latter in Mexico. The United States government designated A'al Waipia a bird sanctuary and stopped all cultivation there in 1957. Over in Mexico, Ki:towak is still being farmed in traditional style by a group of Indians. Nabhan (1982) reports visiting the two oases, accompanied by ornithologists, on back-to-back days three times during one year. At the A'al Waipia bird sanctuary they counted thirty-two species of birds; at the Ki:towak settlement they counted sixty-five. A resident of Ki:towak explained this irony: "When the people live and work in a place, and plant their seeds and water their trees, the birds go live with them. They like those places. There's plenty to eat and that's when we are friends to them" (Nabhan, 1982, p. 98).

Conservation biologist David Ehrenfeld (1989, p. 9) concludes from this "parable of conservation" that "the presence of people may enhance the species richness of an area rather than exert the effect that is more familiar to us." Here, of course we must be cautious. Species richness is not the only measure of ecosystem health. The quality, so to speak, of species is also important. But in general, Nabhan (1982) suggests, the whole desert ecosystem in which they live, not just the Ki:towak oasis, is as adapted to and dependent upon the Papago as they are upon it. Their little charco fields, built to catch and hold the runoff from ephemeral desert rains, are home to a wide variety of coevolved uncultivated plants (some of which the Papago eat) and unfenced animals ("field meat" as the Papago think of them). Undoubtedly the desert ecosystem has been enriched rather than impoverished by millennia of Papago habitation and exploitation.

Arturo Gómez-Pompa and Andrea Kaus (1988), on the basis of the higher incidence of fruit-bearing trees in the remnants of rainforest in southern Mexico, suggests that what appear to the untutored eye to be pristine patches of wilderness, rich in animal as well as plant life, are actually surviving fragments of an extensive lowland Maya permaculture.

Ecologists working in the Amazon have come to similar conclusions about the vast South American rainforest. Darrell A. Posey has studied the methods of living in the Amazon rainforest without destroying it devised by the Kayapó Indians (Stevens, 1990). The Kayapó fish, hunt, gather, and cultivate swiddens. In sharp contrast to the displaced Euro-Brazilian peasants who are entering the region, the Kayapó, through a complex cycle of planting, manage to cultivate a forest clearing for nearly ten years, instead of merely three or four. But after a decade of cultivation, neither is a Kayapó plot simply abandoned. Instead, the Kayapó manage the regeneration of the forest by planting useful native species—first, fast-growing short-lived early succession plants like banana, and later, longlived canopy trees like Brazil nut trees and coconut and oil palms. Thus their fallows become permanent resource patches from which they obtain fruit, nuts, medicines, thatch, and other materials in perpetuity.

I cite these indigenous New World examples of human-nature symbiosis not to suggest that we give the hemisphere back to the Indians or that we all go native and attempt somehow to recreate American Indian culture in the late twentieth century. The three hundred million of us contemporary denizens of North America could not return to the life styles of the ten million Pre-Columbian inhabitants even if we wanted to without exceeding carrying capacity. I simply wish to point out, rather, that the past affords paradigms aplenty of an active, transformative, managerial relationship of people to nature in which both the human and non-human parties to the relationship benefited. The human-nature relationship is an ongoing, evolving one. We can, I am confident, work out our own, post-modern, technologically sophisticated, scientifically informed, sustainable civilization just as in times past the Minoans in the Mediterranean, the vernacular agriculturists of Western Europe, and the Incas in the Andes worked out theirs.

The symbiotic win-win philosophy of conservation is gradually replacing the bifurcated zero-sum approach as the twentieth century gives way to the twenty-first. For example, one of the most promising conservation stratagems in the Amazon rainforest today is the designation not of nature reserves from which people are excluded to protect the forest and its wildlife, but of so-called "extractive reserves" (Hecht and Cockburn, 1989). An extractive reserve is an area where traditional patterns of human-nature symbiosis—such as those evolved by the Amazonian Indians and more recently by the rubber tappers—are protected from loggers, cattle ranchers, miners, and hydroelectric engineers.

Writing in *Nature,* Charles M. Peters, Alwyn H. Gentry, and Robert O. Mendelsohn (1989) report that the nuts, fruits, oils, latex, fiber, and medicines annually harvested from a representative hectare of standing Amazon rainforest in Peru is of greater economic value than the saw logs and pulpwood stripped from a similar hectare—greater even than if, following clear-cutting and slash-burning, the land is, in addition, converted either to a forest monoculture or to a cattle pasture. From a painstaking econometric study they conclude that "without question, the sustainable exploitation of non-wood forest resources represents the most immediately profitable method for integrating the use and conservation of the Amazonian forests" (Peters et al., 1989, p. 656).

But is it possible to export this Third World approach to conservation to the First World?

Here is a modest proposal that I think would have a greater chance of realization than Reed Noss's (1991) suggestion that 50 percent of the lower 48 states be allowed to return to a wilderness condition, a condition which as I have here pointed out has not existed in 50 percent of what is now the lower 48 states for the 100 centuries since the original discovery of the Americas by eastward migrating pedestrian Homo sapiens. It would have a greater chance of realization because it offers a forward looking, sustainable economic alternative to the present pattern of destructive exploitation and because it is not implicitly misanthropic.

First, I agree with Noss (1990b) that we do need big wilderness to help conserve biodiversity, but is 50 percent a reasonable amount? A more practicable alternative might be to start with presently designated wilderness areas and, guided by information on aboriginal settlement and use, expand their political boundaries to coincide with their ecological boundaries. Such expanded wild reserves might be connected where possible by wild corridors, permitting the migration of animals and gene flow between populations as Noss (1990b) also suggests. These core regions and interconnective corridors would serve as sanctuaries for the large carnivores and other species, like the spotted owl, that need old growth or interior habitat to thrive.

One-third of the lower 48 is publicly owned land. These public lands are managed mostly by the USDA Forest Service, the Bureau of Land Management, and the National Park Service, but only by the latter in anything resembling the public interest. In designing a utopia we can stipulate that public agencies function as ideally they ought. Even in the real world there are some winds of discontent blowing in the Forest Service and some stirrings of reform (DeBonis, 1989; USDA Forest Service, 1989). I shall assume that it is theoretically possible for the Forest Service to manage the national forests for ecological integrity and function as well as for a sustainable supply of forest products.

For the western rangelands, however, I propose a complete overhaul of current management philosophy. There is little economic and no ecological justification for grazing domestic animals like cattle and sheep on the western ranges (Conaway, 1987). George Wuerthner (1991, p. 30) reports that

> Grazing [is] permitted on 89 percent of Bureau of Land Management lands and 69 percent of Forest Service lands. . . . Altogether, more than 265 million acres of federal lands . . . are leased under federal grazing programs. Despite the huge acreage involved, the Department of Agriculture estimates that these federal lands provide forage for less than two percent of the cattle and sheep produced annually in the United States. This trifling production comes at considerable ecological cost.

Adding insult to injury, the public treasury actually subsidizes this pathetic livestock industry on these public lands: Grazing fees are significantly lower on public than on private lands and taxpayers pay for a variety of "improvements," most of which, like water impoundment and predator control, are ecologically benighted (Wuerthner, 1991). So what is the justification for this "cowboy welfare" paid for by the public not only in dollars but in a degraded natural heritage? A very few citizens—ranchers—benefit. According to T. H. Watkins (1991), the larger justification is essentially historic and mythic. Europeans brought their domestic stock with them and set up shop in the west and a popular romance grew out of that phenomenon. The present western livestock industry, such as it is, is a relic of that period and because the animals ranched are not suited to the climate, it will, sooner or later, inevitably wither away.

Why not sooner, before more damage is done to the soils, waters, plants, and animals? It's just like the old growth controversy in the Pacific Northwest. If we stop the cutting of primary forests, jobs will be lost. The same jobs will be lost after the ten to twenty years it will take to log out the old growth. But then the forest will be gone too. So why not bite the economic bullet now rather than later while we still have significant old growth left?

The remaining old growth forests are, in my view, good candidates for biosphere reserve style conservation with the local economy confined to the periphery of the core sanctuaries and geared to ecotourism and other nonconsumptive uses of the "resource." The vast rangeland of the west, both public and private, might be economically exploited and conserved quite differently. According to Wuerthner (1991, p. 29), "The current condition of our rangelands is as much a factor of using the wrong animal in the wrong place as it is of lax management." Cattle are inherently destructive, especially in arid country.

So why not ranch the right animals in the right place? Could we take a page out of the approach to conservation advocated by Norman Myers (1981) for wild African fauna and crop the indigenous ungulates of the American west? If I have expressed doubts here about the wilderness condition of most places in the New World before the European conquest, I certainly have no doubt that most New World ecosystems were then in a condition of robust health. The vast Pre-Columbian herds of bison and antelope, and large populations of mule deer, white tailed deer, and elk did not denude stream banks and gully slopes like the cattle and sheep that have succeeded them. And they supported a sizable population of carnivorous primates as well as four-legged predators. A return of these species in comparable numbers might support a new and very different western ranching system and significantly contribute to the diet of a much larger contemporary human population of North America that draws on many other food resources than could the hunting/gathering/dry farming indigenes.

But how can we get from here to this range utopia? I don't know. Very real political and economic obstacles, to say nothing of the sheer inertia of habit and tradition, stand in the way. But before we can figure out how to get from here to anywhere, we have to have a vision to guide us. Utopias may be unattainable in reality, but they are not impractical. They help move us off dead center. And falling short of an ideal is still movement in the right direction.

So far, so much can be said: The idea has been tried successfully in Africa despite dismal infrastructural support, resulting in significant spoilage losses (Casebeer, 1978). With the capital and technical resources available in North America, the efficient processing and distribution of game meat should be well within reason. And demand for game meat has increased in recent years in part because of its low fat content (White, 1987). Wild game meat is also "organic," a quality in both vegetables and meats commanding a premium price in many upscale retail outlets. (Personally, I might add, I find the conventional choice of animal foods boring. While the surf side of a surf and turf entree can be quite varied—anything from sea scallops to thresher shark—the turf fare is dismally limited to beef, pork, mutton, chicken, and turkey.)

Some ranchers, moreover, are already shifting from raising cattle to raising deer and selling permission to hunt on their lands to sportspersons, because it is more profitable (White, 1987). The concept I am suggesting here is not a universalization of the private hunting preserve. I am, rather, suggesting the total elimination of livestock from the western ranges, ripping out unnecessary roads, a massive restoration of native wildlife populations, and a conversion of cowboys to airborne, satellite guided game managers and professional market hunters. But the trend toward game ranching might be a foot in the door for a serious discussion about the biological, economic, political, legal, and social possibility of establishing an ecological and economic regime on the western ranges that is continuous with and an extension of those that prevailed before the advent of the cow and the horseback, and later the pick-up truck, cowboy.

Can we generalize this particular example of the sustainable development alternative to either conventional development, like intensive stock grazing, or wilderness designation and restoration (where none existed before)? Can we envision and work to create an eminently livable, systemic, post-industrial technological society well adapted to and at peace and in harmony with its organic environment? If illiterate, unscientific peoples can do it, can't a civilized, technological society also live, not merely in peaceful coexistence, but in benevolent symbiosis with nature? Is our current industrial civilization the only one imaginable? Aren't there more appropriate, alternative technologies? Can't we be good citizens of the biotic community, like the birds and the bees, drawing an honest living from nature and giving back as much or more than we take?

NOTES

I thank Eugene C. Hargrove for providing an opportunity to broach the central ideas contained in this paper to the Faculty of Environmental Ethics at the University of Georgia's Institute of Ecology. I have benefited from discussion of the role of wilderness in Aldo Leopold's philosophy of conservation with Susan L. Flader, Curt Meine, and Bryan Norton. John Lemons and an anonymous referee for the *Environmental Professional* generously and critically commented on an earlier draft of this paper and rescued it from other errors of fact and doctrine than those that remain.

1. Fully aware that it is gender-biased, I use the term "man" both deliberately and apologetically to refer globally and collectively to the species Homo sapiens because no other term carries the same connotation and flavor, a connotation and flavor that I wish to evoke in the course of this critique, including its decidedly sexist connotation and flavor.

BIBLIOGRAPHY

Allin, C. 1987. The Leopold Legacy and the American Wilderness. In *Aldo Leopold: The Man and His Legacy,* Thomas Tanner, ed. Soil Conservation Society of America, Ankeny, IA, pp. 25–38.

Barrett, S. W. 1980. Indian Fires in the Presettlement Forests of Western Montana. In *Proceedings of the Fire History Workshop, Tucson AZ.* USDA Forest Service General Technical Report RM-81, Fort Collins, CO.

Birch, T. 1990. The Incarceration of Wilderness: Wilderness Areas as Prisons. *Environmental Ethics* 12:3–26.

Botkin, D. B. 1990. *Discordant Harmonies: A New Ecology for the Twenty-First Century.* Oxford University Press, New York.

Brundtland, G. H., chair, World Commission on Environment and Development. 1987. *Our Common Future.* Oxford University Press, New York.

Casebeer, R. L. 1978. Coordinating Range and Wildlife Management in Kenya. *Journal of Forestry* 76:374–375.

Conaway, J. 1987. *The Kingdom in the Country.* Houghton Mifflin, Boston, MA.

Cordell, H. K., and P. C. Reed. 1990. Untrammeled by Man: Preserving Diversity through Wilderness. In *Preparing to Manage Wilderness in the 21st Century: Proceedings of the Conference.* Southeastern Forest Experiment Station, U.S. Department of Agriculture, Forest Service, Asheville, NC, pp. 30–33.

Costanza, R. 1991. Toward an Operational Definition of Ecosystem Health. AAAS annual meeting, February 15, 1991, Washington, DC.

Cronon, W. 1983. *Changes in the Land: Indians, Colonists, and the Ecology of New England.* Hill and Wang, New York.

Dasmann, R. F. 1988. Conservation, Land Use, and Sustainable Development. In *For the Conservation of Earth,* Vance Martin, ed. Fulcrum, Inc., Golden, CO, pp. 68–70.

Day, G. M. 1953. The Indian as an Ecological Factor in Northeastern Forests. *Ecology* 34:329–346.

DeBonis, J. 1989. Speaking Out: A Letter to the Chief of the US Forest Service. *Inner Voice* 1:1, 3.

Devall, B. n.d. Personal communication.

Dobyns, H. F. 1966. Estimating Aboriginal American Population: An Appraisal of Techniques with a New Hemispheric Estimate. *Current Anthropology* 7:395–412.

Dysart III, B. C., and M. Clawson. 1988. *Managing Public Lands in the Public Interest.* Praeger, New York.

Ehrenfeld, D. 1989. Life in the Next Millennium: Who Will Be Left in the Earth's Community? *Orion Nature Quarterly* 8 (spring): 4–13.

Emerson, R. W. 1989. Nature. Beacon Press, Boston, MA.

Flader, S. L. 1974. *Thinking Like a Mountain: Aldo Leopold and the Evolution of an Ecological Attitude toward Deer, Wolves, and Forests.* University of Missouri Press, Columbia, MO.

Gómez-Pompa, A., and A. Kaus. 1988. Conservation by Traditional Cultures in the Tropics. In *For the Conservation of the Earth,* Vance Martin, ed. Fulcrum Inc., Golden, CO, pp. 183–194.

Gordon, J. C., Chair, Commission on Research and Resource Management Policy. 1989. *National Parks: From Vignettes to a Global View.* National Parks and Conservation Association, Washington, DC.

Guha, R. 1989a. Radical American Environmentalism: A Third World Critique. *Environmental Ethics* 11:71–83. [Included in this volume.]

———. 1989b. *The Unquiet Woods: Ecological Change and Peasant Resistance in the Himalaya.* University of California Press, Berkeley, CA.

Hays, S. P. 1959. *Conservation and the Gospel of Efficiency: The Progressive Conservation Movement.* Harvard University Press, Boston, MA.

Hecht, S., and A. Cockburn. 1989. *The Fate of the Forest: Developers, Destroyers, and Defenders of the Amazon.* Verso, New York.

Heizer, R. F. 1955. Primitive Man as an Ecologic Factor. Kroeber Anthropological Society Papers, No. 13, Berkeley, CA.

Hott, L., and D. Garey. 1989. The Wilderness Idea: John Muir, Gifford Pinchot, and the First Great Battle for Wilderness. Florentine Films, Haydenville, MA.

Hughes, J. D. 1983. *American Indian Ecology.* Texas Western Press, El Paso, TX.

Kroeber, A. L. 1939. *Cultural and Natural Areas of Native North America.* University of California Press, Berkeley, CA.

Lemons, J. 1987. United States' National Park Management: Values, Policy, and Possible Hints for Others. *Environmental Conservation* 14:329–328.

Leopold, A. 1918. Forestry and Game Conservation. *Journal of Forestry* 19:404–411.

Leopold, A. 1921. Wilderness and Its Place in Forest Recreation Policy. *Journal of Forestry* 19:718–721.

Leopold, A. 1925a. Conserving the Covered Wagon. *Sunset Magazine* 54 (March): 21, 56.

Leopold, A. 1925b. The Last Stand of the Wilderness. *American Forests and Forest Life* 31:599–604.

Leopold, A. 1925c. Wilderness as a Form of Land Use. *Journal of Land and Public Utility Economics* 1:398–404. [Included in this volume.]

Leopold, A. 1925d. A Plea for Wilderness Hunting Grounds. *Outdoor Life* 56:348–350.

Leopold, A. 1936. Threatened Species: A Proposal to the Wildlife Conference for an Inventory of the Needs of Near-Extinct Birds and Animals. *American Forests* 42:116–119. [Included in this volume.]

Leopold, A. 1939a. A Biotic View of Land. *Journal of Forestry* 37:727–730.

Leopold, A. 1939b. The Farmer as a Conservationist. *American Forests* 45:294–299, 316, 323.

Leopold, A. 1941. Wilderness as a Land Laboratory. *The Living Wilderness* 6 (July): 3.

Leopold, A. 1942. Land Use and Democracy. *Audubon* 44 (Sept./Oct.): 259–265.

Leopold, A. 1945. The Outlook for Farm Wildlife. *Transactions of the 10th North American Wildlife Conference*, pp. 165–168.

Leopold, A. 1949. *A Sand County Almanac: And Sketches Here and There.* Oxford University Press, New York.

Leopold, A. 1991a. The Popular Wilderness Fallacy. In *The River of the Mother of God and Other Essays by Aldo Leopold,* S. L. Flader and J. B. Callicott, eds. University of Wisconsin Press, Madison, WI, pp. 49–52.

Leopold, A. 1991b. Wilderness. In *The River of the Mother of God and Other Essays by Aldo Leopold.* S. L. Flader and J. B. Callicott, eds. University of Wisconsin Press, Madison, WI, pp. 226–229. [Included in this volume.]

Lewis, H. T. 1973. *Patterns of Indian Burning in California: Ecology and Ethnohistory.* Ballena Press Anthropological Papers, No. 1.

Martin, C. 1973. Fire and Forest Structure in the Aboriginal Eastern Forest. *Indian Historian* 6:38–42, 54.

Martin, C. 1978. *Keepers of the Game: Indian-Animal Relationships and Fur Trade.* University of California Press, Berkeley, CA.

Meine, C. 1988. *Aldo Leopold: His Life and Work.* University of Wisconsin Press, Madison, WI.

Merchant, C. 1989. *Ecological Revolutions: Nature, Gender and Science in New England.* University of North Carolina Press, Chapel Hill, NC.

Muir, J. 1901. *Our National Parks.* Houghton Mifflin, Boston, MA.

Myers, N. 1981. A Farewell to Africa. *International Wildlife* 11 (Nov./Dec.): 36–46.

Nabhan, G. P. 1982. *The Desert Smells Like Rain: A Naturalist in Papago Country.* North Point Press, San Francisco, CA.
Nash, R. 1967. *Wilderness and the American Mind.* Yale University Press, New Haven, CT.
Noss, R. F. 1990a. Can We Maintain Our Biological and Ecological Integrity? *Conservation Biology* 4:241–243.
Noss, R. F. 1990b. What Can Wilderness Do for Biodiversity? In *Preparing to Manage Wilderness in the 21st Century—Proceedings of the Conference.* P. Reed, ed. Southeastern Forest Experiment Station, U.S. Department of Agriculture, Forest Service, Asheville, NC, pp. 49–61.
Noss, R. F. 1991. Sustainability and Wilderness. *Conservation Biology* 5:120–122. [Included in this volume.]
Oelschlaeger, M. 1991. *The Idea of Wilderness: From Prehistory to the Age of Ecology.* Yale University Press, New Haven, CT.
Parsons, D. J., D. M. Graber, J. K. Age, and J. W. V. Wagtendonk. 1986. Natural Fire Management in National Parks. *Environmental Management* 10:21–24.
Passell P. 1990. Rebel Economists Add Ecological Costs to Price of Progress. *New York Times* (Nov. 27): B5–B6.
Peters, C. M., A. H. Gentry, and R. O. Mendelsohn. 1989. Valuation of an Amazonian Rainforest. *Nature* 339:656–657.
Pinchot, G. 1947. *Breaking New Ground.* Harcourt, Brace, and Co., New York.
Pyne, S. J. 1982. *Fire in America: A Cultural History of Wildland and Rural Fire.* Princeton University Press, Princeton, NJ.
Randall, A. 1988. What Mainstream Economists Have to Say about the Value of Biodiversity. In *Biodiversity,* E. O. Wilson, ed. National Academy Press, Washington, DC.
Reed, P., ed. 1990. *Preparing to Manage Wilderness in the 21st Century: Proceedings of the Conference.* Southeastern Forest Experiment Station, U.S. Department of Agriculture, Forest Service, Asheville, NC.
Reed, P., and L. Merigliano. 1990. Managing for Compatibility Between Recreational and Nonrecreational Wilderness Purposes. In *Preparing to Manage Wilderness in the 21st Century: Proceedings of the Conference.* P. Reed, ed. Southeastern Forest Experiment Station, U.S. Department of Agriculture, Forest Service, Asheville, NC, pp. 95–107.
Snyder, G. 1990. *The Practice of the Wild.* North Point Press, San Francisco, CA.
Soulé, M. E. 1990. The Onslaught of Alien Species, and Other Challenges in the Coming Decades. *Conservation Biology* 4:233–239.
Stevens, W. K. 1990. Research in "Virgin" Amazon Uncovers Complex Farming. *New York Times* (April 3): B5–B6.
Thompson, T. S. 1988. Vegetation Dynamics in the Western United States: Modes of Response to Climate Fluctuations. In *Vegetation History.* B. Huntly and T. Webb, eds. Kluwer Academic Publishers, Boston, MA.

Thoreau, H. D. 1962. *Excursions.* Corinth Books, New York.

Ulanowicz, R. 1991. Ecosystem Health in Terms of Trophic Flow Networks. AAAS annual meeting, February 15, 1991, Washington, DC.

USDA Forest Service. 1989. *New Perspectives: An Ecological Path for Managing Forests.* Pacific Northwest Research Station, Portland, Oregon, and Pacific Southwest Research Station, Redding, California.

von Droste, B. 1988. The Role of Biosphere Reserves at a Time of Increasing Globalization. In *For the Conservation of the Earth,* Vance Martin, ed. Fulcrum, Inc., Golden, CO, pp. 89–93.

Watkins, T. H. 1991. Aspects of Grass. *Wilderness* 54 (Spring): 27.

Westman, W. E. 1990. Park Management of Exotic Species: Problems and Issues. *Conservation Biology* 4:251–260.

Westman, W. E. 1991. Restoration Projects: Measuring Their Performance. *Environmental Professional* 13:207–215.

White, R. J. 1987. *Big Game Ranching in the United States.* Wild Sheep and Goat International Publishing Co., Mesilla, NM.

Witthoft, J. 1965. *Indian Prehistory of Pennsylvania.* Pennsylvania Historical and Museum Commission, Harrisburg, PA.

Wright, A. 1984. Innocents Abroad: American Agricultural Research in Mexico. In *Meeting the Expectations of the Land: Essays in Sustainable Agriculture and Stewardship,* W. Jackson, W. Berry, and B. Colman, eds. North Point Press, San Francisco, CA, pp. 135–151.

Wuerthner, G. 1991. How the West Was Eaten. *Wilderness* 54 (Spring): 28–37.

Holmes Rolston III

The Wilderness Idea Reaffirmed (1991)

INTRODUCTION

REVISITING THE WILDERNESS, Callicott[1] is a doubtful guide; indeed he has gotten himself lost. That is a pity, because he is on the right track about sustainable development and I readily endorse his positive arguments for developing a culture more harmonious with nature. But these give no cause for being negative about wilderness.

The wilderness concept, we are told, is "inherently flawed," triply so. It metaphysically and unscientifically dichotomizes man and nature. It is ethnocentric, because it does not realize that practically all the world's ecosystems were modified by aboriginal peoples. It is static, ignoring change through time. In the flawed idea and ideal, wilderness respects wild communities where man is a visitor who does not remain. In the revisited idea(l), also Leopold's ideal, humans, themselves entirely natural, reside in and can and ought to improve wild nature.

HUMAN CULTURE AND WILD NATURE

Wilderness valued without humans perpetuates a false dichotomy, Callicott maintains. Going back to Cartesian and Greek philosophy and Christian theology, such a contrast between humans and wild nature is a metaphysical confusion that leads us astray and also is unscientific. But this is not so. One hardly needs metaphysics or theology to realize that there are critical differences between wild nature and human culture. Humans now superimpose cultures on the wild nature out of which they once emerged. There is nothing unscientific or non-Darwinian about the claim that innovations in human culture make it radically different from wild nature.

Information in wild nature travels intergenerationally on genes: information in culture travels neurally as persons are educated into transmissible cultures. (Some higher animals learn limited behaviors from parents and conspecifics, but animals do not form transmissible cultures.) In nature, the coping skills are coded on chromosomes. In culture, the skills are coded in craftsman's traditions, religious rituals, or technology manuals. Information acquired during an organism's lifetime is not transmitted genetically; the essence of culture is acquired information transmitted to the next generation. Information transfer in culture can be several orders of magnitude faster and overleap genetic lines. I have but two children; copies of my books and my former students number in the thousands. A human being develops typically in some one or a few of ten thousand cultures, each heritage historically conditioned, perpetuated by language, conventionally established, using symbols with locally effective meanings. Animals are what they are genetically, instinctively, environmentally, without any options at all. Humans have myriads of lifestyle options, evidenced by their cultures; and each human makes daily decisions that affect his or her character. Little or nothing in wild nature approaches this.

The novelty is not simply that humans are more versatile in their spontaneous natural environments. Deliberately rebuilt environments replace spontaneous wild ones. Humans can therefore inhabit environments altogether different from the African savannas in which they once evolved. They insulate themselves from environmental extremes by their rebuilt habitations, with central heat from fossil fuel or by importing fresh groceries from a thousand miles away. In that sense, animals have freedom within ecosystems, but humans have freedom from ecosystems. Animals are adapted to their niches; humans adapt their ecosystems to their needs. The determinants of animal and plant behavior, much less the determinants of climate or nutrient recycling, are never anthropological, political, economic, technological, scientific, philosophical, ethical, or religious. Natural selection pressures are relaxed in culture; humans help each other out compassionately with medicine, charity, affirmative action, or head-start programs.

Humans act using large numbers of tools and things made with tools, extrasomatic artifacts. In all but the most primitive cultures, humans teach each other how to make clothes, thresh wheat, make fires, bake bread. Animals do not hold elections and plan their environmental affairs; they do not make bulldozers to cut down tropical rainforests. They do not fund development projects through the World Bank or contribute to funds to save the whales. They do not teach their religion to their children. They do not write articles revisiting and reaffirming the idea of wilderness. They do not get confused about whether their actions are natural or argue about whether they can improve nature.

If there is any metaphysical confusion in this debate, we locate it in the claim that "man is a natural, a wild, an evolving species, not essentially different in this respect from all the others."[2] Poets like Gary Snyder perhaps are entitled to poetic license. But philosophers are not, especially when analyzing the concept of wildness. They cannot say that "the works of man, however precocious, are as natural as those of beavers," being "entirely natural," and then, hardly taking a breath, say that "the cultural component in human behavior is so greatly developed as to have become more a difference of kind than of degree."[3] If this were only poetic philosophy it might be harmless, but proposed as policy, environmental professionals who operate with such contradictory philosophy will fail tragically.

"Anthropogenic changes imposed upon ecosystems are as natural as any other."[4] Not so. Wilderness advocates know better; they do not gloss over these differences. They appreciate and criticize human affairs, with insight into their radically different characters. Accordingly, they insist that there are intrinsic wild values that are not human values. These ought to be preserved for whatever they can contribute to human values, and also because they are valuable on their own, in and of themselves. Just because the human presence is so radically different, humans ought to draw back and let nature be. Humans can and should see outside their own sector, their species self-interest, and affirm nonanthropogenic, noncultural values. Only humans have conscience enough to do this. That is not confused metaphysical dichotomy; it is axiological truth. To think that human culture is nothing but natural system is not discriminating enough. It risks reductionism and primitivism.

These contrasts between nature and culture were not always as bold as they now are. Once upon a time, culture evolved out of nature. The early hunter-gatherers had transmissible cultures but, sometimes, were not much different in their ecological effects from the wild predators and omnivores among whom they moved. In such cases, this was as much through lack of power to do otherwise as from conscious decision. A few such aboriginal peoples may remain.

But we Americans do not and cannot live in such a twilight society. Any society that we envision must be scientifically sophisticated, technologically advanced, globally oriented, as well as (we hope) just and charitable, caring for universal human rights and for biospheric values. This society will try to fit itself in intelligently with the ecosystemic processes on which it is superposed. It will, we plead, respect wildness. But none of these decisions shaping society are the processes of wild nature. There is no inherent flaw in our logic when we are discriminating about these radical discontinuities between culture and nature. The dichotomy charge is a half-truth, and, taken for the whole, becomes an untruth.

HUMANS IMPROVING WILD NATURE

Might a mature, humane civilization improve wild nature? Callicott thinks that it is a "fallacy" to think that "the best way to conserve nature is to protect it from human inhabitation and utilization."[5] But, continuing the analysis, surely the fallacy is to think that a nature allegedly improved by humans is anymore real nature at all. The values intrinsic to wilderness cannot, on pain of both logical and empirical contradiction, be "improved" by deliberate human management, because deliberation is the antithesis of wildness. That is the sense in which civilization is the "antithesis"[6] of wilderness, but there is nothing "amiss" in seeing an essential difference here. Animals take nature ready to hand, adapted to it by natural selection, fitted into their niches; humans rebuild their world through artifact and heritage, agriculture and culture, political and religious decisions.

On the meaning of "natural" at issue here, that of nature proceeding by evolutionary and ecological processes, any deliberated human agency, however well intended, is intention nevertheless and interrupts these spontaneous processes and is inevitably artificial, unnatural. (There is another meaning of "natural" by which even deliberated human actions break no laws of nature. Everything, better or worse, is natural in this sense, unless there is the supernatural.) The architectures of nature and of culture are different, and when culture seeks to improve nature, the management intent spoils the wildness. Wilderness management, in that sense, is a contradiction in terms—whatever may be added by way of management of humans who visit the wilderness, or of restorative practices, or monitoring, or other activities that environmental professionals must sometimes consider. A scientifically managed wilderness is conceptually as impossible as wildlife in a zoo.

To recommend that *Homo sapiens* "reestablish a positive symbiotic relationship with other species and a positive role in the unfolding of evolutionary processes"[7] is, so far as wilderness preservation is involved, not just bad advice, it is impossible advice. The cultural processes by their very "nature" interrupt the evolutionary process: there is no symbiosis, there is antithesis. Culture is a post-evolutionary phase of our planetary history: it must be superposed on the nature it presupposes. To recommend, however, that we should build sustainable cultures that fit in with the continuing ecological processes is a first principle of intelligent action, and no wilderness advocate thinks otherwise.

If there are inherent conceptual flaws dogging this debate, we have located another: Callicott's allegedly "improved" nature. In such modified nature, the different historical genesis brings a radical change in value type. Every wilderness enthusiast knows the difference between a pine plantation in the Southeast and an old-growth grove in the Pacific Northwest. Even if the "improvement" is more or less harmonious with the ecosystem, it is fundamentally of a different order. Asian ring-tailed pheasants are rather well naturalized on the contemporary Iowa landscape. But they are there by human introduction, and they remain because farmers plow the fields, plant corn, and leave shelter in the fencerows. They are really as much like pets as like native wild species, because they are not really on their own.

BIODIVERSITY AND WILDERNESS

As an example of his recommended symbiosis where human culture enriches natural systems, Callicott cites a study[8] of two nearby communities, Quitovac (= Ki:towak) in Mexico, where sixty-five bird species were found, and Quitobaquito Springs (or A'al Waipia) in Organ Pipe National Monument, with only thirty-two species. His conclusion is that biodiversity is greater in such rural communities than in wild natural systems.

But this is an unusual case; the locale is desert where water is the limiting factor. If you artificially water the desert, some things will come in that could not live there before. Similarly, if you heat up the tundra, where cold is the limiting factor. We will not be surprised if there are more birds around feeders offering food, water, and shelter than elsewhere. But bird feeders actually may not be increasing biodiversity. We will have to look more closely at what is meant by biodiversity and what is going on in the two communities.

A species count, uninterpreted, doesn't tell us much. In more sophisticated analyses, ecologists use up to a dozen and a half indices of diversity.[9] These include within habitat diversity (alpha diversity), between habitat diversity (beta diversity) and regional diversity (gamma diversity). They include diversity of processes and heterogeneity of fauna and flora, and on and on. If all you do is count species, there are more animal species in the Denver zoo than in the rest of Colorado. Never mind that the processes of nature are entirely gone. Callicott knows that and wants ecosystem health as well as diversity.

Whether there is ecosystem health at Quitovac is less clear. Callicott thinks so, but Nabhan et al.[10] are more circumspect. Though the bird species count was always higher at Quitovac, by a heterogeneity diversity index the avifauna at Quitovac has no advantage over Organ Pipe. (This asks what proportions of the birds are of what species, such as grackles, doves, English sparrows, pigeons.) They also find that Quitovac is "not nearly as diverse in mammals," that ever-present dogs, horses, and cattle, limit the presence of wild animals. Deer and javelina drink and browse frequently at Organ Pipe, seldom at Quitovac. Even rodents are more abundant at Organ Pipe.

They also found more plant diversity at Quitovac, one hundred and thirty-nine species there against eighty at Organ Pipe. It is hardly surprising that if you add some irrigated cultivated fields and orchards, new plant species will appear, and some insects will follow, and birds in turn follow the seeds and insects. They also note that seventeen of these plant species were planted intentionally and that of the fifty-nine species in fields and orchards, many were adventitious species, weeds of disturbed sites. Many were the Old World waifs that, like dandelions, have tagged along after civilization willy-nilly. Is this being offered as a wise symbiosis of nature and culture? Is that enhanced richness in biodiversity?

A species count, offered as evidence of biodiversity without further ado, assumes that if we have the species, we have what we want conserved. But we may have the parts, even extra, artificial parts, but no longer the composition of the former whole. Maybe Quitovac is about as much "ersatz world"[11] as idyllic, humanized ecosystem health with optimized biodiversity. Even a new whole would not have the integrity of the once wild ecosystem. We can and ought to have rural nature, and we will be glad to have rural nature with a high birdcount. But we can have a rural nature with a high species count and not have anywhere on the landscape the radical values of wild, pristine nature. That loss would not be compensated for by the stepped-up species count in agriculturally disturbed lands. In wilderness, we value the interactions as a fundamental component of biodiversity.

The predation pressures, for instance, are never the same on agricultural lands as they are on wildlands. Agriculture means an increase of disturbed soil, with most of these disturbances different in kind from those in wild nature. Different kinds of things grow in such soil, more r-selected species, fewer k-selected ones. Underground, the fungi and soil bacteria are different, so the decomposition regime is different, and that results in differences above ground. The energy flow and the nutrient cycling is different. It is often the case that the highest number of species are found in intermediately disturbed environments, but that considers species counts and alpha diversity alone. If all the environments are kept intermediately disturbed, we lose beta and gamma diversity. Indeed over the landscape as a whole, we lose even species counts, since in disturbed environments the sensitive species go extinct. We are not likely to retain the large carnivores.

Both these oases are water magnets for migrant birds. Quitovac, with its cultivated fields and orchards, draws more migrants into close proximity. Muddy shorelines attract some waders less frequent at Quitobaquito. All this tells us little about whether these migrating birds are safe in their wintering or breeding grounds. In fact, Central American agricultural development, destroying winter grounds, threatens many bird species. Quitovac may draw some breeding species that cannot survive at Quitobaquito or in the unwatered desert. But there are no bird species flourishing in Quitovac that were not flourishing already in their native habitats elsewhere. It is hard to think that much important bird conservation is going on there.

Quitobaquito Springs, far from being depauperate in birds, is one of the best-known sites for observing birds in that region of Arizona, and birders go there from all over the United States to see the migrants and to find the desert species. The oasis is but a small area. Organ Pipe Monument is designated to preserve many other kinds of habitats. Enlarging to consider beta and gamma diversity, the official Monument checklist contains 277 bird species, of which 63 are known to breed in various habitats there, and five more believed to breed as well.[12] Only three are nonnative. Even if the diversity at Quitovac is greater, the diversity preserved by having both a rural area and a wilderness is higher than if we had two rural areas.

Also, whatever the possibilities, we do not want to forget the probabilities, which are that this (allegedly) idyllic picture will be upset by development pressures. Quitovac had been used for centuries, steadily but not intensively. When the comparisons here were made, only two or three dozen persons were using the area. The study concentrates on only a five hectare site, and the natives had only used ten percent of this for cultivated fields and orchards. Before the study could be completed, 125 hectares there were bulldozed to be used for intensive agriculture, including most of the study area, with disastrous results.[13] There may be fewer species at Organ Pipe, but such a disaster is not likely to occur, owing to its sanctuary designation.

WILDERNESS AND CHANGE

Another alleged flaw in the concept of wilderness is that its advocates do not know the fourth dimension, time. That is a strange charge; my experience has been just the opposite. In wilderness, the day changes from dawn to dusk, the seasons pass, plants grow, animals are born, grow up, and age. Rivers flow, winds blow, even the rocks erode; change is pervasive. Indeed, wilderness is that environment in which one is most likely to experience geological time. Try a raft trip through the Grand Canyon.

On the scale of deep time, some processes continue on and on, so that the perennial givens—wind and rain, soil and photosynthesis, life and death and life renewed—can seem almost forever. Species survive for millions of years: individuals are ephemeral. Life persists in the midst of its perpetual perishing. Mountains are reliably there generation after generation. The water cycles back, always moving. In wilderness, time mixes with eternity; that is one reason we value it so highly.

Callicott writes as if wilderness advocates had studied ecology and never heard of evolution. But they know that evolution is the control of development by ecology, and what they value is precisely natural history. They do not object to natural changes. They may not even object to artificial changes in rural landscapes. But, since they know the difference between nature and culture, they know that cultural changes may be quite out of kilter with natural changes. Leopold uses the word "stability" when he is writing in the time frame of land-use planning. On that scale, nature typically does have a reliable stability and farmers do well to figure in the perennial givens.

In an evolutionary time frame, Leopold knows that relative stability mixes with change. "Paleontology offers abundant evidence that wilderness maintained itself for immensely long periods: that its component species were rarely lost, and neither did they get out of hand; that weather and water built soil as fast or faster than it was carried away." That is why "wilderness . . . assumes unexpected importance as a laboratory for the study of land health."[14] Wilderness is the original sustainable development.

With natural processes, "protect" is perhaps a better word than either "preserve" or "conserve." Wilderness advocates do not seek to prevent natural change. There is nothing illusory, however, about appreciating today in wilderness processes that have a primeval character. There, the natural processes of 1992 do not differ much from those of 1492, half a millennium earlier. We may enjoy that perennial character, constancy in change, in contrast with the rapid pace of cultural changes, seldom as dramatic as those on the American landscape of the last few centuries.

A management program in the U.S. Forest Service seeks to evaluate the "limits of acceptable change." This emphasis worries about the rapid pace of cultural change as this contrasts with the natural pace on landscapes. Cordell and Reed[15] are trying to decide the limits of acceptable humanly-introduced changes, artificial changes, since these are of such radically different kind and pace that they disrupt the processes of wilderness. They do not oppose natural changes. At this point, we have an example of how and why environmental professionals will make disastrous decisions, if confused by what is and is not natural. Callicott warns them that they do have to worry about "accelerating rates of environmental change"[16] No one can begin to understand these rates of changes if the changes are thought of as being introduced by a species that is "entirely natural."

When we designate a desert wilderness in Nevada, there really isn't any problem deciding that mustangs are feral animals in contrast with desert bighorns, which are indigenous. There might have been ancestral horses in Paleolithic times in the American West, but they went extinct naturally. The present mustangs came from animals that the Europeans brought over in ships, originally from the plains of Siberia. Bighorns are what they are where they are by natural selection. Mustangs are not so. There is nothing conceptually problematic about that—unless one has never gotten clearly in mind the difference between nature and culture in the first place.

ABORIGINAL PEOPLES AND WILDERNESS

What of the argument that we cannot have any wilderness, because there is none to be had? This is a much stronger claim than that there is no real wilderness left on the American landscape after the European cultural invasion. Even the aboriginals had already extinguished wilderness. Now we have a somewhat different account of the human presence from that earlier advocated. The claim is no longer that the Indians were just another wild species, "entirely natural," but that they actively managed the landscape, so dramatically altering it that there was no wilderness even when Columbus arrived in 1492. It is ethnocentric to think otherwise. This is because we Caucasians exaggerate our own power to modify the landscape and diminish their power. This is a judgment based on prejudice, not on facts.

How much did the American Indians modify the landscape? That is an empirical question in anthropology and ecology. We do not disagree that where there was Indian culture, this altered the locales in which they resided, so that these locales were not wilderness in the pure sense. In that respect, Indian culture is not different in kind from the white man's culture. What we need to know is the degree. Had the Indians, when the white man arrived, already transformed the pre-Indian wilderness beyond the range of its spontaneous self-restoration?

Callicott concedes,[17] rightly, that most of what has been presently designated as wilderness was infrequently used by the aborigines, since it is high, rough, or arid. We have no reason to think that in such areas the aboriginal modifications are irreversible. Were the more temperate regions modified so extensively and irreversibly that so little naturalness remains as to make wilderness designation an illusion? Callicott has "no doubt that most New World ecosystems were in robust health."[18] That suggests that they were not past self-regeneration.

The American Indians on forested lands had little agriculture; what agriculture they had tended to reset succession, and, when agriculture ceases, the subsequent forest regeneration will not be particularly unnatural. The Indian technology for larger landscape modification was bow and arrow, spear, and fire. The only one that extensively modifies landscapes is fire. Fire is—we have learned well by now—also quite natural. Fire suppression is unnatural, but no one argues that the Indians used that as a management tool, nor did they have much capacity for fire suppression. The argument is that they deliberately set fires. Does this make their fires radically different from natural fires? It does in terms of the source of ignition: the one is a result of environmental policy deliberation, the other of a lightning bolt.

But every student of fire behavior knows that on the scale of regional forest ecosystems, the source of ignition is not a particularly critical factor. The question is whether the forest is ready to burn, whether there is sufficient ground fuel to sustain the fire, whether the trees are diseased, how much duff there is, and so on. If conditions are not right, it will be difficult to get the fire going and it will burn out soon. If conditions are right, a human can start a regional fire this year. If some human does not, lightning will start it next year, or the year after that. On a typical summer day, the states of Arizona and New Mexico are each hit by several thousand bolts of lightning, mostly in the higher, forested regions. Doubtless the Indians started some fires too, but it is hard to think that their fires so dramatically and irreversibly altered the natural fire regime in the Southwest that meaningful wilderness designation is impossible today.

We do not want to be ethnocentric, but neither do we want to be naive about the technological prowess of the American Indian cultures. They had no motors, indeed no wheels, no domestic animals, no horses (before the Spanish came), no beasts of burden. The Indians had a hard time getting so simple a thing as hot water. They had to heat stones and drop them in skins or tightly woven baskets. They lived on the landscape with foot and muscle, and in that sense, though they had complex cultures, they had culture with very reduced alternative power. Even in European cultures, in recent centuries the power of civilization to redo the world has accelerated logarithmically.

In Third World nations, perhaps areas that seem "natural" now are often the result of millennia of human modifications through fire, hunting, shifting cultivation, and selective planting and removal of species. This will have to be examined on a case-by-case basis, and we cannot prejudge the answers. We do not know yet how intensively the vast Brazilian rainforests were managed and whether no wilderness designation there is ecologically practicable, even if we desired it. Nor do wilderness enthusiasts advocate that such peoples be removed to accomplish this, where it is possible. What is protested is modern forms of development. Extractive reserves may be an answer, but extractive reserves for latex sold in world markets and manufactured into rubber products can hardly be considered aboriginal wisdom.

Sometimes we will have to make do with what wildness remains in the nooks and crannies of civilization. Meanwhile, where wilderness designation is possible and where there is an exploding population, what should we do? No one objects to trying to direct that explosion into more harmonious forms of human-nature encounter. But constraining an explosion takes some strong measures. One of these ought to be the designation of wilderness.

Perhaps the American Indians did not have enough contrast between their culture and the nature that surrounded them to produce the wilderness idea. It was not an idea that, within their limited power to remake nature, could occur to them. If you have only foot, muscle, bows, arrows, and fire, you do not think much about wilderness conservation. But we, in the twentieth century, do have the wilderness idea: it has crystallized with the possibility, indeed the impending threat, of destroying the last acre of primeval wilderness. It also has crystallized with our deepening, scientific knowledge of how wild nature operates, of DNA, genes, and natural selection, and how dramatically different in kind, pace, and power the processes of culture can be. The Indians knew little of this; they lived still in an animistic, enchanted world.

And we need the wilderness idea desperately. When you have bulldozers that already have blacktopped more acreage than remains pristine, you can and ought to begin to think about wilderness. Such an idea, when it comes, is primitive in one sense: it preserves primeval nature, as much as it can. But it is morally advanced in another sense: it sees the intrinsic value of nature, apart from humans.

Ought implies can; the Indians could not, so they never thought much about the ought. We in the twentieth century can, and we must think about the ought. When we designate wilderness, we are not lapsing into some romantic atavism, reactionary and nostalgic to escape culture. We are breaking through culture to discover, nonanthropocentrically, that fauna and flora can count in their own right (an idea that Indians also might have shared). We realize that ecosystems sometimes can be so respected that humans only visit and do not remain (an idea that the Indians did not need or achieve). A "can" has appeared that has generated a new "ought."

Even some modern American Indians concur. In western Montana, the Salish and Kootenai tribes have set aside 93,000 acres of their reservation as the Mission Mountains Tribal Wilderness; in addition, they have designated the South Fork of the Jocko Primitive Area. In both areas, the Indian too is "a visitor who does not remain"; they want these areas "to be affected primarily by the forces of nature with the imprint of man's work substantially unnoticeable."[19] Indeed, in deference to the grizzly bears, in the summer season, the Indians do not permit any humans at all to visit 10,000 acres that are prime grizzly habitat. In both areas they can claim even more restrictive environmental regulation on what people can do there than in the white man's wilderness. What, when, and how they hunt is an example.

Not a word of the above discussion disparages aboriginal Indian culture. To the contrary, that they survived with the bare skills they had is a credit to their endurance, courage, resolution, and wisdom. A wilderness enthusiast, if he or she has spent much time in the woods armed with only muscle and a few belongings in a backpack, is in an excellent position to appreciate the aboriginal skills.

SUSTAINABLE DEVELOPMENT AND WILDERNESS

Finding out how to remake civilization so that nature is conserved in the midst of sustainable development is indeed a more difficult and important task than saving wild remnants. Little wilderness can be safe unless the sustainable development problem is solved also. I can only endorse Callicott's desire to conserve nature in the midst of human culture. "Human economic activities should at least be compatible with the ecological health of the environments in which they occur."[20] No party to the debate contests that. But this does not mean that wilderness ought not to be saved for what it is in itself.

"The farmer as a conservationist" is quite a good thing and Leopold does well to hope that "land does well for its owner, and the owner does well by his land"; perhaps where a farmer begins, as did Leopold, with lands long abused, "both end up better by reason of their partnership." In that context, "conservation is a state of harmony between men and land."[21] But none of that asks whether there also should be wilderness. Leopold tells us what he thinks about that after his trip to Germany. There was "something lacking. . . . I did not hope to find in Germany anything resembling the great 'wilderness areas' which we dream about and talk about." That was too much to hope; he could dream that only in America. But he did hope to find "a certain quality [—wildness—] which should be, but is not found" in the rural landscape, and, alas, not even that was there.[22] "In Europe, where wilderness now has retreated to the Carpathians and Siberia, every thinking conservationist bemoans its loss."[23] That loss would not be restored if every farmer were a restoration ecologist. All that Leopold says about sustainable development is true, but there is no implication that wilderness cannot or ought not to be saved. Affirming sustainable development is not to deny wilderness.

MONASTIC WILDERNESS AND CIVILIZED COMPLACENCY

Nor is affirming wilderness to deny sustainable development. Callicott alleges, "Implicit in the most passionate pleas for wilderness preservation is a complacency about what passes for civilization."[24] Not so. I cannot name a single wilderness advocate who cherishes wilderness "as an alibi for the lack of private reform," any who "salve their consciences" by pointing to "the few odds and ends" of wilderness and thus avoid facing up to the fact that the ways and means of industrial civilization lie at the root of the current global environmental crisis.[25] The charge is flamboyant; the content runs hollow. Wilderness advocates want wilderness and they also want, passionately, to "re-envision civilization"[26] so that it is in harmony with the nature that humans do modify and inhabit. There is no tension between these ideas in Leopold, nor in any of the other passionate advocates of wilderness that Callicott cites, nor in any with whom I am familiar.

The contrast of monastic sanctuaries with the wicked everyday world risks a flawed analogy. Unless we are careful, we will make a category mistake, because both monastery and lay world are in the domain of culture, while wilderness is a radically different domain. Monastery sets an ideal unattainable in the real civil world (if we must think of it that way), but both worlds are human, both moral. We are judging human behavior in both places, concerned with how far it can be godly. By contrast, the wilderness world is neither moral nor human; the values protected there are of a different order. We are judging evolutionary achievements and ecological stability, integrity, beauty—not censuring or praising human behavior.

Confusion about nature and culture is getting us into trouble again. We are only going to get confused if we think that the issue of whether there should be monasteries is conceptually parallel to the issue of whether there should be wilderness. The conservation of value in the one is by the cultural transmission of a social heritage, including a moral and religious heritage, to which the monastery was devoted. The conservation of value in the other is genetic, in genes subject to natural selection for survival value and adapted fit. There is something godly in the wilderness too, or at least a creativity that is religiously valuable, but the contrast between the righteous and the wicked is not helpful here. The sanctuary we want is a world untrammelled by man, a world left to its own autonomous creativity, not an island of saintliness in the midst of sinners.

We do not want the whole Earth without civilization, for we believe that humans belong on Earth; Earth is not whole without humans and their civilization, without the political animal building his *polis* (Socrates), without peoples inheriting their promised lands (as the Hebrews envisioned). Civilization is a broken affair, and in the long struggle to make and keep life human, moral, even godly, perhaps there should be islands, sanctuaries, of moral goodness within a civilization often sordid enough. But that is a different issue from whether, when we build our civilizations for better or worse, we also want to protect where and as we can those nonhuman values in wild nature that preceded and yet surround us. An Earth civilized on every acre would not be whole either, for a whole domain of value—wild spontaneous nature—would have vanished from this majestic home planet.

INTRINSIC WILDERNESS VALUES

I fear that we are seeing in Callicott's revisiting wilderness the outplay of a philosophy that does not think, fundamentally, that nature is of value in itself. Such a philosophy, though it may protest to the contrary, really cannot value nature for itself. All value in nature is by human projection; it is anthropogenic, generated by humans, though sometimes not anthropocentric, centered on humans. Callicott has made it clear that all so-called intrinsic value in nature is "grounded in human feelings" and "projected" onto the natural object that "excites" the value. "Intrinsic value ultimately depends upon human valuers." "Value depends upon human sentiments."[27]

He explains, "The source of all value is human consciousness, but it by no means follows that the locus of all value is consciousness itself.... An intrinsically valuable thing on this reading is valuable for its own sake, for itself, but it is not valuable in itself, i.e., completely independently of any consciousness, since no value can in principle ... be altogether independent of a valuing consciousness.... Value is, as it were, projected onto natural objects or events by the subjective feelings of observers. If all consciousness were annihilated at a stroke, there would be no good and evil, no beauty and ugliness, no right and wrong; only impassive phenomena would remain." This, Callicott says, is a "truncated sense" of value where " 'intrinsic value' retains only half its traditional meaning."[28]

Talk about dichotomies! Only humans produce value; wild nature is valueless without humans. All it has without humans is the potential to be evaluated by humans, who, if and when they appear, may incline, sometimes, to value nature in noninstrumental ways. "Nonhuman species ... may not be valuable in themselves, but they may certainly be valued for themselves.... Value is, to be sure, humanly conferred, but not necessary homocentric."[29] The language of valuing nature for itself may be used, but it is misleading; value is always and only relational, with humans one of the relata. Nature in itself (a wilderness, for example) is without value. There is no genesis of wild value by nature on its own. Such a philosophy can value nature only in association with human habitation. But that —not some elitist wilderness conservation for spiritual meditation—is the view that many of us want to reject as "aristocratic bias and class privilege."[30]

Sustainable development is, let's face it, irremediably anthropocentric. That is what we must have most places, and humans too have their worthy values. But must we have it everywhere? Must we have more of it and less wilderness? Maybe the value theory here is where the arrogance lies, not in some alleged ethnocentrism or misunderstood doctrine of the dominion of man.

A truncated value theory is giving us a truncated account of biodiversity. Callicott hardly wants wildernesses as "sanctuaries," only as "refugia."[31] A refugia is a seedbed from which other areas get restocked. That is one good reason for wilderness conservation, but we do not want wilderness simply as a place from which the game on our rural lands can be restocked, or even, if we have a more ample vision of wildlife recreation, from which the wildlife that yet persists on the domesticated landscape can be resupplied steadily. Wildernesses are not hatcheries for rural or urban wildlife. Nor are they just "laboratories"[32] for baseline data for sound scientific management. Nor are they raw materials on which we can work our symbiotic enhancements. Nor are they places that can excite us into projecting truncated values onto them. Some of these are sometimes good reasons for conserving wilderness. Leopold sums them up as "the cultural value of wilderness."[33] But they are not the best reasons.

LEOPOLD AND WILDERNESS

Leopold pleads in the "Upshot," in his last book in the penultimate essay, entitled "Wilderness": "Wilderness was an adversary to the pioneer. But to the laborer in repose, able for the moment to cast a philosophical eye on his world, that same raw stuff is something to be loved and cherished, because it gives definition and meaning to his life."[34] He does not mean that wilderness is only a resource for personal development, though it is that. He means that we never know who we are or where we are until we know and respect our wild origins and our wild neighbors on this home planet. We never get our values straight until we value wilderness appropriately. The definition of the human kinds of values is incomplete until we have this larger vision of natural values.

Concluding his appeal for "raw wilderness,"[35] Leopold turns to the "Land Ethic": "The land ethic simply enlarges the boundaries of the community to include soils, waters, plants, and animals, or collectively: the land. . . . A land ethic of course cannot prevent the alteration, management, and use of these 'resources,' but it does affirm their right to continued existence, and, at least in spots, their continued existence in a natural state." We may certainly assert that the founder of the Wilderness Society believed that wilderness conservation is essential in this right to continued existence in a natural state.

"I am asserting that those who love the wilderness should not be wholly deprived of it, that while the reduction of wilderness has been a good thing, its extermination would be a very bad one, and that the conservation of wilderness is the most urgent and difficult of all the tasks that confront us."[36] We must take it as anomalous (else it would be amusing or even tragic) to see Leopold's principal philosophical interpreter, himself a foremost environmental philosopher who elsewhere has said many wise things, now trying to revisit the wilderness idea and de-emphasize it in Leopold.

Just before Leopold plunges into his passionate plea for the land ethic, he calls for "wilderness-minded men scattered through all the conservation bureaus." "A militant minority of wilderness-minded citizens must be on watch throughout the nation, and available for action in a pinch."[37] Alas! His trumpet call is replaced by an uncertain sound. Robert Marshall saluted Leopold as "The Commanding General of the Wilderness Battle."[38] How dismayed he would be by this dissension within his ranks.

On Earth, man is not a visitor who does not remain: this is our home planet and we belong here. Leopold speaks of man as both "plain citizen" and as "king." Humans too have an ecology, and we are permitted interference with, and rearrangement of, nature's spontaneous course: otherwise there is no culture. When we do this there ought to be some rational showing, that the alteration is enriching, that natural values are sacrificed for greater cultural ones. We ought to make such development sustainable. But there are, and should be, places on Earth where the nonhuman community of life is untrammeled by man, where we only visit and spontaneous nature remains. If Callicott has his way, revisiting wilderness, there soon will be less and less wilderness to visit at all.

NOTES

1. J. B. Callicott, "The Wilderness Idea Revisited: The Sustainable Development Alternative," *The Environmental Professional* 13 (1991): 235–247. [Included in this volume.]

2. Ibid., p. 241.

3. Ibid., p. 241.

4. J. B. Callicott, "Standards of Conservation: Then and Now," *Conservation Biology* 4 (1990): 229–232.

5. Callicott, "The Wilderness Idea Revisited," p. 236.

6. Ibid., p. 236.

7. Ibid., p. 240.

8. G. P. Nabhan, A. M. Rea, K. L. Reichhardt, E. Mellink, and C. F. Hutchinson, "Papago Influences on Habitat and Biotic Diversity: Quitovac Oasis Ethnoecology," *Journal of Ethnobiology* 2 (1982): 124–143.

9. A. E. Magurran, *Ecological Diversity and Its Measurement* (Princeton, N.J.: Princeton University Press, 1988); E. C. Pielou, *Ecological Diversity* (New York: John Wiley, 1975).

10. Nabhan et al., "Papago Influences on Habitat and Biotic Diversity."

11. Callicott, "The Wilderness Idea Revisited," p. 243.

12. K. Groschupf, B. T. Brown, and R. R. Johnson, *A Checklist of the Birds of Organ Pipe Cactus National Monument* (Tucson, Ariz.: Southwest Parks and Monument Association, 1987).

13. Nabhan et al. "Papago Influences on Habitat and Biotic Diversity."

14. A. Leopold, *A Sand County Almanac* (New York: Oxford University Press, [1949] 1968), p. 196.

15. H. K. Cordell and P. C. Reed, "Untrammeled by Man: Preserving Diversity through Wilderness," in *Preparing to Manage Wilderness in the 21st Century: Proceedings of the Conference* (Asheville, N.C.: Southeastern Forest Experiment Station, U.S. Department of Agriculture, Forest Service, 1990), pp. 30–33.

16. Callicott, "The Wilderness Idea Revisited," p. 242.

17. Ibid., pp. 241–242.

18. Ibid., p. 244.

19. *Tribal Wilderness Ordinance of the Governing Body of the Confederated Salish and Kootenai Tribes,* 1982.

20. Callicott, "The Wilderness Idea Revisited," p. 239.

21. Leopold, quoted by Callicott, ibid., p. 238.

22. Leopold, quoted by Callicott, ibid.

23. A. Leopold, *A Sand County Almanac,* p. 200.

24. Callicott, "The Wilderness Idea Revisited," p. 236.

25. Ibid., p. 239.

26. Ibid., p. 236.

27. J. B. Callicott, "Non-Anthropocentric Value Theory and Environmental Ethics," *American Philosophical Quarterly* 21 (1984): 299–309.

28. Callicott, "On the Intrinsic Value of Nonhuman Species." In *The Preservation of Species,* B. G. Norton, ed. (Princeton, N.J.: Princeton University Press, 1986), pp. 142–43, 156, 143.

29. Ibid., p. 160.

30. Callicott, "The Wilderness Idea Revisited," p. 237.

31. Ibid., p. 236.

32. Ibid., p. 238.

33. Leopold, *A Sand County Almanac,* p. 200.

34. Ibid., p. 188.

35. Ibid., p. 201.

36. Leopold quoted in C. Meine, *Aldo Leopold: His Life and Work* (Madison: University of Wisconsin Press, 1988), p. 245.

37. Leopold, *A Sand County Almanac,* p. 200.

38. Marshall quoted in C. Meine, *Aldo Leopold: His Life and Work,* p. 248.

J. Baird Callicott

That Good Old-Time Wilderness Religion (1991)

I KNEW THAT CRITICALLY REVISITING the wilderness idea would provoke a reaction about as pleasant as poking a stick into a hornet's nest. My friend and colleague Holmes Rolston is the first wasp out of the hive to sting me.[1] I fear a swarm of others will follow.

So let me reiterate a vital point that Rolston has distorted in his first sentence and his last. I find fault with the wilderness *idea,* not the places so called. In "The Wilderness Idea Revisited" I write, "I am as ardent an advocate of those patches of the planet called 'wilderness areas' as any other environmentalist. . . . Such areas serve the cause of biological conservation most importantly as refugia for species not tolerant of or tolerated by people." My use of the term "refugia" throughout may have been imprudent since the designation of "game refuges" was advocated by turn-of-the-century sportspersons—for purely selfish reasons. I advocate, passionately, the provision of *habitat* so that large predators and those species that require old growth forest, extensive grasslands, wetlands, and so on, may survive and thrive—for their inherent value. Indeed, the more wildlife sanctuaries (a term Rolston prefers and one I also use), the better—as I am on record as insisting.

Rolston's clinic on Sonoran avifauna conservation is a much appreciated elaboration of my caution that "the quality, so to speak, of species is also important."

Rolston attributes my unorthodox reflections on the wilderness idea to theories about the value of nature published elsewhere. As these were taken out of context, it is necessary for me to add an explanation. Most philosophers "cry 'Heresy' on everyone whose sympathies reach a single hair's breadth beyond the boundary epidermis of our own species," to quote Muir.[2] The standard view in both science and philosophy is that value is consciousness-dependent. Assuming the standard view, I argue (in the places Rolston cites) that our sympathies may reach well beyond the boundary epidermis of our own species—to all the others, to ecosystems, even to the biosphere as a whole. Because my argument was with the anthropocentrists, my language may have suggested that I do not think that other conscious beings also value things. But I do in fact think that they may. Value is not only anthropogenic; it is perhaps "vertebragenic."

If it is fair for Rolston to diagnose my dissidence within the ranks of Bob Marshall's wilderness militia by reference to what I have elsewhere written, he cannot cry foul if I diagnose his orthodoxy in similar fashion. However much value Rolston may find in nature, he firmly holds that "Morality appears in humans alone and is not, and never has been, present on the natural scene"—experimental evidence of thoughtful altruism among rhesus monkeys, and anecdotal evidence of voluntary celibacy among wolves, notwithstanding.[3] Rolston is a former Presbyterian minister, as were his father and grandfather before him, and remains committed to Christian ideology, a major tenet of which is that man is a case apart from (the rest of) nature, as his book *Science and Religion* amply testifies.[4] Here culture serves Rolston as a secular surrogate for the image of God cleanly segregating man from nature. The crux (no pun intended) of the debate between us concerns the relation of culture to nature. I follow Darwin in thinking that human culture is continuous with primate and mammalian protoculture and that, no matter how hypertrophic it may lately have become, contemporary human civilization remains embedded in nature. To claim, as Rolston does, that no cultural modification of nature can be ecologically (not anthropocentrically) beneficial, measured by objective criteria, is merely question-begging dogmatism.

Environmental historian Donald Worster once remarked to me that, in his opinion, one could not be an environmentalist unless one enjoyed a

Protestant heritage. Christianity in all its forms, however, is supposed to be antithetical to environmentalism. Or so Lynn White, Jr., led all us environmentalists to believe in his incalculably influential "Historical Roots of Our Ecologic Crisis."[5] Worster didn't explain what he meant by that remark, which flew so directly in the face of common environmental wisdom. His statement was so singularly contrarian that I couldn't forget it. And Worster is so perspicacious a historian of environmental ideas that I couldn't ignore it. Geotheologically speaking, Worster seems right. The wilderness idea plays no significant role in the intellectual environmental history of Catholic Latin America, only in that of Protestant Anglo-America (and, revealingly, Protestant Anglo-Australia). Sometime later, environmental philosopher Mark Sagoff bluntly told me that in his opinion, which is much influenced by the work of historian Perry Miller, the wilderness preservation movement was, more particularly, a Puritan legacy.[6]

Here again, common environmental wisdom, shaped this time by Roderick Nash's incalculably influential book on the wilderness idea, is baldly defied.[7] The Puritans certainly thought long and hard about wilderness, according to Nash—not, however, about how to appreciate it and preserve it, but how to eradicate it and replace it with those shining cities on hills. Nash's representation of Puritan thought about wilderness is among the most vivid and memorable in his canonical history. To the Puritans, according to Nash, the North American "wilderness" was the brooding "antipode" of the Garden of Eden. In the lively Puritan imagination, America was the "stronghold of Satan." Its feather-waving, bone-rattling denizens were "active disciples of the Devil." Their unimproved wilderness abode was "hideous and howling." The American "wilderness" was not only a physical obstacle to English enterprise, but a religious symbol. Hence, transforming New England into fertile farms and fair villages was not, for the Puritans, a mere practical undertaking, it was, to echo Perry Miller, also a divinely appointed "errand into the wilderness."

Puritan thrift and dedication to task soon led to security in the Anglo-American colonies and material wealth. A second generation of Puritans focused their godly zeal not on the religious and moral perversity of the hapless and defeated indigenes, but on faithlessness and self-indulgence in their own midst. A fundamental Calvinist tenet is the sinful condition of man in general. In the subtle and creative mind of Jonathan Edwards, the rest of creation, by contrast, was innocent and unfallen. Indeed, foreshadowing the development of American Transcendentalism, Edwards saw "the images or shadows of divine things" in nature. The wilderness idea—in its positive as well as negative valence—is truly a Puritan creation. The image of God sharply segregates man from nature. For Cotton Mather, that divine spark entitled, even obliged man to master nature and subdue it—as clearly stated in Genesis 1:28. But for Jonathan Edwards, the subsequent Fall, the advent of original sin, and God's curses made of man a kind of environmental pollutant. Man's unnatural, unholy works sullied and fouled the beauty and glory of God's creation—which was so purely manifest in the pristine American landscape. Let me reduce the present argument to its starkest terms: For Christians, man and nature are two, not one; and man was intended to be God's quasi-divine steward whose job it was benignly to govern the creation; but Calvinist Christians especially were preoccupied with man's fall into sin and depravity; thus it was a short step for a Puritan thinker like Edwards to regard man as a scourge on God's innocent creation.

There is more continuity than meets the eye, as Perry Miller convincingly argues, between Calvinist orthodoxy and American Transcendentalism. Jonathan Edwards is the transitional figure. And the received wilderness idea was fathered, as everyone agrees, by the Transcendentalists—Emerson and, especially, Thoreau. While Thoreau's philosophy is hardly Calvinist or even Christian, his temperament is thoroughly Puritan. He was celibate, vegetarian, abstemious, ascetic, self-righteous, judgmental. At points it seems that wild nature serves Thoreau more as a bearing for criticizing his fellow citizens than as something valuable in itself.

John Muir is, perhaps, the most ardent advocate ever of wilderness preservation. He was born in Calvinist Scotland, raised a Presbyterian by a father who could, without hyperbole, be called a religious psychopath, and force-fed the Bible. Finally liberated from his father's tyranny, upon coming of age, Muir was introduced, at the University of Wisconsin, to botany and geology, and, by Jeanne Carr, to the writings of Emerson and Thoreau. Thereupon, it seems, he apostatized. Not so, according to Worster: "Some writers have seen in this emergent Muir . . . a repudiation . . . of his Scotch Presbyterian background."[8] But he remained true to his Puritan roots: "'I am tempted at times,' he admitted in a letter to a friend, 'to adopt the Calvinist doctrine of total depravity.' He did not maintain that human beings had all been born stinking and hideous in the eyes of God, as John Calvin had done," continues Worster, "but he did insist that civilization had managed to corrupt them all, a view that was as bleak and pessimistic. There was a harshly negative side to Muir's vision, a disgust for human pretensions and pride that ran very close to misanthropy. Lord man, the self-proclaimed master of creation, was for him an ugly blot on the face of the Earth."[9] The shift in the valuation of nature—from negative to positive—begun by Edwards, in the Puritan strain of American thought, was completed by Muir. Worster, again, puts this point with both authority and clarity: "There is a positive side to Muir's religious stance, offsetting his pessimism, and it came from his discovery that there is perfect goodness and beauty to be found in the natural world."[10]

Note the Puritan temper and fervor of Muir's words when, in *A Thousand Mile Walk to the Gulf,* he proclaims—for the first time, to my knowledge, in English letters—the rights of nature: "How narrow we selfish conceited creatures are in our sympathies! how blind to the rights of all the rest of creation! With what dismal irreverence we speak of our fellow mortals! . . . They . . . are part of God's family, unfallen, undepraved, and cared for with the same species of tenderness and love as is bestowed on angels in heaven or saints on earth."[11]

Those to whom the wilderness standard was passed from the hands of Muir, after his death in 1914, have more tenuous connections to Puritanism. Aldo Leopold was of German descent and born into a Lutheran household that was not particularly religious. On the other hand, he entered government service in the morally charged Progressive era, Progressivism being arguably a secular permutation of Puritanism. However, Leopold, though a life-long champion of wilderness preservation, devoted most of his time and effort to the integration of "beauty and utility" in the middle, rural landscape, not to wilderness work. And his writing shows no traces of the misanthropic undercurrent running in Muir's. Robert Marshall was a New Dealer and a socialist, with no Puritan influences evident in his writing. Something similar could be said about Sigurd Olson.

But the Calvinist fire blazes up anew in the current generation of wilderness advocates, Rolston most evidently. Interestingly Dave Foreman [in the essay "Wilderness Areas for Real," which follows next in this volume] finds it expedient to mention his Scotch-Irish ancestry. And during his Earth First! period, Foreman's pugnacious misanthropy often got him in hot water. Conservation biologist Reed Noss [in the essay "Sustainability and Wilderness," which follows Foreman's in this volume] staunchly resists the effort to overcome the man/nature dualism at the core of the wilderness idea by way of the alternative human-harmony-with-nature approach to conservation first advanced by Aldo Leopold and now problematically labeled *sustainable development.* Conservation biologist Don Waller [in the essay "Getting Back to the Right Nature," in Part IV of this volume] sternly censures human overpopulation and anthropogenic species extinction with a fervor reminiscent of Edwards's commendation of Hell to sinners. While Rolston does not represent nature conservation to entail a zero-sum, Earth-First!-versus-People-First! choice, as Foreman and Noss are wont to do, he adamantly insists, nevertheless, on maintaining the radical distinction between things human and things natural.

Rolston also epitomizes the faithful who remain committed to a parallel dichotomy—between civilization and savagery—that relegates aboriginal peoples to the status of two-legged wildlife. Here, fortunately, I can defend two points at once: that the difference between native species and exotics is temporally scale-dependent and that the environmental impact of Pre-Columbian American peoples was significant. Are feral mustangs in the American west native or exotic?, I asked. Exotic, Rolston answers: "ancestral horses . . . in the American West went extinct naturally." Naturally for sure, as I understand naturally. But was a contributing agent to this natural extinction cultural Homo sapiens? *Equus* evolved in North America and roamed here up to and shortly after Siberian big game hunters wandered across the Bering land bridge into the western hemisphere.[12] Two species of elephant and 30 other genera of hemispherically extinct large mammals were also here when the spearmen appeared.[13] Did atlatl-launched missiles armed with Clovis warheads conspire with global warming to push these animals over the brink? If, as Paul S. Martin has argued, the sudden extinction of the American megafauna is attributable to a newly arrived predator overkilling prey totally unprepared for projectile-hurling carnivorous apes, then the environmental impact of the Siberian pioneers, measured in terms of biodiversity loss, greatly exceeds that of their technologically superior European counterparts some 100 centuries later. What would the New World biota have been like in 1492 if Columbus had really been its discoverer? Very different, I would suppose; much the same, Rolston seems to think.

After the spasm of (anthropogenic?) extinctions in the western hemisphere at the beginning of the Holocene, Martin speculates that "Major cultural changes would begin."[14] This is the lesson that I wish to draw from American prehistory: Eastward migrating pedestrian Homo sapiens found a genuinely pristine and virgin wilderness. There can be little doubt about that. Though controversial, some archaeologists are inclined to believe that these first pioneers raped it. If so, subsequent generations of Indians apparently learned to live symbiotically with the depauperate natural heritage that their immigrant ancestors bequeathed them. They could because cultural adaptation is more rapid and versatile than genetic adaptation, as Rolston so eloquently and forcefully argues. Perhaps we post-modern Euro-, Afro-, and Asian-Americans can do likewise. Our contemporary overpopulation and technological sophistication may preclude that possibility, in which case all we can do about conservation is defensively protect relatively undisturbed patches of nature—which I'm all for—and occasionally visit them to salve our spirits while living in a world of wounds. But a new generation of post-industrial technologies may make it possible for us to pursue many of our economic activities without compromising ecosystem health. If so, we may envision an alternative, and *complementary,* approach to the conservation of biodiversity as I tried to do in "The Wilderness Idea Revisited."

NOTES

1. Holmes Rolston III, "The Wilderness Idea Reaffirmed," *The Environmental Professional* 13:370–377. [Included in this volume.]
2. J. Muir, *A Thousand Mile Walk to the Gulf* (Boston: Houghton-Mifflin, 1916), p. 139.
3. J. H. Masserman, S. Wechkin, and W. Terris, "Altruistic Behavior in Rhesus Monkeys," *American Journal of Psychiatry* 121 (1964): 584–585; F. Mowat, *Never Cry Wolf* (Boston: Little, Brown, 1963); H. Rolston, *Environmental Ethics: Duties to and Values in the Natural World* (Philadelphia: Temple University Press, 1988), p. 38.
4. H. Rolston, *Science and Religion: A Critical Survey* (New York: Random House, 1987).
5. Lynn White, Jr., "The Historical Roots of Our Ecologic Crisis," *Science* 155 (1967): 1203–1207.
6. See Perry Miller, *Errand into the Wilderness* (New York: Harper and Row, 1964).

7. Roderick Nash, *Wilderness and the American Mind* (New Haven, Conn.: Yale University Press, 1967).

8. Donald Worster, review of Michael P. Cohen, The Pathless Way: John Muir and American Wilderness, *Environmental Ethics* 10 (1988): 268.

9. Ibid., p. 268.

10. Ibid., p. 269.

11. Muir, *A Thousand Mile Walk*, p. 98.

12. G. G. Simpson, *Horses: The Story of the Horse Family in the Modern World through Sixty Million Years of Evolution* (New York: Scribner's, 1956).

13. H. E. Wright and D. G. Frey, eds., *The Quaternary of the United States* (Princeton, N.J.: Princeton University Press, 1965).

14. Ibid., p. 972.

Dave Foreman

Wilderness Areas for Real (1998)

FOR A QUARTER CENTURY, I've been neck-deep (and sometimes over my head) in the Wilderness preservation donnybrook. We conservationists have faced well-heeled timber barons, big ranchers, mining magnates, and other politically powerful captains of industry. Arguing with a friend like J. Baird Callicott over the modern viability of the Wilderness Idea is a curious and often frustrating tussle off to the side.

We have faced off over his criticism of the Wilderness Idea before (*Wild Earth,* Winter 94/95), and I've taken on the general "progressive" critique of Wilderness Areas in "Where Man Is a Visitor" in David Clarke Burks's *Place of the Wild* (Island Press, 1994). Here I will try to respond more specifically to Callicott's approach in "The Wilderness Idea Revisited"; but, since I'm a good little environmentalist, I'll recycle many of the arguments I've used in previous papers. Callicott's criticisms of Wilderness are often roundups of ideas more fully developed by himself and others elsewhere; in such cases, I will contend with the broader criticism of Wilderness. (I'll not respond to Callicott on those points where Holmes Rolston III, in "The Wilderness Idea Reaffirmed," has effectively whacked him about the head and shoulders.)

Scattered through "The Wilderness Idea Revisited," I find eight general areas where I have bones to pick with Callicott. Instead of a point/counterpoint debate, I will use these areas as springboards for a general discussion of the ideas he raises.

WILDERNESS AREAS ARE OKAY, THE WILDERNESS IDEA IS NOT

Callicott says he wants only to criticize the idea of Wilderness but not on-the-ground Wilderness Areas. He even says that Wilderness Areas need to be multiplied and expanded. Why then, criticize them for what they are? Philosophers might call this a logical inconsistency, or some other silver-plated term.

Callicott also says he does not want to discredit Wilderness Areas or make them more vulnerable to development. But this is exactly what he is doing! He is discrediting them by attacking the idea behind them, and others will reap the whirlwind he is sowing to try to open existing Wilderness Areas to clear-cutting, roads, motorized vehicles, and "ecosystem management," and, more dangerously, to argue against the designation of new Wilderness Areas.

Although Callicott criticizes the Wilderness Idea for creating a dualism of Man and Nature, in truth, throughout "The Wilderness Idea Revisited" he creates dualisms. Sustainable development (at least in its ideal form, not in the common land mismanagement that calls itself sustainable) is not an alternative to Wilderness Areas. Indeed, Wilderness Areas and sustainable-use zones are complementary as different regions on a land spectrum running from those places where "man is a visitor who does not remain" to those places where men, women, and children are piled on top of each other by the thousands per square mile. What most frustrates me with Callicott's criticism of Wilderness is the way he characterizes his sustainable development alternative. He does not need to criticize the Wilderness Idea at all. He could simply say, "We need to protect existing Wilderness Areas and expand their size and number; we also need to manage the matrix around them in a way informed by ecology and based on maintaining biodiversity and sustainable human communities." (This, by the way, is the approach of The Wildlands Project.) In other words, Callicott's argument could be entirely positive and not negative—and would not, then, carry the potential for mischief that makes his critique so risky.

ALDO LEOPOLD: WILDERNESS DEFENDER OR ADVOCATE FOR SUSTAINABLE HUMAN USE?

Callicott presents Aldo Leopold as a complex figure, which he was. Aldo Leopold was a big man. He thought deeply, and he thought about a lot of things. His ideas changed during his life as his experience and wisdom grew[1] (we should all be so fortunate). So I just can't understand why Callicott and Susan Flader take a dualistic approach to Leopold. Aldo Leopold advocated the protection of Wilderness Areas. But Aldo Leopold also advocated sustainable agricultural practices that could improve abused land and provide habitat for some wildlife. Which one was the real Aldo Leopold? They both were! There is nothing contradictory between wanting Wilderness Areas and wanting ecologically managed farms, between wanting reserves where no one lives and wanting places where people learn to live in harmony with Nature. True, Leopold's approach to the land blended the approaches of Muir and Pinchot and created a new synthesis. There is nothing new about this way of seeing Leopold, nor is there anything in it that disparages the Wilderness Idea. Why is it necessary for Callicott to make it seem that Leopold was not as supportive of Wilderness as we think he was? There does not need to be a Cartesian dualism here at all; we do not have to pick one or the other. Good lord, I've made a life of defending Wilderness Areas, but I also like four-dollar cigars, good red Bordeaux, French cuisine, and classical music. Do I have to worry that after I'm gone, revisionist historians will come along to argue that Dave Foreman wasn't really a wilderness fanatic at all but was more into refined decadence?

WILDERNESS AREAS AND SACRIFICE ZONES, OR SUSTAINABLE DEVELOPMENT

In supposing wilderness proponents favor an "either/or dichotomy" (zone the land as protected Wilderness Areas, or zone it as sacrifice zones where industrialism can run rampant), Callicott misunderstands the work of the conservation movement. We have fought for Wilderness Areas, yes; we have also fought like hell for sensible, sensitive, sustainable management of other lands. We have fought to protect wild rivers from dams; we've also fought to protect agricultural valleys from dams. We have tried to bring scientific timber-harvesting practices to the National Forests. We have tried to bring scientific livestock management to the public lands. *We have fought for good management of the matrix.* The conservation movement that Callicott criticizes for being totally fixated on Wilderness Areas actually has worked to pass the National Environmental Policy Act (NEPA), the Federal Lands Policy and Management Act (FLPMA), the National Forest Management Act (NFMA), the Resource Planning Act (RPA), and the Endangered Species Act (ESA)—all of which deal mostly with sustainable management of non-Wilderness Area lands. For decades, we have been doing what Callicott now urges us to do.

The reason we keep going back to Wilderness is because every reform measure, from NEPA to NFMA to RPA to FLPMA, gets gutted in practice by agencies controlled by extractive industries. We have tried, god, we have tried to get good management on the land. The reforms usually end up like a Bosnian caught by a bunch of Serbs.

To appreciate why Wilderness Areas must be the centerpiece of conservation strategy, one really needs to spend time in the trenches, fighting Forest Service timber sales, going toe-to-toe with ranchers and loggers and snowmobilers, filing appeals and lawsuits against agency "development" schemes, and lobbying members of Congress to protect a place (and trying to figure out which arguments will work with them).

The Wilderness Act was not so much reform legislation as a monkey wrench in the gears. It says, "We know you (Forest Service, Park Service, other agencies) are incapable of voluntarily protecting these values on the lands you manage. Therefore we are taking the prerogative away from you. We are tying your hands in these ways: no roads, no motorized vehicles or equipment, no logging." Through long years of hard work, experience, and bitter disappointment, effective conservationists have become realists after starting out as idealists who believed their civics textbook's model of how democracy is supposed to work.

I agree we should continue to work for better management of the matrix, and to integrate the Biosphere Reserve idea with Wilderness Areas and National Parks. But I am far less hopeful than Callicott as to the results, for the reasons above. In "The Wilderness Idea Revisited" and elsewhere, he argues that alternatives to industrial agriculture should be encouraged through policy changes; that urban sprawl should be controlled by planning and zoning; that National Forests should be harvested ecologically and sustainably. All that has been the agenda of the conservation movement for decades. We've gotten our faces bloodied from running into swinging ax handles. You think Wilderness is controversial? Try talking zoning, planning, and "alternatives to industrial agriculture" to the property rights militia and agribiz plowboys if you want controversy.

We've been through all this a thousand times before; we're still there as a conservation movement; we'll keep trying in the future. But sustainability is not a new idea, and it sure as hell ain't easy. Through all this, conservation activists have learned that Wilderness Areas, however pared back and compromised they've been, work better than anything else at protecting biodiversity. The fault for abuse of public lands is not with Wilderness; it is with the perversion of Pinchot's progressive utilitarian brand of conservation.

A THIRD WORLD CRITIQUE OF CONSERVATION VIA WILDERNESS PRESERVATION

I fear that Ramachandra Guha and other vociferous critics of American-style Wilderness and National Parks are suffering from Third World jingoism. Guha's critique is "sobering" in the way Pat Buchanan's bombast is sobering. Wilderness is a victim of chronic anti-Americanism. Everything from the United States is bad to some folks. North America and Europe are to blame for all the world's problems.

Some from the United States who approvingly quote Guha are expiating white liberal guilt. (I'm lucky. I come from, at best, a lower middle-class family of Scotch-Irish hillbillies. I have more than my share of moral failings, and one of these days I reckon I should get around to atoning for some of them; but guilt for being pampered I have not.)

Western Civilization, imperialism, and the United States of America deserve plenty of criticism. And I think the United States should be held to higher standards than any other country or society because we have claimed from the beginning to be engaged in a superior social experiment.

But the United States is not wholly evil. We are not the sole source of injustice in the world. The anti-Americanism inherent in Third World criticism of Wilderness and Parks ignores the venality of elites in those countries. To blame white males for all the world's problems is—dare I say it?—racist. Furthermore, the leading Third World critics of Wilderness are Western-educated members of the economic/social elite in their own countries.

None of this is to argue that we should ignore issues of international economic justice. Europe, Japan, and the United States, in cahoots with the robber barons of the Third World, wage conscious economic imperialism against other nations. Certainly we need to safeguard land for use by indigenous peoples and peasants, and to recognize and celebrate their knowledge and stewardship of the land. Wilderness Areas and National Parks need not conflict with the needs and rights of the downtrodden.

Unfortunately, some international social justice proponents who criticize National Parks and Wilderness Areas are just as anthropocentric and development-driven as are fast-buck businessmen and growth-worshipping economists. They merely want to see the supposed economic benefits from the destruction of wilderness go to the poor and socially disenfranchised instead of to the wealthy and politically connected.

To argue that Wilderness is a uniquely American idea and is not internationally universalizable begs the question of whether any single land-management approach is suitable throughout a culturally diverse world. But those who think the Wilderness Idea of places where humans are visitors who do not remain is uniquely American are being ethnocentric or ignorant. Despite Guha's pawing the ground and snorting about imperialism, Wilderness Areas are not a uniquely twentieth-century idea of Americans, Canadians, and Australians. Open-minded research in geography, history, and anthropology shows us that wilderness areas where humans are visitors who do not remain were once widespread throughout the world. (More about this later.)

Can people outside the English frontier colonies appreciate Wilderness for its own sake? I know Native Americans and have met folks in Mexico and Belize who are just as supportive of Wilderness as I am and who believe in the intrinsic value of other species. At an international wilderness mapping conference a few years ago, I met South American biologists as intransigent in their defense of Wilderness as Reed Noss. Jim Tolisano, an ecologist who has worked for the United Nations in many countries, tells of colleagues in Sri Lanka, several African countries, Costa Rica, and the Caribbean who make me look like an old softy.

It is also false to argue that wilderness conservationists have not considered native peoples. Callicott ignores how Wilderness Areas and National

Parks designated by the Alaska National Interest Lands Conservation Act sixteen years ago allowed for subsistence use by Alaskan natives. Today, many tribal peoples in Alaska are among the strongest supporters of Wilderness. We didn't have to throw out the idea of Wilderness Areas and National Parks and replace it with an untested concept like Biosphere Reserves. Conservationists and the Wilderness Idea were dynamic enough to make Wilderness work in the special circumstances of Alaska. Conservationists in Canada are working out similar arrangements there with First Nations.

WILDERNESS IS ETHNOCENTRIC (OR, THE NOBLE SAVAGE)

Callicott argues against the myth of a pristine North America: the America European colonists encountered was heavily managed and modified by Native Americans; indeed, they had "improved" the land and caused the "incredible abundance of wildlife." Other critics of Wilderness have also played variations on this theme.

Anthropology is like the Bible. You can use it to support any claim about humans and Nature you wish. We can argue until we're blue in the face about the level of impact indigenous people had in the Americas. The wisdom until recently was that Native Americans had very little effect on the landscape. New England's Puritans argued so to justify their taking of "unused" land from the Indians. The pendulum has swung the other way in recent years, with crackpots like Alston Chase as well as serious scholars like Callicott claiming that even small populations significantly altered pre-Columbian ecosystems—especially through burning. The "myth of pristine America" is in disrepute. Worshippers of the Noble Savage argue this impact was positive. Some even place the bloody Aztecs, Incas, and Mayans on the ecological pedestal, too.

Many researchers, however, see evidence of ecological collapse in archaeology. Did the Hohokam and Anasazi of the American Southwest overshoot carrying capacity and cause ecosystem failure? Newly mobilized with Spanish mustangs, would the Plains Indians have caused near extinction of the Bison had they been left alone for another hundred years? Did the civilizations of MesoAmerica and the Andes scalp their lands as terribly as did the Assyrians and Greeks? Was the extinction of the Pleistocene megafauna caused by stone-age hunters entering virgin territory?

Deeper questions follow. Is the land ethic of the Hopi a result of a new covenant with the land following the Anasazi ecological collapse seven hundred years ago? Could the hunting ethics of tribes in America (and elsewhere) have been a reaction to Pleistocene overkill?

Where foraging and shifting human horticulture reputedly increase local species diversity, dare we ask about the quality of that increased diversity? Are the additional species common, weedy ones? Are many of these invasive exotics that supplant native species? All biodiversity is not equal. Rare, sensitive, native species are more important than weeds which do well in human-disturbed areas. (Rolston does a fine job of sorting out the Quitobaquito/Quitovac muddle.)

In certain areas of the Americas, high human population density and intensive agriculture led to severely degraded ecosystems. Paul Martin and many other scientists now believe the first wave of skilled hunters entering North America 12,000 years ago caused the extinction of dozens of species of large mammals inexperienced with such a weapon-wielding predator. The "overkill hypothesis" looks to me virtually indisputable, but I question Callicott's suggestion that the North American forests and prairies found by the first European explorers and colonists were primarily the result of burning by native tribes. Certainly, in localized areas North American tribes had an impact on vegetation because of anthropogenic burning. But how extensive could this manipulation have been with a population of only four to eight million[2] north of the Rio Grande in 1500? Reed Noss points out that lightning-caused fires better explain the presence of fire-adapted vegetation than do Indian fires.[3] I do not raise these questions to oppose legitimate land claims of Native Americans and other First Nations. In some cases, tribes are better caretakers of the land than are government agencies. Despite the opposition of some other conservationists in New Mexico, I supported the transfer of Carson National Forest's Blue Lake area to Taos Pueblo in the 1970s. Historically it was their land, and they have done a far better job of protecting its wilderness than the Forest Service would have. However, we must be intellectually honest in investigating human relationships with the land, and we must not pander to the Noble Savage myth and then hold primal peoples up to impossible standards.

Notwithstanding the seesawing over preindustrial societies' role in changing the face of the Earth, there is much evidence that wilderness areas—vast tracts uninhabited by humans—are a familiar concept to many primal cultures. The Gwich'in of the American Arctic talk about going into the bear's or the caribou's home when they go on a hunting expedition away from their villages. Native Hawaiians tell me that before American conquest, some mountains were forbidden to human entry—on pain of death. Jim Tolisano reports that the tribes of Papua New Guinea zone large areas off-limits to villages, horticulture, hunting, and even visitation. "You don't go there. That mountain belongs to the spirits." Like the Papuans, the Yanomami of the Amazon engage in fierce blood feuds (my hillbilly ancestors in eastern Kentucky and earlier in the highlands of Scotland were a lot like them). Between villages is a death zone where one risks one's life by entering. As a result, large areas are left uncultivated, unhunted, and seldom visited. Wildlife thrives. These borderlands are refugia for the animals intensively hunted near settlements.[4] Some anthropologists think that the permanent state of war between some tribes is an adaptation to prevent overshooting carrying capacity, which would result in ecological collapse. (These unused areas on territorial borders are uncannily similar to the places where the territories of wolf packs abut one another and deer occur in higher densities.) My forebears were able to follow Dan'l Boone into the "dark and bloody ground" of Kaintuck because it was uninhabited by Native Americans. The Shawnee north of the Ohio River and the Cherokee from the Tennessee Valley hunted and fought in Kentucky. But none lived there. Wasn't it a Wilderness Area until the Scotch-Irish from Shenandoah invaded?

Geographers, anthropologists, historians, and ecologists need to research these tantalizing threads and others to show that wilderness areas—where humans are visitors who do not remain—were once widespread throughout the world. Wilderness as reality and idea is neither uniquely American nor especially modern. It is widespread and it is ancient. Conservationists have failed to make that point, and we have failed to gather and offer examples of it. (Wilderness needs a few good anthropologists!)

WILDERNESS SEPARATES HUMANS FROM NATURE

I'm sorry that Callicott lends credibility to the old humbug that Wilderness Areas perpetuate a Nature-human dualism. Indeed, other species besides

Homo sapiens can damage their habitats, as Callicott writes. He's right—if there were five billion elephants they would do considerable damage to the Earth; but the important point here is that there are *not* five billion elephants on the planet. There *are* more than five billion humans on the planet, because civilization, modern medicine, agribusiness, and industrial science have allowed us to escape the natural checks on our numbers—*have allowed us to temporarily divorce ourselves from Nature.* The consequences of this are disastrous to all life on Earth, including present and future generations of humans.

It is civilization that has caused a Nature-human dualism. Wilderness Areas are the best idea we've had for healing that breach, for reintegrating people back into Nature in a humble and respectful way. We must realize that we can love something to death, that in possessing a place we destroy it. Wilderness Areas, where we are visitors who do not remain, bring us back into Nature, Nature back into us, without the destruction that permanent habitation would cause. Most of the Earth's surface has been without permanent human habitation for most of our time here. There is nothing dualistic about this, nothing misanthropic. It is normal. It is even—why not say it?—*natural.*

I have spent many, many days and nights in Wilderness Areas from Alaska to Central America. I have not found that these landscapes where I am only a visitor separate me from Nature. When I am backpacking or canoeing, hunting or fishing in a Wilderness, I am home.

Wilderness Areas where humans are visitors who do not remain test us as nothing else can. No other places teach us humility so well—whether we go to them or not. Wilderness asks: Can we show the self-restraint to leave some places alone? Can we consciously choose to share the land with those species who do not tolerate us well? Can we develop the generosity of spirit, the greatness of heart to not be everywhere?

No other challenge calls for self-restraint, generosity, and humility more than Wilderness preservation.

THE WILDERNESS IDEA IS ABOUT STATIC, STABLE LANDSCAPES

Reed Noss, former editor of *Conservation Biology* and the leading theorist of designing nature reserves (including Wilderness Areas) to protect biodiversity, writes, "Callicott erects a straw man of wilderness . . . that is 30 years out of date. No one I know today thinks of wilderness in the way Callicott depicts it."[5]

Callicott claims that Wilderness Areas were established to protect climax communities. Ecology today pooh-poohs the idea of climax communities; ergo, Wilderness Areas are bogus, he believes. In reality, Wilderness Areas are entirely consistent with ecological theories of unstable and changing assemblages of species and seral stages. Wilderness Areas and National Parks, after all, were where modern ideas of fire ecology—that natural fire is a fundamental and vital part of many forest, woodland, and grassland ecosystems—were first translated into "let burn" management policies.

The root for "wilderness" in Old English is *wil-deor-ness:* self-willed land. Self-willed land has fire, storm, and ecosystem change. It has wild beasts who don't cotton to being pushed around by puny hominids. Those who want "snapshot-in-time" Parks are generally the same folks who argue against Wilderness Areas. As our ecological understanding advances, so does our Wilderness philosophy. We conservationists now know that because of profound human-induced ecological changes, we must intervene with science-based management in certain cases, particularly in smaller, isolated areas—as Reed Noss points out.

SUSTAINABLE DEVELOPMENT AND GAME RANCHING

Callicott proposes importing the Third World model of extractive reserves to the western rangelands of the United States and the Third World model of Biosphere Reserves for remaining old-growth forests in the United States. And he says sustainable development is the "new idea in conservation today." Let's look at each of these.

In an optimistic view, sustainable development is unproven; in my cynical view, sustainable development is a fraud—a recorking of the same old vinegar of multiple-use/sustained yield in a new bottle. It's a new guise for multinational corporations and wealthy Third World elites to wring more marks, yen, and dollars out of the land and out of the people.

Something like biosphere reserves were used in President Clinton's plan to protect the ancient forests of the Pacific Northwest. But these "old-growth reserves" (they ain't Wilderness Areas!) are falling to the chain saws, because of a law Clinton signed. Despite all the carping about the "failure" of Wilderness Areas (and National Parks), they have a track record unmatched by any other land designation anywhere in the world. Designated Wilderness Areas have been around for seventy years and they have succeeded in protecting ecological processes and some of the most sensitive species in North America. As conservation activists, scientists, and agency managers develop greater ecological understanding, the concept of Wilderness Areas and their design and management also change. Just as ecosystems are dynamic, the notion of Wilderness is dynamic. It may be a sad commentary on modern humans, but the fact is that only carefully protected core reserves like Wilderness Areas can really maintain the diversity of life. Leading research biologists with experience from the Amazon to the Himalaya have told me that buffer zones, extractive reserves, "sustainable" use zones, and other Biosphere Reserve approaches cannot adequately protect biodiversity without fully protected core reserves. Biosphere Reserves are a fine concept; we will see how they do in practice.

I like Callicott's idea to turn the Western range back over to wildlife and kick out the cattle and sheep. In fact, Howie Wolke and I proposed such an approach in 1980. Game ranching and professional market hunting have grave problems, though. Market hunting is what decimated wildlife populations 100 years ago. Game ranching spins traditional American wildlife philosophy on its head—in the United States you do not own the wildlife that lives on your land. Callicott is correct—there are serious political and economic obstacles to his scheme. To find a workable approach requires us to understand the history of wildlife law and the hunting tradition in the United States and to tease a new "range utopia" out of it.

J. Baird Callicott has done a service to the conservation movement; he has made us think and he has made us better defend our traditional ideas. Some of his suggestions are good and are workable; too many, however, are theoretical and idealistic instead of practical and realistic and ignore human nature, political reality, and the true history of the conservation movement. Aldo Leopold said it best:

"The richest values of wilderness lie not in the days of Daniel Boone, nor even in the present, but rather in the future."

NOTES

1. I am glad Susan Flader and J. Baird Callicott included Aldo Leopold's "The Popular Wilderness Fallacy" in their Leopold anthology, *The River of the Mother of God and Other Essays by Aldo Leopold* (Madison, Wis.: University of Wisconsin Press, 1991). Leopold, more than any other conservation figure, shows how philosophy and action can mature. This essay underscores Leopold's wonderful ability to grow intellectually.

2. The best recent estimates of serious demographers.

3. Reed F. Noss, "Wilderness—Now More Than Ever" in *Wild Earth* 4(4) (Winter 1994/95): 60–63.

4. George Schaller reports that when Amazonian tribes were armed only with blowguns and bows, monkeys could be found half a mile from villages. Now, with the advent of the shotgun, monkeys are not found within five miles of settlements. Jim Tolisano reports similar changes in Papua New Guinea.

5. Noss, "Wilderness—Now More Than Ever," p. 60.

Reed F. Noss

Sustainability and Wilderness (1991)

I AM WRITING in defense of wilderness as a foundation of conservation. The wilderness idea has fallen on hard times of late. Not only does industry fight it at every turn, but even conservation biologists now shun wilderness in favor of the more moderate notion of sustainable management of natural resources. Sustainability is being hailed as a new paradigm for conservation (Salwasser 1990); it seems to represent the perfect middle ground in resource conflicts. Whereas genuine protection of wilderness and biodiversity would demand radical changes in the way we do business as a society, the sustainability notion is safe and nonthreatening. How on earth could anyone be opposed to sustainability?

The paradigm shift from wilderness preservation to sustainable management is easy enough to explain. Biologists have suddenly realized that set-asides are of inadequate size and number to maintain biodiversity. This would be true even under stable environmental conditions; with a changing climate, many reserves will no longer serve the elements they were established to protect (Peters and Darling 1985). If we make the defeatist assumption that substantial new reserves are out of the question, a logical response to our quandary is to manage in a more ecologically sensible manner the "semi-natural matrix" that constitutes most of our land (Brown

1988). Whole-landscape management, we are told, will require an integration of human needs and uses into conservation objectives. Who can argue? A unification of people and nature is something to be applauded, is it not? The Earth first vs. people first polarization has been depressing (Salwasser 1990) and may reflect a dualism in our culture that we could well do without (Callicott 1990). But in our zeal to break down this dualism and come to grips with a dynamic Earth, I worry that we are underrating reserves as a component of conservation strategy. In particular, I urge that we not discount the tremendous value of huge, roadless, essentially unmanaged areas: what has been called "Big Wilderness" (Foreman and Wolke 1989).

I see four primary values of Big Wilderness. One value is scientific: Wilderness provides a standard of healthy, intact, relatively unmodified land. As pointed out by Callicott (1990), this function of wilderness was emphasized by Aldo Leopold in his later years, when his maturity as an ecologist enabled him to move beyond a purely recreational justification for wilderness. Leopold's characterization of wilderness changed from a place "big enough to absorb a two weeks' pack trip" (Leopold 1921) to a "base-datum of normality" for a "science of land health" (Leopold 1941). For ecologists, wilderness areas are controls, baselines against which we may measure the effects of management experiments. Scientists shudder to think of experiments without controls, but this is precisely what happens with most of our land management. Because many effects of land management are expressed at the scale of the landscape, we need control areas large enough to encompass landscape-level processes such as distance propagation, patch dynamics, and between-community fluxes of organisms and materials. Research in natural areas set aside for scientific purposes cannot fill this role; 93 percent of the 213 Forest Service research natural areas are smaller than 1,000 ha and many undoubtedly suffer from edge effects (Noss 1991).

Callicott (1990) discusses some problems with using wilderness as a standard of land health in a dynamic, polluted, and humanized world. Simply put, no place is pristine and every place is continuously changing. But a dynamic and imperfect baseline is better than no baseline at all; the larger the control area, the less it will be affected by many cross-boundary phenomena. We claim that our new approaches to land management emulate nature, but what will be left to emulate if we manipulate everything?

A second value of Big Wilderness is biological. One of the many biological and ecological functions of wilderness is to provide habitat for species that do not get along well with humans. It is no accident that the only places in the world with healthy populations of all native carnivores are Big Wilderness. Roadlessness—or more generally, limited accessibility for predatory humans—is key to wilderness integrity. No matter how ecologically sophisticated our land management practices, how perfectly we think we can mimic natural processes, so long as there are roads and mechanized humans, some species will disappear. Wolves, grizzly bears, black bears, and mountain lions are among the species usually absent in landscapes with high road densities (Noss, in press). Large carnivores are symbolic and authentic indicators of healthy land (Terborgh 1988); when they and the wilderness they depend on are gone, the land is impoverished immeasurably.

The scientific and ecological values of wilderness portrayed so eloquently by Leopold 50 years ago seem to have been forgotten by today's land managers. In the current round of management plans for national forests, the need for wilderness is evaluated on the basis of Recreational Visitor Days (RVDs) or a Wilderness Recreation Opportunity Spectrum (WROS). The latest multiple-use theme of the Forest Service, "New Perspectives" (and its primary component, "New Forestry"), plays down the need for wilderness and other set-asides in its effort to establish "the middle ground between timber production and preservation" (USDA Forest Service 1989). A middle ground of this sort is unlikely to give biodiversity the precedence it deserves. In the first application of New Perspectives accompanied by a Draft Environmental Impact Statement, the Shasta Costa project in the Siskiyou National Forest of Oregon, the Forest Service proposes "using the entire landscape to blend production with protection" (USDA Forest Service 1990a). On the ground, this translates to a 31% reduction of remaining old-growth forest-interior habitat in the watershed by 1998 (USDA Forest Service 1990b). Are conservation biologists willing to accept this consequence of "sustainable" management: further depletion of the last virgin forests in the lower 48 states? If so, we are at odds with national opinion and, I submit, with environmental ethics.

A third value of wilderness is as a source of humility. Wilderness represents self-imposed restraint in a society that generally seeks to dominate and control all of nature. Henry Thoreau, John Muir, Aldo Leopold, Edward Abbey, and other literary defenders of wilderness have expressed the uniquely American idea of wilderness as a sanctuary of freedom, a refuge of sanity in an overcivilized world, and as somewhere to be profoundly humbled (Nash 1982). We need places where we have the liberty to get utterly lost, frozen, starved, or mauled by a grizzly. Less extreme, perhaps, I have found the black flies in the Hudson Bay Lowlands of Ontario to be sufficiently humbling. Conservation biologists and land managers could stand a healthy dose of humility. As Ehrenfeld (1978), Devall and Sessions (1985), Orr (1990), and others have warned, our desire to manage everything is exceedingly arrogant given our ignorance of how nature works. In many cases, what needs to be managed is not nature but rather our own consumptive, manipulative, and destructive behavior.

Finally, some of us believe that wilderness has value for its own sake. No, we cannot prove intrinsic value; it is not a rational inference. Yet, as a scientist, I can find no objective reason for believing that we are fundamentally superior to any other species. What right do we have to subjugate other life and mold the entire Earth to our purposes? Almost everyone will agree that some balance needs to be achieved between land managed primarily for natural values (wilderness or equivalent) and land managed for commodities. The argument is over where that balance lies. The current ratio of protected to exploited land in the United States is a lopsided 5:95; I would suggest 50:50 as a more reasonable compromise.

The multiple values of wilderness and other reserves should temper our enthusiasm for managing whole landscapes for sustainable use. As currently conceived by government agencies, the sustainability notion is hopelessly anthropocentric. What are we sustaining? We hear a lot about sustaining "productive" ecosystems, but the products are unquestioningly assumed to be for human use. Do we really want to sustain our profligate consumption of resources? Why emphasize products and uses, anyhow? Why not instead strive to sustain evolutionary potential and freedom for all creatures? As I see it, the confidence that we can manage landscapes sustainably for multiple uses is no less arrogant than the humanistic assumption that every environmental problem has a technological solution (see Ehrenfeld 1978). Certainly we will need to experiment and learn better ways of living with nature, but it would be wise to have a buffer for our inevitable mistakes. I hope that there is a lot more to conservation biology than the current infatuation with sustainability, and that some of us are still willing to follow Martin Heidegger (1975) and "let beings be."

Set-asides will not fulfill all our conservation objectives; but that realization should not keep us from asking for much more wilderness than we have now, indeed, for reestablishing wilderness where we can by closing roads and eliminating conflicting uses. If our wilderness recovery areas are sufficiently large, multiple, and interconnected, the biota will have a far better chance of adapting to dynamism than if confined to a few isolated scraps. We cannot risk putting all that is left of our native biodiversity into new approaches to multiple-use management; they are only experiments. Wilderness recovery, unlike sustainable management, represents a truly "new perspective" on the human-nature relationship.

Land managers admiringly quote Leopold's statements about harmony between humans and the land, envisioning perhaps a bucolic landscape with working people, working forests, working range, and natural areas happily coexisting. But there is no wilderness in this picture; there are no wolves, grizzlies, or panthers. Leopold, I suspect, would not be happy about the missing pieces. One of my favorite remarks by Leopold I never hear quoted by land managers or bureaucrats, for it is an uncompromising call for defense of America's remaining wildlands: "a militant minority of wilderness-minded citizens must be on watch throughout the nation and vigilantly available for action" (Leopold 1949). Regardless of whether conservation biologists choose to be part of that militant minority, we should at least keep our minds open to the proposition that nature—if given a chance—can still manage land better than we can.

BIBLIOGRAPHY

Brown, J. H. 1988. Alternative conservation priorities and practices. Paper presented at 73rd Annual Meeting, Ecological Society of America, Davis, California. August 1988.

Callicott, J. B. 1990. Standards of conservation: then and now. Conservation Biology 4:229–232.

Devall, B., and G. Sessions. 1985. Deep ecology: living as if nature mattered. Peregrine Smith, Salt Lake City, Utah.

Ehrenfeld, D. 1978. The arrogance of humanism. Oxford University Press, New York.

Foreman, D., and H. Wolke. 1989. The big outside. Ned Ludd Books, Tucson, Arizona.

Heidegger, M. 1975. The end of philosophy. Translated by Joan Stambaugh. Souvenir Press, London, England.

Leopold, A. 1949. A Sand County almanac. Oxford University Press, New York.

Nash, R. 1982. Wilderness and the American mind. 3rd ed. Yale University Press, New Haven, Connecticut.

Noss, R. F. 1991. What can wilderness do for biodiversity? In P. Reed, editor. Proceedings of the conference: Preparing to manage wilderness in the 21st century. USDA Forest Service. Southeastern Forest Experiment Station, Asheville, North Carolina.

Noss, R. F. 1993. Wildlife corridors. In: D. Smith, editor. Ecology of greenways. University of Minnesota Press. Minneapolis.

Orr, D. W. 1990. The question of management. Conservation Biology 4:8–9.

Peters, R. L. and J. D. S. Darling. 1985. The greenhouse effect and nature reserves. BioScience 35:707–717.

Salwasser, H. 1990. Sustainability as a conservation paradigm. Conservation Biology 4:213–216.

Terborgh, J. 1988. The big things that run the world—a sequel to E. O. Wilson. Conservation Biology 2:402–403.

USDA Forest Service. 1989. New Perspectives: an ecological path for managing forests. Pacific Northwest Research Station, Portland, Oregon, and Pacific Southwest Research Station, Redding, California.

USDA Forest Service. 1990a. Shasta Costa from a new perspective. U.S. Government Printing Office, Washington, D.C.

USDA Forest Service. 1990b. Draft Environmental Impact Statement: Shasta Costa timber sales and integrated resource projects. Siskiyou National Forest, Gold Beach, Oregon.

William M. Denevan

The Pristine Myth (1992)

The Landscape of the Americas in 1492[1]

"This is the forest primeval . . ."
Longfellow

WHAT WAS THE NEW WORLD LIKE at the time of Columbus?—"Geography as it was," in the words of Carl Sauer (1971, x).[2] The Admiral himself spoke of a "Terrestrial Paradise," beautiful and green and fertile, teeming with birds, with naked people living there whom he called "Indians." But was the landscape encountered in the sixteenth century primarily pristine, virgin, a wilderness, nearly empty of people, or was it a humanized landscape, with the imprint of native Americans being dramatic and persistent? The former still seems to be the more common view, but the latter may be more accurate.

The pristine view is to a large extent an invention of nineteenth-century romanticist and primitivist writers such as W. H. Hudson, Cooper, Thoreau, Longfellow, and Parkman, and painters such as Catlin and Church.[3] The wilderness image has since become part of the American heritage, associated "with a heroic pioneer past in need of preservation" (Pyne 1982, 17; also see Bowden 1992, 22). The pristine view was restated clearly in 1950 by John Bakeless in his book *The Eyes of Discovery:*

> There were not really very many of these redmen . . . the land seemed empty to invaders who came from settled Europe . . . that ancient, primeval, undis-

> turbed wilderness ... the streams simply boiled with fish ... so much game that one hunter counted a thousand animals near a single salt lick ... the virgin wilderness of Kentucky ... the forested glory of primitive America (13, 201, 223, 314, 407).

But then he mentions that Indian "prairie fires ... cause the often-mentioned oak openings.... Great fields of corn spread in all directions ... the Barrens ... without forest," and that "Early Ohio settlers found that they could drive about through the forests with sleds and horses" (31, 304, 308, 314). A contradiction?

In the ensuing forty years, scholarship has shown that Indian populations in the Americas were substantial, that the forests had indeed been altered, that landscape change was commonplace. This message, however, seems not to have reached the public through texts, essays, or talks by both academics and popularizers who have a responsibility to know better.[4]

Kirkpatrick Sale in 1990, in his widely reported *Conquest of Paradise,* maintains that it was the Europeans who transformed nature, following a pattern set by Columbus. Although Sale's book has some merit and he is aware of large Indian numbers and their impacts, he nonetheless champions the widely-held dichotomy of the benign Indian landscape and the devastated Colonial landscape. He overstates both.

Similarly, *Seeds of Change: A Quincentennial Commemoration,* the popular book published by the Smithsonian Institution, continues the litany of Native American passivity:

> pre-Columbian America was still the First Eden, a pristine natural kingdom. The native people were transparent in the landscape, living as natural elements of the ecosphere. Their world, the New World of Columbus, was a world of barely perceptible human disturbance (Shetler 1991, 226).

To the contrary, the Indian impact was neither benign nor localized and ephemeral, nor were resources always used in a sound ecological way. The concern here is with the form and magnitude of environmental modification rather than with whether or not Indians lived in harmony with nature with sustainable systems of resource management. Sometimes they did; sometimes they didn't. What they did was to change their landscape nearly everywhere, not to the extent of post-Colonial Europeans but in important ways that merit attention.

The evidence is convincing. By 1492 Indian activity throughout the

Americas had modified forest extent and composition, created and expanded grasslands, and rearranged microrelief via countless artificial earthworks. Agricultural fields were common, as were houses and towns and roads and trails. All of these had local impacts on soil, microclimate, hydrology, and wildlife. This is a large topic, for which this essay offers but an introduction to the issues, misconceptions, and residual problems. The evidence, pieced together from vague ethnohistorical accounts, field surveys, and archaeology, supports the hypothesis that the Indian landscape of 1492 had largely vanished by the mid-eighteenth century, not through a European superimposition, but because of the demise of the native population. The landscape of 1750 was more "pristine" (less humanized) than that of 1492.

INDIAN NUMBERS

The size of the native population at contact is critical to our argument. The prevailing position, a recent one, is that the Americas were well-populated rather than relatively empty lands in 1492. In the words of the sixteenth-century Spanish priest, Bartolomé de las Casas, who knew the Indies well:

> All that has been discovered up to the year forty-nine [1549] is full of people, like a hive of bees, so that it seems as though God had placed all, or the greater part of the entire human race in these countries (Las Casas, in MacNutt 1909, 314).

Las Casas believed that more than 40 million Indians had died by the year 1560. Did he exaggerate? In the 1930s and 1940s, Alfred Kroeber, Angel Rosenblat, and Julian Steward believed that he had. The best counts then available indicated a population of between 8–15 million Indians in the Americas. Subsequently, Carl Sauer, Woodrow Borah, Sherburne F. Cook, Henry Dobyns, George Lovell, N. David Cook, myself, and others have argued for larger estimates. Many scholars now believe that there were between 40–100 million Indians in the hemisphere (Denevan 1992). This conclusion is primarily based on evidence of rapid early declines from epidemic disease prior to the first population counts (Lovell 1992).

I have recently suggested a New World total of 53.9 million (Denevan 1992, xxvii). This divides into 3.8 million for North America, 17.2 million for Mexico, 5.6 million for Central America, 3.0 million for the Caribbean,

15.7 million for the Andes, and 8.6 million for lowland South America. These figures are based on my judgment as to the most reasonable recent tribal and regional estimates. Accepting a margin of error of about 20 percent, the New World population would lie between 43–65 million. Future regional revisions are likely to maintain the hemispheric total within this range. Other recent estimates, none based on totaling regional figures, include 43 million by Whitmore (1991, 483), 40 million by Lord and Burke (1991), 40–50 million by Cowley (1991), and 80 million for just Latin America by Schwerin (1991, 40). In any event, a population between 40–80 million is sufficient to dispel any notion of "empty lands." Moreover, the native impact on the landscape of 1492 reflected not only the population then but the cumulative effects of a growing population over the previous 15,000 years or more.

European entry into the New World abruptly reversed this trend. The decline of native American populations was rapid and severe, probably the greatest demographic disaster ever (Lovell 1992). Old World diseases were the primary killer. In many regions, particularly the tropical lowlands, populations fell by 90 percent or more in the first century after contact. Indian populations (estimated) declined in Hispaniola from 1 million in 1492 to a few hundred 50 years later, or by more than 99 percent; in Peru from 9 million in 1520 to 670,000 in 1620 (92 percent); in the Basin of Mexico from 1.6 million in 1519 to 180,000 in 1607 (89 percent); and in North America from 3.8 million in 1492 to 1 million in 1800 (74 percent). An overall drop from 53.9 million in 1492 to 5.6 million in 1650 amounts to an 89 percent reduction (Denevan 1992, xvii–xxix). The human landscape was affected accordingly, although there is not always a direct relationship between population density and human impact (Whitmore et al. 1990, 37).

The replacement of Indians by Europeans and Africans was initially a slow process. By 1638 there were only about 30,000 English in North America (Sale 1990, 388), and by 1750 there were only 1.3 million Europeans and slaves (Meinig 1986, 247). For Latin America in 1750, Sánchez-Albornoz (1974, 7) gives a total (including Indians) of 12 million. For the hemisphere in 1750, the *Atlas of World Population History* reports 16 million (McEvedy and Jones 1978, 270). Thus the overall hemispheric population in 1750 was about 30 percent of what it may have been in 1492. The 1750 population, however, was very unevenly distributed, mainly located in certain coastal and highland areas with little Europeanization elsewhere. In

North America in 1750, there were only small pockets of settlement beyond the coastal belt, stretching from New England to northern Florida (see maps in Meinig 1986, 209, 245). Elsewhere, combined Indian and European populations were sparse, and environmental impact was relatively minor.

Indigenous imprints on landscapes at the time of initial European contact varied regionally in form and intensity. Following are examples for vegetation and wildlife, agriculture, and the built landscape.

VEGETATION

THE EASTERN FORESTS

The forests of New England, the Midwest, and the Southeast had been disturbed to varying degrees by Indian activity prior to European occupation. Agricultural clearing and burning had converted much of the forest into successional (fallow) growth and into semi-permanent grassy openings (meadows, barrens, plains, glades, savannas, prairies), often of considerable size.[5] Much of the mature forest was characterized by an open, herbaceous understory, reflecting frequent ground fires. "The de Soto expedition, consisting of many people, a large horse herd, and many swine, passed through ten states without difficulty of movement" (Sauer 1971, 283). The situation has been described in detail by Michael Williams in his recent history of American forests: "Much of the 'natural' forest remained, but the forest was not the vast, silent, unbroken, impenetrable and dense tangle of trees beloved by many writers in their romantic accounts of the forest wilderness" (1989, 33).[6] "The result was a forest of large, widely spaced trees, few shrubs, and much grass and herbage.... Selective Indian burning thus promoted the mosaic quality of New England ecosystems, creating forests in many different states of ecological succession" (Cronon 1983, 49–51). The extent, frequency, and impact of Indian burning is not without controversy. Raup (1937) argued that climatic change rather than Indian burning could account for certain vegetation changes. Emily Russell (1983, 86), assessing pre-1700 information for the Northeast, concluded that: "There is no strong evidence that Indians purposely burned large areas," but Indians did "increase the frequency of fires above the low numbers caused by lightning," creating an open forest. But then Russell adds: "In most areas climate and soil probably played the major role in determining the precolonial forests." She regards Indian fires as mainly accidental and "merely" augmental to natural fires, and she discounts the reliability of many early accounts of burning.

Forman and Russell (1983, 5) expand the argument to North America in general: "regular and widespread Indian burning (Day 1953) [is] an unlikely hypothesis that regretfully has been accepted in the popular literature and consciousness." This conclusion, I believe, is unwarranted given reports of the extent of prehistoric human burning in North America and Australia (Lewis 1982), and Europe (Patterson and Sassaman 1988, 130), and by my own and other observations on current Indian and peasant burning in Central America and South America; when unrestrained, people burn frequently and for many reasons. For the Northeast, Patterson and Sassaman (1988, 129) found that sedimentary charcoal accumulations were greatest where Indian populations were greatest.

Elsewhere in North America, the Southeast is much more fire prone than is the Northeast, with human ignitions being especially important in winter (Taylor 1981). The Berkeley geographer and Indianist Erhard Rostlund (1957, 1960) argued that Indian clearing and burning created many grasslands within mostly open forest in the so-called "prairie belt" of Alabama. As improbable as it may seem, Lewis (1982) found Indian burning in the subarctic, and Dobyns (1981) in the Sonoran desert. The characteristics and impacts of fires set by Indians varied regionally and locally with demography, resource management techniques, and environment, but such fires clearly had different vegetation impacts than did natural fires owing to differences in frequency, regularity, and seasonality.

FOREST COMPOSITION

In North America, burning not only maintained open forest and small meadows but also encouraged fire-tolerant and sun-loving species. "Fire created conditions favorable to strawberries, blackberries, raspberries, and other gatherable foods" (Cronon 1983, 51). Other useful plants were saved, protected, planted, and transplanted, such as American chestnut, Canada plum, Kentucky coffee tree, groundnut, and leek (Day 1953, 339–40). Gilmore (1931) described the dispersal of several native plants by Indians. Mixed stands were converted to single species dominants, including various pines and oaks, sequoia, Douglas fir, spruce, and aspen (M. Williams 1989, 47–48). The longleaf, slash pine, and scrub oak forests of the Southeast are almost certainly an anthropogenic subclimax created originally by Indian burning, replaced in early Colonial times by mixed hardwoods, and maintained in part by fires set by subsequent farmers and woodlot owners (Garren 1943). Lightning fires can account for some fire-climax vegetation, but Indian burning would have extended and maintained such vegetation (Silver 1990, 17–19, 59–64).

Even in the humid tropics, where natural fires are rare, human fires can dramatically influence forest composition. A good example is the pine forests of Nicaragua (Denevan 1961). Open pine stands occur both in the northern highlands (below 5,000 feet) and in the eastern (Miskito) lowlands, where warm temperatures and heavy rainfall generally favor mixed tropical montane forest or rainforest. The extensive pine forests of Guatemala and Mexico primarily grow in cooler and drier, higher elevations, where they are in large part natural and prehuman (Watts and Bradbury 1982, 59). Pine forests were definitely present in Nicaragua when Europeans arrived. They were found in areas where Indian settlement was substantial, but not in the eastern mountains where Indian densities were sparse. The eastern boundary of the highland pines seems to have moved with an eastern settlement frontier that has fluctuated back and forth since prehistory. The pines occur today where there has been clearing followed by regular burning and the same is likely in the past. The Nicaraguan pines are fire tolerant once mature, and large numbers of seedlings survive to maturity if they can escape fire during their first three to seven years (Denevan 1961, 280). Where settlement has been abandoned and fire ceases, mixed hardwoods gradually replace pines. This succession is likely similar where pines occur elsewhere at low elevations in tropical Central America, the Caribbean, and Mexico.

MIDWEST PRAIRIES AND TROPICAL SAVANNAS

Sauer (1950, 1958, 1975) argued early and often that the great grasslands and savannas of the New World were of anthropogenic rather than climatic origin, that rainfall was generally sufficient to support trees. Even nonagricultural Indians expanded what may have been pockets of natural, edaphic grasslands at the expense of forest. A fire burning to the edge of a grass/forest boundary will penetrate the drier forest margin and push back the edge, even if the forest itself is not consumed (Mueller-Dombois 1981, 164). Grassland can therefore advance significantly in the wake of hundreds of years of annual fires. Lightning-set fires can have a similar impact, but more slowly if less frequent than human fires, as in the wet tropics.

The thesis of prairies as fire induced, primarily by Indians, has its critics (Borchert 1950; Wedel 1957), but the recent review of the topic by Anderson (1990, 14), a biologist, concludes that most ecologists now believe that the eastern prairies "would have mostly disappeared if it had not been for the nearly annual burning of these grasslands by the North American Indians," during the last 5,000 years. A case in point is the nineteenth-century invasion of many grasslands by forests after fire had been suppressed in Wisconsin, Illinois, Kansas, Nebraska, and elsewhere (M. Williams 1989, 46).

The large savannas of South America are also controversial as to origin. Much, if not most of the open vegetation of the Orinoco Llanos, the Llanos de Mojos of Bolivia, the Pantanal of Mato Grosso, the Bolivar savannas of Colombia, the Guayas savannas of coastal Ecuador, the *campo cerrado* of central Brazil, and the coastal savannas north of the Amazon, is of natural origin. The vast *campos cerrados* occupy extremely senile, often toxic oxisols. The seasonally inundated savannas of Bolivia, Brazil, Guayas, and the Orinoco owe their existence to the intolerance of woody species to the extreme alternation of lengthy flooding or waterlogging and severe desiccation during a long dry season. These savannas, however, were and are burned by Indians and ranchers, and such fires have expanded the savannas into the forests to an unknown extent. It is now very difficult to determine where a natural forest/savanna boundary once was located (Hills and Randall 1968; Medina 1980).

Other small savannas have been cut out of the rainforest by Indian farmers and then maintained by burning. An example is the Gran Pajonal in the Andean foothills in east-central Peru, where dozens of small grasslands (*pajonales*) have been created by Campa Indians—a process clearly documented by air photos (Scott 1978). *Pajonales* were in existence when the region was first penetrated by Franciscan missionary explorers in 1733.

The impact of human activity is nicely illustrated by vegetational changes in the basins of the San Jorge, Cauca, and Sinú rivers of northern Colombia. The southern sector, which was mainly savanna when first observed in the sixteenth century, had reverted to rainforest by about 1750 following Indian decline, and had been reconverted to savanna for pasture by 1950 (Gordon 1957, map p. 69). Sauer (1966, 285–88; 1975, 8) and Bennett

(1968, 53–55) cite early descriptions of numerous savannas in Panama in the sixteenth century. Balboa's first view of the Pacific was from a "treeless ridge," now probably forested. Indian settlement and agricultural fields were common at the time, and with their decline the rainforest returned.

ANTHROPOGENIC TROPICAL RAIN FOREST

The tropical rain forest has long had a reputation for being pristine, whether in 1492 or 1992. There is, however, increasing evidence that the forests of Amazonia and elsewhere are largely anthropogenic in form and composition. Sauer (1958, 105) said as much at the Ninth Pacific Science Congress in 1957 when he challenged the statement of tropical botanist Paul Richards that, until recently, the tropical forests have been largely uninhabited, and that prehistoric people had "no more influence on the vegetation than any of the other animal inhabitants." Sauer countered that Indian burning, swiddens, and manipulation of composition had extensively modified the tropical forest.

"Indeed, in much of Amazonia, it is difficult to find soils that are not studded with charcoal" (Uhl et al. 1990, 30). The question is, to what extent does this evidence reflect Indian burning in contrast to natural (lightning) fires, and when did these fires occur? The role of fire in tropical forest ecosystems has received considerable attention in recent years, partly as a result of major wild fires in East Kalimantan in 1982–83 and small forest fires in the Venezuelan Amazon in 1980–84 (Goldammer 1990). Lightning fires, though rare in moist tropical forest, do occur in drier tropical woodlands (Mueller-Dombois 1981, 149). Thunderstorms with lightning are much more common in the Amazon, compared to North America, but in the tropics lightning is usually associated with heavy rain and noncombustible, verdant vegetation. Hence Indian fires undoubtedly account for most fires in prehistory, with their impact varying with the degree of aridity.

In the Río Negro region of the Colombian-Venezuelan Amazon, soil charcoal is very common in upland forests. C-14 dates range from 6260–250 B.P., well within human times (Saldarriaga and West 1986). Most of the charcoal probably reflects local swidden burns; however, there are some indications of forest fires at intervals of several hundred years, most likely ignited by swidden fires. Recent wild fires in the upper Río Negro region were in a normally moist tropical forest (3530 mm annual rainfall) that had experienced several years of severe drought. Such infrequent wild fires in prehistory, along with the more frequent ground fires, could have had significant impacts on forest succession, structure, and composition. Examples are the pine forests of Nicaragua, mentioned above, the oak forests of Central America, and the babassu palm forests of eastern Brazil. Widespread and frequent burning may have brought about the extinction of some endemic species.

The Amazon forest is a mosaic of different ages, structure, and composition resulting from local habitat conditions and disturbance dynamics (Haffer 1991). Natural disturbances (tree falls, landslides, river activity) have been considerably augmented by human activity, particularly by shifting cultivation. Even a small number of swidden farmers can have a widespread impact in a relatively short period of time. In the Río Negro region, species-diversity recovery takes 60–80 years and biomass recovery 140–200 years (Saldarriaga and Uhl 1991, 312). Brown and Lugo (1990, 4) estimate that today about forty percent of the tropical forest in Latin America is secondary as a result of human clearing and that most of the remainder has had some modification despite current low population densities. The species composition of early stages of swidden fallows differs from that of natural gaps and may "alter the species composition of the mature forest on a long-term scale" (Walschburger and von Hildebrand 1991, 262). While human environmental destruction in Amazonia currently is concentrated along roads, in prehistoric times Indian activity in the upland (interflueve) forests was much less intense but more widespread (Denevan, forthcoming).

Indian modification of tropical forests is not limited to clearing and burning. Large expanses of Latin American forests are humanized forests in which the kinds, numbers, and distributions of useful species are managed by human populations. Doubtless, this applies to the past as well. One important mechanism in forest management is manipulation of swidden fallows (sequential agroforestry) to increase useful species. The planting, transplanting, sparing, and protection of useful wild, fallow plants eliminates clear distinctions between field and fallow (Denevan and Padoch 1988). Abandonment is a slow process, not an event. Gordon (1982, 79–98) describes managed regrowth vegetation in eastern Panama, which he believes extended from Yucatán to northern Colombia in pre-European times. The Huastec of eastern Mexico and the Yucatec Maya have similar forms of forest gardens or forest management (Alcorn 1981;

Gómez-Pompa 1987). The Kayapo of the Brazilian Amazon introduce and/or protect useful plants in activity areas ("nomadic agriculture") adjacent to villages or camp sites, in foraging areas, along trails, near fields, and in artificial forest-mounds in savannas (Posey 1985). In managed forests, both annuals and perennials are planted or transplanted, while wild fruit trees are particularly common in early successional growth. Weeding by hand was potentially more selective than indiscriminate weeding by machete (Gordon 1982, 57–61). Much dispersal of edible plant seeds is unintentional via defecation and spitting out.

The economic botanist William Balée (1987, 1989) speaks of "cultural" or "anthropogenic" forests in Amazonia in which species have been manipulated, often without a reduction in natural diversity. These include specialized forests (babassu, Brazil nuts, lianas, palms, bamboo), which currently make up at least 11.8 percent (measured) of the total upland forest in the Brazilian Amazon (Balée 1989, 14). Clear indications of past disturbance are the extensive zones of *terra preta* (black earth), which occur along the edges of the large floodplains as well as in the uplands (Balée 1989, 10–12; Smith 1980). These soils, with depths to 50 cm or more, contain charcoal and cultural waste from prehistoric burning and settlement. Given high carbon, nitrogen, calcium, and phosphorus content, *terra preta* soils have a distinctive vegetation and are attractive to farmers. Balée (1989, 14) concludes that "large portions of Amazonian forests appear to exhibit the continuing effects of past human interference." The same argument has been made for the Maya lowlands (Gómez-Pompa et al. 1987) and Panama (Gordon 1982). There are no virgin tropical forests today, nor were there in 1492.

WILDLIFE

The indigenous impact on wildlife is equivocal. The thesis that "overkill" hunting caused the extinction of some large mammals in North America during the late Pleistocene, as well as subsequent local and regional depletions (Martin 1978, 167–72), remains controversial. By the time of the arrival of Cortéz in 1519, the dense populations of Central Mexico apparently had greatly reduced the number of large game, given reports that "they eat any living thing" (Cook and Borah 1971–79, 3:135, 140). In Amazonia, local game depletion apparently increases with village size and duration (Good 1987). Hunting procedures in many regions seem, however, to have allowed for recovery because of the "resting" of hunting zones intentionally or as a result of shifting of village sites.

On the other hand, forest disturbance increased herbaceous forage and edge effect, and hence the numbers of some animals (Thompson and Smith 1970, 261–64). "Indians created ideal habitats for a host of wildlife species . . . exactly those species whose abundance so impressed English colonists: elk, deer, beaver, hare, porcupine, turkey, quail, ruffed grouse, and so on" (Cronon 1983, 51). White-tailed deer, peccary, birds, and other game increases in swiddens and fallows in Yucatán and Panama (Greenberg 1991; Gordon 1982, 96–112; Bennett 1968). Rostlund (1960, 407) believed that the creation of grassy openings east of the Mississippi extended the range of the bison, whose numbers increased with Indian depopulation and reduced hunting pressure between 1540–1700, and subsequently declined under White pressure.

AGRICULTURE

FIELDS AND ASSOCIATED FEATURES

To observers in the sixteenth century, the most visible manifestation of the Native American landscape must have been the cultivated fields, which were concentrated around villages and houses. Most fields are ephemeral, their presence quickly erased when farmers migrate or die, but there are many eye-witness accounts of the great extent of Indian fields. On Hispaniola, Las Casas and Oviedo reported individual fields with thousands of *montones* (Sturtevant 1961, 73). These were manioc and sweet potato mounds 3–4 m in circumference, of which apparently none have survived. In the Llanos de Mojos in Bolivia, the first explorers mentioned *percheles,* or corn cribs on pilings, numbering up to 700 in a single field, each holding 30–45 bushels of food (Denevan 1966, 98). In northern Florida in 1539, Hernando de Soto's army passed through numerous fields of maize, beans, and squash, their main source of provisions; in one sector, "great fields . . . were spread out as far as the eye could see across two leagues of the plain" (Garcilaso de la Vega 1980, 2:182; also see Dobyns 1983, 135–46).

It is difficult to obtain a reliable overview from such descriptions. Aside from possible exaggeration, Europeans tended not to write about field size, production, or technology. More useful are various forms of relict fields and field features that persist for centuries and can still be recognized, measured, and excavated today. These extant features, including terraces, irrigation works, raised fields, sunken fields, drainage ditches, dams, reservoirs, diversion walls, and field borders number in the millions and are distributed throughout the Americas (Denevan 1980). For example, about 500,000 ha of abandoned raised fields survive in the San Jorge Basin of northern Colombia (Plazas and Falchetti 1987, 485), and at least 600,000 ha of terracing, mostly of prehistoric origin, occur in the Peruvian Andes (Denevan 1988, 20). There are 19,000 ha of visible raised fields in just the sustaining area of Tiwanaku at Lake Titicaca (Kolata 1991, 109) and there were about 12,000 ha of *chinampas* (raised fields) around the Aztec capital of Tenochtitlán (Sanders et al. 1979, 390). Complex canal systems on the north coast of Peru and in the Salt River Valley in Arizona irrigated more land in prehistory than is cultivated today. About 175 sites of Indian garden beds, up to several hundred acres each, have been reported in Wisconsin (Gartner 1992). These various remnant fields probably represent less than 25 percent of what once existed, most being buried under sediment or destroyed by erosion, urbanization, plowing, and bulldozing. On the other hand, an inadequate effort has been made to search for ancient fields.

EROSION

The size of native populations, associated deforestation, and prolonged intensive agriculture led to severe land degradation in some regions. Such a landscape was that of Central Mexico, where by 1519 food production pressures may have brought the Aztec civilization to the verge of collapse even without Spanish intervention (Cook and Borah 1971–79, 3:129–76).[7] There is good evidence that severe soil erosion was already widespread, rather than just the result of subsequent European plowing, livestock, and deforestation. Cook examined the association between erosional severity (gullies, barrancas, sand and silt deposits, and sheet erosion) and pre-Spanish population density or proximity to prehistoric Indian towns. He concluded that "an important cycle of erosion and deposition therefore accompanied intensive land use by huge primitive populations in central Mexico, and had gone far toward the devastation of the country before the white man arrived" (Cook 1949, 86).

Barbara Williams (1972, 618) describes widespread *tepetate,* an indurated substrate formation exposed by sheet erosion resulting from prehistoric agriculture, as "one of the dominant surface materials in the Valley of Mexico." On the other hand, anthropologist Melville (1990, 27) argues that soil erosion in the Valle de Mezquital, just north of the Valley of Mexico, was the result of overgrazing by Spanish livestock starting before 1600: "there is an almost total lack of evidence of environmental degradation before the last three decades of the sixteenth century." The Butzers, however, in an examination of Spanish land grants, grazing patterns, and soil and vegetation ecology, found that there was only light intrusion of Spanish livestock (sheep and cattle were moved frequently) into the southeastern Bajío near Mezquital until after 1590 and that any degradation in 1590 was "as much a matter of long-term Indian land use as it was of Spanish intrusion" (Butzer and Butzer, forthcoming). The relative roles of Indian and early Spanish impacts in Mexico still need resolution; both were clearly significant but varied in time and place. Under the Spaniards, however, even with a greatly reduced population, the landscape in Mexico generally did not recover due to accelerating impacts from introduced sheep and cattle.[8]

THE BUILT LANDSCAPE

SETTLEMENT

The Spaniards and other Europeans were impressed by large flourishing Indian cities such as Tenochtitlán, Quito, and Cuzco, and they took note of the extensive ruins of older, abandoned cities such as Cahokia, Teotihuacán, Tikal, Chan Chan, and Tiwanaku (Hardoy 1968). Most of these cities contained more than 50,000 people. Less notable, or possibly more taken for granted, was rural settlement—small villages of a few thousand or a few hundred people, hamlets of a few families, and dispersed farmsteads. The numbers and locations of much of this settlement will never be known. With the rapid decline of native populations, the abandonment of houses and entire villages and the decay of perishable materials quickly obscured sites, especially in the tropical lowlands.

We do have some early listings of villages, especially for Mexico and Peru. Elsewhere, archaeology is telling us more than ethnohistory. After initially focusing on large temple and administrative centers, archaeologists are now examining rural sustaining areas, with remarkable results. See, for example, Sanders et al. (1979) on the Basin of Mexico, Culbert and Rice (1990) on the Maya lowlands, and Fowler (1989) on Cahokia in Illinois. Evidence of human occupation for the artistic Santarém Culture phase (Tapajós chiefdom) on the lower Amazon extends over thousands of square kilometers, with large nucleated settlements (Roosevelt 1991, 101–2).

Much of the rural precontact settlement was semi-dispersed (*rancherías*), particularly in densely populated regions of Mexico and the Andes, probably reflecting poor food transport efficiency. Houses were both single-family and communal (pueblos, Huron long houses, Amazon malocas). Construction was of stone, earth, adobe, daub and wattle, grass, hides, brush, and bark. Much of the dispersed settlement not destroyed by depopulation was concentrated by the Spaniards into compact grid/plaza style new towns (*congregaciones, reducciones*) for administrative purposes.

MOUNDS

James Parsons (1985, 161) has suggested that: "An apparent mania for earth moving, landscape engineering on a grand scale runs as a thread through much of New World prehistory." Large quantities of both earth and stone were transferred to create various raised and sunken features, such as agricultural landforms, settlement and ritual mounds, and causeways.

Mounds of different shapes and sizes were constructed throughout the Americas for temples, burials, settlement, and as effigies. The stone pyramids of Mexico and the Andes are well known, but equal monuments of earth were built in the Amazon, the Midwest U.S., and elsewhere. The Mississippian period complex of 104 mounds at Cahokia near East St. Louis supported 30,000 people; the largest, Monk's Mound, is currently 30.5 m high and covers 6.9 ha (Fowler 1989, 90, 192). Cahokia was the largest settlement north of the Río Grande until surpassed by New York City in 1775. An early survey estimated "at least 20,000 conical, linear, and effigy mounds" in Wisconsin (Stout 1911, 24). Overall, there must have been several hundred thousand artificial mounds in the Midwest and South. De Soto described such features still in use in 1539 (Silverberg 1968, 7). Thousands of settlement and other mounds dot the savanna landscape of Mojos in Bolivia (Denevan 1966). At the mouth of the Amazon on Marajó Island, one complex of forty habitation mounds contained more than 10,000 people; one of these mounds is 20 m high while another is 90 ha in area (Roosevelt 1991, 31, 38).

Not all of the various earthworks scattered over the Americas were in use in 1492. Many had been long abandoned, but they constituted a conspicuous element of the landscape of 1492 and some are still prominent. Doubtless, many remain to be discovered, and others remain unrecognized as human or prehistoric features.

ROADS, CAUSEWAYS, AND TRAILS

Large numbers of people and settlements necessitated extensive systems of overland travel routes to facilitate administration, trade, warfare, and social interaction (Hyslop 1984; Trombold 1991). Only hints of their former prominence survive. Many were simple traces across deserts or narrow paths cut into forests. A suggestion as to the importance of Amazon forest trails is the existence of more than 500 km of trail maintained by a single Kayapó village today (Posey 1985, 149). Some prehistoric footpaths were so intensively used for so long that they were incised into the ground and are still detectable, as has recently been described in Costa Rica (Sheets and Sever 1991).

Improved roads, at times stone-lined and drained, were constructed over great distances in the realms of the high civilizations. The Inca road network is estimated to have measured about 40,000 km, extending from southern Colombia to central Chile (Hyslop 1984, 224). Prehistoric causeways (raised roads) were built in the tropical lowlands (Denevan 1991); one Maya causeway is 100 km long, and there are more than 1,600 km of causeways in the Llanos de Mojos. Humboldt reported large prehistoric causeways in the Orinoco Llanos. Ferdinand Columbus described roads on Puerto Rico in 1493. Gaspar de Carvajal, traveling down the Amazon with Orellana in 1541, reported "highways" penetrating the forest from river bank villages. Joseph de Acosta (1880, 61:171) in 1590 said that between Peru and Brazil, there were "waies as much beaten as those betwixt Salamanca and Valladolid." Prehistoric roads in Chaco Canyon, New Mexico are described in Trombold (1991). Some routes were so well established and located that they have remained roads to this day.

RECOVERY

A strong case can be made for significant environmental recovery and reduction of cultural features by the late eighteenth century as a result of Indian population decline. Henry Thoreau (1949, 132–37) believed, based on his reading of William Wood, that the New England forests of 1633 were more open, more park-like, with more berries and more wildlife, than Thoreau observed in 1855. Cronon (1983, 108), Pyne (1982, 51), Silver (1990, 104), Martin (1978, 181–82), and M. Williams (1989, 49) all maintain that the eastern forests recovered and filled in as a result of Indian depopulation, field abandonment, and reduction in burning. While probably correct, these writers give few specific examples, so further research is needed. The sixteenth-century fields and savannas of Colombia and Central America also had reverted to forest within 150 years after abandonment (Parsons 1975, 30–31; Bennett 1968, 54). On his fourth voyage in 1502–03, Columbus sailed along the north coast of Panama (Veragua). His son Ferdinand described lands which were well-peopled, full of houses, with many fields, and open with few trees. In contrast, in 1681 Lionel Wafer found most of the Caribbean coast of Panama forest covered and unpopulated. On the Pacific side in the eighteenth century, savannas were seldom mentioned; the main economic activity was the logging of tropical cedar, a tree that grows on the sites of abandoned fields and other disturbances (Sauer 1966, 132–33, 287–88). An earlier oscillation from forest destruction to recovery in the Yucatán is instructive. Whitmore et al. (1990, 35) estimate that the Maya had modified 75 percent of the environment by A.D. 800, and that following the Mayan collapse, forest recovery in the central lowlands was nearly complete when the Spaniards arrived.

The pace of forest regeneration, however, varied across the New World. Much of the southeastern U.S. remained treeless in the 1750s according to Rostlund (1957, 408, 409). He notes that the tangled brush that ensnarled the "Wilderness Campaign of 1864 in Virginia occupied the same land as did Captain John Smith's 'open groves with much good ground between without any shrubs'" in 1624; vegetation had only partially recovered over 240 years. The Kentucky barrens in contrast were largely reforested by the early nineteenth century (Sauer 1963, 30). The Alabama Black Belt vegetation was described by William Bartram in the 1770s as a mixture of forest and grassy plains, but by the nineteenth century, there was only 10 percent prairie and even less in some counties (Rostlund 1957, 393, 401–03). Sections of coastal forests never recovered, given colonist pressures, but Sale's (1990, 291) claim, that "the English were well along in the process of eliminating the ancient Eastern woodlands from Maine to the Mississippi" in the first one hundred years, is an exaggeration.

Wildlife also partially recovered in eastern North America with reduced hunting pressure from Indians; however, this is also a story yet to be worked out. The white-tailed deer apparently declined in numbers, probably reflecting reforestation plus competition from livestock. Commercial hunting was a factor on the coast, with 80,000 deer skins being shipped out yearly from Charleston by 1730 (Silver 1990, 92). Massachusetts enacted a closed season on deer as early as 1694, and in 1718 there was a three-year moratorium on deer hunting (Cronon 1983, 100). Sale (1990, 290) believes that beaver were depleted in the Northeast by 1640. Other fur bearers, game birds, elk, buffalo, and carnivores were also targeted by white hunters, but much game probably was in the process of recovery in many eastern areas until a general reversal after 1700–50.

As agricultural fields changed to scrub and forest, earthworks were grown over. All the raised fields in Yucatán and South America were abandoned. A large portion of the agricultural terraces in the Americas were abandoned in the early colonial period (Donkin 1979, 35–38). In the Colca Valley of Peru, measurement on air photos indicates 61 percent terrace abandonment (Denevan 1988, 28). Societies vanished or declined everywhere and whole villages with them. The degree to which settlement features were swallowed up by vegetation, sediment, and erosion is indicated by the difficulty of finding them today. Machu Picchu, a late prehistoric site, was not rediscovered until 1911.

The renewal of human impact also varied regionally, coming with the Revolutionary War in North America, with the rubber boom in Amazonia, and with the expansion of coffee in southern Brazil (1840–1930). The swamp lands of Gulf Coast Mexico and the Guayas Basin of Ecuador remained hostile environments to Europeans until well into the nineteenth century or later (Siemens 1990; Mathewson 1987). On the other hand, Highland Mexico-Guatemala and the Andes, with greater Indian survival and with the establishment of haciendas and intensive mining, show less evidence of environmental recovery. Similarly, Indian fields in the Caribbean were rapidly replaced by European livestock and sugar plantation systems, inhibiting any sufficient recovery. The same is true of the sugar zone of coastal Brazil.

CONCLUSIONS

By 1492, Indian activity had modified vegetation and wildlife, caused erosion, and created earthworks, roads, and settlements throughout the Americas. This may be obvious, but the human imprint was much more ubiquitous and enduring than is usually realized. The historical evidence is ample, as are data from surviving earthworks and archaeology. And much can be inferred from present human impacts. The weight of evidence suggests that Indian populations were large, not only in Mexico and the Andes, but also in seemingly unattractive habitats such as the rainforests of Amazonia, the swamps of Mojos, and the deserts of Arizona.

Clearly, the most humanized landscapes of the Americas existed in those highland regions where people were the most numerous. Here were the large states, characterized by urban centers, road systems, intensive agriculture, a dispersed but relatively dense rural settlement pattern of hamlets and farmsteads, and widespread vegetation and soil modification and wildlife depletion. There were other, smaller regions that shared some of these characteristics, such as the Pueblo lands in the southwestern U.S., the Sabana de Bogotá in highland Colombia, and the central Amazon floodplain, where built landscapes were locally dramatic and are still observable. Finally, there were the immense grasslands, deserts, mountains, and forests elsewhere, with populations that were sparse or moderate, with landscape impacts that mostly were ephemeral or not obvious but nevertheless significant, particularly for vegetation and wildlife, as in Amazonia and the northeastern U.S. In addition, landscapes from the more distant past survived to 1492 and even to 1992, such as those of the irrigation states of north coast Peru, the Classic Maya, the Mississippian mound builders, and the Tiwanaku Empire of Lake Titicaca.

This essay has ranged over the hemisphere, an enormous area, making generalizations about and providing examples of Indian landscape transformation as of 1492. Ideally, a series of hemispheric maps should be provided to portray the spatial patterns of the different types of impacts and cultural features, but such maps are not feasible nor would they be accurate given present knowledge. There are a few relevant regional maps, however, that can be referred to. For example, see Butzer (1990, 33, 45) for Indian settlement structures/mounds and subsistence patterns in the U.S.; Donkin (1979, 23) for agricultural terracing; Doolittle (1990, 109) for canal irrigation in Mexico; Parsons and Denevan (1967) for raised fields in South America; Trombold (1991) for various road networks; Hyslop (1984, 4) for the Inca roads; Hardoy (1968, 49) for the most intense urbanization in Latin America; and Gordon (1957, 69) for anthropogenic savannas in northern Colombia.

The pristine myth cannot be laid at the feet of Columbus. While he spoke of "Paradise," his was clearly a humanized paradise. He described Hispaniola and Tortuga as densely populated and "completely cultivated like the countryside around Cordoba" (Colón 1976, 165). He also noted that "the islands are not so thickly wooded as to be impassable," suggesting openings from clearing and burning (Columbus 1961, 5).

The roots of the pristine myth lie in part with early observers unaware of human impacts that may be obvious to scholars today, particularly for vegetation and wildlife.[9] But even many earthworks such as raised fields have only recently been discovered (Denevan 1966; 1980). Equally important, most of our eyewitness descriptions of wilderness and empty lands come from a later time, particularly 1750–1850, when interior lands began to be explored and occupied by Europeans. By 1650, Indian populations in the hemisphere had been reduced by about 90 percent, while by 1750 European numbers were not yet substantial and settlement had only begun to expand. As a result, fields had been abandoned, while settlements vanished, forests recovered, and savannas retreated. The landscape did appear to be a sparsely populated wilderness. This is the image conveyed by Parkman in the nineteenth century, Bakeless in 1950, and Shetler as recently as 1991. There was some European impact, of course, but it was localized. After 1750 and especially after 1850, populations greatly expanded, resources were more intensively exploited, and European modification of the environment accelerated, continuing to the present.

It is possible to conclude not only that "the virgin forest was not encountered in the sixteenth and seventeenth centuries; [but that] it was invented in the late eighteenth and early nineteenth centuries" (Pyne 1982, 46). However, "paradoxical as it may seem, there was undoubtedly much more 'forest primeval' in 1850 than in 1650" (Rostlund 1957, 409). Thus the "invention" of an earlier wilderness is in part understandable and is not simply a deliberate creation which ennobled the American enterprise, as suggested by Bowden (1992, 20–23). In any event, while pre-European landscape alteration has been demonstrated previously, including by several geographers, the case has mainly been made for vegetation and mainly for eastern North America. As shown here, the argument is also applicable to most of the rest of the New World, including the humid tropics, and involves much more than vegetation.

The human impact on environment is not simply a process of increasing change or degradation in response to linear population growth and economic expansion. It is instead interrupted by periods of reversal and ecological rehabilitation as cultures collapse, populations decline, wars occur, and habitats are abandoned. Impacts may be constructive, benign, or degenerative (all subjective concepts), but change is continual at variable rates and in different directions. Even mild impacts and slow changes are cumulative, and the long-term effects can be dramatic. Is it possible that the thousands of years of human activity before Columbus created more change in the visible landscape than has occurred subsequently with European settlement and resource exploitation? The answer is probably yes for most regions for the next 250 years or so, and for some regions right up to the present time. American flora, fauna, and landscape were slowly Europeanized after 1492, but before that they had already been Indianized. "It is upon this imprint that the more familiar Euro-American landscape was grafted, rather than created anew" (Butzer 1990, 28). What does all this mean for protectionist tendencies today? Much of what is protected or proposed to be protected from human disturbance had native people present, and environmental modification occurred accordingly and in part is still detectable.

The pristine image of 1492 seems to be a myth, then, an image more applicable to 1750, following Indian decline, although recovery had only been partial by that date. There is some substance to this argument, and it should hold up under the scrutiny of further investigation of the considerable evidence available, both written and in the ground.

NOTES

1. The field and library research that provided the background for this essay was undertaken over many years in Latin America, Berkeley, and Madison. Mentors who have been particularly influential are Carl O. Sauer, Erhard Rostlund, James J. Parsons, and Woodrow Borah, all investigators of topics discussed here.

2. Sauer had a life-long interest in this topic (1963, 1966, 1971, 1980).

3. See R. Nash (1967) on the "romantic wilderness" of America; Bowden (1992, 9–12) on the "invented tradition" of the "primeval forest" of New England; and Manthorne (1989, 10–21) on artists' images of the tropical "Eden" of South America. Day (1953, 329) provides numerous quotations from Parkman on "wilderness" and "vast," "virgin," and "continuous" forest.

4. For example, a 1991 advertisement for a Time-Life video refers to "the unspoiled beaches, forests, and mountains of an earlier America" and "the pristine shores of Chesapeake Bay in 1607."

5. On the other hand, the ability of Indians to clear large trees with inefficient stone axes, assisted by girdling and deadening by fire, may have been overestimated. See W. M. Denevan, "Stone vs. Metal Axes: The Ambiguity of Shifting Cultivation in Prehistoric Amazonia," *Journal of the Steward Anthropological Society,* forthcoming. Silver notes that the upland forests of Carolina were largely uninhabited for this reason. See T. Silver, *A New Face on the Countryside: Indians, Colonists, and Slaves in South Atlantic Forests, 1500–1800* (Cambridge: Cambridge University Press, 1990), p. 51.

6. Similar conclusions were reached by foresters Maxwell (1910) and Day (1953); by geographers Sauer (1963), Brown (1948, 11–19), Rostlund (1957), and Bowden (1992); and by environmental historians Pyne (1982, 45–51), Cronon (1983, 49–51), and Silver (1990, 59–66).

7. B. Williams (1989, 730) finds strong evidence of rural overpopulation (66 percent in poor crop years, 11 percent in average years) in the Basin of Mexico village of Asuncion, ca. A.D. 1540, which was probably "not unique but a widespread phenomenon." For a contrary conclusion, that the Aztecs did not exceed carrying capacity, see Ortiz de Montellano (1990, 119).

8. Highland Guatemala provides another prehistoric example of "severe human disturbance" involving deforestation and "massive" soil erosion (slopes) and deposition (valleys) (Murdy 1990, 186). For the central Andes there is some evidence that much of the *puna* zone (3200–4500 m), now grass and scrub, was deforested in prehistoric times (White 1985).

9. The English colonists in part justified their occupation of Indian land on the basis that such land had not been "subdued" and therefore was "land free to be taken" (Wilson 1992, 16).

BIBLIOGRAPHY

Acosta, Joseph [José] de. 1880 [1590]. *The natural and moral history of the Indies.* Trans. E. Gimston, Hakluyt Society, vols. 60, 61. London.

Alcorn, J. B. 1981. Huastec noncrop resource management: Implications for prehistoric rain forest management. *Human Ecology* 9:395–417.

Anderson, R. C. 1990. The historic role of fire in the North American grassland. In *Fire in North American tallgrass prairies,* ed. S. L. Collins and L. L. Wallace, pp. 8–18. Norman: University of Oklahoma Press.

Bakeless, J. 1950. *The eyes of discovery: The pageant of North America as seen by the first explorers.* New York: J. B. Lippincott.

Balée, W. 1987. Cultural forests of the Amazon. *Garden* 11:12–14, 32.

———. 1989. The culture of Amazonian forests. In *Advances in Economic Botany* 7: 1–21. New York: New York Botanical Garden.

Bennett, C. F. 1968. *Human influences on the zoogeography of Panama.* Ibero-Americana 51. Berkeley: University of California Press.

Borchert, J. 1950. Climate of the central North American grassland. *Annals of the Association of American Geographers* 40:1–39.

Bowden, M. J. 1992. The invention of American tradition. *Journal of Historical Geography* 18:3–26.

Brown, R. H. 1948. *Historical geography of the United States.* New York: Harcourt, Brace.

Brown, S., and Lugo, A. 1990. Tropical secondary forests. *Journal of Tropical Ecology* 6:1–32.

Butzer, K. W. 1990. The Indian legacy in the American landscape. In *The making of the American landscape,* ed. M. P. Conzen, pp. 27–50. Boston: Unwin Hyman.

———, and Butzer, E. K. Forthcoming. The sixteenth-century environment of the central Mexican Bajió: Archival reconstruction from Spanish land grants. In *Culture, form, and place,* ed. K. Mathewson. Baton Rouge, La.: Geoscience and Man.

Colón, C. 1976. *Diario del descubrimiento,* vol. 1, ed. M. Alvar. Madrid: Editorial La Muralla.

Columbus, C. 1961. *Four voyages to the New World: Letters and selected documents,* ed. R. H. Major. New York: Corinth Books.

Cook, S. F. 1949. *Soil erosion and population in Central Mexico.* Ibero-Americana 34. Berkeley: University of California Press.

———, and Borah, W. 1971–79. *Essays in population history.* 3 vols. Berkeley: University of California Press.

Cowley, G. 1991. The great disease migration. In *1492–1992, When worlds collide: How Columbus's voyages transformed both East and West. Newsweek,* Special Issue, Fall/Winter, pp. 54–56.

Cronon, W. 1983. *Changes in the land. Indians, colonists, and the ecology of New England.* New York: Hill and Wang.

Culbert, T. P., and Rice, D. S., eds. 1990. *Pre-Columbian population history in the Maya lowlands.* Albuquerque: University of New Mexico Press.

Day, G. M. 1953. The Indian as an ecological factor in the northeastern forest. *Ecology* 34:329–46.

Denevan, W. M. 1961. The upland pine forests of Nicaragua. *University of California Publications in Geography* 12:251–320.

———. 1966. *The aboriginal cultural geography of the Llanos de Mojos of Bolivia.* Ibero-Americana 48. Berkeley: University of California Press.

———. 1980. Tipología de configuraciones agrícolas prehispánicas. *América Indígena* 40:619–52.

———. 1988. Measurement of abandoned terracing from air photos: Colca Valley, Peru. *Yearbook, Conference of Latin Americanist Geographers* 14:20–30.

———. 1991. Prehistoric roads and causeways of lowland tropical America. In *Ancient road networks and settlement hierarchies in the New World,* ed. C. D. Trombold, pp. 230–42. Cambridge: Cambridge University Press.

———. ed. 1992 [1976]. The native population of the Americas in 1492, 2nd ed. Madison: University of Wisconsin Press.

———. Forthcoming. Stone vs. Metal axes: The ambiguity of shifting cultivation in prehistoric Amazonia. *Journal of the Steward Anthropological Society.*

———, and Padoch, C., eds. 1988. *Swidden-fallow agroforestry in the Peruvian Amazon. Advances in Economic Botany,* vol. 5. New York: New York Botanical Garden.

Dobyns, H. F. 1981. *From fire to flood. Historic human destruction of Sonoran Desert riverine oases.* Socorro, N. M.: Ballena Press.

———. 1983. *Their number become thinned. Native American population dynamics in eastern North America.* Knoxville: University of Tennessee Press.

Donkin, R. A. 1979. *Agricultural terracing in the aboriginal New World.* Viking Fund Publications in Anthropology 56. Tucson: University of Arizona Press.

Doolittle, W. E. 1990. *Canal irrigation in prehistoric Mexico: The sequence of technological change.* Austin: University of Texas Press.

Forman, R. T. T., and Russell, E. W. B. 1983. Evaluation of historical data in ecology. *Bulletin of the Ecological Society of America* 64:5–7.

Fowler, M. 1989. *The Cahokia atlas: A historical atlas of Cahokia archaeology.* Studies in Illinois Archaeology 6. Springfield: Illinois Historic Preservation Agency.

Garcilaso de la Vega, The Inca. 1980 [1605]. *The Florida of the Inca: A history of the Adelantado, Hernando de Soto.* 2 vols. Trans. and ed. J. G. Varner and J. J. Varner. Austin: University of Texas Press.

Garren, K. H. 1943. Effects of fire on vegetation of the southeastern United States. *The Botanical Review* 9:617–54.

Gartner, W. G. 1992. The Hulbert Creek ridged fields: Pre-Columbian agriculture near the Dells, Wisconsin. Master's thesis, Department of Geography, University of Wisconsin, Madison.

Gilmore, M. R. 1931. Dispersal by Indians a factor in the extension of discontinuous distribution of certain species of native plants. *Papers of the Michigan Academy of Science, Arts and Letters* 13:89–94.

Goldammer, J. G., ed. 1990. *Fire in the tropical biota: Ecosystem processes and global challenges.* Ecological Studies, vol. 84. Berlin: Springer-Verlag.

Gómez-Pompa, A. 1987. On Maya silviculture. *Mexican Studies* 3:1–17.

———; Salvador Flores, J.; and Sosa, V. 1987. The "pet kot" : A man-made forest of the Maya. *Interciencia* 12:10–15.

Good, K. R. 1987. Limiting factors in Amazonian ecology. In *Food and evolution: Toward a theory of human food habitats,* ed. M. Harris and E. B. Ross, pp. 407–21. Philadelphia: Temple University Press.

Gordon, B. L. 1957. *Human geography and ecology in the Sinú country of Colombia.* Ibero-Americana 39. Berkeley: University of California Press.

———. 1982. *A Panama forest and shore: Natural history and Amerindian culture in Bocas del Toro.* Pacific Grove: Boxwood Press.

Greenberg, L. S. C. 1991. Garden-hunting among the Yucatec Maya. *Ethnoecologica* 1:30–36.

Haffer, J. 1991. Mosaic distribution patterns of neotropical forest birds and underlying cyclic disturbance processes. In *The mosaic-cycle concept of ecosystems,* ed. H. Remmert, pp. 83–105. Ecological Studies, vol. 85. Berlin: Springer-Verlag.

Hardoy, J. 1968. *Urban planning in pre-Columbian America.* New York: George Braziler.

Hills, T. L., and Randall, R. E., eds. 1968. *The ecology of the forest/savanna boundary.* Savanna Research Series 13. Montreal: McGill University.

Hyslop, J. 1984. *The Inca road system.* New York: Academic Press.

Kolata, A. L. 1991. The technology and organization of agricultural production in the Tiwanaku state. *Latin American Antiquity* 2:99–125.

Lewis, H. T. 1982. Fire technology and resource management in aboriginal North America and Australia. In *Resource managers: North American and Australian hunter-gatherers,* ed. N. M. Williams and E. S. Hunn, pp. 45–67. AAAS Selected Symposia 67. Boulder, Col.: Westview Press.

Lord, L., and Burke, S. 1991. America before Columbus. *U.S. News and World Report,* July 8, pp. 22–37.

Lovell, George. 1992. Essay in *Annals of the Association of American Geographers* 82:369–85.

McEvedy, C., and Jones, R. 1978. *Atlas of world population history.* New York: Penguin Books.

MacNutt, F. A. 1909. *Bartholomew de las Casas: His life, his apostolate, and his writings.* New York: Putnam's.

Manthorne, K. E. 1989. *Tropical renaissance: North American artists exploring Latin America, 1839–1879.* Washington, D.C.: Smithsonian Institution Press.

Martin, C. 1978. *Keepers of the game: Indian-animal relationships and the fur trade.* Berkeley: University of California Press.

Mathewson, K. 1987. Landscape change and cultural persistence in the Guayas wetlands, Ecuador. Ph.D. dissertation, Department of Geography, University of Wisconsin, Madison.

Maxwell, H. 1910. The use and abuse of forests by the Virginia Indians. *William and Mary College Quarterly Historical Magazine* 19:73–103.

Medina, E. 1980. Ecology of tropical American savannas: An ecophysiological approach. In *Human ecology in savanna environments,* ed. D. R. Harris, pp. 297–319. London: Academic Press.

Meinig, D. W. 1986. *The shaping of America. A geographical perspective on 500 years of history,* vol. 1, *Atlantic America, 1492–1800.* New Haven: Yale University Press.

Melville, E. G. K. 1990. Environmental and social change in the Valle del Mezquital, Mexico, 1521–1600. *Comparative Studies in Society and History* 32:24–53.

Mueller-Dombois, D. 1981. Fire in tropical ecosystems. In *Fire regimes and ecosystem properties: Proceedings of the Conference,* Honolulu, 1978, pp. 137–76. General Technical Report WO-26. Washington, D.C.: U.S. Forest Service.

Murdy, C. N. 1990. Prehispanic agriculture and its effects in the valley of Guatemala. *Forest and Conservation History* 34:179–90.

Nash, R. 1967. *Wilderness and the American mind.* New Haven: Yale University Press.

Ortiz de Montellano, B. R. 1990. *Aztec medicine, health, and nutrition.* New Brunswick, N.J.: Rutgers University Press.

Parsons, J. J. 1975. The changing nature of New World tropical forests since European colonization. In *The use of ecological guidelines for development in the American humid tropics,* pp. 28–38. International Union for Conservation of Nature and Natural Resources Publications, n.s., 31. Morges.

———. 1985. Raised field farmers as pre-Columbian landscape engineers: Looking north from the San Jorge (Colombia). In *Prehistoric intensive agriculture in the tropics,* ed. I. S. Farrington, pp. 149–65. International Series 232. Oxford: British Archaeological Reports.

———, and Denevan, W. M. 1967. Pre-Columbian ridged fields. *Scientific American* 217(1):92–100.

Patterson III, W. A., and Sassaman, K. E. 1988. Indian fires in the prehistory of New England. In *Holocene human ecology in northeastern North America,* ed. G. P. Nicholas, pp. 107–35. New York: Plenum.

Plazas, C., and Falchetti, A. M. 1987. Poblamiento y adecuación hidráulica en el bajo Río San Jorge, Costa Atlántica, Colombia. In *Prehistoric agricultural fields in the Andean region,* ed. W. M. Denevan, K. Mathewson, and G. Knapp, pp. 483–503. International Series 359. Oxford: British Archaeological Reports.

Posey, D. A. 1985. Indigenous management of tropical forest ecosystems: The case of the Kayapó Indians of the Brazilian Amazon. *Agroforestry Systems* 3:139–58.

Pyne, S. J. 1982. *Fire in America: A cultural history of wildland and rural fire.* Princeton, N.J.: Princeton University Press.

Raup, H. M. 1937. Recent changes in climate and vegetation in southern New England and adjacent New York. *Journal of the Arnold Arboretum* 1–8:79–117.

Roosevelt, A. C. 1991. *Moundbuilders of the Amazon: Geophysical archaeology on Marajó Island, Brazil.* San Diego: Academic Press.

Rostlund, E. 1957. The myth of a natural prairie belt in Alabama: An interpretation of historical records. *Annals of the Association of American Geographers* 47:392–411.

———. 1960. The geographic range of the historic bison in the southeast. *Annals of the Association of American Geographers* 50:395–407.

Russell, E. W. B. 1983. Indian-set fires in the forests of the northeastern United States. *Ecology* 64:78–88.

Saldarriaga, J. G., and West, D. C. 1986. Holocene fires in the northern Amazon Basin. *Quaternary Research* 26:358–66.

———, and Uhl, C. 1991. Recovery of forest vegetation following slash-and-burn agriculture in the upper Río Negro. In *Rainforest regeneration and management,* ed. A. Gómez-Pompa, T. C. Whitmore, and M. Hadley, pp. 303–12. Paris: UNESCO.

Sale, K. 1990. *The conquest of paradise: Christopher Columbus and the Columbian legacy.* New York: Alfred A. Knopf.

Sánchez-Albornoz, N. 1974. *The population of Latin America: A history.* Berkeley: University of California Press.

Sanders, W. T.; Parsons, J. R.; and Santley, R. S. 1979. *The Basin of Mexico: Ecological processes in the evolution of a civilization.* New York: Academic Press.

Sauer, C. O. 1950. Grassland climax, fire, and man. *Journal of Range Management* 3:16–21.

———. 1958. Man in the ecology of tropical America. *Proceedings of the Ninth Pacific Science Congress, 1957* 20:104–10.

———. 1963 [1927]. The barrens of Kentucky. In *Land and life: A selection from the writings of Carl Ortwin Sauer,* ed. J. Leighly, pp. 23–31. Berkeley: University of California Press.

———. 1966. *The early Spanish Main.* Berkeley: University of California Press.

———. 1971. *Sixteenth-century North America: The land and the people as seen by the Europeans.* Berkeley: University of California Press.

———. 1975. Man's dominance by use of fire. *Geo-science and Man* 10:1–13.

———. 1980. *Seventeenth-century North America.* Berkeley: Turtle Island Press.

Schwerin, K. H. 1991. The Indian populations of Latin America. In *Latin America, its problems and its promise: A multidisciplinary introduction,* ed. J. K. Black, 2nd ed., pp. 39–53. Boulder, Col.: Westview Press.

Scott, G. A. J. 1978. *Grassland development in the Gran Pajonal of eastern Peru.* Hawaii Monographs in Geography 1. Honolulu: University of Hawaii.

Sheets, P., and Sever, T. L. 1991. Prehistoric footpaths in Costa Rica: Transportation and communication in a tropical rainforest. In *Ancient road networks and settlement hierarchies in the New World,* ed. C. D. Trombold, pp. 53–65. Cambridge: Cambridge University Press.

Shetler, S. 1991. Three faces of Eden. In *Seeds of change: A quincentennial commemoration,* ed. H. J. Viola and C. Margolis, pp. 225–47. Washington, D.C.: Smithsonian Institution Press.

Siemens, A. H. 1990. *Between the summit and the sea: Central Veracruz in the nineteenth century.* Vancouver: University of British Columbia Press.

Silver, T. 1990. *A new face on the countryside: Indians, colonists, and slaves in South Atlantic forests, 1500–1800.* Cambridge: Cambridge University Press.

Silverberg, R. 1968. *Mound builders of ancient America: The archaeology of a myth.* Greenwich, Conn.: New York Graphic Society.

Smith, N. J. H. 1980. Anthrosols and human carrying capacity in Amazonia. *Annals of the Association of American Geographers* 70:553–66.

Stout, A. B. 1911. Prehistoric earthworks in Wisconsin. *Ohio Archaeological and Historical Publications* 20:1–31.

Sturtevant, W. C. 1961. Taino agriculture. In *The evolution of horticultural systems in native South America, causes and consequences: A symposium,* ed. J. Wilbert, pp. 69–82. Caracas: Sociedad de Ciencias Naturales La Salle.

Taylor, D. L. 1981. Fire history and fire records for Everglades National Park. Everglades National Park Report T-619. Washington, D.C.: National Park Service, U.S. Department of the Interior.

Thompson, D. Q., and Smith, R. H. 1970. The forest primeval in the Northeast—a great myth? *Proceedings, Tall Timbers Fire Ecology Conference* 10:255–65.

Thoreau, H. D. 1949. *The journal of Henry D. Thoreau,* vol. 7, *September 1, 1854–October 30, 1855,* ed. B. Torrey and F. H. Allen. Boston: Houghton Mifflin.

Trombold, C. D., ed. 1991. *Ancient road networks and settlement hierarchies in the New World.* Cambridge: Cambridge University Press.

Uhl, C.; Nepstad, D.; Buschbacher, R.; Clark, K.; Kauffman, B.; and Subler, S. 1990. Studies of ecosystem response to natural and anthropogenic disturbances provide guidelines for designing sustainable land-use systems in Amazonia. In *Alternatives to deforestation: Steps toward sustainable use of the Amazon rain forest,* ed. A. B. Anderson, pp. 24–42. New York: Columbia University Press.

Walschburger, T., and von Hildebrand, P. 1991. The first 26 years of forest regeneration in natural and man-made gaps in the Colombian Amazon. In *Rain forest regeneration and management,* ed. A. Gómez-Pompa, T. C. Whitmore, and M. Hadley, pp. 257–63. Paris: UNESCO.

Watts, W. A., and Bradbury, J. P. 1982. Paleoecological studies at Lake Patzcuaro on the west-central Mexican plateau and at Chalco in the Basin of Mexico. *Quaternary Research* 17:56–70.

Wedel, W. R. 1957. The central North American grassland: Man-made or natural? *Social Science Monographs* 3:39–69. Washington, D.C.: Pan American Union.

White, S. 1985. Relations of subsistence to the vegetation mosaic of Vilcabamba, southern Peruvian Andes. *Yearbook, Conference of Latin Americanist Geographers* 11:3–10.

Whitmore, T. M. 1991. A simulation of the sixteenth-century population collapse in the Basin of Mexico. *Annals of the Association of American Geographers* 81:464–87.

———, Turner, B. L., II; Johnson, D. L.; Kates, R. W.; and Gottschang, T. R. 1990. Long-term population change. In *The earth as transformed by human action,* ed. B. L. Turner II, et al., pp. 25–39. Cambridge: Cambridge University Press.

Williams, B. J. 1972. Tepetate in the Valley of Mexico. *Annals of the Association of American Geographers* 62:618–26.

———. 1989. Contact period rural overpopulation in the Basin of Mexico: Carrying-capacity models tested with documentary data. *American Antiquity* 54:715–32.

Williams, M. 1989. *Americans and their forests: A historical geography.* Cambridge: Cambridge University Press.

Wilson, S. M. 1992. "That unmanned wild countrey" : Native Americans both conserved and transformed New World environments. *Natural History* May: 16–17.

Wood, W. 1977 [1635]. *New England's prospect,* ed. A. T. Vaughan. Amherst: University of Massachusetts Press.

Thomas H. Birch

The Incarceration of Wildness (1990)

Wilderness Areas as Prisons[1]

BAD FAITH IN WILDERNESS PRESERVATION?

AMERICAN PRESERVATIONISTS CHERISH THE BELIEF that, as Roderick Nash has stated it,

> Wilderness allocation and management is truly a cultural contribution of the United States to the world. Although other nations have established programs to preserve and protect tracts of land, it is only in the United States that a program of broad scope has been implemented, largely because of the fortuitous combination of physical availability, environmental diversity, and cultural receptivity. Despite the continuing ambivalence of American society towards Wilderness, the reserves should be regarded as one of the Nation's most significant contributions.[2]

While wilderness preservation is truly a significant contribution to world civilization, the question whether this contribution, as it is usually understood, is entirely positive ethically is more problematic. As wilderness preservation is generally understood and practiced by mainstream American tradition, and as it often appears to others, particularly those Third and Fourth World peoples who actually live on the most intimate terms with wild nature, it may well be just another stanza in the same old imperialist song of Western civilization.[3]

Nash himself seems close to noticing this problem when he says that

"Civilization created wilderness," and when he points out that "Appreciation of wilderness began in the cities."[4] The urban centers of Western civilization are the centers of imperial power and global domination and oppression. Whatever comes from them, including classic liberalism, is therefore likely to be tainted by the values, ideology, and practices of imperialism, as the mainstream white man (and his emulators) seeks to discharge (impose) his "white man's burden," the burden of his "enlightenment," on all the others, of all sorts, on this planet.

In his most recent book, *The Rights of Nature,* Nash suggests that

> . . . liberty is the single most potent concept in the history of America. The product of both Europe's democratic revolutions and, following Frederick Jackson Turner's hypothesis, the North American frontier, liberalism explains our national origins, delineates our ongoing mission, and anchors our ethics. Natural rights is a cultural given in America, essentially beyond debate as an idea. The liberal's characteristic belief in the goodness and intrinsic value of the individual leads to an endorsement of freedom, political equality, toleration, and self-determination.[5]

This is an accurate statement of what mainstream Western man has taken to be his beneficent burden, as he has sought to bring civilization and liberty (as he conceives it) first to the peoples and the land of North America and then to the entire planet.

Having established the liberal tradition as his starting point, Nash proceeds to subsume or appropriate (reduce) radical or "new" environmentalism into this story:

> Much of the new environmentalists' criticism of American tradition is warranted, but in adopting a subversive, counter-cultural stance, they overlooked one important intellectual foundation for protecting nature that is quintessentially American: natural-rights philosophy, the old American ideal of liberty that they themselves were applying to nature.[6]

Here Nash seems to be suggesting that the American ideals of liberty and natural rights have been overlooked in the new environmentalists' rhetoric but not in their action. Accordingly, Nash presents such diverse thinkers as Paul Shepard, Murray Bookchin, and the deep ecologists as best understood, not as subversive radicals who demand revolutionary changes in our environmental ethics and in our social structures, but as closet champions of the mainstream liberal tradition of natural rights, who would "extend" the benefits and protections of civilization to nature. Just as the once radical appearing abolitionist movement was an extension of our ethics to liberate blacks from exploitation by granting them rights, so the preservation of wilderness is essentially only a next step in the evolution of our liberal tradition, which now would allow even the freedom of self-determination for wild nature.

I suggest that belief in this liberal-tradition story involves self-deception—that it is a cloaking story to cover and legitimate conquest and oppression that needs substantial correction if we are to understand wilderness and the ethics of our relationship with it. Still, if this liberal-tradition story were truly put into practice, if nature were allowed self-determination, then it would be transformed into a radically different story for us to inhabit. That is, this old story does contain some germs for its own transcendence. As things stand, however, self-determination is not permitted for nature, even in legally established wilderness reserves, in spite of much rhetoric to the contrary. Instead, wild nature is *confined* to official wilderness reserves. Why? Probably, I argue, because it would be self-contradictory for imperial power to allow genuine self-determination for the others it would dominate, since doing so would be an abrogation of its power.

John Rodman has exposed the dangers and limitations of our liberal tradition with regard to the animal liberation movement.[7] I am concerned here to do much the same thing with wilderness preservation and the liberation of nature movement. The nub of the problem with granting or extending rights to others, a problem which becomes pronounced when nature is the intended beneficiary, is that it presupposes the existence and the maintenance of a position of power from which to do the granting. Granting rights to nature requires bringing nature into our human system of legal and moral rights, and this is still a (homocentric) system of hierarchy and domination. The liberal mission is to open participation in the system to more and more others of more and more sorts. They are to be enabled and permitted to join the ranks and enjoy the benefits of power; they are to be absorbed. But obviously a system of domination cannot grant full equality to *all* the dominated without self-destructing. To believe that we can grant genuine self-determination to nature, and let its wildness be wild, without dis-inhabiting our story of power and domination, even in its most generous liberal form, is bad faith.

Bad faith is compounded if we believe that the Turner hypothesis, as cited by Nash, can be invoked to support the story of our culture's mission to bring freedom to the wilds of America. The explanatory power of Turner's frontier hypothesis, which proposes that the frontier produced "a culture of individualism, self-reliance, and diffused power—the culture of American democracy," has now been discredited as far as the history of American culture is concerned. Donald Worster writes that it is "a theory that has no water, no aridity, no technical dominance in it."[8] Patricia Limerick holds that it is "ethnocentric and nationalistic" and that "the history of the West is a study of a place undergoing conquest and never fully escaping its consequences."[9] Once we have demythologized American history, we see that Turner's frontier hypothesis is only an instance of a central myth of Western culture, the story that civilization brings light and order to the wild darknesses of savagery—the legitimizing story that cloaks conquest, colonization, and domination. We deceive ourselves if we think that wilderness preservation can be adequately understood in terms of this suspect mythology. To overcome this self-deception we must attend to the less savory side of our tradition, to imperialism and domination (the subtext of the liberal story as usually told).

It is therefore incumbent upon wilderness preservationists, especially those who are privileged to live in the centers of imperial power, to examine their position critically. Even though the establishment of wilderness reservations may well be the best gesture of respect toward nature that Western culture can offer at its present stage of ethical development, unless we Westerners see and acknowledge the shortcomings of this gesture we will languish in self-congratulatory bad faith. My aim here is to expose the bad faith that taints our mainstream justifications for wilderness preservation and to sting us out of it toward a more ethical relationship with wild nature, with wildness itself, and thereby with one another.

BRINGING THE LAW TO THE LAND

At the center of Western culture's bad faith in wilderness preservation are faulty presuppositions about otherness, about others of all sorts, both human and nonhuman, and consequently about the "practical necessities" of our relationship with others. In speaking of "practical necessities" I am raising questions about how others *must* be related to in the deepest sense of *must.*[10]

Problems arise when the other is understood in the usual Western, and imperialistic, manner: as the enemy. In this sense, mainstream Western culture views the oppositional opportunities that otherness affords as adversarial. It presupposes that opposition is fundamentally conflictive, rather than complementary, or communal, or Taoist, or ecosystemic. At best, others are to be "tolerated," which is close to pitying them for their unfortunate inferiority. The central presupposition is thus Hobbesian: that we exist fundamentally in a state of war with any and all others. This is perhaps the most central tenet of our guiding mythology, or legitimizing story, about the necessary manner of relationship with others. Thus, in practice, others are to be suppressed or, when need be, eradicated. This mythology is typical of, but not, of course, limited to mainstream Western culture. William Kittredge has given us a powerful summary of Western culture's leading story:

> It is important to realize that the primary mythology of the American West is also the primary mythology of our nation and part of a much older world mythology, that of law-bringing. Which means it is a mythology of conquest. . . . Most rudimentarily, our story of law-bringing is a story of takeover and dominance, ruling and controlling, especially by strength.[11]

In the case of wildland preservation, bad faith arises when we believe that the simple creation of legal entities such as wilderness areas (a land-use allocation or "disposition" category) can satisfy the practical necessities of relationship with wild land, and with wildness itself. To create legal entities such as wilderness areas is to attempt to *bring the law to wildness,* to bring the law to the essence of otherness, to impose civic law on nature. And this is *all* that it is as long as the customary story is still presupposed, even though it does reform the system of legal institutions. But this is precisely the same sort of reform as the incarceration of Native Americans (paradigm "others") on reservations, even with the putatively well-intended aim of making them over into "productive citizens," in place of the former practice of slaughtering them. Mere reform that is bound by the terms of the prevailing story not only fails to liberate us from the story, but also tends to consolidate its tyranny over us. The reform that appears to be a step in the right ethical direction backfires and turns out to increase the bad faith we have about the ethical quality of what is done. Finding the practical necessities of relationship with nature as other requires breaking the grip of our culture's imperialistic story. It further requires preserving wild land *in order* to help break the grip of this story. It is finally a matter of ethically resolving what Kittredge calls the struggle "to find a new story to inhabit."[12] Then, but only then, can bad faith be left behind.

As things stand for Western culture, committed as it is to the completion of an *imperium* over nature, wilderness reservations are "lockups."[13] The popular terminology of the opponents of wilderness areas about "locking up" wildland is accurate and insightful. The otherness of wildland is objectified into human resource, or value, categories and allocated by law to specific uses (thus bringing law to the land). Of course, as preservationists have often pointed out, allocations of land to specific hard uses such as intensive timber management, strip mining, motorized recreation, etc., are also "lockups," but the attempt to lock out exploiters by locking up wildland turns out, when properly understood, to be just another move in the imperial resource allocation game. This, in fact, is the language of the Wilderness Act of 1964, which states that its purpose is "to *secure* for the American people of present and future generations the *benefits* of an *enduring resource* of wilderness."[14] Wilderness areas in the United States are meant to serve four of the five main multiple uses for public lands: watershed, wildlife, grazing, and recreation. Only timber harvesting is excluded. Mining is often permitted.[15]

But what about the wildness itself? How could wildness be brought under the rule of law? By definition wildness is intractable to definition, is indefinite, and, although it is at the heart of finding utility values in the first place, wildness itself cannot plausibly be assigned any utility value because it spawns much, very much, that is *useless,* and much that is plain disutility. It is for this reason that it is so puzzling, to the point of unintelligibility, to try to construe wildness (or wilderness) as a resource, though we often hear wilderness called a resource. Since wildness is the source of resources, any attempt to construe it as a resource in terms of the rhetoric of resources reduces it to some set of resources.[16] Wildness itself, to the mind of the law-bringing imperium, is lawless; it is the paradigm of the unintelligible, unrepentant, incorrigible outlaw. How then is the imperium to deal with it?

All the usual attempts to subdue wildness by destroying its manifestations fail, although wildness may be driven into hiding for awhile, or, more accurately, may be lost sight of for awhile. Although the forest and the bison and the Indian may be exterminated, this does not affect wildness itself. In the case of wildness itself, there is nothing there to aim at and shoot. As what we might call the "soul" of otherness, wildness is no usual sort of other. To take the manifestations of wildness for the thing itself is to commit a category mistake. Wildness is still very much there and will not go away. How then is the imperium to deal with it, given that the usual strategies of conquest cannot work? Wildness cannot be ostracized, or exterminated, or chastened into discipline through punishment, reward, or even behavior modification techniques. Yet according to the dictates of the imperium, which claims total control, wildness must be, or at least must seem to be, brought into the system, brought under the rule of law. While the older ("conservative") factotums of the imperium still pursue the strategy of obliterating wildness by destroying its manifestations, the modern ("liberal") reformist factotums see the futility of the older strategy and therefore follow a more subtle strategy of "cooptation," or appropriation, through making a place for wildness within the imperial order and putting wildness in this place. The place that is made is the prison, or the asylum. When this place is made and wildness is incarcerated in it, the imperium is completed. Consider Foucault's observations:

> The carceral network does not cast the unassimilable into a confused hell; there is no outside. It takes back with one hand what it seems to exclude with the other. It saves everything, including what it punishes. It is unwilling to waste even what it has decided to disqualify.[17]

In this way, designated wilderness areas become prisons, in which the imperium incarcerates unassimilable wildness in order to complete itself, to finalize its reign. This is what is meant when it is said that there is no wilderness anymore in the contemporary world, in the technological imperium. There is, or will be soon, only a network of wilderness reservations in which wildness has been locked up.

To press the use of prison terminology, we may say that just as the "lockup" occurs at the end of the prison day, so wildness is locked up at the twilight time of modernism. To press prison terminology a bit more, we may say also that when wildness as prisoner "misbehaves" (by being its spontaneous self) the imperium "locks it down." A "lockdown" involves confining prisoners to their cells, revoking privileges, conducting searches, etc., to root out and correct the maleficence.[18]

Wilderness reservations are not intended or tolerated as places where nature is allowed to get out of control, even though a degree of aberrant behavior is permitted, just as a degree of it is permitted within the edifices of the penal system for humans. Wilderness reservations are not meant to be voids in the fabric of domination where "anarchy" is permitted, where nature is actually liberated. Not at all. The rule of law is presupposed as supreme. Just as wilderness reservations are created by law, so too can they be abolished by law. The threat of annihilation is always maintained. Just as a certain inmate, say a tree fungus, may be confined to the wilderness reservations by law, so too can it be exterminated by law, even within the reservation. The imperium does not, and cannot, abrogate these "rights," although it has arrogated them in the first place.

OTHERNESS AS WILDNESS

At the center of the problem of Western culture's incarceration of wildness is its prevailing (mis)understanding of otherness as adversarial, as recalcitrant toward the law, as therefore irrational, criminal, outlaw, even criminally insane (like the grizzly bear). This understanding of the other is a part and product of Western culture's imperialistic mythology of law bringing. It is the meaning of otherness that the story of the imperium has created. It is an enforced misunderstanding, myth or story. Accordingly, texts such as Joseph Conrad's *Heart of Darkness* and E. M. Forster's *A Passage to India* have been incorporated into our literary canon and are taught in our educational institutions, as part of what is called the process of "the indoctrination of the young" by ruling elites,[19] or as what we may more politely call the inculcation and perpetuation of the mythology of Western culture.

If we are to understand what the creation of legally designated wilderness reservations amounts to ethically, then we must disentangle the threads of Western culture's mythology from the realities of otherness. Let us begin by emphasizing that the essence of otherness is wildness. If any other is to preserve its (his, her) identity as other, as other in relation to another person, society, species, or whatever, then it must at bottom resist accepting any *final* identity altogether. An other cannot *essentially* be what it is objectified, defined, analyzed, legislated, or understood to be if it is to be and remain an other. The maintenance of otherness requires the maintenance of a radical openness, or the maintenance of the sort of unconditioned freedom that permits sheer spontaneity and continuous participation in the emergence of novelty.

The need for others to preserve this sort of freedom is especially pronounced in encounters with imperial power (even in its putatively beneficent liberal form), because imperial power seeks to define and fix identities in order to internalize others into its own system of domination—domination through objectification. But the maintenance of this sort of freedom is also necessary for the dynamics of nonconflictive, complementary relationship with otherness. Unfinalized, contingent, or working identities are, of course, also indispensable, but any *finalization* of identities is an anathema that destroys otherness.

A finalization of the identification of the other is a (self-deceived) absorption or ingestion of the other into the subjectivity of the self, or, on the social level, into the "system." Such an absorption is also a finalization of self or of system definition that takes self or system out of the world into a state of alienation. Self-becoming in and out of dialectical response to others and to other-becoming is then no longer possible. The ultimate end of the imperial project is realized at the point of total finalization. This is why wildness, which contradicts any finalization in identification, is at the heart of otherness, as well, of course, as at the heart of any *living* self or society.

THE VOICE AND RESIDENCE OF THE IMPERIAL OTHER

Perhaps the most strikingly articulate voice of the other to emerge within Western culture is that of Jean Genet. Although his writings are fairly well known, they are not part of our literary canon. They are far too unsettling, insightful, and impolite. A foundling, an illegitimate child, a bastard (note *why* this is a term of derogation), Genet was illegal and unlawful from before his birth, and he became a thief, a homosexual, a prostitute, a convict, and as an author, a celebrant of criminality and of all that imperial society finds intolerable and disgusting. As a celebrant of the wild underside of

Western society, Genet is a devastating critic of the entire system of the imperium. He portrays as most beautiful what imperial culture takes as most ugly and unacceptable. Sartre proposes that Genet's original illegality destined him to become an other to the imperium:

> Genet has neither mother nor heritage—how could he be innocent? By virtue of his mere existence he disturbs the natural order and the social order. . . .
>
> . . . the collectivity doomed him to Evil. They were waiting for him. There was going to be a vacancy: some old convict lay dying on Devil's Island; there has to be new blood among the wicked too. Thus, all the rungs of the ladder which he has to descend have been prepared in advance. Even before he emerged from his mother's womb, they had already reserved beds for him in all the prisons of Europe and places for him in all the shipments of criminals. He had only to go to the trouble of being born; the gentle, inexorable hands of the Law will conduct him from the National Foundling Society to the penal colony.[20]

In strict obedience to its own logic, which will be explored more fully below, the imperium creates a place for its own other, which it must have and therefore must create and maintain.[21]

Consider Genet's own reflections about the prison and the palace, the two most illustrative edifices of imperial power:

> They are the two buildings constructed with the most faith, those which give the greatest certainty of being what they are—which are what they are meant to be, and which they remain. The masonry, the materials, the proportions and the architecture are in harmony with a moral unity which makes these dwellings indestructible so long as the social form of which they are the symbol endures. . . . They are also similar in that these two structures are one the root and the other the crest of a living system circulating between these two poles which contain it, compress it, and which are sheer force.[22]

In this passage, we find the voice of the other, the other speaking both as an artifact of the imperium, defined and produced by the imperium, and speaking as himself. Note that the imperium does not recognize that the other has anything at all to say, and certainly nothing to say on matters of importance. With regard to wildland the situation is much worse, for such land, because it is mute, cannot speak out like Genet from its category of incarceration and is therefore a perfect candidate for oppression as paradigm other. In both cases, the other is meant to be silent. In Joseph Conrad's

Heart of Darkness, for example, Africans virtually do not speak at all, Marlow finds the wilderness inarticulate, "dumb," a "thing that couldn't talk and perhaps was deaf as well," and Kurtz turns out to be merely

> A voice. He was little more than a voice. And I heard—him—it—this voice—other voices—all of them were so little more than voices—and the memory of that time itself lingers around me, impalpable, like a dying vibration of one immense jabber, silly, atrocious, sordid, savage, or simply mean, without any kind of sense.[23]

Similarly, consider the "terrifying echo" of the Marabar caves, of E. M. Forster's other, the essential India under the British raj:

> The echo in a Marabar cave . . . is entirely devoid of distinction. Whatever is said, the same monotonous noise replies, and quivers up and down the walls until it is absorbed into the roof. "Boum" is the sound as far as the human alphabet can express it, or "bou-oum," or "ou-boum," utterly dull. . . . And if several people talk at once, an overlapping howling noise begins, echoes generate echoes, and the cave is stuffed with a snake composed of small snakes, which writhe independently.[24]

To Forster's kind and liberal Mrs. Moore this echo had "managed to murmur, 'Pathos, piety, courage—they exist, but are identical, and so is filth. Everything exists, nothing has value.' "[25] So goes our culture's story about the wild essence of the other. For the imperium, the other, and wildness, has nothing sensible to say, and is intolerable when it sometimes noisily, sometimes silently is said to say, "Nothing, Nothing!" It is this same "Nothing" that drives Sartre's Antoine Roquentin to nausea (in *Nausea*), and which for the imperium legitimates law bringing as the establishment of "Meaning," by whatever means are necessary, *ergo,* force and violence.

Genet, however, confronts the imperium with an undeniably articulate voice. Now that Third World and minority literature is beginning to achieve some notice and acceptance in the West, Genet is no longer so alone, although none of this literature has been incorporated into the Western canon. Genet's language equals or even surpasses Proust in its elegance. The highest style, often with what the imperium considers the lowest content, stops and exposes the imperial mind square in the tracks of its bad faith. The voice of the other speaks more beautifully of what the imperium despises than the imperium itself can speak of what it prizes. Genet is intentionally the *revolting* other in both senses of the word, both nauseating and rebellious. In response to the imperium Genet is that "corrosive spirit" (his phrase) of denunciation and revolt that comes out of hiding, refuses to go away, and shows the story of our culture as false, and thereby refutes it.

In concluding her convincing account of *Heart of Darkness* and *A Passage to India* as legitimations of imperialist ideology, Mary Layoun writes:

> Nowhere is the value or meaning-making machinery of the Subject/Colonizer/Text/Ideology brought more into question than when faced with the "reality," or the textual representation of that "reality," of what Jean-Paul Sartre, in a slightly different context, calls "the glance of the Other"—the Other as primeval and savage, as repressed desire, as the unremitting landscape, as the recalcitrant Native, as an incomprehensible language—that not just the "dark continent" or an elusive India are called into question but the white continent and its production of meaning and value—in narratives and outside of them.[26]

If an apology for imperialist ideology is to be at all plausible, then the reality of the other must be represented, and the more this is achieved by attempts to legitimize our false story of the need for law bringing, the more transparently false such attempts become. These legitimations therefore tend toward self-refutation, or at least contain the seeds of their own self-destruction. The glance of the other flashes right through the texts.

If the "glance" of the other shines forth from the texts of Conrad and Forster and calls our culture's mythology into question, then Genet's texts are the *glare* of the other, a glare that starkly illuminates our culture's oppressive structures and practices. In response to the oppression of the imperium Genet adamantly refuses reconstruction or assimilation: "If I had to live . . . in your world, which, nevertheless, does welcome me, it would be the death of me."[27] His aspiration is to ostracism, to the "outside," or, more exactly, to the most marginal, peripheral, the nearest to an "outside" that the imperium permits, which it declares to be outside but which is in reality still very much inside the imperium. He aspires to Guiana, the penal colony, in comparison with which the prisons or reformatories constructed within the centers of power are an obliteration of his otherness:

> I aspire to Guiana. No longer to that geographical place now depopulated and emasculated, but to the proximity, the promiscuity, not in space but in consciousness, of the sublime models, of the great archetypes of misfortune. Guiana is kindly. . . . It suggests and imposes the image of a maternal breast,

> charged, in like manner, with a reassuring power, from which rises a slightly nauseating odor, offering me a shameful peace. I call the Virgin Mother and Guiana the Comforters of the Afflicted.[28]

Genet sees that "The end of the penal colony prevents us from attaining with our living minds the mythical underground regions." The "destruction of the colony corresponds to a kind of punishment of punishment: I am castrated, I am shorn of my infamy."[29] The network of imperial control is completed, and there is no place left where the other can be itself.

Guiana, like roadless and undisposed wildland, was decommissioned because it gave comfort to the other. It made escape possible. As Foucault has pointed out, now even the "criminality" of the other is utilized. Nothing is wasted, not even waste itself. Accordingly, what was once the wasteland of wilderness, the "outside," is brought inside the imperium as legally designated wilderness reserves, thus making a proper place for wildness in the universal system of universal control.

Genet's intractability to assimilation into this system does not permit the usual sorts of rebellion. He has no concern for justice, or for some better system of laws, or even for "morality." He has no concern for any "liberal" reshuffling of the categories of domination. A successful revolt and revolution that is confined by Western culture's presuppositions about otherness would at best amount to reform, to nothing more than some shift in the arrangements of power, as from conservative to liberal, or capitalist to socialist. Genet's revolt cuts far deeper and demands the recognition and respect and liberation of otherness itself. Because the imperium names the other, the other must find a way to insist on its intrinsic namelessness:

> Erotic play discloses a nameless world which is revealed by the nocturnal language of lovers. Such language is not written down. It is whispered into the ear at night in a hoarse voice. At dawn it is forgotten. Repudiating the virtues of your world, criminals hopelessly agree to organize a forbidden universe . . . criminals are remote from you—as in love, they turn away and turn me away from the world and its laws. . . . My adventure, never governed by rebellion or a feeling of injustice, will be merely one long mating, burdened and complicated by a heavy, strange, erotic ceremonial (figurative ceremonies leading to jail and anticipating it).[30]

Genet's path becomes the erotic, that essential element of wildness that the imperium finds most problematic. Recall Conrad's "wild and gorgeous apparition of a woman" and the erotic alienation of Forster's Adela. The erotic side of love remains intractable to reconstruction into anything abstractly and putatively agapic that can be, and is, abused to legitimate oppression for the other's "own good," to relieve the white man of his burden. Because the erotic side of love is so troublesome for the imperium, it is quite properly the territory of the other, and is therefore the place in which Genet feels compelled to take up residence. Both the imperium itself and its other agree that the erotic is a place for the other. The erotic has not yet succumbed to domination, in spite of the imperial attempt to "lock it up" into trivialization, perversion, and sterility (or "simulation," as I argue in the next section), to cast it as evil, and even to use it as an instrument of terror. AIDS, for example, becomes a grizzly bear lurking in the wild otherness of the erotic (and do note how extraordinarily erotic we feel the bear to be).

Like Sid Vicious, Genet sees no alternative to being cast and casting himself as evil, thereby allowing the imperium to define him as *its* sort of other. Although his deepest desire is to deconstruct the imperium, Genet is trapped in the terms of the debate with any opposition that the imperium sets and controls. Thus Genet focuses his love on the criminal, with the liberation of the other as his goal, so far as this is at all possible in the face of, and in spite of, the imperium's definition, and in the only terms remaining for the other to claim—a liberation of the other into its own eroticism and into its own beauty:

> *. . . there is a close relationship between flowers and convicts.* The fragility and delicacy of the former are of the same nature as the brutal insensitivity of the latter. Should I have to portray a convict—or a criminal—I shall so bedeck him with flowers that, as he disappears beneath them, he will himself become a flower, a gigantic new one.[31]

Although Genet strives for liberation mainly through a kind of secular beatification *within* the given framework of the law-bringing imperium, his real liberation requires abolishing the imperium. What is required is a different world, a world in which the imperial order and the imperial other no longer exist, a world in which others can be themselves and not products defined by the imperial order, a world in which the wildness of others of all sorts is respected in the only way this can really be done—by not trying to subject their wildness to the totalizing rule of law, for the bringing of law to wildness, or to the wildness of others, is ineluctably imposition and domination by force. What Western culture needs is an entirely different story about wildness and otherness, a story that does not produce the sort of "criminal" otherness that is Genet's only residence and refuge, and that, given the terms of the prevailing story, is the only fit condition for any other, so defined, to occupy and to celebrate.

WILDERNESS AREAS AS "SIMULATIONS" OF WILDNESS, AND THE RISK OF THE REAL

Jean Baudrillard's brilliant and alarming analysis of modern Western culture starkly illuminates the uses to which imperial culture puts its wild others, both human and nonhuman. Baudrillard's analysis also explains the imperium's need to manufacture and maintain an adversarial sort of other to serve these uses. Briefly, these uses are to provide meaning and legitimation for the institution of imperial power and to enforce its reign with the threat of terror and chaos. For Baudrillard, modern Western culture is headed toward, and to a great extent has already reached, a condition of "hyperreality," and has taken up residence in a world of "simulation," a simulated world of "simulacra," with no remaining contact with reality, including ecological reality. In one of Baudrillard's more noted statements, "The very definition of the real has become: *that of which it is possible to give an equivalent reproduction*. . . . The real is not only what can be reproduced, *but that which is always already reproduced.* The hyperreal . . . which is entirely in simulation."[32]

In order to solidify its reign, to realize its goal of total control, power must create its own world, defined in its own terms, by means of models that are simulations of realities (of all sorts). Total control of such, and only such, simulacra is possible because they are reproducible and therefore fungible. Should any one of them stray from the grip of control, it can be eradicated and replaced with another. Appropriation into this throwaway world involves throwing away the former, and other, reality in favor of a simulation that is illusory: "We live everywhere already in an 'aesthetic' hallucination of reality."[33] Nevertheless, imperial power cannot afford to throw reality away entirely, because it needs some reality or semblance of reality to save its own meaningfulness and legitimacy.

Legally designated wilderness reserves thus become simulacra insofar as it is possible for the imperium to simulate wildness. The pressure on the imperium is to institute simulations of wildness in order to appropriate wildness into the imperium under the rubric of the model. Simulacra are produced according to the dictates of models, and we come to inhabit a modelling of reality that is purported to be all the reality there is. Otherness is incarcerated in simulacra, in models of otherness. But why must the imperium go to the trouble of preserving wild otherness, even if only as simulacra, rather than totally destroying this adversarial opposition and then forgetting about it altogether? Why does the imperium need to create and preserve its Genets and its wilderness preserves? At bottom, it is a matter of the imperium's need to preserve its own meaningfulness, to protect itself from "vanishing into the play of signs":

> Power . . . for some time now produces nothing but signs of its resemblance. And at the same time, another figure of power comes into play: that of a collective demand for *signs* of power—a holy union which forms around the disappearance of power. . . . When it has totally disappeared, logically we will be under the total spell of power—a haunting memory already foreshadowed everywhere, manifesting at one and the same time the compulsion to get rid of it (nobody wants it anymore, everybody unloads it on others) and the apprehensive pining over its loss. . . .[34]

The whole point, purpose, and meaning of imperial power, and its most basic legitimation, is to give humans control over otherness. Once this is totally achieved, or perceived to be achieved, the game is over, and continuing to play it is meaningless. This holds true as much in the case of the manufactured hyperreality, which generates what is really an illusion or "hallucination" of total control, as it does for the real thing (which could be achieved only by some imperial eighteenth-century God the Father). The imperium must therefore attempt to keep its game, itself, alive by preserving its "reality principle," but preferably to the greatest extent possible by simulating this too. Thus, we are given Disneyland and the fantasy fare of television:

> Disneyland is there to conceal the fact that it is the "real" country, all of "real" America, which *is* Disneyland (just as prisons are there to conceal the fact that it is the social in its entirety, in its banal omnipresence, which is carceral). Disneyland is presented as imaginary in order to make us believe that the rest is real, when in fact all of Los Angeles and the America surrounding it are no longer real, but of the order of the hyperreal and of simulation. It is no

> longer a question of a false representation of reality (ideology), but of concealing the fact that the real is no longer real, and thus of saving the reality principle.[35]

This Disneyland sort of simulation of fantasy is not, however, by itself, enough to meet the threat to the meaningfulness of power that is constituted by power's own success in simulation, and is not by itself enough to rejuvenate hyperreality or the imperial game. Thus, "When it is threatened today by simulation (the threat of vanishing in the play of signs), *power risks the real,* risks crisis, it gambles on remanufacturing artificial, social, economic, political stakes. This is a question of life or death for it. But it is too late."[36] Consequently, we are presented periodically with "scandals" such as Watergate and Irangate, and in the United States with election contests between Democrats and Republicans. According to Baudrillard, Watergate was "a trap set by the system to catch its adversaries—a simulation of scandal to regenerative ends."[37]

Still, even though such scandals are meant to be controlled simulations, there is some risk of generating real challenge or resistance to power. When it comes to the wild others, like Genet and wild land, which must continue to exist or be posited as existing in contrast and opposition to imperial power if power is to save its reality principle, the risk of the real is somewhat amplified. In order to do the job of preserving its reality principle, and in spite of its need to simulate or define the other according to its own models, the imperium must leave at least enough otherness intact to *maintain the glance of the other.* It does not seem possible to simulate this otherness entirely while at the same time also preserving the needed significance of its glance. The other must be able to cast its glance at the imperial enterprise to preserve the meaning of that enterprise, to legitimate its purpose of bringing law and order to wild chaos, and to threaten those who might question its good intentions and overall beneficence. There must remain at least some vestiges of wildness to be kept at bay.

The risk of the real is that in seeing the glance of the other, in reading Genet, in attending to wilderness, one sees, or is likely to see, that the other is more than, other than, independent of, the definitions, models, and simulations that the imperium proposes as exhaustive of it. On reading Genet, one sees that, even though perhaps he cannot quite accept it, because he has acquiesced to the imperium's model of what he is, there is far more to him as a human being than the imperium would have it. One sees his real otherness shining through and overwhelming all the imposed categorizations of him. To scratch the surface, one sees a thoroughly sensitive, loving, ethical human being who is perhaps justified in his forms of resistance to what the imperium has done to him—even though we might hope to find different forms of resistance for ourselves. Likewise, in coming to know wilderness, we see beneath its glance as this has been construed for and purveyed to us. When we see the real otherness that is there beneath the imperium's version of it, beneath all the usual categories of use and value, then we see an otherness that can never be fully described, understood, or appropriated, and the entire edifice of the imperium is called into question to such a degree that it becomes practically necessary to resist and deconstruct it, because it so epitomizes bad faith and delusion.[38]

THE GROUND OF SUBVERSION

When Roderick Nash argues that "Civilization created wilderness," he quotes Luther Standing Bear:

> We did not think of the great open plains, the beautiful rolling hills and the winding streams with tangled growth as "wild." Only to the white man was nature a "wilderness" and . . . the land "infested" with "wild" animals and "savage" people. . . . There was no wilderness; since nature was not dangerous but hospitable; not forbidding but friendly.[39]

The real point, which is not the point that Nash is trying to make out of what Luther Standing Bear has said, is that the wilderness, and now the wilderness reservations that the white imperium has created in obedience to its traditional story of law bringing, is an adversarial other to be subdued and controlled. In response I would argue that the other does not *have to be* an adversary or a simulation of an adversary. Certainly it is not an adversary for Luther Standing Bear. He sees the land in terms of a different story, a story which holds that the fundamental human relationship with nature, and with wildness itself, is participatory, cooperative, and complementary, rather than conflictive. At times, of course, and also in the terms of the other story, there is conflict, but normally wild nature sustains, sponsors, empowers, and makes human existence possible. Nature is wild, always wild, in the sense that it is not subject to human control. In this sense, humans are participants in a wildness that is far larger and more powerful than they can ever be, and to which human law bringing is so radically inappropriate as to be simply absurd. This is the sort of understanding of wilderness and its relationship with culture that we need to retrieve and reconstruct in postmodern society in response to the imperium's desire for total dominion.

It seems fair to take the RARE II (Roadless Area Review and Evaluation, second try) process in the United States as typical of and precedent setting for Western civilization's approach to wildland. The United States is, at the moment, the most powerful center of the imperium of Western culture, and the example of RARE II will be followed on the global level. Note that RARE II was implemented by the relatively liberal Carter administration and was intended as a reform of RARE I, to correct its mistakes. The purpose of the RARE process was to search out and evaluate the utilities of all remaining wildland in the national forests with the goal of determining its allocation or disposition, thereby giving it definition, bringing meaning into its nothingness, so that nothing remains unmanaged waste outside of the imperium. RARE II typifies the final step in the imperium's appropriation of wild nature, its most powerful enemy. The RARE process should be seen as a "search and destroy" mission to discover and appropriate or exterminate the last vestiges of wild land in America, to complete the imperium. The acronym for the key instrument of wilderness evaluation in the field was WARS (Wilderness Attribute Rating System). As in the case of a racist joke, the subtext, or presupposition, of this cynical attempt at humor is near the surface and easily seen.[40]

The RARE II process thus marks the completion of the imperium's imposition of its network of control, bringing all wildland under management for some set of utilities. Whereas in the past there were wildland Guianas, which were ignored and to which wildness and wild nature were either let go or ostracized from civilization, places that were *outside* the system of management, but places where wildness could to some extent flower in its own integrities, with RARE II there are only legally designated wilderness areas or reservations in which wild nature, the ultimate other, has been locked into specific management schemata. Whereas, once upon a time, for example in the time of Homer, Western culture was a cluster of tenuously connected islands surrounded by a sea of wildness, civilization now surrounds (or so goes the deluded story) the last islands of wildness, and puts everything to use, wasting nothing. Even Genet is published—just as one of the recognized reasons for official wilderness is to benefit those "oddballs" who thrive on it, or even to permit the furtherance of their "self-realization." Wilderness and wildness are placed on the supermarket shelf of values along with everything else, and everything is enclosed *inside* the supermarket.

Yet there is a contradiction in the imperium's attempt to appropriate wildness, for, as we have seen above, it is not possible practically for the imperium to silence the subversive voice of the other completely or to stifle its glance. For the imperium, the problem with appropriating wildness by incarcerating it in the prisons of wilderness reservations is that the wildness is still there, and it is still wild, and it does "speak" to us. Genet writes, and he insists on his integrity, no matter that it is tainted by the categories that the imperium has inflicted on it. The prisons and asylums are still rich with other minds. When wildness speaks, it always says more than what the imperium would train it to say or train us to hear because wildness stays adamant in its own integrity, as other in its own unconditioned freedom. Thus, managing wildness is contradictory, even though managing official wilderness areas and prisons is not. There is an insurmountable tension in the notion of managing wildness, and of managing land *for* wildness. How much wildfire, how much insect evolution, for example, is to be permitted? We *cannot* know. Wildness is logically intractable to systemization. There can be no natural laws of wildness.

What follows is that making a place for wildness within the system of the imperium creates, institutionalizes, and even legalizes, a basis, literally a ground, for the subversion of the imperial system—just as prisons and reformatories succeed mainly in generating better "criminals." If it does nothing more than open a path for our gaze to fasten with the stars, with the wildness of the universe in which we inhabit our mote of dust, a wilderness reservation can still give the lie to the imperial story of Western civilization. No matter how the imperium deals with wild nature, whether by extermination or incarceration and the logically impossible fiction of total management, wild otherness will continue to show up the belief in our culture's most formative myth of law bringing for the bad faith it is.

Yet, even so, the establishment of wilderness reserves is not enough to counter the imperium's assault on wildness. First, the battle to save wildland through legal preservation, like scandals or simulated scandals, can be a trap that only serves to further power. In the terms of the struggle as set by the imperium, the imperium wins whether a tract of land is classified as legal wilderness or not, because the imperium has been allowed to get away with setting the terms of the debate. The energies of the champions of wildness are appropriated and exhausted by the legal battles for preservation. If an area of wildland is classified as official wilderness, the imperium wins the other it needs; if it is not preserved, the imperium wins something else it needs. The real issue, the preservation of wildness and of knowing human participation in wildness, is very easily lost in the fogs of legal-political controversy, at least in the forms that imperial society permits. But, secondly, we should never forget that the imperium has the power to manage, invade, declassify, abolish, desanctify the legal wildland entities it has created, and the creation of such entities on its terms does little to diminish this power. Because it probably tends to strengthen this power, it is imperative to dig out and clarify what is basically at stake and what underlies the task of preserving wilderness in order to see how such efforts should contribute to the fundamental task of saving wildness and resisting the imperium.

TOWARD RESIDENCE IN SACRED SPACE

Thinking of legally designated wilderness reserves as "sacred spaces" is not by itself enough to rescue official wilderness spaces from the totalizing grip of imperial power. Although there has always been a place for the sacred in the imperial order, in Western culture secular power has long ago triumphed over the church. As far as the imperium is concerned, the sacred, the mystical, and so forth are just the other side of the coin of criminal wild otherness.[41] Wilderness, like religion and morality, is fine for weekends and holidays, but during the working week it may in no way inform business as usual. Thus, the imperium incarcerates its sacred other in churches, convents, and ministries, but if its functionaries (like the Berrigans) take their sacred obligations out of the assigned area and, say, into the streets, they are imprisoned, or otherwise neutralized. It is perfectly fine with the imperium if, on weekends and holidays, some of its citizens wish to follow

John Muir to the temples of the wilderness areas, rather than the usual churches—and this will probably hold for American Indian religions as well.[42] By making room for sacred space, the imperium confirms its tolerance, generosity, its rectitude, its beneficence, and it does so without having to abrogate *its* other (thus maintaining its bad faith). However, actually to inhabit, to live in, sacred space is an anathema, absolutely incompatible with the imperium:

> The idea that holiness inheres in the place where one lives is alien to the European tradition, for in that tradition sacred space is sundered, set aside, a place one goes only to worship. But to live in sacred space is the most forceful affirmation of the sacredness of the whole earth.[43]

For the imperium, only that which is other can be sacred, because all of the usual world, the mundane and the not-so-mundane, is taken to be profane, secular, objective. The imperium is committed to cordoning off sacred space, to separating it as other, effectively keeping it out of the center of our practical lives, and keeping us out of it and thus safe from its subversive effect. Wildness as wilderness land is incarcerated as sacred space. This is perhaps one of the main uses to which the imperial order puts wilderness. It consigns sacred space to the museum of holy relics, as one of the prime manifestations of the wildness it is compelled to incarcerate in order to demonstrate its total triumph.

The point, then, is that even the preservation of wilderness as sacred space must be conceived and practiced as part of a larger strategy that aims to make all land into, or back into, sacred space, and thereby to move humanity into a conscious reinhabitation of wildness. As Gary Snyder has pointed out:

> Inspiration, exaltation, insight do not end . . . when one steps outside the doors of the church. The wilderness as a temple is only a beginning. That is: one should not . . . leave the political world behind to be in a state of heightened insight . . . [but] be able to come back into the present world to see all the land about us, agricultural, suburban, urban, as part of the same giant realm of processes and beings—never totally ruined, never completely unnatural.[44]

Wilderness reserves should be understood as simply the largest and most pure entities in a continuum of sacred space that should also include, for example, wilderness restoration areas of all sizes, mini-wildernesses, pocket-wildernesses in every schoolyard, old roadbeds, wild plots in suburban yards, flower boxes in urban windows, cracks in the pavement, field, farm, home, and workplace, all the ubiquitous "margins." As Wendell Berry puts it:

> . . . lanes, streamsides, wooded fence rows . . . freeholds of wildness . . . enact, within the bounds of human domesticity itself, a human courtesy toward the wild that is one of the best safeguards of designated tracts of true wilderness. This is the landscape of harmony . . . democratic and free.[45]

Wilderness reserves make an indispensable contribution to establishing and inhabiting Berry's "landscape of harmony" writ large, which is how we should write it, a landscape that is thoroughly predicated upon and infused with wildness. The larger wilderness reserves, where the essence of otherness as wildness is most powerfully evident, continuously freshen, enliven, and empower this infusion of wildness, on the analogy of water from mountain watershed sources. The ideal goal, however, is a landscape that is self-sustaining and everywhere self-sufficient in wildness. Enough margins in some locales (perhaps including some Third World locales) could bring this about, or serve as the starting point toward reaching the goal of a larger harmony.

Because the landscape of harmony is an inhabited harmony with otherness and with others, respected in their own integrities, and thus a landscape, a "land" in Leopold's sense, and a form of human life that cooperates with others as complementary to us, it constitutes hope for an implacable counterforce to the momentum of totalizing imperial power. Furthermore, to a great extent, the margins still exist, although we seldom notice them and neglect them. To achieve Berry's landscape of harmony we must, as it were, demarginalize the margins, including the legal wilderness reserves, and come to see and practice their continuing and sustaining primacy to all that we humans can value and construct. Then we can take up residence in wildness, where Luther Standing Bear lived, reinhabiting it now, of course, with different appropriate technologies and social forms. Then we can recover our endangered knowledge of reality and disempower the bad faith that the imperium puts upon us.

THE JUSTIFICATION OF LEGAL WILDLAND ENTITIES

Wilderness reservations are best viewed as holes and cracks, as "free spaces" or "liberated zones," in the fabric of domination and self-deception that fuels and shapes our mainstream contemporary culture. Working to preserve wild nature, in wilderness reservations, or anywhere, is primarily, at this historical moment, an essential holding action, to stop the complete triumph of the bad faith of our culture, especially in regard to ecological reality, and to save us from ineluctably destructive self-deception. Although the culture may deceive itself and believe that wilderness reservations are successful appropriations of the wild and/or sacred and ethical opposition, in fact, their existence, properly understood, helps preserve and foster the possibility of liberation from our imperialist tradition. This subversive potential is what justifies their establishment.

From the ethical standpoint, the purpose and the only justification of laws is to help us fulfill our obligations, or to meet the practical necessities that are incumbent upon us. If legally created wilderness areas do, or can be made to, serve the subversive role I have pointed out, then the laws that create them are thereby ethically justified. Then wilderness reservations serve as a crucial counterfriction to the machine of total domination, slowing it down and creating a window through which a postmodern landscape of harmony may be found. But insofar as wilderness reservations, as they are so often (mis)understood, only serve the completion of the imperium, they are not justifiable. Wilderness must be preserved for the right reasons—to help save the possibility and foster the practice of conscious, active, continuing human participation in wildness, as well as to preserve others for their own sakes. The institution of legally designated wilderness reserves does make an essential contribution toward meeting this larger necessity. However, it is crucial to remember that, important as they are, legal wildland entities are not always and everywhere either the ethically sound or most effective means for meeting this larger necessity, especially if they are imperialistically understood and exported and colonially imposed, either domestically or internationally. I have suggested above what some of the other means may be. But on this question there is very much culturally and economically imaginative and sensitive work waiting to be done.

Of course, all of this looks to the possibility of a more ideal time, to a vision toward which we can struggle, when our practice of respect for nature has become so refined that preservationist laws would no longer be needed, a time when we have moved out of the imperium and taken up residence in wildness. The realization of this vision would mean recovering the sort of relation between humans and others, including human others, that Luther Standing Bear, for instance, sees as basic. It would mean realizing in contemporary practice what Leslie Marmon Silko has called "the requisite balance between human and other."[46] Others would then be seen and lived with as complementary to us, as we all live together in the wild and continuous composition of the world.

NOTES

1. The author wishes to thank his wife Joy DeStefano, Fred McGlynn, Jim Aton, Jim Cheney, Donald Worster, Richard Watson, and Holmes Rolston III for their many helpful comments and suggestions.

2. Roderick Nash, "International Concepts of Wilderness Preservation," in Hendee, Stankey, and Lucas, *Wilderness Management,* Miscellaneous Publication no. 1365 (Washington, D.C.: U.S. Forest Service, 1977), p. 58.

3. For a forceful account of how First World wilderness preservation can appear to Third World peoples see Ramachandra Guha, "Radical American Environmentalism and Wilderness Preservation: A Third World Critique," *Environmental Ethics* 11 (1989): 71–83 [included in this volume]. There are obvious analogies (which cannot be pressed too far) between Third World countries and western American states, like Idaho and Montana, that are rich in wildland areas subject to "lockup" into the wilderness preservation system by colonial and neocolonial powers. Of course, these areas are also subject to lockup into other uses.

4. Roderick Nash, *Wilderness and the American Mind,* 3d ed. (New Haven: Yale University Press, 1982), pp. xiii, 44.

5. Roderick Nash, *The Rights of Nature* (Madison: University of Wisconsin Press, 1989), p. 10.

6. Ibid., p. 11.

7. See John Rodman, "The Liberation of Nature?" *Inquiry* 20 (1977): 83–145, and "Four Forms of Ecological Consciousness Reconsidered," in Donald Scherer and Tom Attig, eds., *Ethics and the Environment* (Englewood Cliffs: Prentice-Hall, 1983), pp. 89–92.

8. Donald Worster, *Rivers of Empire* (New York: Pantheon Books, 1985), p. 11.

9. Patricia Nelson Limerick, *The Legacy of Conquest* (New York: Norton & Company, 1987), pp. 21, 26.

10. I am using the expression "practical necessities" in the sense offered by Bernard Williams: "When a deliberative conclusion embodies a consideration that

has the highest deliberative priority and is also of the greatest importance (at least to the agent), it may take a special form and become a conclusion not merely that one should do a certain thing, but that one *must,* and that one cannot do anything else. We may call this a conclusion of practical necessity . . . a 'must' that is unconditional and *goes all the way down.*" Bernard Williams, *Ethics and the Limits of Philosophy* (Cambridge: Harvard University Press, 1985), pp. 197–98.

11. William Kittredge, *Owning It All* (Saint Paul: Graywolf Press, 1987), pp. 156–57.

12. Ibid., p. 64.

13. I am using the term *imperium* in the sense given by the *Oxford English Dictionary:* "command; absolute power; supreme or imperial power: Empire."

14. Section 2(a) of the Wilderness Act of 1964; emphasis added.

15. See Section 4(a)1 of the Wilderness Act of 1964, where consistency with the Multiple-Use Sustained-Yield Act is stipulated. The Wilderness Act permits the mining of claims established until 1984 for lands covered by the act. Roughly speaking, the legality of mining for other designated wilderness land has been decided on a case by case basis. For an excellent account of wilderness values, see Holmes Rolston III, *Philosophy Gone Wild* (Buffalo: Prometheus Books, 1986), pp. 180–205. Note further that the Wilderness Act, at Section 4(d)4, explicitly reserves the right to further resource uses within designated wilderness areas: ". . . prospecting for water resources, water-conservation works, power projects, transmission lines, and other facilities needed in the public interest, including the road construction and maintenance. . . ." [Included in this volume.]

16. For a sound discussion of wilderness as the source of resources, and not a resource itself, see Holmes Rolston III, "Values Gone Wild," in *Philosophy Gone Wild,* pp. 118–42.

17. Michel Foucault, *Discipline & Punish* (New York: Vintage Books, 1979), p. 301. Also see Foucault's *Madness and Civilization* (New York: Vintage Books, 1973).

18. See the Wilderness Act at Section 4(d)1: ". . . such measures may be taken as may be necessary in the control of fire, insects and diseases, subject to such conditions as the Secretary deems desirable." For an interpretation of the Wilderness Act, see Hendee et al., *Wilderness Management,* p. 82. Note the current halt and reconsideration of the let-burn policy for wilderness fire management, as the result of the huge Yellowstone and Canyon Creek (Scapegoat Wilderness Area) fires in the summer of 1988.

19. See Noam Chomsky, *The Culture of Terrorism* (Boston: South End Press, 1988), p. 32. Chomsky quotes the first major publication of the Trilateral Commission as saying that our educational institutions are responsible for the "indoctrination of the young."

20. Jean-Paul Sartre, *Saint Genet: Actor & Martyr* (New York: Pantheon Books, 1963), pp. 7, 31.

21. Although there is probably some real criminality in all human cultures, in the sense of the monstrous, or criminal insanity, a huge amount of what imperial society sees as criminal is created by its own laws and social structures. Thus in the U.S. (where the per capita prison population approaches that of South Africa, the world's highest) most incarcerated "criminals" are there for economic crimes, for example, stealing tires. Very few are monsters. The strategy of an oppressive culture is to enlarge the category of the really criminal, the monstrous, to include the "criminality" it has fabricated.

22. Jean Genet, *The Thief's Journal* (New York: Bantam Books, 1965), pp. 76–77. The prison to which Genet alludes, Fontevrault, was once in fact a palace.

23. Joseph Conrad, *Heart of Darkness,* ed. Robert Kimbrough (New York: Norton, 1988), pp. 29, 48–49.

24. E. M. Forster, *A Passage to India* (New York: Harcourt Brace, 1924), pp. 147–48.

25. Ibid., p. 149.

26. Mary Layoun, "Production of Narrative Value: The Colonial Paradigm," *North Dakota Quarterly* 55 (1987): 202–03. I have replaced "represented" in Layoun's published text with "repressed" from her original manuscript.

27. Genet, *Thief's Journal,* p. 232.

28. Ibid., p. 230.

29. Ibid., p. 5.

30. Ibid., p. 4.

31. Ibid., p. 3.

32. Jean Baudrillard, *Simulations* (New York: Semiotext[e], 1983), pp. 146 and 147.

33. Ibid., pp. 147–48.

34. Ibid., pp. 44–45.

35. Ibid., p. 25.

36. Ibid., p. 44; emphasis added.

37. Ibid., p. 30. Also see Chomsky, *The Culture of Terrorism,* chap. 4, "The Limits of Scandal." Chomsky argues convincingly that, in the course of treating scandals, the power structure is always careful to avoid asking the real questions, such as, "What moral right do we have to create and finance a proxy army to terrorize Nicaragua?"

38. This is an opportune point to notice just how natural, appropriate, and even plausible it is for Roderick Nash, in his chapter on "The International Perspective" in *Wilderness and the American Mind* [included in this volume], to subsume wild nature and wilderness reservations into the rhetoric of international export-import commercialism. In this vein, he suggests that "national parks and wilderness systems might be thought of as institutional 'containers' that developed nations send to underdeveloped ones for the purpose of 'packaging' a fragile resource" (p. 344). Such a packaging in containers, defined by *its* model of the wild

other, and the experience of it (the "wilderness experience") is precisely what the imperium tries to achieve.

39. Nash, *Wilderness and the American Mind,* p. xiii.

40. In the same vein, the acronym for the latest wilderness management practices is LAC (Limits of Acceptable Change).

41. See Guha, "Radical American Environmentalism and Wilderness Preservation," for development of this point. The imperium is thoroughly "Orientalist," in Edward Said's sense. See his *Orientalism* (New York: Pantheon, 1978). Its other is thus either criminal and diabolical or mystically enlightened, like the noble savage or the guru of the East, but irrational and benighted in either case.

42. Eventually either the recent negative Supreme Court decision on Indian religious rights to preserve sacred lands (*Lyng v. Northwest Indian Cemetery Protective Association*) will be somehow softened by the courts or Congress will (slightly) strengthen the American Indian Religious Freedom Act. The logic of imperial power requires this sort of liberality. Of course, the imperium could never afford the liberality of classifying all land as sacred in any meaningful sense. But some designated and narrowly defined sacred areas will be allowed, or, to use the language of rights, "granted."

43. J. Donald Hughes and Jim Swan, "How Much of the Earth Is Sacred Space?" *Environmental Review* 10 (1986): 256.

44. Gary Snyder, "Good, Wild, Sacred," in Wes Jackson et al., eds., *Meeting the Expectations of the Land* (San Francisco: North Point Press, 1984), p. 205.

45. Wendell Berry, "Preserving Wildness," in *Home Economics* (San Francisco: North Point Press, 1987), p. 151. The antithesis of the "landscape of harmony" is that of industrial monoculture—the landscape of the imperium.

46. Leslie Marmon Silko, "Landscape History, and the Pueblo Imagination," *Antaeus* 57 (1986): 92. The "balance" we need to find is that which Silko says the Pueblo people had to find, and did, in order to become a culture. It is not the balance of cost-benefit analysis, but that of the dance, which requires loving, graceful integration of self and society within the wild whole of an otherness we revere.

William Cronon

The Trouble with Wilderness, or, Getting Back to the Wrong Nature (1995)

THE TIME HAS COME to rethink wilderness.

This will seem a heretical claim to many environmentalists, since the idea of Wilderness has for decades been a fundamental tenet—indeed, a passion—of the environmental movement, especially in the United States. For many Americans, wilderness stands as the last remaining place where civilization, that all too human disease, has not fully infected the earth. It is an island in the polluted sea of urban-industrial modernity, the one place we can turn for escape from our own too-muchness. Seen in this way, wilderness presents itself as the best antidote to our human selves, a refuge we must somehow recover if we hope to save the planet. As Henry David Thoreau once famously declared, "In Wildness is the preservation of the World."[1]

But is it? The more one knows of its peculiar history, the more one realizes that wilderness is not quite what it seems. Far from being the one place on earth that stands apart from humanity, it is quite profoundly a human creation—indeed, the creation of very particular human cultures at very particular moments in human history. It is not a pristine sanctuary where the last remnant of an untouched, endangered, but still transcendent nature can for at least a little while longer be encountered without the contaminating taint of civilization. Instead, it is a product of that civilization, and could hardly be contaminated by the very stuff of which it is made. Wilderness hides its unnaturalness behind a mask that is all the more beguiling because it seems so natural. As we gaze into the mirror it holds up for us, we too easily imagine that what we behold is Nature when in fact we see the reflection of our own unexamined longings and desires. For this reason, we mistake ourselves when we suppose that wilderness can be the solution to our culture's problematic relationships with the nonhuman world, for wilderness is itself no small part of the problem.

To assert the unnaturalness of so natural a place will no doubt seem absurd or even perverse to many readers, so let me hasten to add that the nonhuman world we encounter in wilderness is far from being merely our own invention. I celebrate with others who love wilderness the beauty and power of the things it contains. Each of us who has spent time there can conjure images and sensations that seem all the more hauntingly real for having engraved themselves so indelibly on our memories. Such memories may be uniquely our own, but they are also familiar enough to be instantly recognizable to others. Remember this? The torrents of mist shoot out from the base of a great waterfall in the depths of a Sierra canyon, the tiny droplets cooling your face as you listen to the roar of the water and gaze up toward the sky through a rainbow that hovers just out of reach. Remember this too: looking out across a desert canyon in the evening air, the only sound a lone raven calling in the distance, the rock walls dropping away into a chasm so deep that its bottom all but vanishes as you squint into the amber light of the setting sun. And this: the moment beside the trail as you sit on a sandstone ledge, your boots damp with the morning dew while you take in the rich smell of the pines, and the small red fox—or maybe for you it was a raccoon or a coyote or a deer—that suddenly ambles across your path, stopping for a long moment to gaze in your direction with cautious indifference before continuing on its way. Remember the feelings of such moments, and you will know as well as I do that you were in the presence of something irreducibly nonhuman, something profoundly Other than yourself. Wilderness is made of that too.

And yet: what brought each of us to the places where such memories became possible is entirely a cultural invention. Go back 250 years in American and European history, and you do not find nearly so many people wandering around remote corners of the planet looking for what today we would call "the wilderness experience." As late as the eighteenth century, the most common usage of the word "wilderness" in the English language referred to landscapes that generally carried adjectives far different from the ones they attract today. To be a wilderness then was to be "deserted," "savage," "desolate," "barren"—in short, a "waste," the word's nearest synonym. Its connotations were anything but positive, and the emotion one was most likely to feel in its presence was "bewilderment"—or terror.[2]

Many of the word's strongest associations then were biblical, for it is used over and over again in the King James Version to refer to places on the margins of civilization where it is all too easy to lose oneself in moral confusion and despair. The wilderness was where Moses had wandered with his people for forty years, and where they had nearly abandoned their God to worship a golden idol.[3] "For Pharaoh will say of the children of Israel," we read in Exodus, "They are entangled in the land, the wilderness hath shut them in."[4] The wilderness was where Christ had struggled with the devil and endured his temptations: "And immediately the Spirit driveth him into the wilderness. And he was there in the wilderness forty days, tempted of Satan; and was with the wild beasts; and the angels ministered unto him."[5] The "delicious Paradise" of John Milton's Eden was surrounded by "a steep Wilderness, whose hairy sides / Access denied" to all who sought entry.[6] When Adam and Eve were driven from that garden, the world they entered was a wilderness that only their labor and pain could redeem. Wilderness, in short, was a place to which one came only against one's will, and always in fear and trembling. Whatever value it might have arose solely from the possibility that it might be "reclaimed" and turned toward human ends—planted as a garden, say, or a city upon a hill.[7] In its raw state, it had little or nothing to offer civilized men and women.

But by the end of the nineteenth century, all this had changed. The wastelands that had once seemed worthless had for some people come to seem almost beyond price. That Thoreau in 1862 could declare wildness to be the preservation of the world suggests the sea of change that was going on. Wilderness had once been the antithesis of all that was orderly and good—it had been the darkness, one might say, on the far side of the garden wall—and yet now it was frequently likened to Eden itself. When John Muir arrived in the Sierra Nevada in 1869, he would declare, "No description of Heaven that I have ever heard or read of seems half so fine."[8]

He was hardly alone in expressing such emotions. One by one, various corners of the American map came to be designated as sites whose wild beauty was so spectacular that a growing number of citizens had to visit and see them for themselves. Niagara Falls was the first to undergo this transformation, but it was soon followed by the Catskills, the Adirondacks, Yosemite, Yellowstone, and others. Yosemite was deeded by the United States government to the State of California in 1864 as the nation's first wildland park, and Yellowstone became the first true national park in 1872.[9]

By the first decade of the twentieth century, in the single most famous episode in American conservation history, a national debate had exploded over whether the city of San Francisco should be permitted to augment its water supply by damming the Tuolumne River in Hetch Hetchy Valley, well within the boundaries of Yosemite National Park. The dam was eventually built, but what today seems no less significant is that so many people fought to prevent its completion. Even as the fight was being lost, Hetch Hetchy became the battle cry of an emerging movement to preserve wilderness. Fifty years earlier, such opposition would have been unthinkable. Few would have questioned the merits of "reclaiming" a wasteland like this in order to put it to human use. Now the defenders of Hetch Hetchy attracted widespread national attention by portraying such an act not as improvement or progress but as desecration and vandalism. Lest one doubt that the old biblical metaphors had been turned completely on their heads, listen to John Muir attack the dam's defenders. "Their arguments," he wrote, "are curiously like those of the devil, devised for the destruction of the first garden—so much of the very best Eden fruit going to waste; so much of the best Tuolumne water and Tuolumne scenery going to waste."[10] For Muir and the growing number of Americans who shared his views, Satan's home had become God's own temple.

The sources of this rather astonishing transformation were many, but for the purposes of this essay they can be gathered under two broad headings: the sublime and the frontier. Of the two, the sublime is the older and more pervasive cultural construct, being one of the most important expressions of that broad transatlantic movement we today label as romanticism; the frontier is more peculiarly American, though it too had its European antecedents and parallels. The two converged to remake wilderness in their own image, freighting it with moral values and cultural symbols that it carries to this day. Indeed, it is not too much to say that the modern environmental movement is itself a grandchild of romanticism and post-frontier ideology, which is why it is no accident that so much environmentalist discourse takes its bearings from the wilderness these intellectual movements helped create. Although wilderness may today seem to be just one environmental concern among many, it in fact serves as the foundation for a long list of other such concerns that on their face seem quite remote from it. That is why its influence is so pervasive and, potentially, so insidious.

To gain such remarkable influence, the concept of wilderness had to become loaded with some of the deepest core values of the culture that created and idealized it: it had to become sacred. This possibility had been present in wilderness even in the days when it had been a place of spiritual danger and moral temptation. If Satan was there, then so was Christ, who had found angels as well as wild beasts during His sojourn in the desert. In the wilderness the boundaries between human and nonhuman, between natural and supernatural, had always seemed less certain than elsewhere. This was why the early Christian saints and mystics had often emulated Christ's desert retreat as they sought to experience for themselves the visions and spiritual testing He had endured. One might meet devils and run the risk of losing one's soul in such a place, but one might also meet God. For some that possibility was worth almost any price.

By the eighteenth century this sense of the wilderness as a landscape where the supernatural lay just beneath the surface was expressed in the doctrine of the *sublime,* a word whose modern usage has been so watered down by commercial hype and tourist advertising that it retains only a dim echo of its former power.[11] In the theories of Edmund Burke, Immanuel Kant, William Gilpin, and others, sublime landscapes were those rare places on earth where one had more chance than elsewhere to glimpse the face of God.[12] Romantics had a clear notion of where one could be most sure of having this experience. Although God might, of course, choose to show Himself anywhere, He would most often be found in those vast, powerful landscapes where one could not help feeling insignificant and being reminded of one's own mortality. Where were these sublime places? The eighteenth-century catalog of their locations feels very familiar, for we still see and value landscapes as it taught us to do. God was on the mountaintop, in the chasm, in the waterfall, in the thundercloud, in the rainbow, in the sunset. One has only to think of the sites that Americans chose for their first national parks—Yellowstone, Yosemite, Grand Canyon, Rainier, Zion—to realize that virtually all of them fit one or more of these categories. Less sublime landscapes simply did not appear worthy of such protection; not until the 1940s, for instance, would the first swamp be honored, in Everglades National Park, and to this day there is no national park in the grasslands.[13]

Among the best proofs that one had entered a sublime landscape was the emotion it evoked. For the early romantic writers and artists who first began to celebrate it, the sublime was far from being a pleasurable experience. The classic description is that of William Wordsworth as he recounted climbing the Alps and crossing the Simplon Pass in his autobiographical poem *The Prelude*. There, surrounded by crags and waterfalls, the poet felt himself literally to be in the presence of the divine—and experienced an emotion remarkably close to terror:

> The immeasurable height
> Of woods decaying, never to be decayed,
> The stationary blasts of waterfalls,
> And in the narrow rent at every turn
> Winds thwarting winds, bewildered and forlorn,
> The torrents shooting from the clear blue sky,
> The rocks that muttered close upon our ears,
> Black drizzling crags that spake by the way-side
> As if a voice were in them, the sick sight
> And giddy prospect of the raving stream,
> The unfettered clouds and region of the Heavens,
> Tumult and peace, the darkness and the light—
> Were all like workings of one mind, the features
> Of the same face, blossoms upon one tree;
> Characters of the great Apocalypse,
> The types and symbols of Eternity,
> Of first, and last, and midst, and without end.[14]

This was no casual stroll in the mountains, no simple sojourn in the gentle lap of nonhuman nature. What Wordsworth described was nothing less than a religious experience, akin to that of the Old Testament prophets as they conversed with their wrathful God. The symbols he detected in this wilderness landscape were more supernatural than natural, and they inspired more awe and dismay than joy or pleasure. No mere mortal was meant to linger long in such a place, so it was with considerable relief that

Wordsworth and his companion made their way back down from the peaks to the sheltering valleys.

Lest you suspect that this view of the sublime was limited to timid Europeans who lacked the American know-how for feeling at home in the wilderness, remember Henry David Thoreau's 1846 climb of Mount Katahdin in Maine. Although Thoreau is regarded by many today as one of the great American celebrators of wilderness, his emotions about Katahdin were no less ambivalent than Wordsworth's about the Alps.

> It was vast, Titanic, and such as man never inhabits. Some part of the beholder, even some vital part, seems to escape through the loose grating of his ribs as he ascends. He is more lone than you can imagine.... Vast, Titanic, inhuman Nature has got him at a disadvantage, caught him alone, and pilfers him of some of his divine faculty. She does not smile on him as in the plains. She seems to say sternly, why came ye here before your time? This ground is not prepared for you. Is it not enough that I smile in the valleys? I have never made this soil for thy feet, this air for thy breathing, these rocks for thy neighbors. I cannot pity nor fondle thee here, but forever relentlessly drive thee hence to where I *am* kind. Why seek me where I have not called thee, and then complain because you find me but a stepmother?[15]

This is surely not the way a modern backpacker or nature lover would describe Maine's most famous mountain, but that is because Thoreau's description owes as much to Wordsworth and other romantic contemporaries as to the rocks and clouds of Katahdin itself. His words took the physical mountain on which he stood and transmuted it into an icon of the sublime: a symbol of God's presence on earth. The power and the glory of that icon were such that only a prophet might gaze on it for long. In effect, romantics like Thoreau joined Moses and the children of Israel in Exodus when "they looked toward the wilderness, and, behold, the glory of the Lord appeared in the cloud."[16]

But even as it came to embody the awesome power of the sublime, wilderness was also being tamed—not just by those who were building settlements in its midst but also by those who most celebrated its inhuman beauty. By the second half of the nineteenth century, the terrible awe Wordsworth and Thoreau regarded as the appropriately pious stance to adopt in the presence of their mountaintop God was giving way to a much more comfortable, almost sentimental demeanor. As more and more tourists sought out the wilderness as a spectacle to be looked at and enjoyed for its great beauty, the sublime in effect became domesticated. The wilderness was still sacred, but the religious sentiments it evoked were more those of a pleasant parish church than those of a grand cathedral or a harsh desert retreat. The writer who best captures this late romantic sense of a domesticated sublime is undoubtedly John Muir, whose descriptions of Yosemite and the Sierra Nevada reflect none of the anxiety or terror one finds in earlier writers. Here he is, for instance, sketching on North Dome in Yosemite Valley:

> No pain here, no dull empty hours, no fear of the past, no fear of the future. These blessed mountains are so compactly filled with God's beauty, no petty personal hope or experience has room to be. Drinking this champagne water is pure pleasure, so is breathing the living air, and every movement of limbs is pleasure, while the body seems to feel beauty when exposed to it as it feels the campfire or sunshine, entering not by the eyes alone, but equally through all one's flesh like radiant heat, making a passionate ecstatic pleasure glow not explainable.

The emotions Muir describes in Yosemite could hardly be more different from Thoreau's on Katahdin or Wordsworth's on the Simplon Pass. Yet all three men are participating in the same cultural tradition and contributing to the same myth: the mountain as cathedral. The three may differ about the way they choose to express their piety—Wordsworth favoring an awe-filled bewilderment, Thoreau a stern loneliness, Muir a welcome ecstasy—but they agree completely about the church in which they prefer to worship. Muir's closing words on North Dome diverge from his older contemporaries only in mood, not in their ultimate content:

> Perched like a fly on this Yosemite dome, I gaze and sketch and bask, oftentimes settling down into dumb admiration without definite hope of ever learning much, yet with the longing, unresting effort that lies at the door of hope, humbly prostrate before the vast display of God's power, and eager to offer self-denial and renunciation with eternal toil to learn any lesson in the divine manuscript.[17]

Muir's "divine manuscript" and Wordsworth's "Characters of the great Apocalypse" were in fact pages from the same holy book. The sublime wilderness had ceased to be a place of satanic temptation and become instead a sacred temple, much as it continues to be for those who love it today.

But the romantic sublime was not the only cultural movement that helped transform wilderness into a sacred American icon during the nineteenth century. No less important was the powerful romantic attraction of primitivism, dating back at least to Rousseau—the belief that the best antidote to the ills of an overly refined and civilized modern world was a return to simpler, more primitive living. In the United States, this was embodied most strikingly in the national myth of the frontier. The historian Frederick Jackson Turner wrote in 1893 the classic academic statement of this myth, but it had been part of American cultural traditions for well over a century. As Turner described the process, easterners and European immigrants, in moving to the wild unsettled lands of the frontier, shed the trappings of civilization, rediscovered their primitive racial energies, reinvented direct democratic institutions, and thereby reinfused themselves with a vigor, an independence, and a creativity that were the source of American democracy and national character. Seen in this way, wild country became a place not just of religious redemption but of national renewal, the quintessential location for experiencing what it meant to be an American.

One of Turner's most provocative claims was that by the 1890s the frontier was passing away. Never again would "such gifts of free land offer themselves" to the American people. "The frontier has gone," he declared, "and with its going has closed the first period of American history."[18] Built into the frontier myth from its very beginning was the notion that this crucible of American identity was temporary and would pass away. Those who have celebrated the frontier have almost always looked backward as they did so, mourning an older, simpler, truer world that is about to disappear forever. That world and all of its attractions, Turner said, depended on free land—on wilderness. Thus, in the myth of the vanishing frontier lay the seeds of wilderness preservation in the United States, for if wild land had been so crucial in the making of the nation, then surely one must save its last remnants as monuments to the American past—and as an insurance policy to protect its future. It is no accident that the movement to set aside national parks and wilderness areas began to gain real momentum at precisely the time that laments about the passing frontier reached their peak. To protect wilderness was in a very real sense to protect the nation's most sacred myth of origin.

Among the core elements of the frontier myth was the powerful sense among certain groups of Americans that wilderness was the last bastion of rugged individualism. Turner tended to stress communitarian themes when writing frontier history, asserting that Americans in primitive conditions had been forced to band together with their neighbors to form communities and democratic institutions. For other writers, however, frontier democracy for communities was less compelling than frontier freedom for individuals.[19] By fleeing to the outer margins of settled land and society—so the story ran—an individual could escape the confining strictures of civilized life. The mood among writers who celebrated frontier individualism was almost always nostalgic; they lamented not just a lost way of life but the passing of the heroic men who had embodied that life. Thus Owen Wister in the introduction to his classic 1902 novel *The Virginian* could write of "a vanished world" in which "the horseman, the cow-puncher, the last romantic figure upon our soil" rode only "in his historic yesterday" and would "never come again." For Wister, the cowboy was a man who gave his word and kept it ("Wall Street would have found him behind the times"), who did not talk lewdly to women ("Newport would have thought him old-fashioned"), who worked and played hard, and whose "ungoverned hours did not unman him."[20] Theodore Roosevelt wrote with much the same nostalgic fervor about the "fine, manly qualities" of the "wild rough-rider of the plains." No one could be more heroically masculine, thought Roosevelt, or more at home in the western wilderness:

> There he passes his days, there he does his life-work, there, when he meets death, he faces it as he has faced many other evils, with quiet, uncomplaining fortitude. Brave, hospitable, hardy, and adventurous, he is the grim pioneer of our race; he prepares the way for the civilization from before whose face he must himself disappear. Hard and dangerous though his existence is, it has yet a wild attraction that strongly draws to it his bold, free Spirit.[21]

This nostalgia for a passing frontier way of life inevitably implied ambivalence, if not downright hostility, toward modernity and all that it represented. If one saw the wild lands of the frontier as freer, truer, and more natural than other, more modern places, then one was also inclined to see the cities and factories of urban-industrial civilization as confining, false, and artificial. Owen Wister looked at the post-frontier "transition" that had followed "the horseman of the plains," and did not like what he saw: "a shapeless state, a condition of men and manners as unlovely as is that moment in the year when winter is gone and spring not come, and the face of Nature is ugly."[22] In the eyes of writers who shared Wister's distaste for modernity, civilization contaminated its inhabitants and absorbed them into the faceless, collective, contemptible life of the crowd. For all of its troubles and dangers, and despite the fact that it must pass away, the frontier had been a better place. If civilization was to be redeemed, it would be by men like the Virginian who could retain their frontier virtues even as they made the transition to post-frontier life.

The mythic frontier individualist was almost always masculine in gender: here, in the Wilderness, a man could be a real man, the rugged individual he was meant to be before civilization sapped his energy and threatened his masculinity. Wister's contemptuous remarks about Wall Street and Newport suggest what he and many others of his generation believed—that the comforts and seductions of civilized life were especially insidious for men, who all too easily became emasculated by the femininizing tendencies of civilization. More often than not, men who felt this way came, like Wister and Roosevelt, from elite class backgrounds. The curious result was that frontier nostalgia became an important vehicle for expressing a peculiarly bourgeois form of antimodernism. The very men who most benefited from urban-industrial capitalism were among those who believed they must escape its debilitating effects. If the frontier was passing, then men who had the means to do so should preserve for themselves some remnant of its wild landscape so that they might enjoy the regeneration and renewal that came from sleeping under the stars, participating in blood sports, and living off the land. The frontier might be gone, but the frontier experience could still be had if only wilderness were preserved.

Thus the decades following the Civil War saw more and more of the nation's wealthiest citizens seeking out wilderness for themselves. The elite passion for wild land took many forms: enormous estates in the Adirondacks and elsewhere (disingenuously called "camps" despite their many servants and amenities), cattle ranches for would-be rough riders on the Great Plains, guided big game hunting trips in the Rockies, and luxurious resort hotels wherever railroads pushed their way into sublime landscapes. Wilderness suddenly emerged as the landscape of choice for elite tourists, who brought with them strikingly urban ideas of the countryside through which they traveled. For them, wild land was not a site for productive labor and not a permanent home; rather, it was a place of recreation. One went to the wilderness not as a producer but as a consumer, hiring guides and other backcountry residents who could serve as romantic surrogates for the rough riders and hunters of the frontier if one was willing to overlook their new status as employees and servants of the rich.

In just this way, wilderness came to embody the national frontier myth, standing for the wild freedom of America's past and seeming to represent a highly attractive natural alternative to the ugly artificiality of modern civilization. The irony, of course, was that in the process wilderness came to reflect the very civilization its devotees sought to escape. Ever since the nineteenth century, celebrating wilderness has been an activity mainly for well-to-do city folks. Country people generally know far too much about working the land to regard *un*worked land as their ideal. In contrast, elite urban tourists and wealthy sportsmen projected their leisure-time frontier fantasies onto the American landscape and so created wilderness in their own image.

There were other ironies as well. The movement to set aside national parks and wilderness areas followed hard on the heels of the final Indian wars, in which the prior human inhabitants of these areas were rounded up and moved onto reservations. The myth of the wilderness as "virgin," uninhabited land had always been especially cruel when seen from the perspective of the Indians who had once called that land home. Now they were forced to move elsewhere, with the result that tourists could safely enjoy the illusion that they were seeing their nation in its pristine, original state, in the new morning of God's own creation.[23] Among the things that most marked the new national parks as reflecting a post-frontier consciousness was the relative absence of human violence within their boundaries. The actual frontier had often been a place of conflict, in which invaders and invaded fought for control of land and resources. Once set aside within the fixed and carefully policed boundaries of the modern bureaucratic state, the wilderness lost its savage image and became safe: a place more of reverie than of revulsion or fear. Meanwhile, its original inhabitants were kept out by dint of force, their earlier uses of the land redefined as inappropriate or even illegal. To this day, for instance, the Blackfeet continue to be accused of "poaching" on the lands of Glacier National Park that originally belonged to them and that were ceded by treaty only with the proviso that they be permitted to hunt there.[24]

The removal of Indians to create an "uninhabited wilderness"—uninhabited as never before in the human history of the place—reminds us just how invented, just how constructed, the American wilderness really is. To return to my opening argument: there is nothing natural about the concept of wilderness. It is entirely a creation of the culture that holds it dear, a product of the very history it seeks to deny. Indeed, one of the most striking proofs of the cultural invention of wilderness is its thoroughgoing erasure of the history from which it sprang. In virtually all of its manifestations, wilderness represents a flight from history. Seen as the original garden, it is a place outside of time, from which human beings had to be ejected before the fallen world of history could properly begin. Seen as the frontier, it is a savage world at the dawn of civilization, whose transformation represents the very beginning of the national historical epic. Seen as the bold landscape of frontier heroism, it is the place of youth and childhood, into which men escape by abandoning their pasts and entering a world of freedom where the constraints of civilization fade into memory. Seen as the sacred sublime, it is the home of a God who transcends history by standing as the One who remains untouched and unchanged by time's arrow. No matter what the angle from which we regard it, wilderness offers us the illusion that we can escape the cares and troubles of the world in which our past has ensnared us.[25]

This escape from history is one reason why the language we use to talk about wilderness is often permeated with spiritual and religious values that reflect human ideals far more than the material world of physical nature. Wilderness fulfills the old romantic project of secularizing Judeo-Christian values so as to make a new cathedral not in some petty human building but in God's own creation, Nature itself. Many environmentalists who reject traditional notions of the Godhead and who regard themselves as agnostics or even atheists nonetheless express feelings tantamount to religious awe when in the presence of wilderness—a fact that testifies to the success of the romantic project. Those who have no difficulty seeing God as the expression of our human dreams and desires nonetheless have trouble recognizing that in a secular age Nature can offer precisely the same sort of mirror.

Thus it is that wilderness serves as the unexamined foundation on which so many of the quasi-religious values of modern environmentalism rest. The critique of modernity that is one of environmentalism's most important contributions to the moral and political discourse of our time more often than not appeals, explicitly or implicitly, to wilderness as the standard against which to measure the failings of our human world. Wilderness is the natural, unfallen antithesis of an unnatural civilization that has lost its soul. It is a place of freedom in which we can recover the true selves we have lost to the corrupting influences of our artificial lives. Most of all, it is the ultimate landscape of authenticity. Combining the sacred grandeur of the sublime with the primitive simplicity of the frontier, it is the place where we can see the world as it really is, and so know ourselves as we really are—or ought to be.

But the trouble with wilderness is that it quietly expresses and reproduces the very values its devotees seek to reject. The flight from history that is very nearly the core of wilderness represents the false hope of an escape from responsibility, the illusion that we can somehow wipe clean the slate of our past and return to the tabula rasa that supposedly existed before we began to leave our marks on the world. The dream of an unworked natural landscape is very much the fantasy of people who have never themselves had to work the land to make a living—urban folk for whom food comes from a supermarket or a restaurant instead of a field, and for whom the wooden houses in which they live and work apparently have no meaningful connection to the forests in which trees grow and die. Only people whose relation to the land was already alienated could hold up wilderness as a model for human life in nature, for the romantic ideology of wilderness leaves precisely nowhere for human beings actually to make their living from the land.

This, then, is the central paradox: wilderness embodies a dualistic vision in which the human is entirely outside the natural. If we allow ourselves to believe that nature, to be true, must also be wild, then our very presence in nature represents its fall. The place where we are is the place where nature is not. If this is so—if by definition wilderness leaves no place for human beings, save perhaps as contemplative sojourners enjoying their leisurely reverie in God's natural cathedral—then also by definition it can offer no solution to the environmental and other problems that confront us. To the extent that we celebrate wilderness as the measure with which we judge civilization, we reproduce the dualism that sets humanity and nature at opposite poles. We thereby leave ourselves little hope of discovering what an ethical, sustainable, *honorable* human place in nature might actually look like.

Worse: to the extent that we live in an urban-industrial civilization but at the same time pretend to ourselves that our *real* home is in the wilderness, to just that extent we give ourselves permission to evade responsibility for the lives we actually lead. We inhabit civilization while holding some part of ourselves—what we imagine to be the most precious part—aloof from its entanglements. We work our nine-to-five jobs in its institutions, we eat its food, we drive its cars (not least to reach the wilderness), we benefit from the intricate and all too invisible networks with which it shelters us, all the while pretending that these things are not an essential part of who we are. By imagining that our true home is in the wilderness, we forgive ourselves the homes we actually inhabit. In its flight from history, in its siren song of escape, in its reproduction of the dangerous dualism that sets human beings outside of nature—in all of these ways, wilderness poses a serious threat to responsible environmentalism at the end of the twentieth century.

By now I hope it is clear that my criticism in this essay is not directed at wild nature per se, or even at efforts to set aside large tracts of wild land, but rather at the specific habits of thinking that flow from this complex cultural construction called "wilderness." It is not the things we label as wilderness that are the problem—for nonhuman nature and large tracts of the natural world *do* deserve protection—but rather what we ourselves mean when we use that label. Lest one doubt how pervasive these habits of thought actually are in contemporary environmentalism, let me list some of the places where wilderness serves as the ideological underpinning for environmental concerns that might otherwise seem quite remote from it. Defenders of biological diversity, for instance, although sometimes appealing to more utilitarian concerns, often point to "untouched" ecosystems as the best and richest repositories of the undiscovered species we must certainly try to protect. Although at first blush an apparently more "scientific" concept than wilderness, biological diversity in fact invokes many of the same sacred values, which is why organizations like the Nature Conservancy have been so quick to employ it as an alternative to the seemingly fuzzier and more problematic concept of wilderness. There is a paradox here, of course. To the extent that biological diversity (indeed, even wilderness itself) is likely to survive in the future only by the most vigilant and self-conscious management of the ecosystems that sustain it, the ideology of wilderness is potentially in direct conflict with the very thing it encourages us to protect.[26]

The most striking instances of this have revolved around "endangered species," which serve as vulnerable symbols of biological diversity while at the same time standing as surrogates for wilderness itself. The terms of the Endangered Species Act in the United States have often meant that those hoping to defend pristine wilderness have had to rely on a single endangered species like the spotted owl to gain legal standing for their case—thereby making the full power of sacred land inhere in a single numinous organism whose habitat then becomes the object of intense debate about appropriate management and use. The ease with which anti-environmental forces like the Wise Use Movement have attacked such single-species preservation efforts suggests the vulnerability of strategies like these.

Perhaps partly because our own conflicts over such places and organisms have become so messy, the convergence of wilderness values with concerns about biological diversity and endangered species has helped produce a deep fascination for remote ecosystems, where it is easier to imagine that nature might somehow be "left alone" to flourish by its own pristine devices. The classic example is the tropical rain forest, which since the 1970s has become the most powerful modern icon of unfallen, sacred land—a veritable Garden of Eden—for many Americans and Europeans. And yet protecting the rain forest in the eyes of First World environmentalists all too often means protecting it from the people who live there. Those who seek to preserve such "wilderness" from the activities of native peoples run the risk of reproducing the same tragedy—being forcibly removed from an ancient home—that befell American Indians. Third World countries face massive environmental problems and deep social conflicts, but these are not likely to be solved by a cultural myth that encourages us to "preserve" peopleless landscapes that have not existed in such places for millennia. At its worst, as environmentalists are beginning to realize, exporting American notions of wilderness in this way can become an unthinking and self-defeating form of cultural imperialism.[27]

Perhaps the most suggestive example of the way that wilderness thinking can underpin other environmental concerns has emerged in the recent debate about "global change." In 1989 the journalist Bill McKibben published a book entitled *The End of Nature,* in which he argued that the prospect of global climate change as a result of unintentional human manipulation of the atmosphere means that nature as we once knew it no longer exists.[28] Whereas earlier generations inhabited a natural world that remained more or less unaffected by their actions, our own generation is uniquely different. We and our children will henceforth live in a biosphere completely altered by our own activity, a planet in which the human and the natural can no longer be distinguished, because the one has overwhelmed the other. In McKibben's view, nature has died, and we are responsible for killing it. "The planet," he declares, "is utterly different now."[29]

But such a perspective is possible only if we accept the wilderness premise that nature, to be natural, must also be pristine—remote from humanity and untouched by our common past. In fact, everything we know about environmental history suggests that people have been manipulating the natural world on various scales for as long as we have a record of their passing. Moreover, we have unassailable evidence that many of the environmental changes we now face also occurred quite apart from human intervention at one time or another in the earth's past.[30] The point is not that our current problems are trivial, or that our devastating effects on the earth's ecosystems should be accepted as inevitable or "natural." It is rather that we seem unlikely to make much progress in solving these problems if we hold up to ourselves as the mirror of nature a wilderness we ourselves cannot inhabit.

To do so is merely to take to a logical extreme the paradox that was built into wilderness from the beginning: if nature dies because we enter it, then the only way to save nature is to kill ourselves. The absurdity of this proposition flows from the underlying dualism it expresses. Not only does it ascribe greater power to humanity than we in fact possess—physical and biological nature will surely survive in some form or another long after we ourselves have gone the way of all flesh—but in the end it offers us little more than a self-defeating counsel of despair. The tautology gives us no way out: if wild nature is the only thing worth saving, and if our mere presence destroys it, then the sole solution to our own unnaturalness, the only way to protect sacred wilderness from profane humanity, would seem to be suicide. It is not a proposition that seems likely to produce very positive or practical results.

And yet radical environmentalists and deep ecologists all too frequently come close to accepting this premise as a first principle. When they express, for instance, the popular notion that our environmental problems began with the invention of agriculture, they push the human fall from natural grace so far back into the past that all of civilized history becomes a tale of ecological declension. Earth First! co-founder Dave Foreman captures the familiar parable succinctly when he writes,

> Before agriculture was midwifed in the Middle East, humans were in the wilderness. We had no concept of "wilderness" because everything was wilderness and *we were a part of it.* But with irrigation ditches, crop surpluses, and permanent villages, we became *apart from* the natural world.... Between the wilderness that created us and the civilization created by us grew an ever-widening rift.[31]

In this view the farm becomes the first and most important battlefield in the long war against wild nature, and all else follows in its wake. From such a starting place, it is hard not to reach the conclusion that the only way human beings can hope to live naturally on earth is to follow the hunter-gatherers back into a wilderness Eden and abandon virtually everything that civilization has given us. It may indeed turn out that civilization will end in ecological collapse or nuclear disaster, whereupon one might expect to find any human survivors returning to a way of life closer to that celebrated by Foreman and his followers. For most of us, though, such a debacle would be cause for regret, a sign that humanity had failed to fulfill its own promise and failed to honor its own highest values—including those of the deep ecologists.

In offering wilderness as the ultimate hunter-gatherer alternative to civilization, Foreman reproduces an extreme but still easily recognizable version of the myth of frontier primitivism. When he writes of his fellow Earth First!ers that "we believe we must return to being animal, to glorying in our sweat, hormones, tears, and blood" and that "we struggle against the modern compulsion to become dull, passionless androids," he is following in the footsteps of Owen Wister.[32] Although his arguments give primacy to defending biodiversity and the autonomy of wild nature, his prose becomes most passionate when he speaks of preserving "the wilderness experience." His own ideal "Big Outside" bears an uncanny resemblance to that of the frontier myth: wide open spaces and virgin land with no trails, no signs, no facilities, no maps, no guides, no rescues, no modern equipment. Tellingly, it is a land where hardy travelers can support themselves by hunting with "primitive weapons (bow and arrow, atlatl, knife, sharp rock)."[33] Foreman claims that "the primary value of wilderness is not as a proving ground for young Huck Finns and Annie Oakleys," but his heart is with Huck and Annie all the same. He admits that "preserving a quality wilderness experience for the human visitor, letting her or him flex Paleolithic muscles or seek visions, remains a tremendously important secondary purpose."[34] Just so does Teddy Roosevelt's rough rider live on in the greener garb of a new age.

However much one may be attracted to such a vision, it entails problematic consequences. For one, it makes wilderness the locus for an epic struggle between malign civilization and benign nature, compared with which all other social, political, and moral concerns seem trivial. Foreman writes, "The preservation of wildness and native diversity is *the* most important issue. Issues directly affecting only humans pale in comparison."[35] Presumably so do any environmental problems whose victims are mainly people, for such problems usually surface in landscapes that have already "fallen" and are no longer wild. This would seem to exclude from the radical environmentalist agenda problems of occupational health and safety in industrial settings, problems of toxic waste exposure on "unnatural" urban and agricultural sites, problems of poor children poisoned by lead exposure in the inner city, problems of famine and poverty and human suffering in the "overpopulated" places of the earth—problems, in short, of environmental justice. If we set too high a stock on wilderness, too many other corners of the earth become less than natural and too many other people become less than human, thereby giving us permission not to care much about their suffering or their fate.

It is no accident that these supposedly inconsequential environmental problems mainly affect poor people, for the long affiliation between wilderness and wealth means that the only poor people who count when wilderness is *the* issue are hunter-gatherers, who presumably do not consider themselves to be poor in the first place. The dualism at the heart of wilderness encourages its advocates to conceive of its protection as a crude conflict between the "human" and the "nonhuman"—or, more often, between those who value the nonhuman and those who do not. This in turn tempts one to ignore crucial differences *among* humans and the complex cultural and historical reasons why different peoples may feel very differently about the meaning of wilderness.

Why, for instance, is the "wilderness experience" so often conceived as a form of recreation best enjoyed by those whose class privileges give them the time and resources to leave their jobs behind and "get away from it all"?

Why does the protection of wilderness so often seem to pit urban recreationists against rural people who actually earn their living from the land (excepting those who sell goods and services to the tourists themselves)? Why in the debates about pristine natural areas are "primitive" peoples idealized, even sentimentalized, until the moment they do something unprimitive, modern, and unnatural, and thereby fall from environmental grace? What are the consequences of a wilderness ideology that devalues productive labor and, the very concrete knowledge that comes from working the land with one's own hands?[36] All of these questions imply conflicts among different groups of people, conflicts that are obscured behind the deceptive clarity of "human" vs. "nonhuman." If in answering these knotty questions we resort to so simplistic an opposition, we are almost certain to ignore the very subtleties and complexities we need to understand.

But the most troubling cultural baggage that accompanies the celebration of wilderness has less to do with remote rain forests and peoples than with the ways we think about ourselves—we American environmentalists who quite rightly worry about the future of the earth and the threats we pose to the natural world. Idealizing a distant wilderness too often means not idealizing the environment in which we actually live, the landscape that for better or worse we call home. The majority of our most serious environmental problems start right here, at home, and if we are to solve these problems, we need an environmental ethic that will tell us as much about *using* nature as about *not* using it. The wilderness dualism tends to cast any use as *ab*-use, and thereby denies us a middle ground in which responsible use and non-use might attain some kind of balanced, sustainable relationship. My own belief is that only by exploring this middle ground will we learn ways of imagining a better world for all of us: humans and nonhumans, rich people and poor, women and men, First Worlders and Third Worlders, white folks and people of color, consumers and producers—a world better for humanity in all of its diversity and for all the rest of nature too. The middle ground is where we actually live. It is where we—all of us, in our different places and ways—make our homes.

That is why, when I think of the times I myself have come closest to experiencing what I might call the sacred in nature, I find myself remembering not some remote wilderness but places much closer to home. I think, for instance, of a small pond near my house where water bubbles up from limestone springs to feed a series of pools that rarely freeze in winter and so play home to waterfowl that stay here for the protective warmth even on the coldest of winter days, gliding silently through steaming mists as the snow falls from gray February skies. I think of a November evening long ago when I found myself on a Wisconsin hilltop in rain and dense fog, only to have the setting sun break through the clouds to cast an otherwordly golden light on the misty farms and woodlands below, a scene so unexpected and joyous that I lingered past dusk so as not to miss any part of the gift that had come my way. And I think perhaps most especially of the blown-out, bankrupt farm in the sand country of central Wisconsin where Aldo Leopold and his family tried one of the first American experiments in ecological restoration, turning ravaged and infertile soil into carefully tended ground where the human and the nonhuman could exist side by side in relative harmony. What I celebrate about such places is not *just* their wildness, though that certainly is among their most important qualities; what I celebrate even more is that they remind us of the wildness in our own backyards, of the nature that is all around us if only we have eyes to see it.

Indeed, my principal objection to wilderness is that it may teach us to be dismissive or even contemptuous of such humble places and experiences. Without our quite realizing it, wilderness tends to privilege some parts of nature at the expense of others. Most of us, I suspect, still follow the conventions of the romantic sublime in finding the mountaintop more glorious than the plains, the ancient forest nobler than the grasslands, the mighty canyon more inspiring than the humble marsh. Even John Muir, in arguing against those who sought to dam his beloved Hetch Hetchy valley in the Sierra Nevada, argued for alternative dam sites in the gentler valleys of the foothills—a preference that had nothing to do with nature and everything with the cultural traditions of the sublime.[37] Just as problematically, our frontier traditions have encouraged Americans to define "true" wilderness as requiring very large tracts of roadless land—what Dave Foreman calls "The Big Outside." Leaving aside the legitimate empirical question in conservation biology of how large a tract of land must be before a given species can reproduce on it, the emphasis on big wilderness reflects a romantic frontier belief that one hasn't really gotten away from civilization unless one can go for days at a time without encountering another human being. By teaching us to fetishize sublime places and wide open country, these peculiarly American ways of thinking about wilderness encourage us to adopt too high a standard for what counts as "natural." If it isn't hundreds of square miles big, if it doesn't give us God's-eye views or grand vistas, if it doesn't permit us the illusion that we are alone on the planet, then it really isn't natural. It's too small, too plain, or too crowded to be *authentically* wild.

In critiquing wilderness as I have done in this essay, I'm forced to confront my own deep ambivalence about its meaning for modern environmentalism. On the one hand, one of my own most important environmental ethics is that people should always be conscious that they are part of the natural world, inextricably tied to the ecological systems that sustain their lives. Any way of looking at nature that encourages us to believe we are separate from nature—as wilderness tends to do—is likely to reinforce environmentally irresponsible behavior. On the other hand, I also think it no less crucial for us to recognize and honor nonhuman nature as a world we did not create, a world with its own independent, nonhuman reasons for being as it is. The autonomy of nonhuman nature seems to me an indispensable corrective to human arrogance. Any way of looking at nature that helps us remember—as wilderness also tends to do—that the interests of people are not necessarily identical to those of every other creature or of the earth itself is likely to foster *responsible* behavior. To the extent that wilderness has served as an important vehicle for articulating deep moral values regarding our obligations and responsibilities to the nonhuman world, I would not want to jettison the contributions it has made to our culture's ways of thinking about nature.

If the core problem of wilderness is that it distances us too much from the very things it teaches us to value, then the question we must ask is what it can tell us about *home,* the place where we actually live. How can we take the positive values we associate with wilderness and bring them closer to home? I think the answer to this question will come by broadening our sense of the otherness that wilderness seeks to define and protect. In reminding us of the world we did not make, wilderness can teach profound feelings of humility and respect as we confront our fellow beings and the earth itself. Feelings like these argue for the importance of self-awareness and self-criticism as we exercise our own ability to transform the world around us, helping us set responsible limits to human mastery—which without such limits too easily become human hubris. Wilderness is the place where, symbolically at least, we try to withhold our power to dominate.

Wallace Stegner once wrote of

> the special human mark, the special record of human passage, that distinguishes man from all other species. It is rare enough among men, impossible to any other form of life. *It is simply the deliberate and chosen refusal to make any marks at all.* . . . We are the most dangerous species of life on the planet, and every other species, even the earth itself, has cause to fear our power to exterminate. But we are also the only species which, when it chooses to do so, will go to great effort to save what it might destroy.[38]

The myth of wilderness, which Stegner knowingly reproduces in these remarks, is that we can somehow leave nature untouched by our passage. By now it should be clear that this for the most part is an illusion. But Stegner's deeper message then becomes all the more compelling. If living in history means that we cannot help leaving marks on a fallen world, then the dilemma we face is to decide what kinds of marks we wish to leave. It is just here that our cultural traditions of wilderness remain so important. In the broadest sense, wilderness teaches us to ask whether the Other must always bend to our will, and, if not, under what circumstances it should be allowed to flourish without our intervention. This is surely a question worth asking about everything we do, and not just about the natural world.

When we visit a wilderness area, we find ourselves surrounded by plants and animals and physical landscapes whose otherness compels our attention. In forcing us to acknowledge that they are not of our making, that they have little or no need of our continued existence, they recall for us a creation far greater than our own. In the wilderness, we need no reminder that a tree has its own reasons for being, quite apart from us. The same is less true in the gardens we plant and tend ourselves: there it is far easier to forget the otherness of the tree.[39] Indeed, one could almost measure wilderness by the extent to which our recognition of its otherness requires a conscious, willed act on our part. The romantic legacy means that wilderness is more a state of mind than a fact of nature, and the state of mind that today most defines wilderness is *wonder.* The striking power of the wild is that wonder in the face of it requires no act of will, but forces itself upon us—as an expression of the nonhuman world experienced through the lens of our cultural history—as proof that ours is not the only presence in the universe.

Wilderness gets us into trouble only if we imagine that this experience of wonder and otherness is limited to the remote corners of the planet, or that it somehow depends on pristine landscapes we ourselves do not inhabit. Nothing could be more misleading. The tree in the garden is in reality no less other, no less worthy of our wonder and respect, than the tree in an ancient forest that has never known an ax or a saw—even though the tree in the forest reflects a more intricate web of ecological relationships. The tree in the garden could easily have sprung from the same seed as the tree in the forest, and we can claim only its location and perhaps its form as our own. Both trees stand apart from us; both share our common world. The special power of the tree in the wilderness is to remind us of this fact. It can teach us to recognize the wildness we did not see in the tree we planted in our own backyard. By seeing the otherness in that which is most unfamiliar, we can learn to see it too in that which at first seemed merely ordinary. If wilderness can do this—if it can help us perceive and respect a nature we had forgotten to recognize as natural—then it will become part of the solution to our environmental dilemmas rather than part of the problem.

This will only happen, however, if we abandon the dualism that sees the tree in the garden as artificial—completely fallen and unnatural—and the tree in the wilderness as natural—completely pristine and wild. Both trees in some ultimate sense are wild; both in a practical sense now depend on our management and care. We are responsible for both, even though we can claim credit for neither. Our challenge is to stop thinking of such things according to a set of bipolar moral scales in which the human and the nonhuman, the unnatural and the natural, the fallen and the unfallen, serve as our conceptual map for understanding and valuing the world. Instead, we need to embrace the full continuum of a natural landscape that is also cultural, in which the city, the suburb, the pastoral, and the wild each has its proper place, which we permit ourselves to celebrate without needlessly denigrating the others. We need to honor the Other within and the Other next door as much as we do the exotic Other that lives far away—a lesson that applies as much to people as it does to (other) natural things. In particular, we need to discover a common middle ground in which all of these things, from the city to the wilderness, can somehow be encompassed in the word "home." Home, after all, is the place where finally we make our living. It is the place for which we take responsibility, the place we try to sustain so we can pass on what is best in it (and in ourselves) to our children.[40]

The task of making a home in nature is what Wendell Berry has called "the forever unfinished lifework of our species." "The only thing we have to preserve nature with," he writes, "is culture; the only thing we have to preserve wildness with is domesticity."[41] Calling a place home inevitably means that we will use the nature we find in it, for there can be no escape from manipulating and working and even killing some parts of nature to make our home. But if we acknowledge the autonomy and otherness of the things and creatures around us—an autonomy our culture has taught us to label with the word "wild"—then we will at least think carefully about the uses to which we put them and even ask if we should use them at all. Just so can we still join Thoreau in declaring that "in Wildness is the preservation of the World," for *wild*ness (as opposed to wilderness) can be found anywhere: in the seemingly tame fields and woodlots of Massachusetts, in the cracks of a Manhattan sidewalk, even in the cells of our own bodies. As Gary Snyder has wisely said, "A person with a clear heart and open mind can experience the wilderness anywhere on earth. It is a quality of one's own consciousness. The planet is a wild place and always will be."[42] To think ourselves capable of causing "the end of nature" is an act of great hubris, for it means forgetting the wildness that dwells everywhere within and around us.

Learning to honor the wild—learning to remember and acknowledge the autonomy of the other—means striving for critical self-consciousness in all of our actions. It means that deep reflection and respect must accompany each act of use, and means too that we must always consider the possibility of non-use. It means looking at the part of nature we intend to turn toward our own ends and asking whether we can use it again and again and again—sustainably—without its being diminished in the process. It means never imagining that we can flee into a mythical wilderness to escape history and the obligation to take responsibility for our own actions that history inescapably entails. Most of all, it means practicing remembrance and gratitude, for thanksgiving is the simplest and most basic of ways for us to recollect the nature, the culture, and the history that have come together to make the world as we know it. If wildness can stop being (just) out there and start being (also) in here, if it can start being as humane as it is natural, then perhaps we can get on with the unending task of struggling to live rightly in the world—not just in the garden, not just in the wilderness, but in the home that encompasses them both.

NOTES

1. Henry David Thoreau, "Walking," *The Works of Thoreau,* ed., Henry S. Canby (Boston: Houghton-Mifflin, 1937), p. 672.

2. *Oxford English Dictionary,* s.v. "wilderness"; see also Roderick Nash, *Wilderness and the American Mind,* 3rd ed. (New Haven: Yale University Press, 1967, 1982), pp. 1–22; and Max Oelschlaeger, *The Idea of Wilderness: From Prehistory to the Age of Ecology* (New Haven: Yale University Press, 1991).

3. Exodus, 32:1–35, KJV.

4. Exodus, 14:3, KJV.

5. Mark 1:12–13 KJV; see also Matthew, 4:1–11; and Luke, 4:1–13.

6. John Milton, "Paradise Lost," *John Milton: Complete Poems and Major Prose,* ed. Merritt Y. Hughes (New York: Odyssey Press, 1957), pp. 280–81, lines 131–142.

7. I have discussed this theme at length in "Landscapes of Abundance and Scarcity," in Clyde Milner et al., eds., *Oxford History of the American West* (New York: Oxford University Press, 1994), pp. 603–37. The classic work on the Puritan "city on a hill" in colonial New England is Perry Miller, *Errand Into the Wilderness* (Cambridge, Mass.: Harvard University Press, 1956).

8. John Muir, *My First Summer in the Sierra* (1911), reprinted in *John Muir: The Eight Wilderness Discovery Books* (London: Diadem; Seattle: The Mountaineers, 1992), p. 211.

9. Alfred Runte, *National Parks: The American Experience,* 2nd ed. (Lincoln: University of Nebraska Press, 1987).

10. John Muir, *The Yosemite* (1912), reprinted in *John Muir: Eight Wilderness Discovery Books,* p. 715.

11. Scholarly work on the sublime is extensive. Among the most important studies are Samuel Monk, *The Sublime: A Study of Critical Theories in XVIII-Century England* (New York, 1935); Basil Willey, *The Eighteenth-Century Background: Studies on the Idea of Nature in the Thought of the Period* (London: Chatto and Windus, 1949); Marjorie Hope Nicolson, *Mountain Gloom and Mountain Glory: The Development of the Aesthetics of the Infinite* (Ithaca: Cornell University Press, 1959); Thomas Weiskel, *The Romantic Sublime: Studies in the Structure and Psychology of Transcendence* (Baltimore: Johns Hopkins University Press, 1976); Barbara Novak, *Nature and Culture: American Landscape Painting, 1825–1875* (New York: Oxford University Press, 1980).

12. The classic works are Immanuel Kant, *Observations on the Feeling of the Beautiful and Sublime* (1764), trans. John T. Goldthwait (Berkeley: University of California Press, 1960); Edmund Burke, *A Philosophical Enquiry into the Origin of Our Ideas of the Sublime and Beautiful*, ed. James T. Boulton (1958; Notre Dame: University of Notre Dame Press, 1968); William Gilpin, *Three Essays: On Picturesque Beauty, On Picturesque Travel, and on Sketching Landscape* (London, 1803).

13. See Ann Vileisis, "From Wastelands to Wetlands," unpublished senior essay, Yale University, 1989; Runte, *National Parks*.

14. William Wordsworth, "The Prelude," Book 6, in Thomas Hutchinson, ed., *The Poetical Works of Wordsworth* (London: Oxford University Press, 1936), p. 536.

15. Henry David Thoreau, *The Maine Woods* (1864), in *Henry David Thoreau* (New York: Library of America, 1985), pp. 640–41.

16. Exodus 16:10, KJV.

17. John Muir, *My First Summer in the Sierra,* p. 238. Part of the difference between these descriptions may reflect the landscapes the three authors were describing. Kenneth Olwig notes that early American travellers experienced Yosemite as much through the aesthetic tropes of the pastoral as through those of the sublime. The ease with which Muir celebrated the gentle divinity of the Sierra Nevada had much to do with the pastoral qualities of the landscape he described.

18. Frederick Jackson Turner, *The Frontier in American History* (New York: Henry Holt, 1920), pp. 37–38.

19. Richard Slotkin has made this observation the linchpin of his comparison between Turner and Theodore Roosevelt. See Slotkin, *Gunfighter Nation: The Myth of the Frontier in Twentieth-Century America* (New York: Atheneum, 1992), pp. 29–62.

20. Owen Wister, *The Virginian: A Horseman of the Plains* (New York: Macmillan, 1902), pp. viii–ix.

21. Theodore Roosevelt, *Ranch Life and the Hunting Trail* (1888; New York: Century, 1899), p. 100.

22. Wister, *Virginian,* p. x.

23. On the many problems with this view, see William M. Denevan, "The Pristine Myth: The Landscape of the Americas in 1492," *Annals of the Association of American Geographers* 82 (1992), pp. 369–85. [Included in this volume.]

24. Louis Warren, "The Hunter's Game: Poachers, Conservationists, and Twentieth-Century America," (Ph.D. diss., Yale University, 1994).

25. Wilderness also lies at the foundation of the Clementsian ecological concept of the climax.

26. On the many paradoxes of having to manage wilderness in order to maintain the appearance of an unmanaged landscape, see John C. Hendee et al., *Wilderness Management,* USDA Forest Service Miscellaneous Publication No. 1365 (Washington, D.C.: Government Printing Office, 1978).

27. This argument has been powerfully made by Ramachandra Guha, "Radical American Environmentalism and Wilderness Preservation: A Third World Critique," *Environmental Ethics* 11 (1989), pp. 71–83. [Included in this volume.]

28. Bill McKibben, *The End of Nature* (New York: Random House, 1989).

29. Ibid., p. 49.

30. Even comparable extinction rates have occurred before, though we surely would not want to emulate the Jurassic-Cretaceous boundary extinctions as a model for responsible manipulation of the biosphere!

31. Dave Foreman, *Confessions of an Eco-Warrior* (New York: Harmony Books, 1991), p. 69 (italics in original). For a sampling of other writings by followers of deep ecology and/or Earth First!, see Michael Tobias, ed., *Deep Ecology* (San Diego: Avant Books, 1984); Bill Devall and George Sessions, *Deep Ecology* (Salt Lake City: Gibbs M. Smith, 1985); Michael Tobias, *After Eden: History, Ecology, and Conscience* (San Diego: Avant Books, 1985); Dave Foreman and Bill Haywood, eds., *Ecodefense: A Field Guide to Monkey Wrenching* (2nd ed., Tucson: Ned Ludd Books, 1987); Bill Devall, *Simple in Means, Rich in Ends: Practicing Deep Ecology* (Salt Lake City: Gibbs Smith, 1988); Steve Chase, ed., *Defending the Earth: A Dialogue Between Murray Bookchin & Dave Foreman* (Boston: South End Press, 1991); John Davis, ed., *The Earth First! Reader: Ten Years of Radical Environmentalism* (Salt Lake City: Gibbs Smith, 1991); Bill Devall, *Living Richly in an Age of Limits: Using Deep Ecology for an Abundant Life* (Salt Lake City: Gibbs Smith, 1993); and Michael E. Zimmerman et al., eds., *Environmental Philosophy: From Animal Rights to Deep Ecology* (Englewood Cliffs, N.J.: Prentice-Hall, 1993). A useful survey of the different factions of radical environmentalism can be found in Carolyn Merchant, *Radical Ecology: The Search for a Livable World* (New York: Routledge, 1992). For a very interesting critique of this literature (first published in the anarchist newspaper *Fifth Estate*), see George Bradford, *How Deep Is Deep Ecology?* (Ojai, Calif.: Times Change Press, 1989).

32. Foreman, *Confessions of an Eco-Warrior,* p. 34.

33. Ibid., p. 65. See also Dave Foreman and Howie Wolke, *The Big Outside: A Descriptive Inventory of the Big Wilderness Areas of the U.S.* (Tucson: Ned Ludd Books, 1989).

34. Foreman, *Confessions of an Eco-Warrior,* p. 63.

35. Ibid., p. 27.

36. It is not much of an exaggeration to say that the wilderness experience is essentially consumerist in its impulses.

37. Cf. Muir, *Yosemite,* in *John Muir: Eight Wilderness Discovery Books,* p. 714.

38. Wallace Stegner, ed., *This Is Dinosaur: Echo Park Country and Its Magic Rivers* (New York: Alfred A. Knopf, 1955), p. 17. Emphasis in original.

39. Katherine Hayles helped me see the importance of this argument.

40. Analogous arguments can be found in John Brinckerhoff Jackson, *Beyond "Wilderness," A Sense of Place, A Sense of Time* (New Haven: Yale University Press, 1994), pp. 71–91; and in the wonderful collection of essays by Michael Pol-

lan, *Second Nature: A Gardener's Education* (New York: Atlantic Monthly Press, 1991).

41. Wendell Berry, *Home Economics* (San Francisco: North Point, 1987), pp. 138, 143.

42. Gary Snyder, quoted in the *New York Times,* "Week in Review," Sept. 18, 1994, p. 6.

Marvin Henberg

Wilderness, Myth, and American Character (1994)

THERE IS A JOKE among U.S. Forest Service employees—many of whom opposed the 1964 Wilderness Act—that prior to 1964 only God could make wilderness but now only the U.S. Congress can. The joke refers to the act's prohibiting release of potential wilderness land to other use or designation until Congress has judged its suitability for inclusion in the National Wilderness Preservation System. That language has resulted in great fuss and fury over public lands. In the coterminous United States, some 57 million acres still await congressional determination of their potential wilderness value. That is considerably more than the approximately 34 million acres, excluding Alaska, now officially designated American wilderness. Most of the disputed land lies west of the Mississippi River; in my own state of Idaho, for instance, approximately 9 million acres await release from the language of the Wilderness Act.

These figures indicate the extent to which wilderness designation is a political hot potato. It is also a philosophical hot potato, replete with paradox. Some philosophers hold the idea of wilderness to be purely an invention of the mind, a time-bound product of humanity's triumph in successfully inhabiting all but the most inhospitable portions of the earth's landmass. Others hold it to be something real and palpable, as suggested by its Old English etymology—"wildeorness," a place of wild beasts.

Much of the difficulty in conceiving of wilderness stems from the paradox that is human nature. How do we account for ourselves? Are we the dark angels of our various religious conceptions or the natural bodies of Darwinian evolution—bodies that, through a fluke of gambling nature, happened to stumble upon consciousness? To what extent are our activities and actions "natural"? If *Homo sapiens* is, as Jared Diamond argues, simply a third species of chimpanzee, we have, no matter what we do, wilderness all about us. If we are one kind of beast, the literal "place of the beasts" contains us and all we have wrought—our art and poetry no less than our skyscrapers and sewage systems.

Our kinship with other animals aside, most conceptions of wilderness distinguish sharply between humans and the other "beasts." This line of thinking resolves one paradox only to create another. Wilderness lands become, in the words of the Wilderness Act, areas "untrammeled by man, where man himself is a visitor who does not remain." Thanks to the attractions of this idea, some areas are so popular as sanctuaries from the hazards and trials of urban life that it is difficult to find solitude, a prime value of wilderness recreation.

The four federal agencies responsible for administering wilderness lands have been forced into "wilderness management"—a paradox if ever there was one. It doesn't take a philosopher to point out that "managing" a wild species risks its eventual domestication. For instance, winter feeding of elk and deer—a widespread policy of many state wildlife agencies—may, over time, tame animals whose present attraction is that they are wild. Someday perhaps the sole large mammal to be genetically wild—that is, whose procreation is left to the spontaneity of nature—will be *Homo sapiens*. Perhaps, though, not even we will remain genetically wild, given our increasing ability to intervene in the human genome.

In mentioning the air of paradox surrounding both the idea of wilderness and the practice of "managing" wilderness, I invite you to think of the role of metaphor in articulating diverse conceptions of wilderness. The phrase "wilderness as" comes naturally to our lips: wilderness as a wasteland, as a gymnasium, as a playground, as a prison, and as a pharmacy—to name but a few of the images defended in wilderness literature. Often wilderness is conceived as a proving ground to test for valuable personal and social traits. I shall here consider a specific case under the idea of wilderness as proving ground, examining the claim that the received wilderness idea bears a special connection to American character. That connection is expressed by Wallace Stegner:

> Something will have gone out of us as a people if we ever let the remaining wilderness be destroyed; if we permit the last virgin forests to be turned into comic books and plastic cigarette cases; if we drive the few remaining members of the wild species into zoos or to extinction; if we pollute the last clear air and dirty the last clean streams and push our paved roads through the last of the silence, so that never again will Americans be free in their own country from the noise, the exhausts, the stinks of human and automotive waste. . . . We need wilderness preserved—as much of it as is still left, and as many kinds—because it was the challenge against which our character as a people was formed.[1]

As with most claims related to wilderness, this one generates its share of controversy. Supporters of what I shall henceforth call the *character thesis* point to the pride we in the United States take in our wilderness heritage. They point to our leadership in conservation and preservation—our historic firsts in establishing the National Park System and, later, the National Wilderness Preservation System. They point to the spread of the wilderness idea to countries ranging from New Zealand to Zimbabwe. Finally, they find in contemporary wilderness experience manifold echoes of good character—honesty, self-reliance, and simplification of wants, to name but a few.

CHANGING ATTITUDES

Critics of connecting the wilderness idea to our national character point first to historical relativism in America's attitudes toward untamed nature. Less than a century and a half before Stegner waxed eloquent on the importance of wilderness to American character, Alexis de Tocqueville told a different story:

> In Europe people talk a great deal about the wilds of America, but the Americans themselves never think about them; they are insensible to the wonders of inanimate nature and they may be said not to perceive the mighty forests that surround them till they fall beneath the hatchet.[2]

Ultimately, historical differences between contemporary Americans and their ancestors need not trouble defenders of the character thesis. National character, like individual character, takes on the craggy lines of wisdom because of, rather than in spite of, turmoil and reversal of fortune. Ideals shift—that which is lost (or nearly so) gets appreciated when we no longer have it: innocence for one, wilderness for another.

In addition, de Tocqueville's remarks are generalized and composite; could he, for instance, have had the privilege of meeting Virginia's own William Byrd II nearly a hundred years earlier, he would have found a man in whom wilderness sensibility was highly developed. The reversal in American appreciation of wilderness, a story so ably told by historian Roderick Nash, was not created *ex nihilo.* It had seeds, most dying on the hard granitic soil of public indifference, but a few nurtured against extinction until a field could be cultivated for them.

WHOSE CHARACTER?

A second and more intractable problem for the character thesis lies in its ethnic exclusivity. Whom do we conceive as having their characters formed by the "challenge" of wilderness? Not American Indians; according to Standing Bear, an Oglala Sioux, the land of North America was never wild in conception, but tame. Not African Americans, first enslaved on plantations of the New World and later confined, many of them, to urban ghettos—"city wildernesses" in the parlance of Robert A. Woods' turn-of-the-century book, *The City Wilderness.* Not Polly Beamis, a young woman kidnapped in her native China and carried off to Oregon Territory. Her character was formed by fending off lustful drunks in saloons, where she served as a hostess and eventually purchased her way to freedom by surreptitiously sweeping and collecting gold dust from the floors.

For these and other diverse peoples of America, wilderness as conceived in the mainstream preservation movement played little role in shaping character. Relatively few people experienced the frontier, whose "closing" Frederick Jackson Turner turned into a powerful metaphor for America's first inward glance—our first hint that we might have to reinvent ourselves by, among other things, protecting wildlands and wildlife. Fewer people still—at least according to Patricia Limerick, Patrick White, and other revisionist historians currently challenging the Turner thesis—had reason to care about the frontier, its wilderness edge, or its supposed vanishing. According to the revisionists, the idea that wilderness was a strong force in shaping the American character is a "myth" as in a false and misleading tale.

WHICH VIRTUES?

If the ethnocentrism of the character thesis is a second problem, its vagueness as to the traits engendered by wilderness experience is a third. Theodore Roosevelt, for instance, thought of the wilds as a proving ground for virility, male camaraderie, and the honing of a warrior caste. Such a view is less than palatable in these decades of deep ecology and ecofeminism. Why virility and aggressiveness over placidity and nurture?

CHARACTER AND SELF-CONCEPT

Fundamentally, then, the character thesis is in serious philosophical trouble. The main difficulty lies in selective readings both of character and of wild nature. Human character runs the gamut from the virtuous to the vicious, with numerous shades of each. In addition, some of our favored virtues may be inconsistent with each other. As Isaiah Berlin observes, the honor of Achilles cannot be harmonized with the mercy of Christ. These two species of virtue are incommensurable, as only disturbing figures like Machiavelli and Nietzsche have dared to proclaim.

When our own dark image is glimpsed in the supposed mirror of wild nature, the difficulty is compounded. Wild nature may be, with Tennyson, "red in tooth and claw" or, with Annie Dillard, gentle as a spring day on Tinker Creek, the epitome of harmony and symbiosis. We search in wildness for what we want and, unsurprisingly, find it exactly as it "ought" to have been.

Yet for all its cultural exclusivity, its vagueness as to what constitutes a virtue, and its tendency to shape wild nature after its own favored image, there remains something to be said for the character thesis. It has a ring that many Americans harken to—a ring, if not of truth, at least of innocently faithful self-conception. Conceive the point this way: Suppose the Liberty Bell were to be rung and we as a people were to hear it. Thanks to the bell's famous crack, the sound would not be faithful to its original—its "true" sound. But would that matter? Would it even be relevant to the spirit the bell represents? We have, of course, detractors of the ideal of liberty represented by the bell, and many of their criticisms are apt, pointed. Freedom has not been equally extended to all within the fabric of our nation, and that is a criticism whose measure we must take. It is not to be ignored, but neither is it to be made into the whole story.

Let us ask the critic this: With what would you rather take your chances, a political system whose ideal is sounded by the cracked knell of the Liberty Bell or a political system with no such symbol? I, for one, fervently believe in the positive power of ideals. Their appeal is nonrational, even ritualistic; but as an aspirant to philosophy, I have concluded that concepts alone mainly divide rather than unite human beings. We need symbols and their emotive associations. Among the symbols we need most is wilderness.

Stegner's words thus emphasize that, culturally rather than ethnically or personally, to be American is to conceive of ourselves as a wilderness people. The flaws in this thesis are both as prominent and as irrelevant as the crack in the Liberty Bell. Review the political struggle leading to passage of the Wilderness Act and you will find a robust populism stirred from the depths of our national self-conception.

Sometimes, thankfully, our ideals—erroneous and unflattering as they may appear under some lights—stir us to prefer the social good to the getting and spending by which we lay waste our powers. Because we must believe something about ourselves, I submit that belief in ourselves as a people shaped by wilderness is productive of greater good than of ill. In this fight the character thesis becomes a different kind of myth—not a false and misleading tale, but a symbolic means of uniting us in celebration of something larger than ourselves.

My defense of this thesis, however, is fideistic rather than rationalistic. Reason alone is incompetent to penetrate and sufficiently articulate the mysteries of wild nature. Reason concocts its arguments with judicious concern for the other side. For every Wallace Stegner lamenting production of plastic cigarette cases, we have a Martin Krieger making the case for plastic trees. Reason can carry us into the realm of computer-generated virtual realities to ask, "Why not extinguish the real thing so long as the wilderness experience can be provided in surrogate?" Quickly forgotten when reason exerts its generalizing, abstracting sway is the joy of particularity—a dimension of experience open only to the "inherent imbecility of feeling," in George Eliot's wonderful phrase. It is the joy of knowing a specific place or person, as opposed to grasping a generalized category or purpose.

Here is Rockbridge County native and wilderness philosopher Holmes Rolston III, on the importance of particularity:

> Wildness is nature in what philosophers call idiographic form. Each wilderness is one of a kind, so we give it a proper name—the Rawahs, the Dismal Swamp. We climb Mount Ida or canoe on the Congaree River. Even when exploring some nameless canyon or camping at a spring, one experiences a concrete locus never duplicated in idiosyncratic detail. In culture, there is but one Virginia and each Virginian has a proper name. The human differences include conscious self-affirmations and heritages for which nature provides little precedent. But nature first is never twice the same. Always in the understory there are distinctive landscape features—the Shenandoah Valley or the Chesapeake Bay—with which the Virginians interact, each with a unique genetic set. Before culture emerges, nature is already endlessly variable. This feature is crucial to what we mean by wildness.[3]

Thanks to its endless variability, the best way of capturing the particularity of wilderness lands is through narrative. As Rolston observes, "There is no narrative in biology text, but a trip into wildness is always a story."[4] Each parcel in the National Wilderness Preservation System features stories with multiple plots and restless casts of plant, animal, and human characters wandering through a unique geography. Drama lies in the pure contingency of relations.

Wilderness understanding depends on emotional singularity and kinesthetic presence more than on abstract generalization. It has more the imprint of natural history than of molecular biology. Please do not misunderstand: We need molecular biology too, for, among other things, it tests the contingent relations described in wild nature, just as it does similar work in paleontology to test the integrity of claims in natural history. But the history itself—for instance, the evolutionary mixing some 70 million years ago of fauna from North and South America known to paleontologists as the Great American Interchange—cannot be replicated. We trace it in the fossil record, speculating about cause and effect, but our understanding is always in the form of a story, a narrative. Narratives of natural history abound in reconstructed details of climate, predation, birth and rearing of young, migration, cataclysm, evolutionary branching, and extinction. As a complement, narratives of wilderness offer up miniature slices within the grander narratives of natural history.

INTRODUCING 'THE FRANK'

Let me observe the particularity of wilderness by acquainting you with Idaho's Frank Church River of No Return area. Begin with the name and its particularity. Frank Church was floor manager of the 1964 Wilderness Act in the U.S. Senate. "River of No Return" dates to the Lewis and Clark expedition's exploration of the Salmon River. After following the river into one of its spectacular canyons, William Clark pronounced the steep cliffs and fierce rapids to be impassable.

Continue to the particularity of the area. "The Frank," as it is called by its partisans, is the largest official wilderness area outside Alaska—over 2.3 million acres. Located in central Idaho, it is contiguous with two other wilderness areas, the 200,000-acre Gospel Hump and the 1.1-million-acre Selway-Bitterroot. Sheer size gives the River of No Return area outstanding wilderness qualities of remoteness and isolation. Its ecosystem is as undisturbed as can be found in the continental United States. Ecologists and wildlife biologists regard The Frank and its environs as unique for containing wholly within its borders both the summer and the winter ranges of all its large mammals.

Following procedures outlined in the Wilderness Act, Congress created The Frank in 1980. Special conditions apply, creating special peculiarities of management. For example, although banned in most wilderness areas outside Alaska, planes may fly into The Frank, using any of 18 primitive airstrips. Power boats, also generally banned from wilderness, are allowed on the Main Salmon River, which forms the 86-mile-long northern boundary. On the pristine Middle Fork of the Salmon—105 miles through the heart of the wilderness—only nonmotorized watercraft are allowed. There are many small inholdings of private land, most along the two rivers and in the larger creek drainages. Finally, there is a special mining reserve where, in a national emergency, extraction of cobalt, a strategic mineral, may be authorized.

Lest you are tempted to accept the view perpetuated by foes of wilderness designation that wilderness areas "lock up" the land so it receives no use, let me reassure you to the contrary. In 1992 more than 20,000 people rafted the combined Middle Fork and Main Salmon rivers. Thousands more traveled the Main Salmon in their jet boats—the only craft powerful enough to ply the rapids of the river. There were 10,000 registered backcountry users, traveling by foot or pack animal (llamas are now common), while in September and October alone—hunting season—there were some 9,000 plane landings.

The periphery of the wilderness is growing rapidly, with thousands of people moving each year into the Treasure and Magic valleys of Idaho and the Bitterroot Valley of Montana. Many of these people are attracted precisely because of the proximity of The Frank and other wilderness lands. A recent survey by Gundars Rudzitis and Harley Johansen shows that migration into counties containing wilderness is heavily influenced by environmental quality and opportunities for outdoor recreation. Seventy-two percent of recent immigrants cited proximity to wilderness as a special amenity influencing their relocation. In contrast, only 55 percent of long-time residents thought the nearby wilderness a special amenity. Presuming that the new arrivals act on their expressed preferences, wilderness lands will be more heavily used as in-migration continues.

All these particularities create management headaches for the Forest Service. The wilderness portion of the agency's budget is minuscule. For instance, the North Fork Ranger District, responsible for management of one-fifth of The Frank, has a total wilderness budget, including overhead, of only $100,000. Even this paltry sum is considerably larger than the wilderness budget for the other four ranger districts with responsibility in The Frank, for the North Fork district patrols the Main Salmon River, where most human impact is concentrated.

Somehow, despite pressure from those who love it too well and from others who abuse it by poaching or littering, The Frank remains a magnificent political achievement, because as a place it is exactly that—magnificent.

Defenders of wilderness often point to the millions of people who, though they will never visit a particular place, are happy simply knowing it is there, protected. This argument would be toothless if nonvisitors were reduced to knowing a place by name alone. Even 10,000 scenic photographs, each worth its proverbial thousand words, would make little difference. Names and pictures come fully to life only when animated by the storyteller's art. Wild places, from Thomas Hardy's Egdon Heath to Jack London's Yukon, are known best when woven into narrative. So also for the beasts, mythical and real: They too are known best when endowed with character and related to each other by incident.

In winter, the Frank Church wilderness is virtually empty of human presence, yet the animals remain. Every creek drainage is a scene for timeless stories of birth and death, play and predation, nurture and starvation. In February 1993 I was witness in The Frank to the mute testimony of fresh blood alternately lavished and trickled across a mile or more of glistening white snow. A 400-foot slide down a steep canyon, four wide, bloody, fur-strewn depressions, and a trail of crimson cougar prints told a dramatic story.

One male cougar, lurking in the rocks above snowline, had pounced on a second male traversing the slope. They slid down together, biting and clawing for a hundred feet, at which point one of them sank tooth or claw into an artery of the other. The bleeding began, dyeing the snow, spurting blood a full five meters downwind. The cougars crashed into the bottom of the canyon. The wounded one lay below; the other, its belly matted with blood from its rival, lay 10 feet above. They hissed and snarled, keeping their distance. The wounded male fled upstream along the trail. The rival followed.

They battled once again at a precipice a mile upstream from their first encounter. Both fell onto the frozen shore of Big Creek, 60 feet below. They paced uphill alongside each other, snarling but apart. One of them shredded a mountain mahogany bush, seeking to intimidate the other. Again they fought, again both fell—this time into the creek. Only one of them, as the tracks clearly showed, swam out. The other, dazed, perhaps already dead from lack of blood, slid under the ice to await the spring thaw. The victor returned to the scene of the two battles above the precipice, then headed upslope to resume its furtive, solitary existence.

At the time I encountered these tracks and pieced together their meaning with the able assistance of two wildlife biologists, I was one of only four people within a 50-mile radius. Surrounding me in each of hundreds of other creek drainages were the tracks, the evidence, the leavings from thousands of similar stories. All of them will be gone with the thaw, as will the cougar carcass in Big Creek, washed into the Middle Fork by April torrents. But what is magnificent and enduring is the wilderness itself, complete with the ennobling thought that somewhere in this great continent of ours, nature can still be so prodigal as to waste the life of a lithe, fierce, full-grown cougar.

Such stories as may be woven together from silent cougar tracks are more magnificent for being imagined than observed, for imagination is unobtrusive. It extends human experience while keeping for wild animals the essential condition of their wildness—freedom from thoughtless incursion.

If wilderness is about the mythic American character, it is equally about characters who live their natural lives apart from us. Stories told on errands outside the wilderness are scripted by powers larger than human. Nature and narrative fit hand and glove. If we will let nature abide wildly in some few remaining portions of the earth, we will be immeasurably richer for it. We will not only secure a future for coming generations, we will secure stories for them—stories of wonder, of kinship with other living beings, and of richness and fecundity from a prodigal source.

While love of narrative for its own sake may be exclusively a human trait, the characters we enjoy are drawn from life. Animals are unequivocally the favorite characters of our young. (Imagine, for a moment, children's literature without animals.) Will we have wild animals for our storylines of the future? Not unless we protect their habitat and freedom.

NOTES

1. Wallace Stegner, "The Wilderness Idea," in David Brower, ed., *Wilderness: America's Living Heritage* (San Francisco: Sierra Club Books, 1961), p. 97.

2. Alexis de Tocqueville, *Journey to America,* J. P. Mayer, ed., George Lawrence, trans. (New Haven: Yale University Press, 1960), p. 335.

3. Holmes Rolston III, "Values Gone Wild," in *Philosophy Gone Wild* (Buffalo, N.Y.: Prometheus Books, 1986), pp. 137–38.

4. Ibid., p. 140.

PART FOUR

Beyond the Wilderness Idea

SELECTIONS FROM Aldo Leopold

Threatened Species (1936)

A Proposal to the Wildlife Conference for an Inventory of the Needs of Near-Extinct Birds and Animals

THE VOLUME OF EFFORT expended on wildlife conservation shows a large and sudden increase. This effort originates from diverse courses, and flows through diverse channels toward diverse ends. There is a widespread realization that it lacks coordination and focus.

Government is attempting to secure coordination and focus through reorganization of departments, laws, and appropriations. Citizen groups are attempting the same thing through reorganization of associations and private funds.

But the easiest and most obvious means to coordination has been overlooked: explicit definition of the immediate needs of particular species in particular places. For example: Scores of millions are being spent for land purchase, C.C.C. labor, fences, roads, trails, planting, predator-control, erosion control, poisoning, investigations, water developments, silviculture, irrigation, nurseries, wilderness areas, power dams, and refuges, within the natural range of the grizzly bear.

Few would question the assertion that to perpetuate the grizzly as a part of our national fauna is a prime duty of the conservation movement. Few would question the assertion that any one of these undertakings, at any time and place, may vitally affect the restoration of the grizzly, and make it either easy or impossible of accomplishment. Yet no one has made a list of the specific needs of the grizzly, in each and every spot where he survives, and in each and every spot where he might be reintroduced, so that conservation projects in or near that spot may be judged in the light of whether they *help or hinder* the perpetuation of the noblest of American mammals.

On the contrary, our plans, departments, bureaus, associations, and movements are all focused on abstract categories such as recreation, forestry, parks, nature education, wildlife research, more game, fire control, marsh restoration. Nobody cares anything for these except as means toward ends. What ends? There are of course many ends which cannot and many others which need not be precisely defined at this time. But it admits of no doubt that the immediate needs of threatened members of our fauna and flora must be defined now or not at all.

Until they are defined and made public, we cannot blame public agencies, or even private ones, for misdirected effort, crossed wires, or lost opportunities. It must not be forgotten that the abstract categories we have set up as conservation objectives may serve as alibis for blunders, as well as ends for worthy work. I cite in evidence the C.C.C. crew which chopped down one of the few remaining eagle's nests in northern Wisconsin, in the name of "timber stand improvement." To be sure, the tree was dead, and according to the rules, constituted a fire risk.

Most species of shootable non-migratory game have at least a fighting chance of being saved through the process of purposeful manipulation of laws and environment called management. However great the blunders, delays, and confusion in getting management of game species under way, it remains true that powerful motives of local self-interest are at work in their behalf. European countries, through the operation of these motives, have saved their resident game. It is an ecological probability that we will evolve ways to do so.

The same cannot be said, however, of those species of wilderness game which do not adapt themselves to economic land-use, or of migratory birds which are owned in common, or of non-game forms classed as predators, or of rare plant associations which must compete with economic plants and livestock, or in general of all wild native forms which fly at large or have only an esthetic and scientific value to man. These, then, are the special and immediate concern of this inventory. Like game, these forms depend for their perpetuation on protection and a favorable environment. They need "management"—the perpetuation of good habitat—just as game does, but the ordinary motives for providing it are lacking. They are the threatened element in outdoor America—the crux of conservation policy. The new organizations which have now assumed the name "wildlife" instead of "game," and which aspire to implement the wildlife movement, are I think obligated to focus a substantial part of their effort on these threatened forms.

This is a proposal, not only for an inventory of threatened forms in each of their respective places of survival, but an inventory of the information, techniques, and devices applicable to each species in each place, and of local human agencies capable of applying them. Much information exists, but it is scattered in many minds and documents. Many agencies are or would be willing to use it, if it were laid under their noses. If for a given problem no information exists, or no agency exists, that in itself is useful inventory.

For example, certain ornithologists have discovered a remnant of the Ivory-billed Woodpecker—a bird inextricably interwoven with our pioneer tradition—the very spirit of that "dark and bloody ground" which has become the locus of the national culture. It is known that the Ivory-bill requires as its habitat large stretches of virgin hardwood. The present remnant lives in such a forest, owned and held by an industry as reserve stumpage. Cutting may begin, and the Ivory-bill may be done for at any moment. The Park Service has or can get funds to buy virgin forests, but it does not know of the Ivory-bill or its predicament. It is absorbed in the intricate problem of accommodating the public which is mobbing its parks. When it buys a new park, it is likely to do so in some "scenic" spot, with the general objective of making room for more visitors, rather than with the specific objective of perpetuating some definite thing to visit. Its wildlife program is befogged with the abstract concept of inviolate sanctuary. Is it not time to establish particular parks or their equivalent for particular "natural wonders" like the Ivory-bill?

You may say, of course, that one rare bird is no park project—that the Biological Survey should buy a refuge, or the Forest Service a National Forest, to take care of the situation. Whereupon the question bounces back: the Survey has only duck money; the Forest Service would have to cut the timber. But is there anything to prevent the three possible agencies concerned from getting together and agreeing whose job this is, and while they are at it, a thousand other jobs of like character? And how much each would cost? And just what needs to be done in each case? And can anyone doubt that the public, through Congress, would support such a program? Well this is what I mean by an inventory and plan.

Some sample lists of the items which need to be covered are wilderness and other game species, such as grizzly bear, desert and bighorn sheep, caribou, Minnesota remnants of spruce partridge, masked bobwhite, Sonora deer, peccary, sagehen; predator and allied species, such as the wolf, fisher, otter, wolverine and Condor; migratory birds, including the trumpeter swan, curlews, sandhill crane, Brewster's warbler; plant associations, such as prairie floras, bog floras, Alpine and swamp floras.

In addition to these forms, which are rare everywhere, there is the equally important problem of preserving the attenuated edges of species common at their respective centres. The turkey in Colorado, or the ruffed grouse in Missouri, or the antelope in Nebraska, are rare species within the meaning of this document. That there are grizzlies in Alaska is no excuse for letting the species disappear from New Mexico.

It is important that the inventory represent not merely a protest of those privileged to think, but an agreement of those empowered to act. This means that the inventory should be made by a joint committee of the conservation bureaus, plus representatives of the Wildlife Conference as representing the states and the associations. The plan for each species should be a joint commitment of what is to be done and who is to do it. The bureaus, with their avalanche of appropriations, ought to be able to loan the necessary expert personnel for such a committee, without extra cost. To sift out any possible imputation of bureaucratic, financial, or clique interest, the interbureau committee should feed its findings to the public through a suitable group in the National Research Council, and subject to the Council's approval. The necessary incidental funds for a secretary, for expense of gathering testimony and maps, and for publications might well come from the Wildlife Institute, or from one of the scientific foundations.

There is one cog lacking in the hoped-for machine: a means to get some kind of responsible care of remnants of wildlife remote from any bureau or its field officers. Funds can hardly be found to set up special paid personnel for each such detached remnant. It is of course proved long ago that closed seasons and refuge posters without personnel are of no avail. Here is where associations with their far-flung chapters, state officers or departments, or even private individuals can come to the rescue. One of the tragedies of contemporary conservation is the isolated individual or group who complains of having no job. The lack is not of jobs, but of eyes to see them.

The inventory should be the conservationist's eye. Every remnant should be definitely entrusted to a custodian-ranger, warden, game manager, chapter, ornithologist, farmer, stockman, lumberjack. Every conservation meeting—national, state, or local—should occupy itself with hearing their annual reports. Every field inspector should contact their custodians—he might often learn as well as teach. I am satisfied that thousands of enthusiastic conservationists would be proud of such a public trust, and many would execute it with fidelity and intelligence.

I can see in this set-up more conservation than could be bought with millions of new dollars, more coordination of bureaus than Congress can get by new organization charts, more genuine contacts between factions than will ever occur in the war of the inkpots, more research than would accrue from many gifts, and more public education than would accrue from an army of orators and organizers. It is, in effect, a vehicle for putting Jay Darling's concept of "ancestral ranges" into action on a quicker and wider scale than could be done by appropriations alone.

Wilderness (1935)

TO AN AMERICAN CONSERVATIONIST, one of the most insistent impressions received from travel in Germany is the lack of wildness in the German landscape.

Forests are there—interminable miles of them, spires of spruce on the skyline, glowering thickets in ravines, and many a quick glimpse "where the yellow pines are marching straight and stalwart up the hillside where they gather on the crest." Game is there—the skulking roebuck or even a scurrying *Rudel* of red-deer is to be seen any evening, even from a train-window. Streams and lakes are there, cleaner of cans and old tires than our own, and no worse beset with hotels and "bide-a-wee" cottages. But yet, to the critical eye, there is something lacking that should not be lacking in a country which actually practices, in such abundant measure, all of the things we in America preach in the name of "conservation." What is it?

Let me admit to begin with the obvious difference in population density, and hence in population pressure on the economic mechanisms of land-use. I knew of that difference before coming over, and think I have made allowance for it. Let it further be clear that I did not hope to find in Germany anything resembling the great "wilderness areas" which we dream and talk about, and sometimes briefly set aside, in our National Forests and Parks. Such monuments to wilderness are an esthetic luxury which Germany with its timber deficit and the evident land-hunger of its teeming millions, cannot afford. I speak rather of a certain quality which should be but is not found in the ordinary landscape of producing forests and inhabited farms, a quality which still in some measure persists in some of the equivalent landscapes of America, and which we I think tacitly assume will be enhanced by rather than lost in the hoped-for practice of conservation. I speak specifically to the question of whether and under what limitations that assumption is correct.

It may be well to first inquire whether the Germans themselves, who know and love their rocks and rills with an intensity long patent to all the world, admit any such esthetic deficit in their countryside. "Yes" and "no" are of course worthless as criteria of such a question. I offer in evidence, first, the existence of a very vigorous esthetic discontent, in the form of a "Naturschutz" (nature-protection) movement, the equivalent of which preceded the emergence of the wilderness idea in America. This impulse to save wild remnants is always, I think, the forerunner of the more important and complex task of mixing a degree of wildness with utility. I also submit that the Germans are still reading Cooper's "Leatherstocking" and Parkman's "Oregon Trail," and still flock to the wild-west movies. And when I asked a forester with a philosophical bent why people did not flock to his forest to camp out, as in America, he shrugged his shoulders and remarked that perhaps the tree-rows stood too close together for convenient tenting! All of which, of course, does not answer the question. Or does it?

And this calls to mind what is perhaps the first element in the German deficit: their former passion for unnecessary outdoor geometry. There is a lag in the affairs of men—the ideas which were seemingly buried with the cold hard minds of the early-industrial era rise up out of the earth today for us to live with. Most German forests, for example, though laid out over a hundred years ago, would do credit to any cubist. The trees are not only in rows and all of a kind, but often the various age-blocks are parallelograms, which only an early discovery of the ill-effects of wind saved from being rectangles. The age-blocks may be in ascending series—1, 2, 3—like the proverbial stepladder family. The boundary between wood and field tends to be sharp, straight, and absolute, unbroken by those charming little indecisions in the form of draw, coulee, and stump-lot, which, especially in our "shiftless" farming regions, bind wood and field into an harmonious whole. The Germans are now making a determined effort to get away from cubistic forestry—experience has revealed that in about the third successive crop of conifers in "pure" stands the microscopic flora of the soil becomes upset and the trees quit growing, but it will be another generation before the new policy emerges in landscape form.

Not so easily, though, will come any respite from what the geometrical mind has done to the German rivers. If there were only room for them, it would be a splendid idea to collect all the highway engineers in the world, and also their intellectual kith and kin the Corps of Army Engineers, and settle them for life upon the perfect curves and tangents of some "improved" German river. I am aware, of course, that there are weighty commercial reasons for the canalization of the larger rivers, but I also saw many a creek and rivulet laid out straight as a dead snake, and with masonry banks to boot. I am depressed by such indignities, and I have black misgivings over the swarm of new bureaus now out to improve the American countryside. It is, I think, an historical fact that no American bureau equipped with money, men, and machines ever refused *on principle* to straighten a river, save only one—the Soil Conservation Service.

Another more subtle (and to the average traveller, imperceptible) element in the deficit of wildness is the near-extirpation of birds and animals of prey. I think it was Stewart Edward White who said that the existence of one grizzly conferred a flavor to a whole county. From the German hills that flavor has vanished—a victim to the misguided zeal of the game-keeper and the herdsman. Even the ordinary hawks are nearly gone—in four months travel I counted only ____. And the great owl or "Uhu"—without whose vocal austerity the winter night becomes a mere blackness—persists only in the farthest marches of East Prussia. Before our American sportsmen and game-keepers and stockmen have finished their self-appointed task of extirpating our American predators, I hope that we may begin to realize a truth already written bold and clear on the German landscape: that success in most over-artificialized land-uses is bought at the expense of the public interest. The game-keeper buys an unnatural abundance of pheasants at the expense of the public's hawks and owls. The fish-culturist buys an unnatural abundance of fish at the expense of the public's herons, mergansers, and terns. The forester buys an unnatural increment of wood at the expense of the soil, and in that wood maintains an unnatural abundance of deer at the expense of all palatable shrubs and herbs.

This effect of too many deer on the ground flora of the forest deserves special mention because it is an illusive burglary of esthetic wealth, the more dangerous because unintentional and unseen. Forest undergrowth consists of many species, some palatable to deer, others not. When too dense a deer population is built up, and there are no natural predators to trim it down, the palatable plants are grazed out, whereupon the deer must be artificially fed by the game-keeper, whereupon next year's pressure on the palatable species is still further increased, etc. ad infinitum. The end result is the extirpation of the palatable plants—that is to say an unnatural simplicity and monotony in the vegetation of the forest floor, which is still further aggravated by the too-dense shade cast by the artificially crowded trees, and by the soil-sickness already mentioned as arising from conifers. One is put in mind of Shakespeare's warning that "virtue, grown into a pleurisy, dies of its own much." Be that as it may, the forest landscape is deprived of a certain exuberance which arises from a rich variety of plants fighting with each other for a place in the sun. It is almost as if the geological clock had been set back to those dim ages when there were only pines and ferns. I never realized before that the melodies of nature are music only when played against the undertones of evolutionary history. In the German forest—that forest which inspired the *Erlkönig*—one now hears only a dismal fugue out of the timeless reaches of the carboniferous.

Reed F. Noss

Wilderness Recovery (1991)

Thinking Big in Restoration Ecology

INTRODUCTION

I AM NOT A RESTORATION ECOLOGIST. My practical experience in restorative management consists of a few times helping with prairie burns in the midwest and pine land burns in Florida. My overriding interest has been in developing ambitious conservation strategies at regional to continental scales. Through this work, a mixture of science and advocacy, I have come to realize that no big conservation project is adequate in today's world without a major restoration component. There is simply too little land left in near-prime condition, human influences are everywhere, and some ecosystem types are virtually gone.

Restoration projects and techniques, large and small, are aptly discussed in the pages of *Restoration and Management Notes.* In perusing those pages, the reader quickly notes that the vast majority of restoration projects are tiny. They typically involve a few acres of prairie, a forest stand, or a small wetland. The diminutive nature of most restoration projects is understandable. Full-blown ecological restoration at a watershed or landscape scale, often relying on high-tech engineering, can be incredibly expensive. Restoration of the Kissimmee River in South Florida, converted to a canal by the U.S. Army Corps of Engineers between 1961 and 1971, is expected to cost $300 million—10 times the cost of the original channelization

(Duplaix, 1990). Another exceptionally large restoration project, the rehabilitation of 36,000 heavily logged acres out of 48,000 acres added to Redwood National Park in 1978, was budgeted for $33 million over a 10–15 year period (Belous, 1984). As of 1990, only about $12–13 million has been spent. Although the project is considered a success so far, at least another decade of work will be needed (Lee Purkerson, National Park Service, personal communication).

The point of this article is that landscape restoration need not be prohibitively expensive, even when applied at scales that dwarf the Kissimmee River and Redwood National Park projects. Landscape-scale restoration can rely largely on the natural recovery processes of ecosystems, aided by human labor. Road closures alone can work wonders. The billions of dollars that federal agencies spend annually degrading natural ecosystems, often through such exorbitant programs as below-cost timber sales, subsidized grazing, dams, and road construction, can be diverted to restoration projects. Labor-intensive restoration, in turn, can employ many former timber workers, road engineers, and ranchers, whose prior activities have created the need for restoration in the first place. The net benefit to biodiversity and human society will be tremendous.

In the following essay, I outline a strategy and recipe for wilderness recovery based on a land ethic ("sensu" Leopold, 1949). The re-establishment of huge, wild, functional ecosystems replete with large carnivores and their prey is the pinnacle of restoration ecology and human re-harmonization with nature (Sayen, 1989). In order to achieve human dignity, we must fully respect the dignity of other life forms and pay retribution for past offenses against them; this can be achieved by assisting in the healing of ecosystems (Maser, 1990). Restoration ecology must be expanded from the local and community level to the regional landscape level in order to accomplish these goals.

A WILDERNESS VISION

Why wilderness? Wilderness is defined in the Wilderness Act of 1964 as "an area where the earth and its community of life are untrammeled by man, where man himself is a visitor who does not remain," and which "generally appears to have been affected primarily by the forces of nature, with the imprint of man's work substantially unnoticeable." Hence, wilderness is a place where humans do not dominate, a place to be humbled.

ECOLOGICAL VALUES OF WILDERNESS

Wilderness also has ecological values. It is no accident that the only ecosystems that include all native carnivores are very large roadless areas. In the lower 48 states, the only ecosystem that still regularly contains both grizzly bears and wolves is the Northern Continental Divide complex in northern Montana and adjacent Canada, the "healthiest big mountain ecosystem in America south of Canada" (Foreman and Wolke, 1989). (The North Cascades of Washington and adjacent Canada also have grizzlies and wolves, but in very tenuous populations.) The presence of large carnivore populations indicates a relatively healthy ecosystem; the predators themselves may play a fundamental role in maintaining the diversity of the system through indirect effects on the food web (Terborgh, 1988). Roadlessness defines wilderness and is the key to its ecological health. Probably no single feature of human-dominated landscapes is more threatening to biodiversity (aquatic and terrestrial) than roads. Direct effects of roads include fragmentation and isolation of populations, roadkill, pollution and sedimentation of streams and wetlands, and exotic species invasions (Diamondback, 1990; Bennett, 1991). Many species of small mammals rarely or never cross roads, even two-lane roads closed to public traffic (Oxley et al., 1974; Mader, 1984; Swihart and Slade, 1984). Roads, therefore, reduce effective population sizes and gene flow, and will be significant dispersal barriers during climate change. Another set of species—largely weeds and pests—uses roadsides as dispersal corridors. In the Northwest, Port Orford cedar root rot fungus, black-stain root disease fungus, spotted knapweed, and the gypsy moth disperse and invade natural habitats via roads (Schowalter, 1988). Sedimentation of streams and ruination of fisheries is often associated with roads, especially in steep terrain subject to landslides and debris flows. Erosion from logging roads in an Idaho study was 220 times greater than from undisturbed sites (Megahan and Kidd, 1972).

Indirect effects of roads are many, but the most important to consider here are those related to human access. Many species associated with wilderness are there not because the habitat is ideal (it usually is not), but rather because they do not get along well with humans. In northern Wisconsin and Minnesota, road density is the best predictor of wolf habitat suitability, not because wolves avoid roads (in fact, they often use roads as travelways) but because roads bring people with guns, snares, and traps. Above 0.9 miles of road per square mile of habitat, wolves in this region cannot maintain populations (Thiel, 1985; Mech et al., 1988). Black bears and grizzly bears show similar mortality responses with increasing road density and access to legal and illegal hunters (Brody, 1984; McLellan and Mace, 1985). Elk habitat capability also drops rapidly with increasing density of roads open to public traffic (Lyon, 1983).

Roadless areas offer refugia to those species, from wildruns of anadromous fish to large carnivores and ungulates, sensitive to the impacts of roads. Because large mammals require enormous amounts of habitat to maintain viable populations (see below), roadless areas must be large to offer sufficient security. Hence the need for Big Wilderness (Foreman and Wolke, 1989; Noss, 1991). If Americans want the symbols of American wilderness—large carnivores and herds of ungulates—to remain on their public lands (and every indication is that they do), then they should be fully informed about roads. If properly informed, many might support the concept of road closures for the restoration of Big Wilderness.

Postage-stamp nature reserves have protected some important elements of biodiversity, but they are not whole. They fail to maintain populations of area-dependent animals, do not represent complete biological communities, do not perpetuate the ecological processes necessary to assure landscape-level diversity, and are heavily influenced by phenomena beyond their borders (Noss, 1987). Furthermore, because small reserves usually require a considerable amount of manipulative management in order to maintain what diversity they have (White and Bratton, 1980), they fail the naturalness test and do little to promote humility.

Many argue that, by applying ecological principles, we can manage landscapes to maintain biodiversity and still build roads, harvest timber, drill for oil, mine ore, graze cows and sheep, and roar around in off-road vehicles. Lawmakers have put this multiple-use philosophy on the books. Federal laws, most notably the National Forest Management Act (NFMA) of 1976, tell land-managing agencies to provide for many human uses while at the same time maintaining diversity and viable populations. But the record of multiple-use management is a sorry one; common and weedy species have prospered at the expense of rare and sensitive species and entire ecosystems are being degraded (Noss, 1983; Norse et al., 1986; Wilcove, 1988; Grumbine, 1990). Much of the damage is connected with road-building and other forms of habitat fragmentation (Noss 1987). Although innovative approaches to forest management, under the banner "New Forestry" or "New Perspectives" (Franklin, 1989; Gillis, 1990), are theoretically attractive, as presently applied in the Pacific Northwest they are resulting in continued liquidation of old-growth forests. Destruction of endangered ecosystems is not justified under any banner, no matter how scientifically sophisticated.

INTRINSIC VALUE

Biologists can argue wilderness versus non-wilderness, large versus small reserves, set-asides versus managed lands, forever. In the end, they will surely agree that many conservation functions (for example, protection of rare plant populations) are served by small nature reserves, other functions (such as protection of species not overly sensitive to human disturbance) are served by multiple-use lands, while still other species (big carnivores) and processes (big wildfires) require mega-wilderness. But besides its ecological values, besides even the amenity values of inspiring humility and contemplation, Big Wilderness is essential. It has intrinsic value.

Some people, for reasons quite beyond the rational, believe that huge, wild areas are valuable for their own sake. These areas should be as free as possible from human influence in order to satisfy the criterion of sacred otherness (Reed, 1989; Birch, 1990). Science cannot prove or disprove intrinsic value in wilderness or any other entity. But if we accept that humans have intrinsic value, as do almost all ethical traditions, then only anthropocentric prejudice would keep us from recognizing such value in others. In the final analysis, intrinsic values are the only values that stand on their own, unfettered by tenuous links to utility. What good is wilderness? One might as well ask the rocks in the canyon. "Wilderness," Ed Abbey wrote, "needs no defense, only more defenders."

BALANCE

Implicit in conservation strategy is that a balance should be achieved between land devoted to primarily natural values and land developed for human purposes. How balanced is the ratio of wild to developed land in the United States? Conservationists often look to the tropics with greatest alarm, but many North American ecosystems are in equally critical condition and stand to lose as great a proportion of their biotas (Moyle and Williams, 1990). By 1920, the northeastern and central states had already lost 96 percent of their virgin forests (Reynolds and Pierson, 1923). Today, the Pacific Northwest holds most of the old-growth forests in the lower 48 states, yet less than 13 percent remains of the ancient forest in western Washington, western Oregon, and northwestern California (Norse, 1990). The longleaf pine forests of the Southeastern Coastal Plain, once the dominant regional ecosystem, have declined by at least 98 percent (Noss, 1989). Very little of our land is strictly protected, despite the claims of commodity interests. Designated wilderness represents only 1.8 percent of the 48 states, or 4 percent of the U.S. including Alaska, and many of these areas are open to grazing, mining and other disruptive uses (Watkins, 1989). Only about 3 percent of the 48 states is protected in wilderness, parks, or equivalent reserves and conservationists generally do not expect to more than double that figure (Scott et al., 1990-a).

To set aside only 3 percent of the 48 states in reserves, and only 1.8 percent as wilderness, does not seem very balanced. No one can say how much land is "needed" to maintain biodiversity; where we draw the line is a reflection of our values. My values tell me that an order of magnitude increase—to 50 percent—is a reasonable compromise, and that large-scale wilderness recovery is needed to restore that balance. Many people will think that asking for 50 percent of our land as wilderness is either utopian or insane (or worse). But then again, most people (and nearly all elected officials) believe in infinite economic growth. Few accept the inevitability of catastrophe if we stay on our present course. I rest my case.

RECOVERY PRINCIPLES

Wherever we draw the line between wilderness and development, the degraded condition of most ecosystems means that preservation, strictly speaking, is seldom a tenable option. Clear-cuts and abandoned fields must be reforested, cows and fences removed, roads obliterated, natural fire and hydrological regimes restored, native carnivores and other species returned to their former range. Wilderness must be allowed to recover.

If we accept that some level of wilderness recovery is a worthy conservation goal, where do we begin? First, it must be recognized that preservation and restoration are complementary elements of any regional conservation strategy (Noss, 1985-a; Sayen, 1989). A common environmentalist criticism of restoration is that it will divert attention or resources from preserving the last pristine areas, and may even be used to justify further habitat degradation if a "don't worry, we can fix it" attitude prevails. This unfortunate situation must not be allowed to develop; restoration must be seen as a healing art, not as "an environmental license to kill" (Jordan, 1990). Using "no net loss" of habitat or biodiversity as a guiding principle, all existing, relatively unaltered natural areas must be protected in addition to restoring habitats critical to regional biodiversity. The following are some general principles to consider in designing a wilderness recovery project.

ECOSYSTEM REPRESENTATION

Habitat destruction, like biodiversity, is not uniformly or randomly distributed over the land. The most productive and often most diverse habitats were the first to be settled and converted to intensive production. Forty years ago, Kendeigh et al. (1950–51) noted that no areas of "virgin vegetation . . . of sufficient size to contain all the animal species in the self-maintaining populations historically known to have occurred in the area" remained in deciduous forest, prairie, or lower elevations in the Rocky Mountains. Opportunities for creating large reserves were limited to inaccessible southern swamps, boreal forests, higher elevations in the western mountains, desert, and tundra. Today of 261 major terrestrial ecosystems in the United States and Puerto Rico, defined by a combination of Bailey's ecoregions and Kuchler's potential natural vegetation, 104 (40%) are not protected in designated wilderness areas (Davis, 1988). Wilderness boundaries often coincide with timberline: the "rock and ice" phenomenon.

Representation of all ecosystem types in wilderness and other protected areas is one of the most important and widely accepted goals of conservation. In 1987, delegates of 62 nations at the Fourth World Wilderness Conference voted unanimously for a resolution to preserve "representative examples of all major ecosystems of the world to ensure the preservation of the full range of wilderness and biological diversity" (Davis, 1988). Because any major ecosystem type will vary across its geographic range in such attributes as species composition, vegetation structure, and natural disturbance regime, multiple examples of each major ecosystem should be protected or restored. Restoration priorities can be established by determining which ecosystem types in the region have declined most markedly from pre-settlement condition.

BIGNESS

The desirability of large reserves, "bigness," is one of the few generally accepted principles of conservation (Soulé and Simberloff, 1986). Protected examples of ecosystem types must be large enough to maintain viable populations of all native species (Kendeigh et al., 1950–51) and to persist in concert with natural disturbances (Pickett and Thompson, 1978). Large reserves are easier to defend against encroachment from outside, suffer less intensive edge or boundary effects, and require less management per unit area (White and Bratton, 1980; Noss, 1983).

How large must a wilderness area or other reserve be to maintain native biodiversity? Estimates by conservation biologists of minimum viable populations and corresponding reserve sizes are alarmingly high. A recent review of empirical studies (Thomas, 1990) concluded that an average population of 1,000 individuals may be adequate for species of normal population variability, but 10,000 individuals may be needed for long-term persistence of highly variable birds and mammals. How these minimum population estimates translate into area requirements depends on factors such as habitat quality and social behavior, that determine population density and dispersion. Hubbell and Foster (1986) suggested that tens to hundreds of km^2 are required to maintain populations of tropical trees, which occur in low densities. Thiollay (1989) concluded that rainforest reserves in French Guiana must be between 1 and 10 million ha to maintain a complete bird community, including viable populations of diurnal raptors and large game birds. Schonewald-Cox (1983) estimated that reserves of 1,000 to 100,000 ha might maintain viable populations of small herbivorous and omnivorous mammals, but reserves of 1 to 10 million ha are needed for long-term viable populations of large carnivores and ungulates.

Natural disturbances must also be taken into consideration, as reserves that are small relative to the scale of natural disturbance may experience radical fluctuations in the proportions of different serial stages, which in turn will endanger populations dependent on particular stages. Shugart and West (1981) estimated that landscapes 50–100 times the largest disturbance patch may approach a steady state in habitat diversity. Because boreal forests experience natural fires covering up to 10,000 ha, 1 million ha might be necessary to achieve a steady state in boreal regions. In the Greater Yellowstone Ecosystem, landscape stability by some measures (constant heterogeneity) is reached at about 1 million ha (Romme and Despain, 1990). For ecosystems that experience small disturbances, such as eastern deciduous forests characterized by treefall gaps and occasional watershed-sized fires, existing national parks and forests (at 100,000 ha or so) may be in approximate steady state (Shugart and West, 1981). The current system of protected areas in this region fails to represent most ecosystem types, however (Davis, 1988). Furthermore, the most space-demanding species in these forests, eastern cougar and wolves (if re-introduced), would require larger reserves, again on the order of 1 million ha.

How well do existing reserves meet these ambitious size criteria? Research natural areas (RNAs), designated for their ecological and scientific values, are far too small by our criterion. Of the 213 Forest Service RNAs (as of 1990), 93% are smaller than 1000 ha; the remaining 7% are smaller than 5000 ha (Noss, 1991). The 320 units in the National Park system (as of 1983) are distributed more evenly among size classes, though only 10% are larger than 100,000 ha and 3% (10 areas) are larger than 1 million ha (Schonewald-Cox, 1983). Most (81%) of the 474 units in the National Wilderness Preservation System (as of 1989) are between 1000 and 100,000 ha, and only 1% (6 areas) are larger than 1 million ha. Thus, few protected areas approach the approximately 1 million ha needed to maintain natural disturbance regimes (for some ecosystem types) and viable populations of large mammals. Society has to face the fact that its existing protected areas are too small (Grumbine, 1990).

Hence the need for wilderness recovery. In almost all cases, representing ecosystems in protected areas of sufficient size to assure viability is possible today only through restoration. For future parks or wilderness areas to represent the diversity that greeted the first European visitors, they will have to be "grown rather than decreed" (Janzen, 1988). The tall grass prairie of the American midwest, for example, has been reduced by at least 98 percent. In Missouri, only 0.5 percent remains, all of it in fragments too small to maintain the native flora or fauna (Rissler, 1988). Preserving or restoring small pieces of prairie has produced depauperate "museum pieces" (Noss and Harris, 1986), not viable ecosystems. An unprecedented project of the Nature Conservancy seeks to restore an entire prairie landscape in the Osage Hills of Oklahoma, starting with an initial 30,000-acre (12,000 ha) core area (Madson, 1990). The prairie preserve will someday contain a herd of bison and a fire regime that replicates the original. But what about wolves? To be complete, restoration needs to be considered at a scale of millions of hectares.

IMPLEMENTATION

Accepting that all ecosystem types, across their range of natural variation, should be represented in reserves large enough to assure long-term viability, what steps can the wilderness restorationist take to put a project on the ground? The answer lies in looking to the landscape and seeing where the conservation needs and opportunities are.

SPATIAL CONTEXT

Spatial context is a key to determining where to locate landscape-scale restoration projects. The goal is a regional landbase where protected areas representing all ecosystem types are enclosed or otherwise linked by continuous natural habitat. Such a system of interconnected reserves can form a whole greater than the sum of its parts (Noss and Harris, 1986). Although each reserve individually might be too small to assure viability of a carnivore population, for example, the network as a whole might suffice. Small, satellite reserves protecting local biodiversity hotspots, such as a rock outcrop with an endemic plant, complement the regional network but cannot be allowed to substitute for it.

A map of all public and private managed areas in the region of interest is a good place to start. Parcels can be rated in terms of the degree of protection given to them, as is done for all managed areas by the Nature Conservancy. Such a map portrays the skeleton to which you add flesh: designate critical areas on public land in protected status, fill in gaps (such as inholdings) in public ownership, acquire surrounding lands to increase reserve size, identify restoration priorities, and establish linkages between reserves.

Next, consult the best available vegetation maps of remote sensing information to determine the current distribution of ecosystem types in the region. The land ownership and protection map can then be overlaid on the vegetation map, ideally using a Geographic Information System (GIS), to determine how well existing protected areas and other public lands represent ecosystem types. The "gap analysis" project of Mike Scott and colleagues (Scott et al., 1990-b) is ideally suited to this type of analysis. At a state or regional scale, Scott's group is mapping the distribution of current vegetation and associated species (based on habitat suitability), determining centers of species richness and endemism, and identifying gaps in the coverage of ecosystems and species ranges in protected areas. Hot spots of high species richness outside of existing reserves become priorities for protection. Conservationists can apply this basic approach at any geographic scale, and at whatever level of resolution is desired, to identify protection and restoration needs.

Once the ecosystems most in need of representation in the wilderness recovery system are identified, proposed reserve boundaries must be delineated on a map. Boundaries should be drawn to encompass areas of adequate size; conform to topography, watersheds, and natural geomorphic features; and, take advantage of existing undeveloped areas with low road density. Within your idealistic boundaries, depict (perhaps by color coding) on the map which private land should be acquired, which existing public lands need upgraded protective designation, which roads need to be closed (perhaps all!), and other restoration needs. Draw linkages or broad habitat corridors between reserves and clusters of reserves (Harris, 1984; Noss and Harris, 1986; Noss, 1987; Noss, in press) to allow for seasonal wildlife migration, dispersal of plants and animals, and long-range movements in response to climate change.

At this stage, you have mapped a prospective wilderness recovery system. It covers perhaps 50 percent of your region, more or less depending on the extent of landscape modification and the existing public land base. Now you need to buffer the preserves from surrounding, intensive-use land (clear-cuts, corn fields, cities, etc.). A gradation of buffer zones, with intensity of use increasing outwards, can surround each core wilderness preserve and comprise a "multiple-use module" (MUM) (Harris, 1984; Noss and Harris, 1986), patterned after the biosphere reserve model (UNESCO, 1974). Corridors can be zoned in a similar fashion (Noss, 1987). Such zoning can indeed provide for multiple human uses, but its primary function must be to insulate core areas from adverse human influences. The core area must provide sufficient roadless acreage to protect the most sensitive species (generally, large carnivores) that live or will be reintroduced there.

DESIGN AND MANAGEMENT CONSIDERATIONS

A wilderness recovery proposal is obviously idealistic. It is not something you take to your friendly public lands managers and legislators, and expect to be greeted with enthusiasm. Many will balk at its apparent unreasonableness. But nothing is more unreasonable than the willful destruction of biodiversity. For conservationists to put forth something less than what is really needed, to compromise nature, is foolhardy. You will never get all you ask for, but the higher you shoot, the higher you will score. If anything, it is human nature to underestimate real needs; perhaps million-hectare reserves are too small.

My wilderness recovery proposal for Florida (Noss, 1985a, 1987), portrayed on a map that encompassed half the state in protected areas, has probably inspired more public enthusiasm than backlash (although the Farm Bureau happily provided some of the latter). A gubernatorial candidate supported the proposal, it was endorsed in principle by members of the Florida Panther Technical Advisory Council and the Florida Department of Natural Resources in its nongame plan, and some of the key linkages have been purchased by state and federal agencies, assisted by the Nature Conservancy. Despite the rapid pace of habitat destruction in Florida, piece by piece much of the system is coming together.

The greatest failure of the Florida proposal so far relates to the refusal of public land-managing agencies to close roads and restore natural conditions on the lands entrusted to them. National forests and other existing public lands constitute most of the "core preserves" in the diagram, yet most are not being managed in a way that protects biodiversity. As an illustration of the limitations of the current system, the large core preserve in the northeastern corner of Florida, overlapping Georgia, is the Okefenokee National Wildlife Refuge–Osceola National Forest complex. Pinhook Swamp, a linkage between Okefenokee and Osceola, was purchased in 1988 and added to the National Forest, a significant conservation achievement (Noss and Harris, 1990).

This complex of nearly 500,000 ha is the first priority reintroduction site for the Florida panther, an endangered species with an 85 percent probability of extinction within 25 years, according to population viability analysis (Ballou et al., 1989). To test the feasibility of panther reintroduction into the Okefenokee-Osceola complex, 5 neutered and radio-collared Texas puma were released within the complex in 1988. The cats did not fare well; within 6 months, three died, two of gunshot. The two remaining puma discovered domestic animals (on private lands) and had to be recaptured (Noss and Harris, 1990). Clearly, some management changes—especially road closures to minimize human access and removal of domestic animals and inholdings—will have to be made before Okefenokee-Osceola can regain its top carnivore and be considered a recovered ecosystem.

PUTTING IT ON THE GROUND

On-the-ground implementation of a wilderness recovery strategy is contingent on factors specific to each regional landscape. Some general guidelines apply:

1. close and revegetate roads;
2. remove fences and other human structures;
3. eradicate exotic species whenever feasible, including (perhaps especially) livestock;
4. reintroduce populations of extirpated native species, including large predators;
5. restore hydrological regimes and soils; and
6. reintroduce or mimic natural disturbance regimes. Many American ecosystems are naturally shaped by fire. Prescribed burning, preceded where necessary by manual removal of artificially dense vegetation that developed due to fire suppression, may be needed to restore habitat structure to a condition where wildfires will have effects approximating those in a presettlement system.

As emphasized throughout this paper, closure and rehabilitation of roads is perhaps the single most important ingredient in a recipe for wilderness recovery. Many land-managing agencies close roads, at least seasonally, to protect wildlife such as elk or bears. Barriers erected by land managers are often inadequate, however. A study on the Flathead National Forest in Montana found that some 80 percent of "obliterated" roads inventoried by the Forest Service were driveable by ordinary passenger vehicles and 38 percent of the barriers erected for road closures involve 'ripping' (physical obliteration) and revegetation (Hammer, 1990). Obliteration of roads can be relatively expensive (but not as expensive as building new roads). In the Redwood National Park restoration, rehabilitation of logging roads has proceeded at an average rate of 15 miles per year. Costs for "worst-case" roads have reached $40,000 per mile, but ordinarily have ranged between $8,000 and $25,000 per mile (Belous, 1984). Many roads that pose minor erosion threats can be allowed to recover through natural revegetation, at zero cost, so long as their entrances are effectively blocked to vehicles, including off-road vehicles.

Much of the restoration of ecosystems at a landscape scale can rely on natural revegetation; the exceptions are mostly sites with severe soil destruction or dominance by exotic species, which must be dealt with case by case. In most human-modified landscapes being considered for wilderness recovery, fragments of remnant natural habitat are embedded in a matrix of agriculture, clear-cuts, or tree farms. Restoration, then, is primarily "the initiation and coalescence of *growing habitat fragments*" (Janzen, 1988, emphasis in the original). The degraded habitat between fragments must be returned to a physical condition where it can be reseeded by native plant propagules. This will often require clearing sites of foreign vegetation (including tree plantations) and will require manual planting of native species if seed sources are too distant or seed dispersal too slow to overcome the invasion of exotics.

One of the most challenging tasks in restoration is determining the target, the natural patterns and processes you are trying to replicate. Reconstructions of presettlement vegetation and disturbance regimes (Noss, 1985-b) and comparison with existing wilderness baselines, though not perfect, are all there is to work from. The target, of course, is always moving; any point in time specified as "natural" is just one frame in a very long movie. Furthermore, because of long-range transport of pollutants, enhanced UV-B levels due to ozone depletion, and global warming, an ecosystem can never be fully restored as long as industrial humans dominate the planet. Discussion of other elements in a wilderness recovery strategy can be found in Mueller (1985), Noss (1985-a), Noss and Harris (1986), Sayen (1987), Friedman (1988), and Foreman and Wolke (1989).

CONCLUSION

Preservation and restoration are essential partners in a comprehensive conservation strategy. Conservationists must insist that every wild and natural area be saved, and that many degraded areas be restored to viability by closing roads and reintroducing missing species and processes. Wilderness recovery must not be compromised in an effort to appear reasonable; the time for compromise, if ever, was when North America was still a wilderness continent.

Restoration is a life-affirming art (Jordan, 1990; Maser, 1990). It offers the only hope of true victory in conservation. Normally, conservation battles are never won; defeat is simply postponed. Designation under the Wilderness Act creates no new wilderness; for every acre officially designated, many more acres of wilderness are destroyed. The Oregon Wilderness Act of 1984, as a case in point, protected a mere 853,062 acres out of 4 million acres of eligible roadless areas on national forests; the remaining acres were "released" for multiple-abuse management (Foreman and Wolke, 1989). Wilderness recovery seeks simply to bring back ecosystems that contain all of their parts and to keep them healthy.

> *Without enough wilderness America will change. Democracy, with its myriad personalities and increasing sophistication, must be fibred and vitalized by regular contact with outdoor growths—animals, trees, sun warmth and free skies—or it will dwindle and pale.*
> Walt Whitman

NOTE

I have been greatly inspired by the wilderness recovery ideas of Jasper Carlton, Bill Devall, Ed Abbey, Dave Foreman, Mitch Friedman, Ed Grumbine, Keith Hammer, Bob Mueller, Tony Povilitis, Jamie Sayen, David Wheeler, Howie Wolke, George Wuerthner, Bob Zahner, and other unrealistic dreamers.

BIBLIOGRAPHY

Ballou, J. D., T. J. Foose, R. C. Lacy and U. S. Seal. 1989. "Florida panther (*Felis concolot coryi*): Population viability analysis and recommendations." Captive Breeding Specialist Group, Species Survival Commission, IUCN, Apple Valley, MN.

Belous, R. 1984. "Restoration among the redwoods." *Restoration and Management Notes* 2: 57–65.

Bennett, A. F. 1991. "Roads, roadsides, and wildlife conservation: A review." In *Nature Conservation: The Role of Corridors,* D. A. Saunders and R. J. Hobbs, eds. Surrey Beatty and Sons, Chipping Norton, NSW, Australia. (In press.)

Birch, T. H. 1990. "The incarceration of wilderness: Wilderness areas as prisons." *Environmental Ethics* 12: 3–26. [Included in this volume.]

Brody, A. J. 1984. "Habitat use by black bears in relation to forest management in Pisgah National Forest, North Carolina." M. S. Thesis. University of Tennessee, Knoxville.

Davis, G. D. 1988. "Preservation of natural diversity: The role of ecosystem representation within wilderness." Paper presented at National Wilderness Colloquium, Tampa, Florida, January, 1988.

Diamondback. 1990. "Ecological effects of roads" (or, "The road to destruction"). Pages 1–5 in *Killing Roads: A Citizens' Primer on the Effects and Removal of Roads,* J. Davis, ed. Earth First! Biodiversity Project Special Publication. Tucson, AZ, pp. 1–5.

Duplaix, N. 1990. "South Florida water: Paying the price." *National Geographic* 178(l): 89–112.

Foreman, D., and H. Wolke. 1989. *The Big Outside.* Ned Ludd Books, Tucson, AZ.

Franklin, J. F. 1989. "Toward a new forestry." *American Forests,* Nov./Dec.: 37–44.

Friedman, M., ed. 1988. *Forever Wild. Conserving the Greater North Cascades Ecosystem.* Mountain Hemlock Press, Bellingham, WA.

Gillis, A. M. 1990. "The new forestry." *BioScience* 40: 558–562.

Grumbine, R. E. 1990. "Viable populations, reserve size, and federal lands management: A critique." *Conservation Biology* 4: 127–134.

Hammer, K. J. 1986. "An on-site study of the effectiveness of the U.S. Forest Service Road Closure Program in Management Situation One Grizzly Bear Habitat, Swan Lake Ranger District, Flathead National Forest, Montana." Unpublished report.

Hammer, K. J. 1988. "Roads revisited: A travelway inventory of the Upper Swan and Lower Swan geographic units, Swan Lake Ranger District, Flathead National Forest." Stage 2, Report No. 4.

Hammer, K. J. 1990. "A road ripper's guide to the national forests." In *Killing*

Roads: A Citizens' Primer on the Effects and Removal of Roads, J. Davis, ed. Earth First! Biodiversity Project Special Publication. Tucson, AZ, pp. 6–8.

Harris, L. 1984. *The Fragmented Forest.* University of Chicago Press, Chicago, IL.

Hubbell, S. P., and R. B. Foster. 1986. "Commonness and rarity in a neotropical forest: Implications for tropical tree conservation." In *Conservation Biology: The Science of Scarcity and Diversity,* M. E. Soulé, ed. Sinauer Associates, Sunderland, MA, pp. 205–231.

Janzen, D. H. 1988. "Management of habitat fragments in a tropical dry forest: Growth." *Annals of the Missouri Botanical Garden* 75: 105–116.

Jordan, W. R., III. 1990. "Two psychologies." *Restoration and Management Notes* 8: 2.

Kendeigh, S. C., H. I. Baldwin, V. H. Cahalane, C. H. D. Clarke, C. Cottam, W. P. Cottam, I. McT. Cowan, P. Dansereau, J. H. Davis, F. W. Emerson, L. T. Haig, A. Hayden, C. L. Hayward, J. M. Linsdale, J. A. MacNab, and J. E. Potzger. 1950–51. "Nature sanctuaries in the United States and Canada: A preliminary inventory." *Living Wilderness* 15(35): 1–45.

Leopold, A. 1949. *A Sand County Almanac.* Oxford University Press, New York.

Lyon, L. J. 1983. "Road density models describing habitat effectiveness for elk." *Journal of Forestry* 81: 592–595.

McLellan, B. N., and R. D. Mace. 1985. *Behavior of Grizzly Bears in Response to Roads, Seismic Activity, and People.* British Columbia Ministry of Environment, Cranbrook, B.C.

Mader, H. J. 1984. "Animal habitat isolation by roads and agricultural fields." *Biological Conservation* 29: 81–96.

Madson, J. 1990. "On the Osage." *Nature Conservancy* 40(3): 7–15.

Maser, C. 1990. "On the 'naturalness' of natural areas: A perspective for the future." *Natural Areas Journal* 10: 129–133.

Mech, L. D., S. H. Fritts, G. L. Radde, and W. J. Paul. 1988. "Wolf distribution and road density in Minnesota." *Wildlife Society Bulletin* 16: 85–87.

Megahan, W. F., and W. J. Kidd. 1972. "Effects of logging and logging roads on erosion and sediment deposition from steep terrain." *Journal of Forestry* 70: 136–141.

Moyle, P. B., and J. E. Williams. 1990. "Biodiversity loss in the temperate zone: Decline of the native fish fauna of California." *Conservation Biology* 4: 275–284.

Mueller, R. F. 1985. "Ecological preserves for the eastern mountains." *Earth First!* 5(8): 20–21.

Norse, E. A. 1990. *Ancient Forests of the Pacific Northwest.* The Wilderness Society and Island Press, Washington, D.C.

Norse, E. A., K. L. Rosenbaum, D. S. Wilcove, B. S. Wilcox, W. H. Romme, E. W. Johnson, and M. L. Stout. 1986. *Conserving Biological Diversity in Our National Forests.* The Wilderness Society, Washington, D.C.

Noss, R. F. 1983. "A regional landscape approach to maintain diversity." *BioScience* 33: 700–706.

Noss, R. F. 1985-a. "Wilderness recovery and ecological restoration: An example for Florida." *Earth First!* 5(8): 18–19.

Noss, R. F. 1985-b. "On characterizing presettlement vegetation: How and why." *Natural Areas Journal* 5(1): 18–19.

Noss, R. F. 1987. "Protecting natural areas in fragmented landscapes." *Natural Areas Journal* 7: 2–13.

Noss, R. F. 1989. "Longleaf pine and wiregrass: Keystone components of an endangered ecosystem." *Natural Areas Journal* 9: 211–213.

Noss, R. F. 1991. "What can wilderness do for biodiversity?" In *Proceedings of the Conference: Preparing to Manage Wilderness in the 21st Century,* P. Reed, ed. USDA Forest Service, Southeastern Forest Experiment Station, Asheville, NC (in press).

Noss, R. F. In press. "Wildlife corridors." In D. Smith, ed. *Ecology of Greenways* (Publisher to be determined).

Noss, R. F., and L. D. Harris. 1986. "Nodes, networks, and MUMS: Preserving diversity at all scales." *Environmental Management* 10: 299–309.

Noss, R. F., and L. D. Harris. 1990. "Habitat connectivity and the conservation of biological diversity: Florida as a case study." In *Proceedings of the 1989 Society of American Foresters National Convention.* Spokane, WA. Sept. 24–27. Pp. 131–135.

Oxley, D. J., M. B. Fenton, and G. R. Carmody. 1974. "The effects of roads on populations of small mammals." *Journal of Applied Ecology* 11: 51–59.

Pickett, S. T. A., and J. N. Thompson. 1978. "Patch dynamics and the design of nature reserves." *Biological Conservation* 13: 27–37.

Reed, P. 1989. "Man apart: An alternative to the self-realization approach." *Environmental Ethics* 11: 53–69.

Reynolds, R. V., and A. H. Pierson. 1923. "Lumber Cut of the United States, 1870–1920." USDA Bulletin No. 1119, Washington, D.C.

Rissler, P. G. 1988. "Diversity in and among grasslands." In *Biodiversity,* E. O. Wilson, ed. National Academy Press, Washington, D.C., pp. 176–180.

Romme, W. H. and D. G. Despain. 1990. "Effects of spatial scale on fire history and landscape dynamics in Yellowstone National Park." Paper presented at Fifth Annual Landscape Ecology Symposium: The Role of Landscape Ecology in Public-Policy Making and Land-use Management. March 21, 1990. Oxford, OH.

Sayen, J. 1987. "The Appalachian Mountains: Vision and Wilderness." *Earth First!* 7(5): 26–30.

Sayen, J. 1989. "Notes towards a restoration ethic." *Restoration and Management Notes* 7: 57–59.

Schonewald-Cox, C. M. 1983. Conclusions. "Guidelines to management: A beginning attempt." In *Genetics and Conservation: A Reference for Managing Wild Plant and Animal Populations,* C. M. Schonewald-Cox, S. M. Chambers, B. MacBryde, and W. L. Thomas, eds. Benjamin-Cummings, Menlo Park, CA, pp. 141–145.

Schowalter, T. D. 1988. "Forest pest management A synopsis." *Northwest Environmental Journal* 4: 313–318.

Scott, J. M., B. Csuti, and K. A. Smith. 1990-a. "Playing Noah while playing the devil." *Bulletin of the Ecological Society of America* 71: 156–159.

Scott, J. M., B. Csuti, K. Smith, J. E. Estes, and S. Caicco. 1990-b. "Gap analysis of species richness and vegetation cover: An integrated conservation strategy for the preservation of biological diversity." In *Balancing on the Brink: A Retrospective on the Endangered Species Act,* K. Kohn, ed. Island Press, Washington, D.C.

Shugart, H. H., and D. C. West. 1981. "Long-term dynamics of forest ecosystems." *American Scientist* 69: 647–652.

Soulé, M. E., and D. Simberloff. 1986. "What do genetics and ecology tell us about the design of nature reserves?" *Biological Conservation* 35: 19–40.

Swihart, R. K., and N. A. Slade. 1984. "Road crossing in Sigmodon hispidus and Microtus ochrogaster." *Journal of Mammalogy* 65: 357–360.

Terborgh, J. 1988. "The big things that run the world—a sequel to E. O. Wilson." *Conservation Biology* 2: 402–403.

Thiel, R. P. 1985. "Relationship between road densities and wolf habitat suitability in Wisconsin." *American Midland Naturalist* 113: 404–407.

Thiollay, J. M. 1989. "Area requirements for the conservation of rain forest raptors and game birds in French Guiana." *Conservation Biology* 3: 128–137.

Thomas, C. D. 1990. "What do real population dynamics tell us about minimum viable population sizes?" *Conservation Biology* 4: 324–327.

UNESCO. 1974. "Task Force on Criteria and Guidelines for the Choice and Establishment of Biosphere Reserves." *Man and the Biosphere Report* No. 22, Paris, France.

Watkins, T. H. (ed.). 1989. "A special report—Wilderness America: A vision for the future of the nation's wildlands." *Wilderness* 52(184): 3–64.

White, P. S., and S. P. Bratton. 1980. "After preservation: Philosophical and practical problems of change." *Biological Conservation* 18: 241–255.

Wilcove, D. S. 1988. "National Forests: Policies for the Future." *Vol. 2. Protecting Biological Diversity.* The Wilderness Society, Washington, D.C.

Donald M. Waller

Getting Back to the Right Nature (1998)

A Reply to Cronon's "The Trouble with Wilderness"

> *Ability to see the cultural value of wilderness boils down, in the last analysis, to a question of intellectual humility. . . . It is only the scholar who understands why the raw wilderness gives definition and meaning to the human enterprise.*
>
> Aldo Leopold (1949, pp. 200–201)

In contrast to Leopold's assertion, Bill Cronon (1995a, b)—and J. Baird Callicott (1991; 1994a) before him—argue forcefully that our ideas about wilderness are so culturally and historically mired as to have become an albatross around the neck of contemporary conservationists. Like Callicott, Cronon presents us with an erudite and wide-ranging analysis of the historical bases for our concepts of wilderness and concludes that such ideas have become a liability hindering what might otherwise be a broader and more constructive debate regarding how humans relate to their environment. In particular, Cronon proposes that historic romantic notions of awe and admiration for wild nature fused with Turner's myth of the frontier in America to produce the icon of wilderness we now accept. Cronon further argues that our artificial and now outmoded notions of wilderness have become a cultural liability by obscuring what should be a substantial debate regarding how better to mesh our human culture and activities with natural environments.

Most of us can readily accept Cronon's and Callicott's initial arguments that our ideas about wilderness are laden with cultural and historical assumptions and biases. But Cronon goes much further to argue that environmentalists improperly use wilderness as a standard for judging both nature and ourselves to the point that "wilderness poses a serious threat to responsible environmentalism in the late 20th century." For example, by defining wilderness as a place without people, we are left with a standard that only fosters continuing alienation and a destructive dualism between areas judged appropriate for development versus areas reserved for the rest of nature. Cronon argues both that our peculiarly American view of wilderness is too culture-bound to serve as such a standard and that this standard has become counterproductive in the sense that it leads us to overvalue the remote and grand western national parks and exotic tropical forests while undervaluing the opportunities we have nearer at hand to reconstruct our cities and countryside in a more environmentally benign manner. The implication is that if we could only learn to better appreciate the small and contrived pieces of nature that surround us, we would gain the will to care more for our environment as a whole.

What is noteworthy about these critiques is that they come from serious and respected scholars deeply concerned with our culture's inability to successfully and sustainably mesh its social and economic activities with the biological world that sustains it. Their argument is not with existing physical wilderness areas (for which they avow concern and support despite Cronon's unfortunate title). Rather, they seek to expand our vision of how we might care for nature more responsibly while we pursue our own welfare. This is, of course, the central conundrum of sustainable development. While few conservationists would object to this general and important goal, many might object to their decision to use a critique of wilderness as a vehicle to achieve this goal. In addition, it seems all too likely that unfriendly critics of protecting wild lands will borrow provocative arguments from these friendly critics to serve quite different ends.

Has wilderness become an albatross preventing a more constructive dialogue about how to address the broader set of environmental crises we face? In this essay, I question the idea that North American concepts of wilderness are historically so static, or so confining, as to stymie further debate or progress on land-use issues. In fact, I will argue the opposite: that concepts of wilderness are currently undergoing a remarkable reformation to encompass a broader set of values and processes and that this broadening is likely to increase rather than diminish the importance of wilderness in future land-use debates. It should be obvious that I write as a scientist rather than as a historian or environmental philosopher. Nevertheless, I will attempt to marshal historical evidence that wilderness has served a broader, and broadening, set of goals, especially in the twentieth century, and to argue that this broadening represents a logical extension of human values. At the end of the essay, I return to the central conundrum to ask how our concepts of wild and wilderness might better inform our approaches to rebuilding human settlements and restoring degraded habitats.

DO CONCERNS FOR WILDERNESS DIMINISH THE PROTECTION OF LOCAL ENVIRONMENTS?

I begin by questioning an initial premise of Cronon's: that by idolizing wilderness and working for its protection we tend to diminish our concern for, and protection of, nearer and more mundane environments such as our cities and farms. Does concern for distant big wilderness areas necessarily decrease our concern for local environmental quality or environmental justice? An implicit assumption here appears to be that our overall efforts to protect the environment represent a zero-sum game so that additional concern for one area diminishes resources available to protect or restore other areas. If these premises are correct, we might be justified in refocusing our attention to more local issues.

What evidence exists that people who care more for remote wilderness care less for nearby or more mundane examples of nature? A qualified social scientist should address this issue in earnest (as well as the zero-sum assumption). My own admittedly subjective experience suggests just the opposite: individuals who care strongly about their nearby oaks and wetlands seem more likely to work passionately to preserve remote wild places such as the Arctic Wildlife Refuge. While ecologically aware and concerned citizens are inevitably torn among a wide set of worthy environmental causes and may decide to allocate their limited resources to remote and wild areas, this should never be taken to imply a lack of concern with local conditions (or for environmental social justice). Those few who do loudly proclaim their "misanthropic" preference for the protection of big wild areas over other human values are often doing so simply to provoke others to consider their point of view. Even Dave Foreman, co-founder of the radical group Earth First! and current Sierra Club Board member, notes that: "Wilderness advocates are not anti-people. Most of us support campaigns for human health and for social and economic justice." (1994, p. 65).

We might also consider the words and actions of early wilderness advocates such as John Muir and Aldo Leopold. Did not Muir plead passionately with his family to protect the small wetland on their farm (Wolfe 1945; Fox 1981)? Aldo Leopold in his essay "Illinois Bus Ride," laments the "success" of modern agriculture:

> There are not hedges, brush patches, fencerows, or other signs of shiftless husbandry. The cornfield has fat steers, but probably no quail. The fences stand on narrow ribbons of sod; whoever plowed that close to barbed wires must have been saying, "Waste not, want not." (Leopold 1949, p. 119)

Far from devaluing local conditions, Leopold pleads for us to extend our ecological sensibility even to these seemingly marginal scraps of habitat.

THE TREE IN THE GARDEN

Cronon raises several interesting points via his metaphor of the tree in the garden. After asking whether, by valuing distant and grand wilderness areas, we don't devalue the "humble places and experiences" nearer to home, he concludes his essay by suggesting that we should all learn to "honor the wild" that surrounds us, including the tree in the garden which he argues is no less wild than a tree standing in an ancient forest. If we are to value wildness, then we should recognize it wherever it occurs and seek to restore it to our local environments rather than reserving it, at a distance, for special occasions.

Is the tree in the garden actually wild? In certain important senses, of course it is. It is derived from an unbroken line of ancestors that ultimately stretches back some 3.5 billion years. Its genetic code embodies and records this evolutionary history, directing the construction of mitochondria, plastids, and ribosomes into cells and ultimately the arrangement of these cells into roots, trunks, leaves, and flowers. The tree in the garden also grows via the same intricate biochemical processes of respiration and photosynthesis that occur in its forest cousins (and, indeed, all other plants). Depending on its species, we might expect it to support a squirrel or two, a few birds, and legions of smaller creatures. Importantly, the tree in the garden also serves to represent the rest of nature and, by proxy, our relationship to it (most obviously in a Zen garden). This is clearly a role Cronon favors for the tree in the garden and one that may alone justify its existence.

Despite these several similarities, however, the tree in the garden differs in important and fundamental ways from the tree in an ancient forest. The tree in the garden has a distinct (and probably impoverished) set of birds in its branches. It also has a different suite of lichens on its bark and branches, or perhaps no lichens at all if our garden is in a polluted city. A different complement of nematodes, fungi, and bacteria thrive around its roots. And, most importantly, its flowers and seeds are likely to face a compromised fate due to a lack of appropriate pollinators and dispersers and/or the scarcity of appropriate "safe-sites" for germination and establishment. In sum, the tree in the garden is not wild because it has been removed from its ancestral ecological and evolutionary context. We should define an organism as tame or wild according to its context rather than its constitution. To paraphrase the poet Goethe's phrase "Ein Mensch ist kein Mensch," one tree (removed from its context) is no tree at all. While it exists as an interesting artifact that will always reflect its remarkable biotic history, its future is evolutionary oblivion. (Unless, of course, it happens to be one of the small number of species well-adapted to growing in urban landscapes, in which case its evolutionary prognosis may be excellent—at least in the short term.)

DUALISTIC FALLACIES AND THE DIFFERENCE BETWEEN WILDNESS AND WILDERNESS

So what makes something wild? We tend to think in terms of things and places as being either wild or tamed, artificial or natural. Yet rarely are things and places so easily categorized. More often, degrees of wildness exist. Part of Cronon's critique centers on our historic tendency to draw dichotomies between wild vs. tame, or natural vs. unnatural. In this epoch of global climate change and the long-distance transport of heavy metals, persistent pesticides, and other pollutants, we must accept the fact that no area on Earth remains pristine or fully free of human influence, as McKibben (1989) persuasively argues. Humans have become a biologic and even geologic force across the globe. Similarly, because humans are one of a related set of primate species, we must also accept the fact that humans are natural and forever a part of, and dependent upon, natural ecosystems.

While Cronon makes these points clearly and effectively, he leaves further questions unresolved. If degrees of wildness exist, and if humans must be accepted as an integral part of the systems we seek to protect and maintain, how are we to establish criteria for evaluating human behavior? What boundaries shall we place upon our own tendency to expand and subvert other biotic systems to our own ends? If no boundaries exist between wild and tame, natural and unnatural, why shouldn't we establish parks to protect rock quarries, dammed rivers, and hog farms? If all areas are considered as natural and wild, or denatured and tamed, as any other, why should we concern ourselves with conserving nature at all? This is the dilemma of environmental relativism raised, yet not resolved, in Cronon's essay.

The tree in the garden metaphor suggests that we should seek to expand on the distinction between wildness and wilderness. Although Cronon's essay begins with Thoreau's famous assertion that "In Wildness is the preservation of the World," it goes on to blur this distinction by focusing on the historical and cultural roots of our ideas about wilderness. If wilderness is, admittedly, a very human construct laden with cultural meaning, wildness is just the opposite: that which is not, and cannot be, a human construct. Wildness existed before human cultures expanded and will exist long after human cultures have vanished. Wildness clearly also persists in many corners of our acculturated cities and farms (for example, the hedgerows mentioned by Leopold in "Illinois Bus Ride"). Because ecological systems change constantly, no single static state can be considered wild. Nevertheless, the rate and extent of human-induced changes now greatly exceed the ability of many organisms to adapt evolutionarily to these changes, meaning that we are robbing them of their wild context. However romanticized and idealized our notions of "wild" and "wilderness" are, there is and always will be a gap separating the artificial from the wild—the "otherness" that Cronon refers to.

If Cronon had been writing about the state of wildness instead of our use of wilderness concepts, he surely would have made a different set of points. Perhaps he would have pointed out that no other large animal has ever existed on Earth in such vast numbers (now 5.75 billion and growing by 95 million each year) with such profound impacts on so many ecosystems. Scientists now estimate that fully 40 percent of the entire world's net primary productivity goes, directly or indirectly, to serve our one species (Vitousek et al. 1986). He surely would have explained how expanded habitat destruction, combined with the ongoing homogenization of the world's biota via the long-distance exchange of weedy plant, animal, and microbe species, is precipitating the greatest extinction spasm in 66 million years, with losses expected to exceed 20 percent of the species on Earth (Wilson 1992). Cronon might also have reported to us the remarkable statistic that only 11 percent of the world's lands remain wild and that only 4.3 percent are legally protected as parks or nature reserves (Wilson 1992, 337). Despite having led the world in establishing national parks and wilderness areas, we in the United States have protected only 3.9 percent of our land and have actually paved more land than we have protected as wilderness (Callicott 1994a). Cronon might have gone on to remind us of the recent report by the National Biological Service documenting our losses, with 96 of the 261 distinct ecosystem types in the United States now covering less than 30 percent of the area they once did and prairie grasslands persisting in less than 1 percent of their original area. Finally, he might have reminded us about just how imperiled even these small remaining scraps of wild nature are in today's political climate where neoconservatives rush to dump decades of legislation protecting rare species and habitats, accelerate mineral resource development in the Arctic Wildlife Refuge, and sell off entire national parks. We even see remote stands of old-growth forest now being logged without the protection of any environmental laws (and with little economic return) under the guise of protecting "forest health."

The reality of catastrophic and accelerating biological change suggests that we should discard the red herring of arguing over what is truly wilderness or "natural" and instead focus on efficiently conserving what remains and developing criteria to evaluate human actions in these terms. To do so, we must retain a clear vision of (and respect for) what is wild. Fortunately, we can define "wildness" in terms that are much less prone to misinterpretation and misuse than our use of "wilderness." In particular, it seems appropriate to define an organism's, or a habitat's, wildness in terms of its ecological and evolutionary context, that is, its habitual relationships to other organisms and the surrounding environment.

For an organism to be considered "wild" (and in most cases for it to persist at all) it must exist in an ecological context essentially similar to the one its ancestors evolved in. That is, wild species need relatively intact habitats that retain historical patterns of disturbance, connectivity, and ecological interaction. Because only "wild" habitats provide these conditions to the species that occur within them, only they can ensure an evolutionary future to the lineage. Note that this definition is explicitly historical in the sense that it stipulates a continuity (but not stasis—see below) in the reigning conditions. The definition also recognizes that the traits we find in every species reflect their evolutionary history. In Dawkin's (1995, p. 8) words: "Any organism is a model of the world in which it lives . . . [or] to be more precise, . . . the worlds in which the animal's ancestors lived." Any species exists only as a population of individuals, the separate survival and reproduction of which represent that species' viability. To assume that organisms can readily adapt to live in the radically altered environments we construct reflects either biological ignorance or remarkable hubris.

Under this definition, isolated organisms removed from their natural context, as we find in any zoo or botanical garden, can never be considered wild. They lack the most crucial aspect of their identity, namely their interactions with their habitat, including other conspecifics and the vast number of other species that constitute the world their ancestors evolved in. Any plant or animal plucked from this context plays a far different ecological and evolutionary role than it would if left in the wild. Such plants and animals demand external feeding and propagation, experience artificial breeding and selection (advertent or inadvertent), and are unlikely to leave any long-term evolutionary legacy. While such organisms, like the tree in the garden, will always retain vestiges of their wild evolutionary past, they are prone also to change fundamentally in character, as has occurred so dramatically in domesticated plants and animals. A Chihuahua is no wolf, even if they share most of their genome.

THE PROBLEM OF ECOLOGICAL CHANGE

Another issue that arises often in recent discussions of what is wild or natural concerns the question of ecological balance or stasis vs. change. Cronon and others have pointed out that we have a long tradition of assuming that nature, left to its own devices, exhibits stasis or equilibrium. Early American plant ecologists such as Frederick Clements, impressed with the phenomenon of ecological succession, stressed the ability of many biotic communities to rebound from disturbance by passing through a predictable sequence of seral stages to converge once again on a stable equilibrium (Colinvaux 1973; Tobey 1981). In addition, theoretical ecology for many years stressed notions of stability (MacArthur 1955; May 1973; Pimm 1991). Such equilibria appear conveniently congruent with classical notions of a primeval divine order prone to being disturbed by intrusive human action. If ecological change occurs constantly, however, what basis do we have for judging those changes induced by humans as unnatural or inappropriate?

In recent decades ecologists have stressed the dynamic and often unpredictable nature of ecological change, in response both to secular change or erratic disturbances in the environment and to the often complex interactions that occur among species. Fires, floods, windstorms, and other forms of natural disturbance are a chronic and inherent part of most ecosystems that greatly influences their composition and the characteristics of many of their component species (see, e.g., contributions in Pickett and White 1985). The ecologist Daniel Botkin (1990) popularized this paradigm shift within ecology and termed the acceptance of non-equilibrium ecological dynamics "the new ecology." Despite this paradigm shift, however, no ecologist would argue that the nature and especially the rates of contemporary anthropogenic change are benign. Species well-adapted to certain types of disturbance and rates of ecological change are often exquisitely vulnerable to disruptions of these patterns. These species face imminent extinction because they cannot adapt ecologically or evolutionarily to the rapid pace and novelty of human-caused change. In sum, it is the nature, rate, and extent of change that are now "unnatural," not the occurrence of change per se (Callicott 1996a; Pickett and Ostfeld 1995).

This paradigm shift within academic ecology might have gone unnoticed except that writers such as Chase (1986), Callicott (1991), Cronon (1995b) and others leapt on this new paradigm to argue that our traditional notions of wilderness are far too static to adequately represent our contemporary understanding of ecological dynamics and to allow proper approaches to managing these areas. Proctor (1995, p. 291) explicitly extends the issue of ecological dynamics into the realm of environmental ethics: "The connection between ethics and reality is not straightforward, because reality itself is not straightforward. For instance, there is no clear balance of nature in Pacific Northwest forests." Such a statement (appearing in the same book as Cronon's essay) implies the misconception that ecological change is somehow so disordered or random as to preclude any coherent or consistent approach to management. This is seriously misleading; while ecological dynamics are an obvious challenge in many situations, natural forms of ecological disturbance and change are commonly rather predictable and an integral part of managing many ecosystems. Thus, the difficulties of managing ecologically dynamic systems have given rise to criticism in a few particular cases (e.g., the National Park Service and fire or their sometimes too-rigid goal of recreating 'vignettes' of primitive America circa 1492). Such critics, however, assume far more importance and generality for this criticism than seems appropriate. In fact, most ecologists, conservationists, and land managers have long observed and accepted the roles of disturbance and ecological change within natural systems and incorporated this understanding into their management. Even two-thirds of a century ago, Adams (1929, p. 10) concisely dispensed with the straw man of stasis in the context of preserving natural values:

> Thus, when ecologists emphasize the need of setting aside reservations for the preservation of natural conditions they do not mean, and certainly do not expect, the conditions to remain indefinitely "balanced," fixed and unchanged or unchanging, because they know that it is utterly impossible, both theoretically and practically . . . to keep it free from all outside influences.

Modern ecologists, in fact, argue that maintaining these historically dominant modes and patterns of disturbance are a fundamental part of effective conservation (e.g., Alverson et al. 1994). Nevertheless, unfriendly critics of wilderness and environmental protection sometimes mis-characterize this discussion to make their own political points. The Mobil Corporation (1995), for example, quoted a similar-sounding but fatuous opinion in Easterbrook (1995) in an op-ed advertisement in the *New York Times*: "the notion of a fragile environment is profoundly wrong. . . . The environment . . . is close to indestructible." While the environment may be indestructible, its inhabitants surely are not.

ECOLOGICAL RELATIVISM AND THE PROBLEM OF ESTABLISHING BIOTIC VALUE

These arguments over the meaning of wild, natural, and wilderness might seem academic and semantic except for the very real implications they carry for how we manage our lands and waters. If we accept that we cannot clearly separate natural from artificial and wild from tame, and that nature lacks stability and internal self-regulation, where does that leave us? Do we have any criteria left for judging our actions? Proctor (1995, p. 288) argues explicitly with regard to conservationists that "the ancient forest they strive to protect is as much a reflection of their own particular view of nature as it is some primeval system under siege by logging." Are the images we hold of ancient forests and broad sweeps of prairie grassland then only an illusion based on culturally tainted artifacts? Is the environmental destruction we see around us also an illusion? Shall we do whatever we please on the land and in our rivers and lakes as long as we agree to label these effects "natural"? Such conclusions gain some credibility if in discarding the dualisms described above we feel compelled to embrace environmental relativism.

No ecologist or conservationist biologist would suggest that wild areas do not deserve protection or that any changes we precipitate in the lands and waters are as natural and acceptable as any other. Botkin (1990), Alverson et al. (1994), and other ecologists argue rather that we must depend upon our best science to judge what kinds and rates of ecological change are acceptable. Real differences in conservation value clearly exist among different areas and among human actions affecting these areas. Lands also plainly differ in the degree to which their original biotic value has been degraded by direct or indirect human influence and the opportunities we have to conserve their remaining biotic value. Once we accept that lands differ in conservation value, then we must also accept that protecting certain areas from further degradation is far more significant than protecting other areas.

But what is this biotic value? Are biological criteria for evaluating lands and human actions merely a substitute or cover for motives that remain at base metaphysical? Are we merely replacing the awe and reverence that Cronon documents as the legacy of wilderness with another mystical concept now cloaked in the garb of scientific legitimacy? While it would hardly seem that scientific justifications would stem from any romantic ideal based on the grandeur of scenery or the opportunity to prove oneself in a wild and remote setting, Cronon (1995b, p. 81) suggests just this: "Although at first blush an apparently more 'scientific' concept than wilderness, biological diversity in fact invokes many of the same sacred values." While an individual's true motives may forever remain obscure, it seems unjustified to so dismiss a coherent and powerful scientific rationale for wild areas. Most conservation biologists would argue instead that clear criteria for evaluating biotic value can be established purely on scientific grounds without recourse to any such "sacred" values. In fact, biologists have plainly laid out scientific systems for evaluating the relative utility of different lands for conservation for more than fifteen years (Diamond and May 1981). Such systems have become routine in conservation organizations such as The Nature Conservancy, which is dedicated to identifying, purchasing, and permanently protecting natural areas with the greatest biotic value. They explicitly rank species and community types according to their conservation value and combine these data in their Natural Heritage databases (Noss 1987). They critically evaluate the biotic value of all the lands they consider for protection in terms of quantitative criteria established specifically for this purpose. Those writing Environmental Impact Statements or attempting to identify critical habitat under the Endangered Species Act must also routinely quantify the biotic value of lands. Such judgments have become ever more important as Habitat Conservation Plans have been advanced in lieu of absolute protection of all suitable habitat (Beatley 1994).

Thus, few conservation biologists are shy about devising scientific criteria to estimate biotic value. We rarely concern ourselves with questions regarding what is truly "natural," "stable," "pristine," or "undisturbed by human activity," and consider debates over these matters rather sterile. This is not to say that developing biotic criteria for conservation is simple or free of controversy. While our doubts and concerns are many, they differ from those that seem to plague environmental relativists and revolve instead around how unsure we are about levels of threat and the diverse and often indirect effects of human activities. In fact, the nascent job of developing biodiversity criteria and standards is technically quite demanding and fraught with frequent scientific debate. Such scientific criteria cannot be purely objective or free of all cultural bias. That they are artifacts of our culture, however, in no way implies that what they seek to describe or quantify is also an artifact or culture-bound. Many species will persist and thrive or decline to extirpation and extinction in response to human activity and there is nothing remotely subjective about the permanence of their extinction. Indeed, it is the epic proportion of these losses and their rapidity that drives conservation biologists to make the many and admittedly imperfect judgments they must in deciding which areas most deserve conservation effort.

SCIENTIFIC VALUES FOR CONSERVING WILD LANDS

Cronon describes the broad set of human values that we ascribe to wilderness including scenic grandeur, the chance to test oneself against primitive conditions, and opportunities to revere sublime or sacred aspects of nature. After analyzing how these values have shifted historically, he argues that our wilderness concepts are now so culturally inappropriate as to be counterproductive. Ironically, this argument comes just at a time when an entirely new set of concerns for wild lands has emerged to reinvigorate the wilderness movement. Increasingly, justifications and criteria for preserving wilderness areas are turning away from scenic and recreational values to address instead the *biotic* values that wild lands sustain (Alverson et al. 1994; Noss and Cooperrider 1994; Waller 1996). While we might agree with Cronon that traditional anthropocentric values for wilderness have become obsolete, his essay does little to inform the reader of how our notions of wild lands are expanding in the late twentieth century to encompass a new realm of more urgent justifications. I will argue that these shifts have given traditional concerns for wilderness a new lease on life and placed the designation of wild lands at the forefront of conservation thinking and action. While these shifts are mostly recent, they also reflect subtle historical trends that Cronon tends to downplay or overlook.

The science of conservation biology recently emerged from the conventional scientific fields of ecology, systematics, population biology, island biogeography, and wildlife management to address more synthetically the many factors that now threaten the persistence of individual species and even whole biological communities. Why is this shift in perspective occurring now? It stems from the growth in scientific knowledge, the growing realization that declines in biological diversity represent the most serious and long-lasting environmental threats of our time, and the realization that these threats now stem largely from how we mismanage our landscapes. Furthermore, unlike simpler physical environmental issues such as clean air and clean water, which responded quickly and dramatically to legislated reductions in pollutants, we encounter ever more threats to biodiversity despite prohibitions on trade in endangered species and the Endangered Species Act. These extinctions directly reflect losses and conversion of habitats throughout the world. Despite this realization, and moves to establish more protected reserves, habitat losses continue to accelerate with continued population and economic growth.

A surprising number of scientific justifications for conserving large and undeveloped lands have emerged from conservation biology in recent years (Alverson et al. 1994). Most conspicuously, we continue to discover how many of the ecological interactions crucial for sustaining plant and animal species depend critically on how large, connected, and intact areas of habitat are. Smaller natural areas, or those subject to human disturbance, are highly prone to losing a substantial fraction of their species and these losses occur via several distinct but related mechanisms. Large wide-ranging carnivores and ungulates are directly sensitive to human activity; their requirement for extensive home ranges causes them to disappear quickly from smaller fragmented habitats. Many smaller species of amphibians, mammals, and plants are also quite sensitive to human disturbance, or to human-altered disturbance regimes. Many species are also incapable of dispersing across open or inhospitable habitats such as clear-cuts or roads, which dissect their populations into smaller subunits that are increasingly vulnerable to genetic and demographic hazards. Similarly, many neotropical migrant songbird species are suffering serious declines across eastern North America in apparent response to progressive habitat losses and fragmentation. With the increase in edge habitats, nest predators and parasites favored by such edges have also increased, drastically reducing nest success. Smaller areas also fail to sustain or survive historically dominant patterns of natural disturbance such as fire and windthrow, which causes further losses of species dependent on these disturbances. In their stead, farming, roads, channelization, and other forms of development enhance opportunities for the weedy and often exotic species that increasingly dominate our landscapes, further displacing many native species. While many of these changes are delayed or occur too slowly to catch our attention, they have already caused dramatic losses. Their eventual effects will surely be catastrophic.

Because habitat loss, fragmentation, and other forms of degradation threaten such a large fraction of our biota, we must pursue strategies that can succeed not only in conserving populations now but also in perpetuating the ecological conditions that will sustain their evolutionary future. Interestingly, all the mechanisms reviewed above strongly support approaches to conservation that emphasize large and/or connected wild areas relatively free of human disturbance. Only large areas support larger, more viable, and interconnected populations of rare and threatened species and perpetuate the ecological processes that sustain other elements of biodiversity. While some elements of diversity can be sustained in small areas, and certain species clearly need localized and particular habitats, conservation biologists agree on the fundamental importance of allocating large blocks of suitable habitat as a first defense against further species losses (Soulé and Simberloff 1986). These findings have recast the agenda of most conservation organizations, including The Nature Conservancy, which now embraces the importance of protecting and restoring areas much larger than the small (50–100 acre) remnants it originally sought to protect.

While conserving wildlife has always existed as an accessory justification for wilderness, few of our wilderness areas and only one of our national parks (Everglades National Park) have been expressly designated to preserve biotic values. In addition, scientific and lay concerns for these biotic values have themselves expanded from a few vertebrate species (the "charismatic megafauna") to embrace the much broader concerns for other species now termed biodiversity. Callicott (1991; 1994a) and Cronon criticize wilderness designations for their "man vs. nature" dichotomy, yet some such separation is essential to sustain the many species that cannot tolerate the more denatured habitats that now dominate our landscapes.

EXPANDING HUMAN VALUES FOR WILD LANDS

As scientific justifications for conserving large wild areas have expanded in recent years, so has the set of human values centered on nature generally and biological diversity in particular (Norton 1986). A broad shift in moral perspective has occurred among many environmentalists away from the anthropocentric goals and values Cronon describes so well toward values that are at least ostensibly concerned with the ecological viability of species and biological communities. Nash (1989, p. 7) describes this development as "arguably the most dramatic expansion of morality in the course of human thought."

In reviewing Western environmental ethics, Callicott (1994b) distinguishes between instrumental or utilitarian values that can be considered anthropocentric in that they serve human goals and intrinsic or inherent values that transcend immediate human utility and so can be legitimately labeled nonanthropocentric. Instrumental values are further broken down into the material benefits we receive from nature (goods, services, and genetic information), and the psycho-spiritual benefits we receive in the form of aesthetic beauty, awe, and scientific knowledge. Aside from these justifications, conserving lands to sustain species diversity may also be viewed as a form of enlightened self-interest in that doing so provides the basis for new foods, drugs, chemicals, and biological control agents. Worster (1994, p. 432) places a similar emphasis on the consequences of human actions in terms of their eventual implications for our welfare:

> The challenge is to determine which changes are in our enlightened self-interest and are consistent with our most rigorous ethical reasoning, always remembering our inescapable dependency on other forms of life.

Thus, instrumental values alone represent a powerful and inclusive set of arguments for conserving biodiversity generally and, by extension, the large areas that this diversity depends upon. Nevertheless, it is the intrinsic worth of other species that has moved to the fore of environmental ethics and captured the attention of many environmentalists. Some authors and those involved in the animal rights movement suggest that we extend intrinsic value to individual organisms, a philosophy termed *biocentric* by Paul Taylor (1986). In its extreme (and ultimately absurd) form, organisms of all sizes and kinds are valued equally. Holmes Rolston III (1988) instead proposes that we grant additional intrinsic value to individuals of larger and more sentient species and to individuals grouped into larger entities such as species and ecosystems. Ecologists and conservation biologists, however, are far more likely to discount the importance of any given individual and value instead the evolutionary lineages we term *species* and the ecological communities and processes that sustain these species. Callicott

(1989; 1994b) terms such views *ecocentric* (to distinguish them from the narrower forms of biocentrism) and traces these ideas back to Darwin's *The Descent of Man* and especially to Aldo Leopold, who elaborated them in "The Land Ethic" and other essays.

What is noteworthy here is how science informs this area of environmental ethics and, in return, how the values so elaborated tend to match those of conservation biology. For instrumental values, science has greatly extended our understanding of how many goods and services we receive from nature, including the vast set of invisible 'ecosystem services' that humans took for granted until quite recently (Ehrlich and Ehrlich 1991). Similarly, science has contributed greatly to the instrumental value we extract from nature. All our efforts to "improve" crops, domesticated animals, and bacteria and their ability to produce our foodstuffs and drugs via genetic engineering depend utterly on the genetic information we extract from nature. (Ironically, much of our recent progress stems from the discovery of a thermophilic bacterium in the wilderness hot springs of Yellowstone National Park. This bacterium produces a thermostable DNA polymerase enzyme that is the basis for the polymerase chain reaction [PCR], a method used to amplify DNA. Although this enzyme and the associated technique won Kerry Mullis a Nobel Prize and millions of dollars via a patent, neither the bacterium's discoverer nor the wilderness that bred this remarkable form of life benefited from the remarkable advances they enabled.)

Thus, conservation science provides the tools to achieve either ecocentric goals or more pragmatic utilitarian goals. Those who recognize the inherent value of conserving nature are joining forces with conservation biologists dedicated to applying scientific methods to maintain species diversity and ecological systems for perhaps more mundane ends. Goals do not matter in this case because both instrumental and intrinsic values for biodiversity generate identical criteria for making environmental decisions. Nevertheless, ecocentric values lend considerable moral force to scientific arguments for conserving diversity. Oddly, while many Christians embrace conservation ethics as an implicit aspect of their religious beliefs (C. DeWitt, personal communication), some disparage those who attempt to grant intrinsic value to nature as members of an anti-Christian New Age cult (e.g., Coffman 1992).

Although conservation science provides the theory and data to guide environmental decisions, this science, like many others, can only make probabilistic predictions in many cases. This uncertainty poses a significant problem for politicians and judges who must set policy via the writing and interpretation of statutes. Wiener (1995) has explicitly explored the legal implications of evolutionary biology and modern views of nature. Like Cronon, Callicott, and Chase, he is interested in the "New Ecology" and how uncertainty regarding ecological change should influence our attitudes and approaches to conserving nature. Unlike those authors, however, he concludes that human actions must be judged by their ecological consequences rather than by the arbitrary terms and categories we tend to use in describing natural systems. Labeling human actions as "natural" does nothing to justify them. Courts, however, have yet to absorb how conservation science and scientific uncertainty should be incorporated into their decisions. This is particularly evident in recent cases where the Forest Service was excused from its failure to apply contemporary conservation science despite the species diversity provision of the 1976 National Forest Management Act (still the only federal statute with any such provision; Alverson et al. 1994; King 1995; Kuhlmann 1995).

HOW COMPATIBLE ARE SCIENTIFIC AND TRADITIONAL VALUES FOR WILDERNESS?

How do scientific approaches based on concerns for biodiversity accord with more traditional approaches to managing wilderness? Contemporary scientific justifications for protecting large areas from major human influence appear largely congruent with traditional arguments for wilderness based on anthropocentric values like aesthetics and reverence for remote uninhabited areas. Classically, a wilderness must be large, relatively undeveloped (e.g., low road density), and correspondingly 'wild' in its aspect. Traditional wilderness management attempts to limit human activities ("where man is a visitor who does not remain"). Conservation biologists also seek large areas relatively free of human disturbance. If they evaluate lands according to size and degree of wildness, lands accorded the highest value will often resemble, or be, classical wilderness. Given such convergence, should we enlarge our concept of wilderness to encompass intrinsic and other biotic values or should we instead coin a new term to make clear our concerns for biotic components that extend beyond more classical values?

While prescribing large areas free of roads and most other forms of major human disturbance often meshes well with conserving biodiversity, there are also cases where traditional wilderness areas have done a poor job of conserving biodiversity or where managing for one set of goals interferes with the other. Heavy recreational use by hikers may disrupt habitat use by wolves or other species sensitive to human presence. Similarly, limited development of certain kinds may be perfectly compatible with sustaining these or other species as suggested by the oil wells in at least one Audubon wildlife sanctuary. Furthermore, managers pursuing biotic goals may need road access and the discretion to burn or apply herbicides when restoring or managing certain types of habitat. Where goals compete or conflict, differences in goals and/or styles of management may strongly differentiate wilderness dedicated to aesthetic and recreational goals from natural areas dedicated to conserving species or ecosystems.

Many traditional wilderness areas are handicapped in their ability to serve biotic ends by the fact that they overwhelmingly tend to be located in habitats of little economic value such as mountains or deserts. While such "rock and ice" locales clearly serve aesthetic and recreational ends, they also avoid many areas of high biological productivity and diversity, including low-altitude eastern and western forests, prairies, and wetlands. This often makes it possible to distinguish wild areas intended to address recreational and aesthetic standards from wild areas intended to sustain biological values by protecting rare species or community types. Rugged coasts or spectacular mountains suggest an area chosen for recreation or scenery whereas floodplain forests, mangrove swamps, or calcareous fens suggest the use of a distinct set of biotic criteria.

Aldo Leopold, chief architect of the nation's first wilderness area in the Gila National Forest in 1924 and co-founder of the Wilderness Society, embraced both recreational and scientific justifications for wilderness (Meine 1988). After initially justifying wilderness primarily for its contributions to recreation (Leopold 1921), Leopold embraced a broader set of aesthetic values (Leopold 1925), but finally arrived in his essay "Wilderness as a Land Laboratory" at the idea that the most critical role for wilderness was a scientific one: that such lands serve as an invaluable control for judging the effects of various human impacts (Leopold 1941). Others joined Leopold early in this century to enunciate scientific values for wilderness (Sumner 1921; Adams 1929).

While Leopold felt no need to devise new terminology to identify new purposes for conserving wild lands, other individuals and groups have (Callicott, 1996b). For example, the United Nations has chosen to recognize a set of "biosphere reserves" characterized by their goal of conserving biodiversity, large area, and the existence of surrounding buffer zones supporting limited economic activity. Similarly, Solheim et al. (1987) and Alverson et al. (1994) also elected to label their proposals to preserve large blocks of forest in the upper Midwest "Diversity Maintenance Areas" (DMAs) in an explicit attempt to distinguish these proposals from traditional wilderness aimed at recreation. Later, however, we also recognized virtues, like Leopold, in broadening the classical wilderness concept:

> The merits of associating ecosystem protection with wilderness lie in their very close interconnection in fact, the centrality of a hands off policy for achieving biotic health, and our proven willingness as a society to take strong legal measures to protect lands of exceptional value, called wilderness, for nonmarket reasons. These hallmarks of wilderness would well serve biodiversity (Alverson et al. 1994, p. 243).

To allow for differences in purpose and management, however, we suggested recognizing distinct classes of wilderness.

Many biologists and environmentalists find ecocentric approaches to conservation fully compatible with the more traditional attitudes based on reverence, aesthetics, and recreation that Cronon describes. It is also interesting to contemplate whether aesthetics may be changing to encompass biotic values more completely, as Callicott (1993) himself argues. For example, although originally denigrated as a worthless mosquito- and alligator-infested swamp, the Everglades has metamorphosed into a valued national park, a World Natural Heritage site, and the center of a thriving tourist industry. Ecocentric values may also be taken as sacred by some, although this is by no means a necessary moral step.

The concordance between conservation science and intrinsic biotic values is particularly evident in what has been termed the "New Conservation Movement" (Foreman 1991a). This movement has arisen in direct response to the emergence of conservation biology and the accompanying realization that conventional approaches to conserving wild lands based on scenic beauty or the potential for rugged recreation are failing to adequately protect species diversity. These twin developments are related, of course, in that conservation biology provides a strong scientific basis for the new conservation movement and that both recognize the special importance of habitat area, landscape context, and biotic interactions for conserving biodiversity. Although some environmentalists have yet to embrace this shift, it is an increasingly dominant theme within many conservation organizations. It is also noteworthy that conservation biologists such as Michael Soulé and Reed Noss have been actively involved in working with environmentalists on activities like the ambitious Wildlands Project.

By suggesting that biodiversity acts as a cover for sacred values, Cronon seems to imply either that biodiversity lacks scientific substance or that moral and scientific justifications for wilderness are somehow incompatible with one another. Such positions seem at odds with the personal history of many of the individuals most directly involved with the wilderness movement in America. Is it simply a coincidence that so many of these individuals were excellent field naturalists with a scientific bent? One could start with Thoreau and recall his patient and perceptive observations of ant warfare, the details of tree dispersal, and the process of ecological succession (a term he apparently coined). One might next consider John Muir who, as the Archbishop of Nature worship in the late nineteenth century, must be considered among those ascribing sacred values to wild lands. Yet this is the same man who began his career as a successful inventor and machinist, who gained college training in botany and geology, and who convincingly demonstrated the glacial origin of high Sierra valleys in California such as Yosemite (Fox 1981; Wolfe 1945). Further, recall that Aldo Leopold was a scientist of considerable stature as well as a prime mover for wilderness and eventually a major spiritual and moral leader of modern environmentalism. He moved seamlessly from justifying wilderness primarily on recreational grounds to more holistic points of view where land and species should be protected for both scientific and ethical reasons. Perhaps more than any other individual, Leopold embodies the philosopher-scientist who sought, and found, deep resonance between scientific and moral principles.

Let us finally consider our contemporary, Dave Foreman, a founder of the radical environmental group "Earth First!" and a chief spokesman for "The Big Outside" consisting of large and remote areas of wilderness free of direct human management. In his autobiography, Foreman (1991b) traces the evolution of his own thinking and moral stances relative to conserving wilderness. Interestingly, despite a personal history rich in wilderness exploration and defiant moral stands on behalf of wilderness, Foreman (1995/96) has settled firmly on biotic justifications for wilderness as being most appropriate in the sense of being both scientifically sound and morally defensible. Indeed, he now spends much of his effort extending and defending concepts of wilderness explicitly in the context of protecting biological diversity (Foreman 1994) and working with biologists to promote an ambitious regional network of wild areas (Foreman 1992).

Thus, many of the chief proponents of wilderness have been expert naturalists who drew on their intimate familiarity with the subtlety and nuance of natural systems to argue persuasively for conservation. Although trained in science, these individuals freely invoked moral or spiritual bases for the protection of wild lands without considering such arguments to be in any way antithetical to their scientific outlook. (One might even be tempted to suggest that their deep understanding of wild species and natural events contributed to their moral perspective rather than the other way around.) While scientific rationales for protecting wilderness are admittedly tied to a particular cultural and historical context, they also provide a rigorous, cohesive, and ethically defensible basis for choosing and managing wild areas. Those most directly involved in the difficult work of protecting wild lands have converged on biotic values for both ethical and scientific reasons. This reflects both the escalating threats to species diversity and the fact that our scientific knowledge regarding these threats is increasing.

Ultimately, it is matter of judgment whether to extend classical concepts of wilderness so as to encompass a wider set of biotic and ethical values or to adopt instead a new terminology to differentiate areas dedicated to conserving biotic values from those established for recreation or aesthetic goals. A Venn diagram would show a considerable area of overlap between classical wilderness areas and areas serving biotic goals, but also separate areas where these distinct goals (or methods of management) distinguished the two. Such choices will reflect both personal preference and pragmatic issues surrounding the conservation of particular areas.

PROSPECTS FOR ECOLOGICAL RESTORATION

Rather than posing a serious threat to responsible environmentalism, I have argued, as a biologist, that wilderness defined as large, connected, and relatively intact ecosystems should form the backbone of any ecologically informed program to conserve our natural heritage. If there is a trouble with wilderness, it is that we do not have enough of it and that we continue to devalue it just as we have historically. Cronon and Callicott propose that we expand upon this meager base by granting greater value to the nature that persists in our nearby fields, forests, and cities and examining how our everyday lives affect this nature. This is, of course, natural and useful. Surely wilderness areas are now too scarce to represent a vastness to be loathed and conquered. But they are far too valuable for ecological and social restoration to be marginalized as representing only arenas for outdoor recreation or romantic icons of an earlier cultural era. Their value stems instead from the ecological integrity they embody. One cannot find this integrity in "the tree in the garden" severed as it is from genetic links to its conspecifics and its ecological links to other species. One has difficulty even finding it in 40- and 80-acre remnant patches of forest or prairie. We must accept the very real differences that exist in the degree of wildness of the lands and waters around us and use the better examples as foci to begin the serious and important work of restoring more nature to our cities and countryside. To blame wilderness for the concepts others have heaped on it (or logical inconsistencies among those concepts) is to blame the victim instead of the aggressor.

The intent of authors like Cronon and Callicott is obviously to clear away the rusty unworkable tools and cobwebs of past thinking and so prepare the way for the serious business of devising more politically feasible ways to live benignly on the land. These authors also worry that, by expending so much time and effort on what they see as the narrow issue of wilderness, American society has compromised its ability to enrich and restore our more mundane habitats. While I have questioned the presumption that worrying about wilderness prevents us from valuing nearer and plainer habitats, we can agree that we must now move forward to consider how better to manage the unprotected matrix. Following Leopold, Meine (1992, p. 170) wisely suggests: "a mature conservation/environmental movement will work across the full spectrum of land types, from the wild to the semi-wild to the cultivated to the settled to the urbanized, and will recognize the relevance of each to all the others."

Evaluating areas according to their location, size, and degree of wildness allows us to evaluate their relative value and potential for restoration, allowing escape from the quicksand of environmental relativism. Our science will also allow us to critically evaluate what will, and what will not, suffice to stem the hemorrhage of biodiversity. Science, however, is only a tool whose success in these efforts will depend utterly on our values. As we approach these difficult tasks we should be catholic in accepting the plurality of values that serve these goals. We should eschew semantic arguments and avoid red herrings such as the argument that all human actions are equally 'natural' or that wild areas follow no inherent or ecologically appropriate dynamics. These distractions divert us from more important issues and lend intellectual support to those who would conclude that natural systems can provide no guidance for judging human actions.

As conservation moves away from its historical job of preserving the best remaining natural habitats to encompass the broader work of managing and restoring a broad spectrum of natural communities, we face further questions regarding active versus passive management. As it may be inappropriate to assume that, left alone, nature will take an appropriate course, various forms of active management and restoration are often necessary. Such actions should always be guided and informed by an understanding of ecological context and historical dynamics. In addition, we must recognize that smaller and/or more degraded areas will usually require more active (and costly) efforts to manage and restore their natural values than larger areas (Alverson and Waller 1992). This itself constitutes a strong argument in favor of conserving large areas free of chronic human disturbance. Here we face questions similar to those posed in the wilderness debate: What is natural? How should ecological change be incorporated? How can we agree on a basis for setting goals? As in the case of wild lands, we should allow precepts of ecology and conservation science to determine appropriate actions and adjudicate among competing tactics. Furthermore, by applying our management and restoration efforts as experimental treatments, we have the opportunity to extend our science via the process now termed "adaptive management."

Ecological restoration has sometimes been compared to gardening. The difference between restoration and gardening, however, lies in their goals:

for gardening, this is clearly the artifice of some aesthetic ideal. For ecological restoration, our goals are more subtle and complex, namely creating habitat for native species and reestablishing the ecological processes inherent in natural communities (recreating, in ecologist G. Evelyn Hutchinson's provocative metaphor, the ecological theater and the evolutionary play). While many restored areas are, and likely will remain, too small or fragmented to ever sustain full complements of the species that originally inhabited such habitats, they serve well to remind us, like the tree in the garden, of the larger biotic context of which they, and we, are a part and upon which we depend so completely. In doing this work, however, we should never confuse the synechdoche with the larger and wilder habitat it symbolizes. Instead, we must accept the fact that we share our planet with myriad other forms of life, many of which demand larger and ecologically more intact habitats if they are to survive. If we accept Cronon's suggestion to embrace all areas, from the city to the wilderness, as "home," let it be a revered home that we treat gently and share generously.

NOTE

I am grateful to the several friends and colleagues who took time to carefully read an early draft of this essay and share their reactions and suggestions: Baird Callicott, Bill Cronon, Jan Dizard, Nancy Langston, Greg Aplet, David Wilcove, Hugh Iltis, Curt Meine, and Ann Shaffer. While I did not always accept their suggestions, their input was critical in allowing me to express myself more clearly.

BIBLIOGRAPHY

Adams, C. C. 1929. "The Importance of Preserving Wilderness Conditions." Report of the Director, New York State Museum, Albany, N.Y.

Alverson, W. S., and D. M. Waller. 1992. "Is It Un-Biocentric to Manage? *Wild Earth* 2(4): 9–10.

Alverson, W. S., W. Kuhlmann, and D. M. Waller. 1994. *Wild Forests: Conservation Biology and Public Policy.* Washington, D.C.: Island Press.

Beatley, T. 1994. *Habitat Conservation Planning: Endangered Species and Urban Growth.* Austin, Tex.: Texas University Press.

Botkin, D. B. 1990. *Discordant Harmonies: A New Ecology for the Twenty-first Century.* New York: Oxford University Press.

Callicott, J. B. 1989. *In Defense of the Land Ethic: Essays in Environmental Philosophy.* Albany: State University of New York Press.

Callicott, J. B. 1991. "The Wilderness Idea Revisited: The Sustainable Development Alternative." *The Environmental Professional* 13: 235–47. [Included in this volume.]

Callicott, J. B. 1993 "The Land Aesthetic." In S. Armstrong and R. Botzler, eds., *Environmental Ethics: Divergence and Convergence.* New York: McGraw-Hill.

Callicott, J. B. 1994a. "A Critique of and an Alternative to the Wilderness Idea." *Wild Earth* 4(4): 54–59.

Callicott, J. B. 1994b. "Conservation Values and Ethics." In G. K. Meffe and C. R. Carroll, eds., *Principles of Conservation Biology,* pp. 24–49. Sunderland, Mass.: Sinauer Associates.

Callicott, J. B. 1996a. "Do Deconstructive Ecology and Sociobiology Undermine Leopold's Land Ethic?" *Environmental Ethics* 18: 353–372.

Callicott, J. B. 1996b. "Should Wilderness Areas Become Biodiversity Reserves?" *The George Wright Forum* 13(2): 32–38. [Included in this volume.]

Chase, A. 1986. *Playing God in Yellowstone: The Destruction of America's First National Park.* Boston, Mass.: Atlantic Monthly Press.

Coffman, M. S. 1992. *Environmentalism! The Dawn of Aquarius or the Twilight of a New Dark Age?* Bangor, Me.: Environmental Perspectives, Inc.

Colinvaux, P. A. 1973. *Introduction to Ecology.* New York: John Wiley & Sons, Inc.

Cronon, W. 1995a. "The Trouble with Wilderness." *New York Times Sunday Magazine,* Aug. 13, pp. 42–43.

Cronon, W. 1995b. "The Trouble with Wilderness, or, Getting Back to the Wrong Nature." In W. Cronon, ed., *Uncommon Ground: Toward Reinventing Nature,* pp. 69–90. New York: W. W. Norton. [Included in this volume.]

Dawkins, R. 1995. "The Evolved Imagination: Animals as Models of Their World." *Natural History* 104(9): 8, 10–11, 22–24.

Diamond, J. M., and R. M. May. 1981. "Island Biogeography and the Design of Nature Reserves." In R. M. May, ed., *Theoretical Ecology: Principles and Applications,* pp. 228–52. Sunderland, Mass.: Sinauer Associates.

Easterbrook, G. 1995. *A Moment on the Earth: The Coming Age of Environmental Optimism.* New York: Viking Penguin.

Ehrlich, P. R., and A. H. Ehrlich. 1991. *Healing the Planet: Strategies for Resolving the Environmental Crisis.* Reading, Mass.: Addison-Wesley.

Foreman, D. 1991a. "The New Conservation Movement." *Wild Earth* 1(2): 6–12.

Foreman, D. 1991b. *Confessions of an Eco-Warrior.* New York: Harmony Books.

Foreman, D. 1992. "Developing a Regional Wilderness Recovery Plan." *Wild Earth* (The Wildlands Project Special Issue): 26–29.

Foreman, D. 1994. "Wilderness Areas Are Vital: A Reply to Callicott." *Wild Earth* 4(4): 64–68.

Foreman, D. 1995/96. "Wilderness: From Scenery to Nature." *Wild Earth* 5(4): 8–16. [Included in this volume.]

Fox, S. 1981. *The American Conservation Movement: John Muir and His Legacy.* Madison: University of Wisconsin Press.

King, P. S. 1995. "Applying Daubert to the 'Hard Look' Requirement of NEPA: Scientific Evidence before the Forest Service in *Sierra Club v. Marita.*" *Wisconsin Environmental Law Journal* 2: 147–71.

Kuhlmann, W. 1995. "Defining the Role of Conservation Biology in the Law of Protecting Ecosystems." In R. E. Grumbine, ed., *Environmental Policy and Biodiversity,* pp. 209–20. Washington, D.C.: Island Press.

Leopold, A. 1921. "The Wilderness and Its Place in Forest Recreational Policy." *Journal of Forestry* 19: 718–21.

Leopold, A. 1925. "Wilderness as a Form of Land Use." *Journal of Land and Public Utility Economics* 1: 398–404. [Included in this volume.]

Leopold, A. 1941. "Wilderness as a Land Laboratory." *Living Wilderness* 6: 3.

Leopold, A. 1949. *A Sand County Almanac and Sketches Here and There.* New York: Oxford University Press.

MacArthur, R. H. 1955. "Fluctuations of Animal Populations, and a Measure of Community Stability." *Ecology* 36: 533–36.

McKibben, B. 1989. *The End of Nature.* New York: Random House.

May, R. M. 1973. *Stability and Complexity in Model Ecosystems.* Princeton, N.J.: Princeton University Press.

Meine, C. 1988. *Aldo Leopold: His Life and Work.* Madison: University of Wisconsin Press.

Meine, C. 1992. "The Utility of Preservation and the Preservation of Utility: Leopold's Fine Line." In M. Oelschlaeger, ed., *The Wilderness Condition: Essays on Environment and Civilization.* San Francisco: Sierra Club Books.

Mobil Corporation. 1995. "More Good News" Advertisement appearing on the op-ed page of the *New York Times,* Oct. 5.

Nash, R. 1989. *The Rights of Nature: A History of Environmental Ethics.* Madison: University of Wisconsin Press.

Norton, B. G., ed. 1986. *The Preservation of Species: The Value of Biological Diversity.* Princeton, N.J.: Princeton University Press.

Noss, R. F. 1987. "From Plant Communities to Landscapes in Conservation Inventories: A Look at The Nature Conservancy (USA)." *Biological Conservation* 41: 11–37.

Noss, R. F., and A. Y. Cooperrider. 1994. *Saving Nature's Legacy: Protecting and Restoring Biodiversity.* Washington, D.C.: Island Press.

Pickett, S. T. A., and R. S. Ostfeld. 1996. "The Shifting Paradigm in Ecology." In R. L. Knight and S. F. Bates, eds., *A New Century for Natural Resources Management.* Washington, D.C.: Island Press.

Pickett, S. T. A., and P. S. White, eds. 1985. *The Ecology of Natural Disturbance and Patch Dynamics.* New York: Academic Press.

Pimm, S. L. 1991. *The Balance of Nature.* Chicago, Ill.: University of Chicago Press.

Proctor, J. D. 1995. "Whose Nature? The Contested Moral Terrain of Ancient Forests." In W. Cronon, ed., *Uncommon Ground: Toward Reinventing Nature,* pp. 269–97. New York: W. W. Norton.

Rolston, H., III. 1988. *Environmental Ethics: Duties to and Values in the Natural World.* Philadelphia: Temple University Press.

Solheim, S. L., W. S. Alverson, and D. M. Waller. 1987. "Maintaining Biotic Diversity in National Forests: The Necessity for Large Blocks of Mature Forest." *Endangered Species Technical Bulletin Reprint* 4(8): 1–3.

Soulé, M., and D. Simberloff. 1986. "What Do Genetics and Ecology Tell Us About the Design of Nature Preserves?" *Biological Conservation* 35: 19–40.

Sumner, F. B. 1921. "The Responsibility of the Biologist in the Matter of Preserving Natural Conditions." *Science,* n.s., 54: 39–43.

Taylor, P. W. 1986. *Respect for Nature: A Theory of Environmental Ethics.* Princeton, N.J.: Princeton University Press.

Tobey, R. C. 1981. *Saving the Prairies: The Life Cycle of the Founding School of American Plant Ecology, 1895–1955.* Berkeley, Calif.: University of California Press.

Vitousek, P., P. Ehrlich, A. Ehrlich, and P. Matson. 1986. "Human Appropriation of the Products of Photosynthesis." *Bioscience* 36: 368–73.

Waller, D. M. 1996. "Biodiversity as a basis for conservation efforts." In W. Snape, ed., *Biodiversity and the Law,* pp. 16–32. Washington, D.C.: Island Press.

Wiener, J. B. 1995. "Law and the New Ecology: Evolution, Categories, and Consequences." *Ecology Law Quarterly* 22: 325–58.

Wilson, E. O. 1992. *The Diversity of Life.* New York: W. W. Norton.

Wolfe, L. M. 1945. *Son of the Wilderness: The Life of John Muir.* Madison: University of Wisconsin Press.

Worster, D. 1994. *Nature's Economy: A History of Ecological Ideas,* 2nd ed. Cambridge: Cambridge University Press.

Dave Foreman

Wilderness (1995)

From Scenery to Nature

TWO SCENES, ONLY MONTHS APART:

October 31, 1994. President Bill Clinton lifts his pen from the California Desert Protection Act and the acreage of the National Wilderness Preservation System soars to over 100 million acres, nearly half of which is outside Alaska, and the acreage of the National Park System jumps to almost 90 million acres, over one-third outside Alaska. American Wilderness Areas and National Parks—the world's finest nature reserve system—are a legacy of citizen conservationists from Barrow to Key West, of courageous federal agency employees, and of farsighted elected officials. One hundred million acres is more than I thought we would ever protect when I enlisted in the wilderness wars (and I'm far from a hoary old warhorse like Dave Brower or Ed Wayburn—I've only been fighting for a quarter of a century).

February 14, 1995. The *New York Times* reports on a National Biological Service study done by three distinguished biologists. Reed Noss, editor of the widely-cited scientific journal *Conservation Biology* and one of the report's authors, says, "We're not just losing single species here and there, we're losing entire assemblages of species and their habitats." The comprehensive review shows that ecosystems covering half the area of the 48 states are endangered or threatened. Longleaf Pine Ecosystem, for example, once the dominant vegetation of the coastal plain from Virginia to Texas and covering more than 60 million acres, remains only in dabs and scraps covering less than 2 percent of its original sprawl. Ninety-nine percent of the native grassland of California has been lost. There has been a 90 percent loss of riparian ecosystems in Arizona and New Mexico. Of the various natural ecosystem types in the United States, 58 have declined by 85 percent or more and 38 by 70 to 84 percent.[1]

The dissonance between these two events is as jarring as chain saws in the forest, dirt bikes in the desert, the exploding of harpoons in the polar sea.

How have we lost so much while we have protected so much?

The answer lies in the goals, arguments, and process used to establish Wilderness Areas and National Parks over the last century.

In his epochal study, *National Parks: The American Experience* (University of Nebraska Press, 1979), Alfred Runte discusses the arguments crafted to support establishment of the early National Parks. Foremost was what Runte terms *monumentalism*—the preservation of inspirational scenic grandeur like the Grand Canyon or Yosemite Valley and the protection of the curiosities of nature like Yellowstone's hot pots and geysers. Later proposals for National Parks had to measure up to the scenic quality of Mt. Rainier or Crater Lake. Even the heavily glaciated Olympic Mountains were denied National Park designation for many years because they weren't deemed up to snuff. Then, after the icy mountains were grudgingly accepted as National Park material, the National Park Service and even some conservation groups bristled over including the lush temperate forests of the Hoh and Quinalt valleys in the new Park, seeing them as mere trees unworthy of National Park designation. National Park status was only for the "crown jewels" of American nature, an award akin to the Congressional Medal of Honor. If a substandard area became a National Park, it would tarnish the *idea* of National Parks as well as diminish all other National Parks. (In our slightly more enlightened age, the stupendous conifers are the most celebrated feature of Olympic National Park.)

A second argument for new National Parks, Runte explains, was based on "*worthless lands.*" Areas proposed for protection, conservationists argued, were unsuitable for agriculture, mining, grazing, logging, and other make-a-buck uses. Yellowstone could be set aside because no one in his right mind would try to grow corn there; no one wanted to mine the glaciers of Mt. Rainier or log the sheer cliffs of the Grand Canyon. The worthless-lands argument often led Park advocates to agree to boundaries gerrymandered around economically valuable forests eyed by timber interests, or simply to avoid proposing timbered lands altogether. Where Parks were designated over industry objections (such as Kings Canyon National Park, which was coveted as a reservoir site by Central Valley irrigators), protection prevailed only because of the dogged efforts of the Sierra Club and allied groups. Such campaigns took decades.

When the great conservationist Aldo Leopold and fellow rangers called for protecting Wilderness Areas in the National Forests in the 1920s and '30s, they adapted the monumentalism and worthless-lands arguments with success. The Forest Service's enthusiasm for Leopold's wilderness idea was, in fact, partly an attempt to head off the Park Service's raid on the more scenic chunks of the National Forests. "Why transfer this land to the Park Service?" the Forest Service asked. "We have our own system to recognize and protect the crown jewels of American scenery!" Wilderness advocates also reiterated the *utilitarian* arguments used decades earlier for land protection. The Adirondack Preserve in New York, for example, had been set aside to protect the watershed for booming New York City. The first Forest Reserves in the West had been established to protect watersheds above towns and agricultural regions. Such utilitarian arguments became standard for Wilderness Area advocacy in the twentieth century.

The most common argument for designating Wilderness Areas, though, touted their *recreational* values. Leopold, who railed against "Ford dust" in the backcountry, feared that growing automobile access to the National Forests would supplant the pioneer skills of early foresters. "Wilderness areas are first of all a series of sanctuaries for the primitive arts of wilderness travel, especially canoeing and packing," said Leopold. He defined Wilderness Areas as scenic roadless areas suitable for pack trips of two weeks' duration without crossing a road. Bob Marshall in the 1930s elaborated on the recreation arguments. Wilderness Areas were reservoirs of freedom and inspiration for those willing to hike the trails and climb the peaks. John Muir, of course, had used similar recreation arguments for the first National Parks.

In the final analysis, most areas in the National Wilderness Preservation

System and the National Park System were (and are) decreed because they had friends. Conservationists know that the way to protect an area is to develop a constituency for it. We create those advocates by getting them into the area. Members of a Sierra Club group or individual hikers discover a wild place on public land. They hike the trails, run the rivers, climb the peaks, camp near its lakes. They photograph the area and show slides to others to persuade them to write letters in its support. We backcountry recreationists fall in love with wild places that appeal to our sense of natural beauty. Conservationists also know the many political compromises made in establishing boundaries by chopping off areas coveted by industry for lumber, forage, minerals, oil and gas, irrigation water, and other natural resources—"worthless" lands coming back to haunt us.

The character of the National Wilderness Preservation and National Park systems is formed by these monumental, worthless, utilitarian, and recreational arguments. Wilderness Areas and National Parks are generally scenic, have rough terrain that prevented easy resource exploitation or lack valuable natural resources (timber and minerals especially), and are popular for non-motorized recreation.

So, in 1995, despite the protection of nearly 50 million acres of Wilderness Areas and about 30 million acres of National Parks in the United States outside of Alaska, we see true wilderness—biological diversity with integrity—in precipitous decline. In 1992, The Wildlands Project cited some of these losses in its mission statement:

- Wide-ranging, large carnivores like Grizzly Bear, Gray Wolf, Mountain Lion, Lynx, Wolverine, and Jaguar have been exterminated from many parts of their pre-European settlement ranges and are in decline elsewhere.
- Populations of many songbirds are crashing.
- Waterfowl and shorebird populations are approaching record lows.
- Native forests have been extensively cleared and degraded, leaving only remnants of most forest types—such as the grand California redwoods and the low-elevation coniferous forests of the Pacific Northwest. Forest types with significant natural acreages, such as those of the Northern Rockies, face imminent destruction.

- Tallgrass and Shortgrass Prairies, once the habitat of the most spectacular large mammal concentrations on the continent, have been almost entirely converted to agriculture or other human uses.

It is important to note, however, that *ecological integrity* has always been at least a minor goal and secondary justification in Wilderness Area and National Park advocacy. At the Sierra Club Biennial Wilderness Conferences from 1949 to 1973, scientists and others presented ecological arguments for wilderness preservation and discussed the scientific values of Wilderness Areas and National Parks. In the 1920s and '30s, the Ecological Society of America and the American Society of Mammalogists developed proposals for ecological reserves on the public lands. The eminent ecologist Victor Shelford was an early proponent of protected wildlands big enough to sustain populations of large carnivores.

Some of this country's greatest conservationists have been scientists, too. One of the many hats John Muir wore was that of a scientist. Aldo Leopold was a pioneer in the sciences of wildlife management and ecology, and argued for Wilderness Areas as ecological baselines. Bob Marshall had a Ph.D. in plant physiology. Olaus Murie, long-time president of The Wilderness Society, was an early wildlife ecologist and one of the first to defend the wolf.

Moreover, not all National Parks were protected primarily for their scenery. Mt. McKinley National Park was set aside in 1917 not for its stunning mountain but as a wildlife reserve. Everglades National Park, finally established in 1947, was specifically protected as a wilderness ecosystem. Even the Forest Service used ecosystem representation to recommend areas for Wilderness in the Second Roadless Area Review and Evaluation (RARE II) in 1977–79.

Somehow, though, professional biologists and advocates for wilderness preservation drifted apart—never far apart, but far enough so that the United States Forest Service lumped its wilderness program under the division of recreation.

That drifting apart was brought to an abrupt halt when the most important—and most depressing—scientific discovery of the twentieth century was revealed some fifteen years ago. During the 1970s, field biologists had grown increasingly alarmed at population losses in a myriad of species and by the loss of ecosystems of all kinds around the world. Tropical rainforests were falling to saw and torch. Coral reefs were dying from god knows what. Ocean fish stocks were crashing. Elephants, Rhinos, Gorillas, Tigers, and other charismatic megafauna were being slaughtered. Frogs everywhere were vanishing. These staggering losses were in oceans and on the highest peaks; they were in deserts and in rivers, in tropical rainforests and Arctic tundra alike.

A few scientists—like Michael Soulé, later founder of the Society for Conservation Biology, and Harvard's famed entomologist E. O. Wilson—put these disturbing anecdotes and bits of data together. They knew, through studies of the fossil record, that in the 500 million years or so of terrestrial evolution there had been five great extinction events—the hard punctuations in the equilibrium. The last occurred 65 million years ago at the end of the Cretaceous when dinosaurs became extinct. Wilson and company calculated that the current rate of extinction was one thousand to ten thousand times the background rate of extinction in the fossil record. That discovery hit with all the subtlety of an asteroid striking Earth:

RIGHT NOW, TODAY, LIFE FACES THE SIXTH
GREAT EXTINCTION EVENT IN EARTH HISTORY.

The cause is just as disturbing: eating, manufacturing, traveling, warring, and breeding by five and a half billion human beings.

The crisis we face is biological meltdown. Wilson warns that one-third of all species on Earth could become extinct in 40 years. Soulé says that the only large mammals that will be left after the turn of the century will be those we consciously choose to protect; that for all practical purposes "the evolution of new species of large vertebrates has come to a screeching halt."

That 1980 realization shook the daylights out of biology and conservation. Biology could no longer be removed from activism, if scientists wished their research subjects to survive. Conservation could no longer be about protecting outdoor museums and art galleries, and setting aside backpacking parks and open-air zoos. Biologists and conservationists began to understand that species can't be brought back from the brink of extinction one by one. Nature reserves had to protect entire ecosystems, guarding the flow and dance of evolution.

A new branch of applied biology was launched. Conservation biology, Michael Soulé declared, is a *crisis* discipline.

Conservation biologists immediately turned their attention to nature reserves. Why hadn't National Parks, Wilderness Areas, and other reserves prevented the extinction crisis? How could reserves be better designed and managed in the future to protect biological diversity? Looking back, we see that four lines of scientific inquiry led to the sort of reserve design now proposed by The Wildlands Project and our allies.

Conservation biologists first drew on an obscure corner of population biology called *island biogeography* for insights. In the 1960s, E. O. Wilson and Robert MacArthur studied colonization and extinction rates in oceanic islands like the Hawaiian chain. They hoped to devise a mathematical formula for the number of species an island can hold, based on factors such as the island's size and its distance from mainland.

They also looked at continental islands. Oceanic islands have never been connected to the continents. Hawaii, for example, is a group of volcanic peaks rising from the sea floor. Any plants or animals had to get there from somewhere else. But continental islands, like Borneo or Vancouver or Ireland, were once part of nearby continents. When the glaciers melted 10,000 years ago and the sea level rose, these high spots were cut off from the rest of the continents and became islands. Over the years, continental islands invariably lose species of plants and animals that remain on their parent continents, a process called *relaxation.* On continental islands, island biogeographers tried to develop formulas for the rate of species loss and for future colonization, and to determine whether equilibrium would someday be reached.

Certain generalities jumped out at the researchers. The first species to vanish from continental islands are the big guys. Tigers. Elephants. Bears. The larger the island, the slower the rate at which species disappear. The farther an island is from the mainland, the more species it loses; the closer, the fewer. An isolated island loses more species than one in an archipelago.

In 1985, as Soulé, David Ehrenfeld, Jared Diamond, William Conway, Peter Brussard, and other top biologists were forming the Society for Conservation Biology, ecologist William Newmark looked at a map of the western United States and realized that its National Parks were also islands. As the sea of development had swept over North America, National Parks had become islands of natural habitat. Did island biogeography apply?

Newmark found that the smaller the National Park and the more isolated it was from other wildlands, the more species it had lost. The first species to go had been the large, wide-ranging critters—Gray Wolf, Grizzly Bear, Wolverine. Faunal relaxation had occurred, *and was still occurring.* Newmark predicted that all National Parks would continue to lose species. Even Yellowstone National Park isn't big enough to maintain viable populations of all the large wide-ranging mammals. Only the complex of National Parks in the Canadian Rockies is substantial enough to ensure their survival.

While Newmark was applying island biogeography to National Parks, Reed Noss and Larry Harris at the University of Florida were using the *metapopulation* concept to design reserves for the Florida Panther, an Endangered subspecies, and the Florida Black Bear, a Threatened subspecies. Metapopulations are populations of subpopulations. A small isolated population of bears or Panthers faces genetic and stochastic threats. With few members of the population, inbreeding is likely, and this can lead to all kinds of genetic weirdness. Also a small population is more vulnerable than a large one to local extinction (*winking out,* in ecological jargon). If the animals are isolated, their habitat can't be recolonized by members of the species from another population. But if habitats are connected so that animals can move between them—even as little as one horny adolescent every ten years—then inbreeding is usually avoided, and a habitat whose population winks out can be recolonized by dispersers from a nearby population.

Noss and Harris designed a nature reserve system for Florida consisting of core reserves surrounded by buffer zones and linked by habitat corridors. Florida is the fastest growing state in the nation. When the Noss proposal, calling for 60 percent of Florida to be protected in such a nature reserve network, was published in 1985, it was considered . . . well, impractical. But over the last decade this visionary application of conservation biology has been refined by the State of Florida, and now state agencies and The Nature Conservancy are using the refinement to set priorities for land acquisition and protection of key areas.

In 1994 the Florida Game and Fresh Water Fish Commission published a 239-page document, *Closing the Gaps In Florida's Wildlife Habitat Conservation System.* Using GIS computer mapping technology, *Closing the Gaps* identified Biodiversity Hot Spots for Florida. The study looked in detail at range occurrences and habitat needs for 33 sensitive species ranging from the Florida Panther to the Pine Barrens Treefrog, and at 25,000 known locations of rare plants, animals, and natural communities. Existing conservation lands in Florida cover 6.95 million acres. The hot spots—called Strategic Habitat Conservation Areas—encompass another 4.82 million acres. Florida is working with private landowners to protect identified areas and has appropriated $3.2 billion to purchase Strategic Habitat Conservation Areas by the year 2000. Once a new Ph.D.'s pie-in-the-sky, a conservation biology-based reserve system is now the master plan for land protection in Florida.

While metapopulation dynamics and island biogeography theory were being applied to nature reserve design, biologists were beginning to recognize the value of large carnivores to their ecosystems. Previously, scientists had tended to see wolves and Wolverines and Jaguars as relatively unimportant species perched on top of the food chain. They really didn't have that much influence on the overall functioning of the natural system, biologists thought. Until the 1930s, in fact, the National Park Service used guns, traps, and poison to exterminate Gray Wolves and Mountain Lions from Yellowstone and other Parks (they succeeded with the wolf). Early in his career, even Aldo Leopold beat the drum for killing predators.

Today, biologists know that lions and bears and wolves are ecologically essential, in addition to being important for a sense of wildness in the landscape. For example, the eastern United States is overrun with White-tailed Deer. Their predation on trees is preventing forest regeneration and altering species composition, according to University of Wisconsin botanists Don Waller, Steve Solheim, and William Alverson. If allowed to return, wolves and Mountain Lions would scatter deer from their concentrated wintering yards and reduce their numbers, thereby allowing the forest to return to more natural patterns of succession and species composition.

Large herds of Elk are overgrazing Yellowstone National Park. Conservation biologists hope that the recent reintroduction of the Gray Wolf will control Elk numbers and keep large herds from loafing in open grasslands.

Michael Soulé has shown that native songbirds survive in suburban San Diego canyons where Coyotes remain; they disappear when Coyotes disappear. Coyotes eat foxes and prowling house cats. Foxes and cats eat quail, cactus wrens, gnatcatchers, and their nestlings. Soulé calls this phenomenon of increasing mid-sized carnivores because of decreasing large carnivores *mesopredator release.*

In the East, David Wilcove, staff ecologist for the Environmental Defense Fund, has found that songbirds are victims of the extirpation of wolves and Cougars. Neotropical migrant songbirds such as warblers, thrushes, and flycatchers winter in Central America and breed in the United States and Canada. The adverse effects of forest fragmentation on songbird populations are well documented; but Wilcove has shown that songbird declines are partly due to the absence of large carnivores in the East. Cougars and Gray Wolves don't eat warblers or their eggs, but raccoons, foxes, and possums do, and the Cougars and wolves eat these midsize predators. When the big guys were hunted out, the populations of the middling guys exploded—with dire results for the birds. Soulé's mesopredator release rears its ugly head again.

On the Great Plains, the tiny Swift Fox is Endangered. Why? Because the wolf is gone. Swift Foxes scavenged on wolf kills but wolves didn't bother their little cousins. Coyotes, however, eat Swift Foxes. Wolves eat Coyotes. Get rid of the wolf and Swift Foxes don't have wolf kills to clean up, and abundant Coyotes eat up the foxes.

John Terborgh of Duke University (in my mind the dean of tropical ecology) is currently studying the ecological effects of eliminating large carnivores from tropical forests. He tells us that large carnivores are major regulators of prey species numbers—the opposite of once-upon-a-time ecological orthodoxy. He has also found that the removal or population decline of large carnivores can alter plant species composition, particularly the balance between large and small-seeded plants, due to increased plant predation by animals normally preyed upon by large carnivores.

In addition to being critical players in various eat-or-be-eaten schemes, large carnivores are valuable as *umbrella species.* Simply put, if enough habitat is protected to maintain viable populations of top predators like Wolverines or Harpy Eagles, then most of the other species in the region will also be protected. Those that aren't, such as rare plants with very restricted habitats, can usually be protected with vest-pocket preserves of the old Nature Conservancy variety.

A final piece in conservation biology's big-picture puzzle is the importance of natural disturbances. Caribbean forests are adapted to periodic hurricanes. Many plant communities in North America evolved with wildfire. Floods are crucial to new trees sprouting in riparian forests. To be viable, habitats must be large enough to absorb major natural disturbances (types of *stochastic events* in ecologist lingo). When Yellowstone burned in 1988, there was a great hue and cry over the imagined destruction; but ecologists tell us that the fire was natural and beneficial. Because Yellowstone National Park covers two million acres and is surrounded by several million acres more of National Forest Wilderness Areas and roadless areas, the extensive fires affected only a portion of the total reserve area.

Things didn't turn out so well when The Nature Conservancy's Cathedral Pines Preserve in Connecticut was hammered by tornadoes in 1989. In this tiny patch of remnant old-growth White Pine forest (with some trees 150 feet tall), 70 percent of the trees were knocked flat, devastating the entire forest patch. Had the tornadoes ripped through an old-growth forest of hundreds of thousands of acres, they instead would have played a positive role by opening up small sections to new forest growth.

These four areas of recent ecological research—island biogeography, metapopulation theory, large carnivore ecology, and natural disturbance dynamics—are the foundation for The Wildlands Project. We used insights from these four fields to set our goals for protecting Nature in a reserve network. For a conservation strategy to succeed, it must have clearly defined goals. These goals should be scientifically justifiable and they should be visionary and idealistic. Reed Noss, science director for the Project, set out the four fundamental goals of The Wildlands Project in 1992:

1. Represent, in a system of protected areas, all native ecosystem types and seral stages across their natural range of variation.
2. Maintain viable populations of all native species in natural patterns of abundance and distribution.
3. Maintain ecological and evolutionary processes, such as disturbance regimes, hydrological processes, nutrient cycles, and biotic interactions, including predation.
4. Design and manage the system to be responsive to short-term and long-term environmental change and to maintain the evolutionary potential of lineages.

With the criteria embodied in these goals, we can look closely at existing Wilderness Areas and National Parks and answer our original question—why has the world's greatest nature reserve system failed to prevent biological meltdown in the United States?

As we have seen, Wilderness Areas and National Parks are generally islands of wild habitat in a matrix of human-altered landscapes. By fragmenting wildlife habitat we imperil species from Grizzlies to warblers who need large, intact ecosystems. Because they have been chosen largely for their scenic and recreational values, and to minimize resource conflicts with extractive industries, Wilderness Areas and National Parks are often "rock and ice"—high elevation, arid, or rough areas which are beautiful and are popular for backpacking, but which also are *relatively* unproductive habitats. For the most part, the richer deep forests, rolling grasslands, and fertile river valleys on which a disproportionate number of rare and Endangered species depend have passed into private ownership or, if public, have been "released" for development and resource exploitation. To make matters worse, the elimination of large carnivores, suppression of natural fire, and livestock grazing have degraded even the largest and most remote Wilderness Areas and National Parks in the lower 48 states.

To achieve TWP's four reserve design goals, we must go beyond current National Park, Wildlife Refuge, and Wilderness Area systems. Our ecological model for nature reserves consists of large Wilderness cores, buffer zones, and biological corridors. The core Wilderness Areas would be strictly managed to protect and, where necessary, to restore native biological diversity and natural processes. Traditional Wilderness recreation is entirely compatible, so long as ecological considerations come first. Biological corridors would provide secure routes between core reserves for the dispersal of wide-ranging species, for genetic exchange between populations, and for migration of plants and animals in response to climate change. Surrounding the core reserves, buffer zones would allow increasing levels of compatible human activity away from the cores. Active intervention or protective management, depending on the area, would aid in the restoration of extirpated species and natural conditions.

Admittedly, there has been some debate among scientists about reserve design. Some aspects of corridors have been criticized. Several "scientists" representing the anti-conservation wise use/militia movement have misstated these controversies, ignoring the general consensus that has emerged among reputable scientists on all sides of these discussions.

This emerging consensus has been summarized in several forms during the last five years. In 1990 with the Conservation Strategy for the Northern

Spotted Owl, Jack Ward Thomas, now Chief of the Forest Service, set forth five reserve design principles "widely accepted among specialists in the fields of ecology and conservation biology." In 1992, Reed Noss updated those five and added an important sixth principle:

1. Species well distributed across their native range are less susceptible to extinction than species confined to small portions of their range.
2. Large blocks of habitat containing large populations of a target species are superior to small blocks of habitat containing small populations.
3. Blocks of habitat close together are better than blocks far apart.
4. Habitat in contiguous blocks is better than fragmented habitat.
5. Interconnected blocks of habitat are better than isolated blocks; corridors or linkages function better when habitat within them resembles that preferred by target species.
6. Blocks of habitat that are roadless or otherwise inaccessible to humans are better than roaded and accessible habitat blocks.

Based on his studies of faunal extinctions in fragmented chaparral habitats in San Diego County, Michael Soulé summarized some reserve design principles in a very understandable way for us layfolk:

A. Bigger is better.
B. Single large is usually better than several small.
C. Large native carnivores are better than none.
D. Intact habitat is better than artificially disturbed.
E. Connected habitat is usually better than fragmented.

In a 1995 report for the World Wildlife Fund, *Maintaining Ecological Integrity in Representative Reserve Networks,* Noss added several more fundamental principles:

- Ecosystems are not only more complex than we think, but more complex than we can think (Egler 1977).
- The less data or more uncertainty involved, the more conservative a conservation plan must be (i.e., the more protection it must offer).
- Natural is not an absolute, but a relative concept.

- In order to be comprehensive, biodiversity conservation must be concerned with multiple levels of biological organization and with many different spatial and temporal scales.
- Conservation biology is interdisciplinary, but biology must determine the bottom line (for instance, where conflicts with socio-economic objectives occur).
- Conservation strategy must not treat all species as equal but must focus on species and habitats threatened by human activities (Diamond 1976).
- Ecosystem boundaries should be determined by reference to ecology, not politics.
- Because conservation value varies across a regional landscape, zoning is a useful approach to land-use planning and reserve network design.
- Ecosystem health and integrity depend on the maintenance of ecological processes.
- Human disturbances that mimic or simulate natural disturbances are less likely to threaten ecological integrity than are disturbances radically different from the natural regime.
- Ecosystem management requires cooperation among agencies and landowners and coordination of inventory, research, monitoring, and management activities.
- Management must be adaptive.
- Natural areas have a critical role to play as benchmarks or control areas for management experiments, and as refugia from which areas being restored can be recolonized by native species.

Now what? Where do we go with all this?

Conservation biology has shown us the crisis we face (and it *is* a crisis despite the sugary "What, me worry?" attitude of Eco-Pollyannas like Gregg Easterbrook); conservation biology has developed the theory supporting the protection of biological diversity; and conservation biology has set out a new model of how nature reserves should be designed. It is up to citizen conservationists to apply conservation biology to specific land use decisions and Wilderness Area proposals. We have the political expertise, the love for the land, and the ability to mobilize support that an ambitious Nature protection campaign demands.

There is wide agreement among conservation biologists that existing Wilderness Areas, National Parks, and other federal and state reserves are the building blocks for an ecological reserve network. Inspired by Noss's and Soulé's work, conservationists in the Northern Rockies, led by the Alliance for the Wild Rockies, applied conservation biology principles there as early as 1990. Biologists like pioneer Grizzly Bear researcher John Craighead and conservationists like former Wilderness Society head Stewart Brandborg reckoned that if Yellowstone is not large enough to maintain viable populations of Grizzlies and Wolverines, then we need to link Yellowstone with the big Wilderness Areas of central Idaho, the Glacier National Park/Bob Marshall Wilderness complex in northern Montana and on into Canada's Banff/Jasper National Park complex. Maintaining metapopulations of wide-ranging species means landscape connectivity must be protected throughout the entire Northern Rockies. The Northern Rockies Ecosystem Protection Act (NREPA), which would designate 20 million acres of new Wilderness Areas in the United States and protect corridors between areas, has been introduced into Congress and drew over 60 cosponsors in 1994. The proposal is now being refined by scientists and conservationists in Canada and the United States for a Yellowstone to Yukon reserve system. Scores of grassroots Wilderness groups have helped advance the legislation. The Sierra Club was the first major national conservation organization to endorse NREPA.

Other conservation groups are using conservation biology to develop alternative proposals for the next generation of National Forest Management Plans. They are seeking to identify biological hot spots including habitat for sensitive species, remaining natural forest, and travel corridors for wide-ranging species. With such maps they will argue for expanding existing Wilderness Areas into ecologically rich habitats and for protecting wildlife linkages. In many areas roads need to be closed in sensitive ecosystems, once present species like wolves and Mountain Lions reintroduced, and damaged watersheds restored. The Southern Rockies Ecosystem Project is coordinating several groups in a comprehensive conservation biology approach to new National Forest Plans in Colorado. The Southern Appalachian Forest Coalition is developing a conservation biology management strategy for all National Forests in its region. SREP and SAFC are the best examples of regional coalitions working from conservation biology principles.

One of the central messages of conservation biology is that ecosystems and wildlife ranges do not follow political boundaries. Many nature reserves will need to cross international borders. The best application of this so far is in Central America where a consortium of government agencies, scientists, and private groups are working with Wildlife Conservation International to link existing National Parks and other reserves from Panama to Mexico's Yucatan. This proposed nature reserve network, called *Paseo Pantera* (Path of the Panther), would allow Jaguars and Mountain Lions to move between core reserves throughout Central America.

To the north, the Canadian Parks and Wilderness Society and World Wildlife Fund Canada are incorporating conservation biology in their Endangered Spaces campaign throughout Canada. In every province and territory, scientists and activists are working to identify core reserves and connecting corridors based on the needs of large carnivores, biological hot spots, and "enduring features" on the landscape. The Canadians are working with conservationists in Alaska and the northern part of the lower 48 states on cross-border reserves and linkages.

National conservation groups in the United States like the Sierra Club, Wilderness Society, Defenders of Wildlife, World Wildlife Fund, and American Wildlands have been influenced by The Wildlands Project and are seeking to incorporate conservation biology into their work. The Sierra Club's Ecoregions Campaign could become a promising initiative for bringing conservation biology to conservation policy. Early this year the Sierra Club brought together Noss and Soulé with Club activists and public opinion, political, and marketing experts to explore how to "sell" biodiversity to politicians and the public.

In fifteen years, conservation biology has wrought a revolution. The goal for nature reserves has moved beyond protecting scenery to protecting all Nature—the diversity of genes, species, ecosystems, and natural processes. No longer are conservationists content with protecting remnant and isolated roadless areas; more and more biologists have come to agree with Reed Noss, who says, "Wilderness recovery, I firmly believe, is the most important task of our generation." Recycling, living more simply, and protecting human health through pollution control are all important. But it is only by encouraging wilderness recovery that we can learn humility and respect, that we can come home, at last. And that the grand dance of life will continue in all its beauty, integrity, and evolutionary potential.

NOTE

1. See Reed Noss's article, "What Do Endangered Ecosystems Mean to The Wildlands Project?" *Wild Earth* (Winter 1995/96).

J. Baird Callicott

Should Wilderness Areas Become Biodiversity Reserves? (1996)

THE TWENTY-FIRST CENTURY is just around the corner. That calls for reflection and reassessment of the conservation philosophy that has governed the disposition of public lands in the United States since the late nineteenth century. As jurisdictions, the national forests, range lands, and parks didn't just happen. They were created for a reason. But, over a century, new thinking can emerge that challenges the raison d'être of old institutions. Here I would like to review, and suggest a way to reformulate, the raison d'être of wilderness areas in the public domain.

During the 1980s the "crisis discipline" called conservation biology emerged. The crisis that it aims to address is the precipitous and accelerating loss of species, or, more generally and abstractly, the loss of biological diversity at every level of organization—of genetic diversity within populations, of diverse populations within species, of species diversity, of diverse assemblages of species populations (biotic communities), landscape-scale diversity, and diverse biomes. Conservation biology has quietly transformed the agenda of conservation *from* either conserving natural resources ("wise use," etc.) or conserving pristine Nature ("wilderness preservation") *to* conserving biological diversity (or "biodiversity" for short) and ecological integrity.

The utilitarian Pinchot philosophy of conservation—summed up in the maxim, "the greatest good of the greatest number for the longest time," and in the general policy of "maximum sustained yield" of "natural resources"—is easy to criticize, but hard to kill. Though anthropocentric, reductive, and based on a pre-ecological scientific paradigm, extractive resourcism is still very much alive in the USDA Forest Service, the BLM, and most other federal and state land management agencies—with the problematic exception of the National Park Service. After I also dedicated a couple of articles to criticism of the once sacrosanct (in environmentalist circles) Muir philosophy of conservation—wilderness preservation—the floodgates have opened and a torrent of criticism has washed over the wilderness idea, finally cresting in a recent *New York Times Sunday Magazine* article, "The Trouble with Wilderness," by environmental historian William Cronon (1995).

And just what is wrong with the wilderness idea? From a general, philosophical point of view, three things are wrong it.

First, no less than Pinchot's utilitarian concept of conservation, classic wilderness preservationism perpetuates the pre-evolutionary strict separation of "man" from "nature." Pinchot had infamously declared, "There are two things on this material earth, people and natural resources." But the same man-nature dichotomy is just as fundamental to Muir's outlook: "I have precious little sympathy for the selfish propriety of ... man," Muir wrote, "and if a war of races should occur between the wild beasts and Lord Man, I would be tempted to sympathize with the bears." Resourcism and preservationism share equally the modern split between the human and natural realms. Preservationism simply puts an opposite spin on the value question, defending bits of innocent, pristine, virgin nature against the depredations of greedy and destructive "man."

In fact, I'm beginning to think that wilderness loving—no less than wilderness hating—has its roots in Puritanism. John Winthrop, Michael Wigglesworth, and Cotton Mather saw the American continent as the stronghold of Satan and the Pilgrims as Christian soldiers onwardly liberating the land from its devil-worshiping denizens. But some among the second generation of Puritans saw the American continent as the garden of Eden sullied by the hand of fallen and originally sinful man. Jonathan Edwards, for example, could see in nature "the shadows and images of divine things," anticipating American Transcendentalism—while, at the same time, vigorously defending the "Christian doctrine of original sin," and scaring the bejesus out of his congregation with an account of "sinners in the hands of an angry God." The philosophy of wilderness preservation was first articulated by Emerson, the Transcendentalist. Thoreau had a thoroughly Puritan temperament, however heretical his theology. Muir was steeped from an early age in Calvinist misanthropy. I have yet to look into the Sunday School records of the leading wilderness advocates in the first half of the twentieth century—the Robert Marshalls, Sigurd Olsons, the Muries, the Howard Zahnizers, and David Browers—but many of the leading contemporary advocates of wilderness have Puritan roots: Dave Foreman boasts about his Scotch-Irish (eo ipso Calvinist) heritage; Bill McKibben was raised a Presbyterian and is now a Methodist; and Holmes Rolston III followed in the footsteps of both his father and grandfather to become a Presbyterian minister (and is still a staunch apologist for the Protestant Christian faith).

Second, the wilderness idea ignores the presence and considerable impact of indigenous peoples in their native ecosystems. North and South America had been fully inhabited and radically affected by Homo sapiens for 10,000 or more years before the European conquest (Denevan 1992).

And third, the received wilderness idea assumes that, if preserved, an ecosystem will remain in a stable steady-state, while current thinking in ecology stresses the importance of constant, but patchy, perturbation and the inevitability of change (Botkin 1990).

We have to be very careful, here, however, not to throw the baby out with the bath water. The *idea* of wilderness that we have inherited from Thoreau, Muir, and their successors may be ill conceived, but there's nothing whatever wrong with the *places* that we call wilderness, except that they are too small, too few and far between, and, as I shall directly explain, mostly mislocated. Those who have long campaigned for wilderness preservation (Noss 1994; Foreman 1994) are concerned lest honest, friendly critics of the wilderness *idea,* such as Cronon and I, give unwitting aid and comfort to the real enemies—the likes of Rush Limbaugh, Ron Arnold, and the "Wise Use Movement," the congressional delegation from Alaska, and the rest of the shock troops in the Newt Gingrich–led Republican Revolution—of designated wilderness *areas.* It is incumbent, therefore, on environmentally well-intentioned critics of the received wilderness idea to offer something positive with which to replace it.

And what might that be? In my earlier discussions of the wilderness idea I emphasized one half of a whole answer to that question. Following the lead of the twentieth century's third towering figure in conservation philosophy, Aldo Leopold, I stressed our need to find ways to inhabit and use nature that are at the same time ecologically benign (Callicott 1991). We need to work on sustainable livelihood, if not sustainable development (Callicott and Mumford 1997). Examples abound of past human cultures that lived sustainably, in harmony with their non-human neighbors (Gómez-Pompa and Kaus 1988). On the other hand, some species—most obviously large predators—do not coexist well with Homo sapiens. If members of such species are to have a place to live, then sustainable inhabitation and use of *most* places must be complemented by setting aside *some* places in which human inhabitation and use are either prohibited or severely restricted.

Such places are designated wilderness *areas*. In addition, however, to the above-noted conceptual problems with the received wilderness *idea,* the system of wilderness areas that we have inherited from our forebears only accidentally serves the vital habitat needs of endangered species—because wilderness areas were created with purposes other than biological conservation in mind (Foreman 1995). A review of the preservationist literature from the mid-nineteenth to the mid-twentieth century indicates that most traditional preservationists were not concerned primarily with providing habitat for members of those species that do not coexist well with people, but with such things as the recreational, scenic, and spiritual values of the human *experience* of wilderness. And just such values informed wilderness preservation policy. Hence, designated wilderness areas were selected, not because they were particularly rich or diverse in species, but for their recreational, scenic, and spiritual potential.

One institution, the zoological garden, that we have inherited from our forebears has been quick to adapt to the changed agenda of conservation —as redefined by conservation biology. Capturing and displaying wild animals appears to be as old as civilization itself; "the first known large collections were assembled in Egypt around 2,500 B.C.E." (Dunlap and Kellert 1995, 184) Modern zoos that display exotic animals from faraway places to a curious public became an urban commonplace in the nineteenth century A.D. Gawking at animals, imprisoned like criminals behind bars, was what, until very recently, zoos were all about. When wildlife cinematography began routinely appearing on television, people could see moving images of animals in the wild. By comparison, the incarcerated zoo animals, in their cramped and barren cages, appeared lethargic, forlorn, unhealthy, and incomplete. Simultaneously, animal welfare and environmental ethics came on the scene. Rather suddenly, the very existence of zoos has become morally problematic (Fox 1990). To survive, zoos had to change.

One response was to simulate the natural habitats of the inmates, displaying mixed-species groups of animals in open, landscaped compounds, secured by moats, rather than bars. But conservation biology was the real godsend for the public relations problems of zoos. Zoos contribute to conservation biology in several ways (Luoma 1987; Hutchins et al. 1995). They provide subjects and facilities for biological research. For some highly endangered species whose natural habitats are engulfed by the deluge of human overpopulation, zoos are arks, havens of last resort. More generally, a consortium of American and European zoos have initiated a Species Survival Plans program that features not only maintaining viable populations of threatened species, but also maintaining genetic variability in captive species populations through scientifically sophisticated captive breeding. The ultimate goal of this program is to reintroduce zoo-bred animals into the wild—if and when enough of their habitat can be reclaimed and restored (Balmford et al. 1996). Finally, taking advantage of the fact that people still flock to zoos in great numbers for family entertainment, zoos are attempting subtly to educate their patrons about the biodiversity crisis and the dire necessity for biological conservation.

Zoos play an important role in what conservation biologists call "ex situ" (off site) conservation. (Other ex situ conservation institutions, such as the International Crane Foundation, also exist—originally created not to exhibit but to help conserve threatened species.) What conservation biologists call "in situ" (on site) conservation is by far the preferred approach, ex situ conservation being to an endangered species somewhat as intensive hospital care is to a gravely ill organism (Snyder et al. 1996). Life in a hospital with no hope of going home is a living death. Similarly, a species' existence solely in zoos, with no hope of a return to the wild, is a living extinction.

Here's another analogy. Designated wilderness areas are, I suggest, to in situ biodiversity conservation, what zoos are to ex situ conservation. But just as zoos had to remake themselves to function as ex situ conservation institutions, so wilderness areas also need a makeover to function as in situ conservation institutions.

How then would conservation biology change wilderness policy? We can get a start on answering that question by reviewing how zoos have been changed by conservation biology.

The first order of business is a name change. The old Bronx Zoo in New York City has been renamed. It is now the "Wildlife Conservation Park." The director of the National Zoo in Washington, D.C., Michael Robinson (1989), has proposed "biological park" as a new generic name for the institutions formerly called zoos. What's in a name? Shakespeare asked. Rather a lot. Names are fraught with all sorts of associations—baggage. The name "zoo" conjures up images of animals in cages—there to be stared at, fed Crackerjacks and other snacks, teased, and such. "Biological conservation park" puts patrons on notice that the place they are visiting has a higher calling than some site for public amusement on the same scurrilous level as a circus tent or dog track. I suggest we rename wilderness areas "biodiversity reserves." That would put patrons on notice that the back country in the national parks and forests doesn't exist primarily for the enjoyment of trekkers, climbers, canoers, campers, and solitude seekers—as wilderness advocates argued from the mid nineteenth to the mid twentieth century—but for the nonhuman inhabitants of such places.

On the other hand, zoos have not closed their gates to the public. Far from it. People still patronize erstwhile zoos—in record numbers—most of them completely oblivious to the fact that they are visiting not zoos but biological conservation parks. Any thoughtful and tasteful visitor to a remodeled biological conservation park will immediately notice that exhibits have become more spacious, natural looking, and ecologically informed. But the conservation agenda of biological conservation parks takes priority over the public entertainment agenda, despite the fact that the vast majority of the public is there to do some good old-fashioned gawking at charismatic megafauna, not to be educated about the biodiversity crisis and such things as the genetic niceties of captive breeding. The silverback mountain gorilla that patrons may have come especially to admire just may be on loan to another facility in hopes that he will romance a female of his species located there. Or, at that other facility, the courting couple may not be on display to the public so as to allow them the privacy they require to consummate their union. If so, too bad. Patrons will be informed of the reason for their disappointment and will have to be content with just knowing that their favorite exhibit is temporarily serving a loftier purpose. Similarly, in the Yellowstone or Glacier biodiversity reserves (or, as they are now still called, wilderness areas), backpackers might just have to be excluded from grizzly bear or gray wolf habitat altogether if their recreational activities prove to be in conflict with the conservation of these beleaguered species. More controversially—though personally I do not see why it should be—if the needs of bears and wolves in national forest biodiversity reserves (a.k.a. wilderness areas) are in conflict with livestock grazing, biological conservation should take precedence.

It would be hard to argue that the old zoos were not located in the right places. If properly designed and managed, London is as good a place as New York for a biological conservation park, and San Diego is as good as Chicago. Unfortunately, designated wilderness areas are not always located in the best places to perform their newfound and overriding conservation function. In the blunt characterization of one unregenerate wilderness advocate, much of the wilderness system in the United States, for all its stupendous glory, is rock and ice (Foreman 1991). But it's a start. The Yellowstone Biodiversity Reserve (as it might be renamed) should be expanded to become coextensive with the Greater Yellowstone Ecosystem and connected up with the Selway-Bitterroot, Frank Church River of No Return, and other proximate wild lands. Politically impossible, you may be thinking. That's part of what's wrong with the wilderness label. It pits the politically anemic historic rationale of wilderness preservation (recreation and aesthetics for an elite few) against the politically more robust claims of jobs and profits. Preserving biodiversity is a more universal and higher-minded conservation aim than the provision of outdoor recreation and monumental scenery—which, with a little help from their spin doctors, Congressional demagogues can make to look like a government-subsidized luxury for social misfits. And, unlike traditional wilderness areas—which are partly defined in terms of the absence of "man" and "his works"—all human economic activity need not be ruled out, by definition, in biodiversity reserves. Under certain circumstances, selective logging, regulated hunting, and careful mineral extraction might be made compatible with in situ Species Survival Plans.

The next step is to establish biodiversity reserves in the places that are biologically rich but scenically poor, and that thus got overlooked by the historic wilderness preservation movement. Three general categories of places appropriate for biodiversity reserve designation come to mind: First, representative biomes with their characteristic species. The biome most neglected by the waning twentieth century's North American wilderness preservation movement is surely the Great Plains. No monumental scenery, no wilderness designation. The plains are sufficiently vast, sparsely populated, and climatically diverse to warrant the establishment of a whole network of biodiversity reserves from Manitoba to Chihuahua. Second, what conservation biologists call "hot spots"—areas of particularly rich biodiversity (which often occur at the intersection of biomes)—are obvious candidates for designation as biodiversity reserves (Lydeard and Mayden 1995). Third, unique ecosystems, such as the Florida Everglades—the most threatened ecosystem in the United States, according to a recent Defenders of Wildlife assessment—are also obvious candidates.

A pipe dream? Maybe, maybe not. The Republican Revolution in the U.S. Congress may fizzle after the next election. You can't fool all the people all the time. The populism of the anti-environment far right is a sham. Who gets represented and who gets their legislative agendas enacted is who contributes big bucks to the campaign coffers. The cynical bet is that those who merely vote can be manipulated. But tax breaks and government subsidies for the rich and ripoffs for everyone else can't play for too long in Peoria or anywhere else in a healthy democracy. A militant minority—big ranching, big mining, big drilling, big logging, big real estate development—are the instigators and beneficiaries of the current effort in Congress to sell out the environment and literally sell off our public domain. I don't think it will fly much longer. In the meantime, hopefully, the current public and academic debate about the fate of endangered species, the wilderness idea, and the environment in general will mix some new and creative thinking with the venerable American traditions of nature conservation and preservation. And, hopefully, the twenty-first century will be characterized by a more serious and coherent conservation agenda than its predecessors.

NOTE

Presented to the Seventh North American Interdisciplinary Wilderness Conference, Reno, Nevada, March 1, 1996.

BIBLIOGRAPHY

Balmford, A., G. Mace, and N. Leader-Williams. 1996. "Setting Priorities for Captive Breeding." *Conservation Biology* 10: 719–27.

Botkin, D. B. 1990. *Discordant Harmonies: A New Ecology for the Twenty-first Century.* New York: Oxford University Press.

Callicott, J. B. 1991. "The Wilderness Idea Revisited: The Sustainable Development Alternative." *Environmental Professional* 13: 235–47. [Included in this volume.]

Callicott, J. B. 1994. "A Critique of and Alternative to the Wilderness Idea." *Wild Earth* 4(4): 54–59.

Callicott, J. B., and K. Mumford. 1997. "Sustainability as a Conservation Concept." *Conservation Biology* 11: 32–40.

Cronon, W. 1995. "The Trouble with Wilderness." *New York Times Sunday Magazine,* August 13, pp. 42–43. [Included in this volume.]

Denevan, W. M. 1992. "The Pristine Myth: The Landscape of the Americas in 1492." *Annals of the Association of American Geographers* 82: 369–85. [Included in this volume.]

Dunlap, J., and S. R. Kellert. 1995. "Zoos and Zoological Parks." In W. R. Reich, ed., *Encyclopedia of Bioethics,* vol. 1, pp. 184–86.

Foreman, D. 1991. "Dreaming Big Wilderness." *Wild Earth* 1 (spring): 10–13.

Foreman, D. 1994. "Wilderness Areas Are Vital: A Response to Callicott." *Wild Earth* 4 (winter): 64–68.

Foreman, D. 1995. "Wilderness: From Scenery to Nature." *Wild Earth* 8(4): 8–15. [Included in this volume.]

Fox, M. W. 1990. *Inhumane Society: The American Way of Exploiting Animals.* New York: St. Martin's Press.

Gómez-Pompa, A., and A. Kaus. 1988. "Conservation by Traditional Cultures in the Tropics." In V. Martin, ed., *For the Conservation of the Earth.* Golden, Col.: Fulcrum, Inc.

Hutchins, M., K. Willis, and R. J. Wiese. 1995. "Strategic Collection Planning: Theory and Practice." *Zoo Biology* 14: 2–22.

Luoma, J. R. 1987. *A Crowded Ark: The Role of Zoos in Wildlife Conservation.* Boston: Houghton Mifflin.

Lydeard, C., and R. L. Mayden. 1995. "A Diverse and Endangered Aquatic Ecosystem of the Southeast United States." *Conservation Biology* 9: 800–805.

Noss, R. F. 1994. "Wilderness—Now More Than Ever: A Response to Callicott." *Wild Earth* 4 (winter): 60–63.

Robinson, J. G. 1993. "The Limits to Caring: Sustainable Living and the Loss of Biodiversity." *Conservation Biology* 7: 20–28.

Robinson, M. 1989. "Zoos Today and Tomorrow." *Anthrozoos* 2: 10–14.

Snyder, N. F. R., S. R. Derrickson, S. R. Beissinger, J. W. Wiley, T. B. Smith, W. D. Toone, and B. Miller. 1996. "Limitations of Captive Breeding in Endangered Species Recovery." *Conservation Biology* 10: 338–48.

R. Edward Grumbine

Using Biodiversity as a Justification for Nature Protection in the US (1996)

INTRODUCTION

TO FATHOM THE LAST TWENTY-FIVE YEARS of growth in awareness of biological diversity in the US, take this simple test. Ask yourself the following questions:

1. What were two primary goals of environmental activists in 1970?
2. What were three species threatened with extinction at the time?
3. Was the term "ecosystem" in your personal lexicon on Earth Day 1970?

Your response to the first question likely includes air and water pollution, as these problems were receiving much attention at the time of the first Earth Day. Congress had already passed several laws to address such concerns and additional legislation was forthcoming. The second question was probably more difficult to answer—the Endangered Species Act as we know it today did not exist. You might have mentioned Whooping Cranes, Bald Eagles, or Bison, but most citizens were just beginning to wake up to the loss of species as a critical problem. As for the final question, you probably would not have had a solid working definition of "ecosystem" unless you had taken a college course in biology. In 1970, few activists in the nascent environmental movement had yet to embrace scientific ecology as an organizing principle.

If you were to ask yourself these same questions for 1996 your answers would be surprisingly different. Though pollution is still perceived as a threat by most Americans, many more environmentalists would now highlight the loss of biological diversity as a key problem. For question #2 you would have no trouble listing numerous species—Northern Spotted Owl, Peregrine Falcon, Grizzly Bear, Snail Darter, Mission Blue Butterfly, Kirtland's Warbler, or any of a dozen other commonly known endangered life-forms. And though your definition of ecosystem might not pass muster with a Ph.D., you would have little trouble describing it as a community in which plants and animals interact with the physical environment.

In the mid 1990s, after the 25th anniversary of Earth Day, loss of biodiversity is at center stage for many concerned citizens, activists, scientists, and managers (Grumbine 1992). This was not the case in 1970. The current emphasis on biodiversity has grown from a complex mix of cultural factors which are easy to highlight but difficult to untangle. First and foremost, there has been since the first Earth Day an unprecedented growth in scientific understanding of biological diversity, the ecological functions that diversity serves, and the biological consequences of environmental deterioration. This new knowledge has in turn been bolstered by trends in U.S. environmentalism that reflect broad changes in American social values.

As historian Samuel Hays (1987) has observed, the first Earth Day marked the high water mark in the metamorphosis from conservation to environmentalism in the US. Americans, with greater amounts of education, disposable income, and leisure time were beginning to view *game* as *wildlife,* value *nonconsumptive* outdoor activities (e.g., photography) equally with *consumptive* pursuits (e.g., hunting), and voice stronger concern about resource *protection* as well as resource *management.* During the 1970s and 1980s, as environmental groups gained members, larger budgets, and lobbying clout, their agenda expanded from countering threats to specific parks and wildernesses to include concern for general environmental problems such as population growth, resource consumption, pollution, and energy policy. Arguments challenging human-centered values also grew stronger, to the point that a new field of a philosophical inquiry—environmental ethics—began to flourish (Nash 1989). Overall, more Americans began to actively question whether *progress,* defined simply as endless material growth, could really be sustained into the future.

Today, another phase in the evolution of American environmental values appears to be taking place, spurred on by new understanding of biodiversity. If the original Earth Day marked the beginning of a more inclusive approach to managing Nature for humans, future Earth Days may come to represent the rise in importance of biodiversity protection as the primary basis for human work with Nature.

In this paper, I trace how the concept of biodiversity has evolved toward its present position at the center of a compelling scientific framework for protecting Nature. I focus on two related trends—the development of the ecological roots of knowledge about biodiversity and the development within environmentalism of scientific justifications for protecting Nature.

DEFINING BIODIVERSITY AND THE BIODIVERSITY CRISIS

"Biodiversity" has become a central rallying cry for a growing portion of the US environmental movement. The term and its relative, "ecosystem management," are referred to so often that the media portrays them as "buzzwords," empty phrases that everyone employs but few understand. Open any recent textbook, however, and biodiversity is easily defined. Noss and Cooperrider (1994, p. 5) provide a standard definition:

> [Biodiversity is] the variety of life and its processes. It includes the variety of living organisms, the genetic differences among them, the . . . ecosystems in which they occur and the ecological and evolutionary processes that keep them functioning, yet ever changing and adapting.

Where did the modern concept of biodiversity come from? Part of the answer is that biodiversity has appeared today because it is disappearing so rapidly. Conservation biology, the science of scarcity and diversity, would not be needed if not for significant loss of lifeforms. A science exploring extinction and habitat fragmentation, it has blossomed since the 1980s as a response to widespread destruction of species and ecosystems.

In the US, thousands of species are either listed or awaiting protection as candidates for listing under the Endangered Species Act. Estimates of species at risk over the next decade range from 2.5–15% of all lifeforms on Earth (Primack 1993). Beyond individual plants and animals, many US ecosystem types have been reduced to critical levels (Noss et al. 1994). Yet only about 6% of the US is in some kind of protected classification. Biologists are beginning to describe not only species and ecosystems at risk but also endangered biophysical processes, including large mammal and song bird migrations, river system flooding and deposition patterns, and forest nutrient cycles (Brower 1994). Some scientists warn that entire faunal groups may "all but disappear" within the next century, including primates, large carnivores, and most hoofed animals (Soulé 1986). In both direct and indirect ways, human activities are causing a biodiversity crisis—the largest mass extinction in 65 million years.

SCIENTIFIC ROOTS OF BIODIVERSITY

The modern definition of biodiversity derives from the development of the science of ecology and its application to conservation issues. To fathom why scientists and environmentalists did not comprehend biodiversity fully in 1970 requires a glance at how ecology has matured as a discipline. There have been at least four developmental stages in the science of ecology: formative, descriptive, quantitative, and predictive non-equilibrium. (For full treatments see Worster 1994, Golley 1993, and McIntosh 1985.)

Several people stand out as major influences on ecological thinking long before the field coalesced into a unified discipline. Charles Lyell, the father of geology, in his book *Principles of Geology* (1830) helped overturn Linnaean concepts of a static nature under strict divine rule. Lyell was among the first to understand that geologic change occurred gradually over eons, that species dispersed actively around the world, and that competition was a driving force in biotic interactions. Lyell was a major influence on Charles Darwin. In *On the Origin of Species* (1859), Darwin built upon Lyell and advanced natural selection as the primary mechanism of evolution. Contemporary with Darwin, but living in the New England woods,

Henry David Thoreau was one of the first naturalists to understand succession as a major pattern of change in ecosystems. Thoreau also was one of the first to glimpse the loss of species and habitat and its cultural ramifications at the onset of the Industrial Revolution. George Perkins Marsh contributed a pioneering global account of humanity's role in reducing the capacity of Earth to support life in *Man and Nature* (1864).

Ecology in the early decades of the 20th century was a descriptive, holistic science. The key themes were the balance of nature and succession toward a stable, climax state. Plant ecologist Frederick Clements dominated the field with his idea of communities as interdependent superorganisms evolving collectively.

By the time A. G. Tansley coined the term "ecosystem" in 1935, Clements's views were falling from favor. Qualitative, descriptive ecology was being superseded by a more quantitative ecology of energy and nutrient flows, food chains, and trophic levels. Natural history was out, mathematical models were in. The science of interrelationships was becoming subject to compartmentalization and reductionism. In the 1940s, Raymond Lindeman developed important theories on energy flows in ecosystems and G. E. Hutchinson refined the concept of feedback and constructed some of the first mathematical models of populations. Later ecologists built on these fundamentals with Eugene Odum (ecosystem characteristics), Frank Bormann and Gene Likens (nutrient flows), and Robert MacArthur (population models) making key contributions.

Since the late 1970s, as knowledge of natural patterns and processes has accumulated and the biodiversity crisis has grown, a new ecological worldview has been emerging (Pickett et al. 1992, Botkin 1990). Ecology is moving away from a reductionist approach toward a more contextual, non-equilibrium perspective. Where in the past scientists (and environmentalists) characterized ecosystems as orderly and relatively balanced, current viewpoints emphasize systems as dynamic, changing at different space and time scales, and full of uncertainty. Nature is episodic as often as it is homeostatic. Nature is not always in "balance"; and changes are difficult, sometimes impossible, to predict.

Definitions of biodiversity reflect these latest changes. No longer is diversity just about numbers of species or types of ecosystems. The new emphasis on non-equilibrium processes (especially natural disturbances such as fires and floods) has resulted in a comprehensive definition that includes not only the diversity of life from genes to landscapes, but also the fundamental patterns and processes of Nature that weave lifeforms over time and in space.

EARLY ECOLOGICAL JUSTIFICATIONS FOR PROTECTING NATURE

In the first decades of this century, the descriptive balance-of-Nature view of ecology reigned supreme for biologists and citizens alike. Two of America's greatest naturalists, Joseph Grinnell and Tracy Storer (1916, p. 377), wrote in an early *Science* article that mammals in National Parks added "the witchery of movement" to the "natural charm of the landscape." These distinguished biologists believed that the National Parks' highest purpose was to "furnish examples of the earth as it was before the advent of the white man" (p. 377).

A few years later, views were beginning to change. Beginning in the 1920s, several professional ecologists published papers calling for Nature protection for the sake of science. In 1920, Victor Shelford criticized the Park Service and the Forest Service for an unecological approach to management. Francis Sumner (1920, 1921) called for "nature conservation" over resource management. Both Shelford and Sumner were members of the new Ecological Society of America and advocated setting aside representative examples of all US ecosystems in a comprehensive national system. Other ecologists joined them, publishing articles such as "The Preservation of Natural Areas in the National Forests" (Pearsons 1922) and "The Importance of Preserving Wilderness Conditions" (Adams 1929). In 1921 the American Academy for the Advancement of Science endorsed the Ecological Society's policies on reserves.

What sparked this outcry from a few leading scientists? Wilderness historian Craig Allin (1982) suggests that it resulted from a massive upsurge in road building on public lands by both the Forest Service and Park Service between 1916 and 1921. Also likely is that Shelford and his colleagues, on the cusp between Clementsian and quantitative ecology, recognized the need for protecting representative ecosystem types as examples of steady state conditions and as baselines for gathering new scientific data.

Shelford's efforts led to the remarkable paper "The Preservation of Natural Biotic Communities" (Shelford 1933). This visionary work (not unlike The Wildlands Project's vision today) outlined a national strategy for preserves that included protection for both species and ecosystems, expansion of park and reserve boundaries to match species habitat needs, managing for ecological "fluctuations" (i.e., natural disturbances), and a core/buffer zoning approach to planning.

Reading Shelford's paper more than 60 years later, one can only dream of what condition US public lands would be in today if policy-makers of the time had embraced Shelford's bold vision. Unfortunately, no sanctuary system was forthcoming. Shelford's work did result, though, in the beginning of the Forest Service's Research Natural Area program, where small examples of different timber types were declared off-limits to commercial logging.

While Victor Shelford developed his Nature sanctuary plan, three other biodiversity pioneers, George Wright, Ben Thompson, and Joseph Dixon, were focusing on the National Parks. As wildlife experts studying the habitat needs of park fauna, Wright and his colleagues discovered that every single park was far too small to sustain large mammal populations over time. At the conclusion of their landmark *Fauna of the National Parks of the United States* (1933, pp. 37–39) they made one of the first statements suggesting biodiversity as the *raison d'être* for parks: ". . . perhaps our greatest natural heritage," rather "than just scenic features . . . is nature itself, with all its complexity and its abundance of life."

Wright and his colleagues were proved correct in both their scientific and policy assessments by the debate that surrounded the creation of Everglades National Park in 1934. Wildlands advocates count Everglades as the first park where biodiversity preservation was used to justify protection. The park was established for the "preservation intact of the unique flora and fauna and the essential primitive conditions" (US Statutes at Large 1934); but this legal language obscures the true justification behind the protection of the park. The record shows that conservationists convinced Congress to accept wildlife as "scenery" since the river of grass had no magnificent mountains or gorges (Runte 1987). Though the Everglades bill does represent a statutory milestone for accepting wilderness and wildlife, grandeur, magnificence, and Romantic ideas of the balance of Nature continued to hold sway.

BIODIVERSITY FROM ALDO LEOPOLD TO EARTH DAY

As ecology developed into a modern science in the 1930s and 1940s, there remained a need to consolidate ecological justifications for protecting Nature into a coherent whole. The person who accomplished this, as much as anyone, was Aldo Leopold. Best known for *A Sand County Almanac* (1949), Leopold wed the science of ecology with a land ethic where humans were "plain members and citizens" of Earth.

Leopold's thinking about ecology and management went through a profound transformation over several decades. In 1921, he was using recreational justifications for protecting wildlands that were radical for the time. But Leopold soon left such arguments behind. Beginning in 1933 he published a series of papers that provide the basis for much of the current definition of biodiversity as well as the ethical foundations of conservation biology. Leopold made four key contributions.

In 1939, he offered one of the first inclusive definitions of biodiversity: ". . . the biota as a whole is useful, and biota includes not only plants and animals, but soils and water as well" (Leopold 1939, p. 727). Leopold expanded on this in 1944 by adding the concept of health to conservation:

> Conservation is a state of health in the land. The land consists of soil, water, plants, and animals, but health is more than a sufficiency of these components. It is a state of vigorous self-renewal in each of them, and in all collectively. . . . In this sense land is an organism and conservation deals with its functional integrity, or health (Leopold 1991, p. 310).

This commingling of biodiversity conservation and land health is the root of current attempts to define ecological health and ecological integrity.

Leopold's second contribution was to use the new ecological concepts of biotic pyramids, energy flows, and food chains to point out defects in prevailing balance-of-nature perspectives on ecosystems. He suggested that balance implies "only one point at which balance occurs, and that balance is normally static" (Leopold 1939, p. 727).

Third, Leopold used his awareness of the dynamics of Nature to provide a scientific rationale for wilderness protection. In 1941, he wrote that "all wilderness areas . . . have a large value to land-science" and that their principal utility was as a "base-datum of normality, a picture of how healthy land maintains itself..." (Leopold 1941, p. 3).

The fourth and most important contribution of Leopold to understanding biodiversity was that he placed people squarely in Nature as "plain members and citizens of the land community" (Leopold 1949, p. 204). He had already recognized this intimate relationship as early as 1933 when he defined civilization as "a state of mutual and interdependent cooperation between human animals, other animals, plants, and soils..." (Leopold 1933, p. 635). With this conception Leopold became the first modern ecologist to link the health of land with the health of culture.

Like Shelford and Wright before him, Leopold had little immediate influence on policy. By the 1960s, however, as the pace of environmental deterioration quickened, other ecologists were beginning to catch up with where Leopold had been. Science came to play an increasingly key role in environmental policy debates. Rachel Carson in *Silent Spring* (1962) built her argument against pesticides by exposing their negative effects on both human health and ecosystem functioning. By 1968, the international scientific community was becoming active. The United Nations Educational, Scientific, and Cultural Organization (UNESCO) sponsored a global conference that year on the use and conservation of the biosphere. This led to the biosphere reserve model of ecosystem protection. All of this work forged links between Leopold's view of science and the environmental problems of the day.

Congress and the Administration, too, were beginning to act for biodiversity. In 1966 the Endangered Species Preservation Act was passed. This prototype of more powerful laws to come protected only vertebrates and contained many other loopholes. A year before Earth Day, in 1969, as the environmental movement gathered strength, Congress extended protection to invertebrates. In 1972, President Nixon stated that "even the most recent act to protect endangered species simply does not provide the kind of management tool needed to act early enough to save a vanishing species" (Nixon 1972). Nixon signed the Marine Mammals Protection Act that year.

The groundswell of presidential and popular support for ecology and endangered species surrounding Earth Day 1970 led to Congress passing the 1973 Endangered Species Act with but 12 no-votes total in both legislative houses. The ESA, still the strongest American environmental law, validated Aldo Leopold's "ecological consciousness" toward species and ecosystems and set the stage for future policy debates.

BIODIVERSITY AND THE ENVIRONMENTAL MOVEMENT

Environmental protection did not gain a lasting place in American values merely as a result of the development of the science of ecology. Nor did biodiversity come to the fore simply because of individual biologists such as Shelford, Leopold, and Carson. It also took the concerted efforts of leaders within the environmental movement to understand the implications of ecology and render these ideas accessible to the American public.

Just as ecology has developed as a science, so have environmentalist arguments evolved for protecting Nature. When Robert Marshall, Aldo Leopold and their colleagues founded The Wilderness Society in 1935, they focused only on wilderness and roadless areas. They believed that wildlands should be protected primarily for the benefits they conferred on people. Marshall's views, in particular, were influential: wilderness offered a respite from civilization, encouraged spiritual contemplation, and offered a unique aesthetic experience (Marshall 1930). During this period Leopold was only beginning to voice his biotic view of land; and the conservation movement had, aside from Shelford's work, little scientific ecology on which to base political prescriptions.

In 1949, the year *A Sand County Almanac* appeared, the Sierra Club convened its first biennial wilderness conference. These conferences were to become the main philosophical and strategic forum for the movement. Through the first four gatherings, there was little mention of ecology as having anything to do with wilderness protection. Recreational, spiritual, and aesthetic justifications prevailed. The *Sierra Club Bulletin* from 1950–1976 had only two references to ecology, four to endangered species, and five to wildlife conservation (Sierra Club 1976).

But the power of scientific ecology in general and Leopold's ideas in particular were beginning to be felt. In 1950 journalist Bernard DeVoto, responding to the Echo Park Dam controversy in Dinosaur National Monument, proclaimed that the park deserved to be protected "as wilderness . . . for the field study of . . . the balance of Nature, the web of life, the interdependence of species . . ." (DeVoto 1950, p. 44). DeVoto's is a classic attempt to incorporate Leopoldian ecology with Romantic ideas of balance in Nature.

In 1951 and again in 1955, at the third and fifth wilderness conferences, Howard Zahniser of The Wilderness Society (TWS) unveiled a national plan for wilderness protection, based partly on Victor Shelford's original Nature sanctuary vision (Kendeigh et al. 1950–51). Zahniser's plan became the precursor to the original Wilderness Act bill in 1956. In Congressional hearings over the new bill, Zahniser mentioned scientific baseline data arguments in favor of the legislation, but these justifications were never highlighted by conservationists during the debate.

As executive director of the Sierra Club, David Brower was as responsible as any leader for bringing science to conservation. Brower controlled the agendas of the wilderness conferences. Beginning in 1959 with Raymond Cowles, he invited several professional ecologists to address the conferences. Cowles spoke of population growth from an ecological perspective. In 1963, James Gilligan, author of the first Ph.D. dissertation on US wilderness policy, described wildlands as "essential habitat for scarce species." Slowly, biodiversity was creeping into conservation arguments.

Conservationists did not find it easy, however, to include ecology along with recreational and spiritual justifications for wilderness. Sharing the podium in 1963 with Gilligan was forest ecologist Stephen Spurr, whose view of ecology challenged Brower and the conferees. Spurr argued strongly against any wilderness preservation strategy grounded in a stable, balance of Nature view. "Stability is only relative, and only superficial," spoke the ecologist, and "natural succession will never recreate old patterns, but will constantly create new patterns!" (Gilligan 1963, p. 60). Spurr used ecological theory to confront the conferees' "nostalgia" for a nature that never existed. Instead of drawing lines around roadless areas and lobbying Congress to designate new wilderness, Spurr argued for greater use of science and technology to manipulate Nature for human ends.

This conflict between ecology and preservation was manifest again in 1963 with the influential *Wildlife Management in the National Parks,* the so-called Leopold Report (A. S. Leopold et al. 1963). At the behest of Interior Secretary Steward Udall, a blue ribbon committee chaired by Aldo Leopold's son, zoologist A. Starker Leopold, was convened to review wildlife in the parks. The committee's report was both revolutionary and paradoxical. Following ecology (and the thirty year old insights of George Wright), the report concluded that "maintaining suitable habitat is the key to sustaining animal populations, and ... protection, though it is important, is not of itself a substitute for habitat" (pp. 1–2). But after verifying Spurr's assessment that ecosystems change over time, the Leopold Committee recommended that "the biotic associations within each park be maintained ... as nearly as possible in the condition that prevailed when the area was first visited by the white man." Each park "should represent a vignette of primitive America!" (p. 4). As historian Alfred Runte (1987) has noted, these scientists could not escape their cultural values. Science required them to portray Nature as dynamic, yet they advocated freezing Nature into pre-European landscapes.

The Sierra Club, Wilderness Society, and National Parks and Conservation Association were all quick to endorse the Leopold Report. These groups supported the committee's *philosophy* while avoiding the committee's *ecology*. The following year the Wilderness Act was passed by Congress. Ecological values rated all of three words in the new law.

Despite these inconsistencies, support for endangered species and broad environmental protection continued to grow. In 1968, the Sierra Club lobbied for a national ecological survey, but the bill died in Congress (McCloskey 1968). At the biennial wilderness conference in 1969, population biologist Paul Ehrlich proclaimed that population growth and resource consumption were inextricably linked to the loss of wilderness.

BIODIVERSITY COMES OF AGE: 1970–1990

During the 1970s and into the 1980s, scientific and policy conceptions of biodiversity continued to converge with environmentalist notions of ecology. Ecologists added to their knowledge of competition, diversity, stability, and community dynamics (see Cody and Diamond 1975; and on biogeography, MacArthur and Wilson 1967). R. H. Whitaker (1972) refined and broadened the concept of diversity to include within-habitat (alpha), between-habitat (beta), and regional (gamma) diversity. The same year that Whitaker published his classic work, the United Nations Conference on the Environment was held in Stockholm. For all the impassioned debate in Sweden, though, few ecologists attended. Scientists were not ready to present their ideas in political forums.

Several national and international conferences and policy documents built upon Stockholm. In 1981 the US Council on Environmental Quality produced the *Global 2000 Report to the President,* which was the first US policy document to attempt a definition of biodiversity. The US State Department, following the Council's lead, sponsored an International Strategy Conference on Biological Diversity in 1981. A World Charter for Nature was ratified by the UN General Assembly in 1982. The charter included recommendations to protect parks and wildernesses, but was especially notable for its preamble which tied protecting diversity to an ethical position: "Every form of life is unique, warranting respect regardless of its worth to man, and, to accord other organisms such recognition, man must be guided by a moral code of action." The US was the only member of the General Assembly to vote against the charter.

Yet by the time UN delegates were voting on the World Charter for Nature, biodiversity protection had already been codified in US law—at least in the National Forests. The 1976 National Forest Management Act (NFMA) today remains the only US law that explicitly requires a federal agency to protect viable populations and ecosystems. As with the Everglades legislation four decades prior, however, the motives of Congress were unclear. The NFMA was a response to excessive clearcut logging in the National Forests. Forest activists and Congress were concerned about stand conversions, the forestry practice of logging native forests to replace them with preferred commercial species, creating industrial monocultures. Yet Congress did not understand biodiversity well enough to act decisively. In an extremely ambiguous section of the NFMA, legislators required the Forest Service to "provide for the diversity of plant and animal communities" (US Code 1982).

The NFMA diversity provision was clarified by a committee of scientists who wrote rules under the Code of Federal Regulations whereby the law would be implemented. These rules, completed in 1979 and revised in 1982, require the Forest Service to preserve existing variety, maintain viable populations, recognize forests as ecosystems, and base management on ecological relationships. Clear as the rules were, it would take many years and numerous appeals and lawsuits to force the agency to begin to implement them.

Along with NFMA, two additional events in the late 1970s brought ecologists and activists closer together. As the pace of development continued, concerns were raised as to how "external threats" would affect protected areas. The National Parks and Conservation Association (1979) published a national report documenting such threats. The Park Service (1980) released its own study highlighting similar problems. The following year, Congress, in response to erosion and watershed degradation on lands surrounding Redwood National Park, amended the Park Service Organic Act to affirm park protection. While judicial interpretations have limited the effectiveness of the Redwood Park Amendments, the issue of external threats served notice that protected areas were in fact embedded in an ecological matrix that required protection as a whole.

Further illuminating biodiversity in the late 1970s were several books by prominent scientists warning of an extinction crisis. Norman Myers's *The Sinking Ark* (1979) was read widely and caused much debate. Paul and Anne Ehrlich titled a 1981 textbook *Extinction*. The Nature Conservancy, ahead of most conservation groups in understanding diversity, began to build a national database that cataloged threatened and endangered species, habitat types, and other elements of biodiversity (The Nature Conservancy 1975).

By the beginning of the 1980s a critical threshold was being reached in scientific comprehension and environmental awareness of biodiversity. The first International Conference on Conservation Biology, held at the University of California, San Diego in 1978, brought together a diverse group of geneticists, population biologists, evolutionists, and biogeographers. The conference resulted in the path-finding anthology *Conservation Biology: An Ecological and Evolutionary Approach* (Soulé and Wilcox 1980). The synthetic discipline of conservation biology was born. Soon thereafter, other books appeared linking conservation with genetics, evolution, and population biology (Frankel and Soulé 1981, Schonewald-Cox et al. 1983, Harris 1984). In 1986, a second conference of the newly formed Society for Conservation Biology was held, followed by the initial publication of a professional journal. In late 1986 in Washington, DC, the Smithsonian Institution and the National Academy of Sciences hosted the first high-profile international gathering of professionals concerned with loss of biodiversity. From this time onward there has been a great outpouring of papers and reports covering all aspects of the new field.

What was new about conservation biology? The discipline was synthetic, with island biogeography, population genetics, and habitat fragmentation studies leading the way. There was an emphasis on applying academic theory to management problems. And conservation biology was explicitly value-laden: diversity, complexity, and evolution were imbued with normative value (Soulé 1985). Many conservation biologists supported shifting the burden of proof in environmental decisions from those who wished to protect diversity to those who desired to develop Nature. While conservation biology was mission-oriented, however, the methods used to gather data were objective, peer reviewed, and as value-free as in any other scientific discipline. The field has contributed these general management goals to conservation (Meffe and Carroll 1994):

1. Critical ecological processes must be maintained.
2. Goals and objectives must come from a deep understanding of the ecological properties of the system.
3. External threats to reserves must be minimized and external benefits maximized.
4. Evolutionary processes must be conserved.
5. Management must be adaptive and minimally intensive.

The Earth First! movement was well out in front of almost all environmental groups in using conservation biology arguments as first principles in protecting Nature. Evident in the earliest volumes of the *Earth First! Journal* (1981–1982), this ecological wilderness perspective was consolidated in *The Big Outside* (Foreman and Wolke 1989): "Protecting *natural diversity,* then, must be the major goal of the wilderness movement . . . natural diversity means all indigenous species must be free to evolve under natural conditions, in as many different natural habitats as possible" (p. 24).

Still, many environmental groups were slow to embrace the new field. It took the Northern Spotted Owl and its old-growth forest habitat to catapult biodiversity toward the forefront of environmentalism. The owl awakened activists (and managers and Congress) to critical aspects of protecting biodiversity. What began in the 1970s as an owl-only issue was transformed by 1990 into an ecosystem protection issue, in part due to EF!'s dramatization of threats to old-growth forests. Once old-growth ecosystems were adopted as a strategy focus, it became easier for activists to appreciate the role that ecosystem patterns and processes played in maintaining biodiversity. As scientific assessments on owls and old growth were joined by reports on Marbled Murrelets, salmonids, and other species, activists were pushed toward another level of sophistication, recognizing the need for regional/landscape-scale protection. In 1994, ecosystem management studies supported (in concept) by environmentalists were initiated by the federal government for the entire Columbia River Basin and the Sierra Nevada Mountains in California.

BIODIVERSITY BEYOND EARTH DAY 1995

Three years after Earth Day 1970, when Congress passed the Endangered Species Act, protecting diversity was perceived as "low-cost no-lose" (Yaffee 1982, p. 57). Today the Act faces efforts to gut its most stringent provisions. Yet it is abundantly clear that the biodiversity crisis has worsened and that the law should be strengthened. The sum of our growing scientific understanding of biodiversity reveals a deep chasm between environmental policy and environmental protection (Grumbine 1994b).

The single major consequence of the revolution in awareness of biodiversity has been to deepen our appreciation of interrelationships. For ecologists, this trend has been manifest in two important ways: the evolving definition of diversity from species to the current inclusive hierarchical view, and the change from static balance to non-equilibrium theories of Nature. In the environmental movement, biodiversity has begun to nudge activists away from viewing Nature as a series of special places (parks and wilderness) embedded in developed landscapes toward the protection of regional landscapes or greater ecosystems (Grumbine 1990) where use and protection are grounded in a sense of limits on human behavior.

In American society at large the concept of protecting biodiversity continues to challenge cherished but outmoded images of people vs. Nature. While anthropocentric values and resourcism still hold sway with the majority, a growing number of citizens are asking provocative questions. Are there limits to private property rights when biodiversity is at risk? Is industrial-scale resource depletion sustainable? What should give when human activities are exposed by conservation science as endangering species, ecosystems, and landscapes? Who should decide what constitutes acceptable levels of risk in losing elements of biodiversity? These questions were not part of the discussion on Earth Day 1970. We had no definition for biodiversity, no comprehensive Endangered Species Act, and property owners were unconcerned about their "rights." "Sustainability" was not part of the environmental lexicon. There were no conservation biologists to decry threats to viable populations, and "plenty" of old growth remained to be cut. Environmental protection was perceived as either/or, cut-and-dried, not replete with uncertainty and multiple levels of risk.

Conservation biologists today confront myriad complex issues. In a society that considers science to be value-neutral, how do you practice objective science and yet advocate for biodiversity (for example, see Noss 1994 and Brussard et al. 1994). And how do you create a tighter link between science and policy? In the 1930s, when Victor Shelford designed a national reserve system using science, the political response was a series of tiny Research Natural Areas. In the 1990s, several scientific panels recommended a moratorium on old-growth logging and the result was President Clinton's Option Nine, and Congress's "forest health" rider—the first of which failed to uphold the scientists' recommendations, the second of which runs directly counter to them.

Activists, too, are adjusting to the new world of diversity. They are becoming less hesitant to employ scientific arguments in their strategies, regardless of the perceived political costs. The history of US environmentalism reflects a tendency by activists to downplay scientific rationales in favor of ethical justifications. The normative standards of conservation biology can help to overcome this tendency.

Activists should not accept science uncritically, however. The history of the concept of biodiversity makes clear that along with evolving scientific "fact," ecological theory is also dependent on cultural context. The balance of Nature steady-state model was partly a product of Romantic values at the last turn of the century, just as chaos, uncertainty, and non-equilibrium theories are tied to current circumstances. The process of science suggests that the current definition of biodiversity provides an improved picture of how Nature works; but the views that Stephen Spurr expressed at the 1969 Sierra Club wilderness conference are alive and well today. The influential ecologist Daniel Botkin believes "We can engineer nature at nature's rate and in nature's way . . ." (Botkin 1990, p. 190). The current debate over defining ecosystem management provides another indication of American ambivalence over new concepts of biodiversity (Grumbine 1994a). Is it people over Nature or people in partnership with Nature? Ecosystem management or ecosystem protection? The goal of biodiversity protection considers all human use of Nature as flowing from ecosystems only after basic patterns and processes are maintained and restored. If ecosystem management for native diversity is to take hold and flourish beyond 1995, the new goal of protecting biodiversity and the old standard of providing resources for human use must be reconciled. This is a values and political question which does not depend exclusively on science for resolution.

One hundred thirty-seven years ago, Charles Darwin ascertained that all life—including humans—is subject to the forces of evolution. Three decades before the first Earth Day, Aldo Leopold, working a cut-over sand county farm in Wisconsin, saw through the delusion that people are separated from and not responsible to Nature. In 1962, Rachel Carson published her blockbuster against pesticides. In the late 1980s, American school children became aware of the Northern Spotted Owl and its old-growth home. Yet a century and a third after the *Origin of Species* many US citizens do not believe in evolution. Twenty-five years passed between the appearance of *A Sand County Almanac* and a law to protect species from extinction. DDT, banned in the US since 1969, is still manufactured by US companies for export. The pace of positive change is painfully slow. Sociologist Bill Devall is correct to note that though Americans have been quick to support environmental reforms, changes that require difficult behavioral and values adjustments remain incomplete (Devall 1995).

Biodiversity protection represents the core idea that may bring Americans to support protecting all species, human and nonhuman together. Noting the tension between scientific and environmentalist views of ecosystems, ecologist Frank Golley (1993, p. 205) remarked, "It is not clear to me where ecology ends and the study of the ethics of nature begins, nor is it clear to me where biological ecology ends and human ecology begins. These divisions become less and less useful." Moving from a 19th century model of preserving Nature toward a 21st century image of protecting biodiversity will help break down further the delusion that people and Nature can be separate. The hope is that adjusting management goals to reduce extinction and habitat destruction will not only end the present biodiversity crisis but also provide the opportunity for people to forge a new relationship with Nature. Hope and time are intertwined—most biologists do not believe that we have the luxury of an additional 25 years to wait for biodiversity to become accepted by society as ecology was from 1970 to 1995.

Long before Earth Day's 50th anniversary, Americans must learn that there can be no alternative to protecting the native sources of life—healthy, functioning wild ecosystems.

NOTE

The author would like to thank Curt Meine, Reed Noss, and John Davis for their helpful comments on earlier drafts of this essay. An earlier version of this essay appeared in the 1995 *Humbolt Journal of Social Relations* 21 (1): 35–59.

BIBLIOGRAPHY

Adams, Charles. 1929. "The Importance of Preserving Wilderness Conditions," *New York State Museum Bulletin* 279: 37–34.

Allin, Craig. 1982. *The Politics of Wilderness Preservation.* Westport, CT: Greenwood Press.

Botkin, Daniel. 1990. *Discordant Harmonies.* New York: Oxford University Press.

Brower, Lincoln. 1994. "A New Paradigm in Conservation Biology. Endangered Biological Phenomena": Pages 104–106 in Gary Meffe and C. R. Carroll, eds., *Principles of Conservation Biology.* Sunderland, MA: Sinauer Associates.

Brussard, Peter, et al. 1994. "Cattle and Conservation Biology—Another View," *Conservation Biology* 8(4): 919–921.

Carson, Rachel. 1962. *Silent Spring.* Boston: Houghton Mifflin.

Cody, Martin L., and J. Diamond. 1975. *Ecology and Evolution of Communities.* Cambridge, MA: Belknap Press of Harvard University Press.

Darling, F. F., and N. Eichhorn. 1969. *Man and Nature in the National Parks: Reflections on Policy.* Washington, DC: The Conservation Foundation.

Darwin, Charles. 1859. *On the Origin of Species.* New York: Modern Library, n.d.

Devall, Bill. 1995. "Twenty Five Years Since Earth Day: Reflections of a Sometimes Social Activist," *Humbolt Journal of Social Relations.*

DeVoto, Bernard. 1950. "Shall We Let Them Ruin Our National Parks?," *Saturday Evening Post* 223: 44.

Ehrenfeld, David. 1970. *Biological Conservation.* New York: Holt Rinehart and Winston.

Ehrlich, Paul, and Anne Ehrlich. 1981. *Extinction: The Causes and Consequences of the Disappearance of Species.* New York: Random House.

Foreman, Dave, and Howie Wolke. 1989. *The Big Outside.* Tucson: Ned Ludd Books.

Frankel, J., and M. Soulé. 1981. *Conservation and Evolution.* Cambridge: Cambridge University Press.

Gilligan, James. 1963. "Summary of the Outdoor Recreation Resource Review Committee," Pages 51, 52 in François Leydet, ed., *Tomorrow's Wilderness.* San Francisco: Sierra Club.

Golley, Frank. 1993. *A History of the Ecosystem Concept in Ecology.* New Haven, CT: Yale University Press.

Grinnell, Joseph, and Tracy Storer. 1916. "Animal Life as an Asset of National Parks," *Science* 44 (September 15, 1916).

Grumbine, E. 1990. "Protecting Biological Diversity Through the Greater Ecosystem Concept," *Natural Areas Journal* 19 (3): 114–120.

Grumbine, R. Edward. 1992. *Ghost Bears: Exploring the Biodiversity Crisis.* Washington, DC: Island Press.

Grumbine, R. Edward. 1994a. "What Is Ecosystem Management?," *Conservation Biology* 8 (1): 27–38.

Grumbine, R. Edward. 1994b. *Environmental Policy and Biodiversity.* Washington, DC: Island Press.

Harris, Larry. 1984. *The Fragmented Forest.* Chicago: University of Chicago Press.

Hays, Samuel. 1987. *Beauty, Health, and Permanence.* Cambridge: Cambridge University Press.

Kendeigh, S. C., et al. 1950–1951. "Nature Sanctuaries in the United States and Canada: A Preliminary Inventory," *The Living Wilderness* 15 (35): 1–45.

Leopold, Aldo. 1933. "The Conservation Ethic," *Journal of Forestry* 31 (6): 634–643.

Leopold, Aldo. 1939. "A Biotic View of Land," *Journal of Forestry* 37 (9): 727–730.

Leopold, Aldo. 1941. "Wilderness as a Land Laboratory," *Living Wilderness* 6:3.

Leopold, Aldo. 1949. *A Sand County Almanac.* New York: Oxford University Press.

Leopold, Aldo. 1991. "Conservation: In Whole or in Part?" Pages 310–319 in Susan L. Flader and J. B. Callicott, eds. *The River of the Mother of God and Other Essays by Aldo Leopold.* Madison, WI: University of Wisconsin Press.

Leopold, A. Starker, et al. 1963. *Wildlife Management in the National Parks.* Washington, DC: US Department of the Interior. [Included in this volume.]

Lyell, Charles. (1830). *Principles of Geology.* Chicago: University of Chicago Press, 1990–1991.

MacArthur, Robert, and E. O. Wilson, 1967. *The Theory of Island Biogeography.* Princeton: Princeton University Press.

McCloskey, Michael. 1968. "A Conservation Agenda for 1969," *Sierra Club Bulletin* 53 (12): 5.

McIntosh, R. P. 1985. *The Background of Ecology: Concept and Theory.* Cambridge, MA: Cambridge University Press.

Marsh, George Perkins. (1864). *Man and Nature: or, Physical Geography as Modified by Human Action.* Cambridge: Cambridge University Press, 1965.

Marshall, Robert. 1930. "The Problem of the Wilderness," *Scientific Monthly* 30: 142–143. [Included in this volume.]

Meffe, Gary, and C. R. Carroll, eds. 1994. *Principles of Conservation Biology.* Sunderland, MA: Sinauer Associates.

Myers, Norman. 1979. *The Sinking Ark.* Oxford: Pergamon Press.

Nash, Roderick. 1989. *The Rights of Nature.* Madison, WI: University of Wisconsin Press.

Nature Conservancy, The. 1975. *The Preservation of Natural Diversity: A Survey and Recommendations.* Washington, DC: The Nature Conservancy.

Nixon, Richard. 1972. Weekly Compilation of Presidential Documents 8 (February 8, 1972), 218–224.

Noss, Reed. 1994. "Cows and Conservation Biology," *Conservation Biology* 8 (3): 613–616.

Noss, Reed, and Alan Cooperrider. 1994. *Saving Nature's Legacy.* Washington, DC: Island Press.

Noss, Reed, et al. 1994. *Endangered Ecosystems of the United States: A Preliminary Assessment of Loss and Degradation.* Washington, DC: US Fish and Wildlife Service.

NPCA. 1979. "Adjacent Lands Survey: No Park Is an Island," *National Parks and Conservation Association Magazine* 53 (3): 4–9.

Pearsons, G. A. 1992. "The Preservation of Natural Areas in the National Forest," *Ecology* 3: 284–287.

Pickett, S. T. A., et al. 1992. "The New Paradigm in Ecology: Implications for Conservation Biology above the Species Level." Pages 65–88 in P. Fiedler and S. Jains, eds., *Conservation Biology.* New York: Chapman and Hall.

Primack, Richard. 1993. *Essentials of Conservation Biology.* Sunderland, MA: Sinauer Associates.

Runte, Alfred. 1987. *National Parks: The American Experience.* Lincoln, NE: University of Nebraska Press.

Schonewald-Cox, Christine, et al., eds. 1983. *Genetics and Conservation: A Reference for Managing Wild Animal and Plant Populations.* Menlo Park, CA: Benjamin/Cummings.

Shelford, Victor. 1920. "Preserves of Natural Conditions," *Transactions of the Illinois State Academy of Science* 13: 37–58.

Shelford, Victor. 1933. "The Preservation of Natural Biotic Communities," *Ecology* 14 (2): 240–245.

Sierra Club Bulletin. 1976. *Sierra Club Bulletin Index 1950–1976.* San Francisco: Sierra Club.

Soulé, Michael. 1985. "What Is Conservation Biology?" *BioScience* 35: 727–734.

Soulé, Michael, et al. 1986. "The Millennium Ark: How Long a Voyage, How Many Staterooms, How Many Passengers?" *Zoo Biology* 5: 101–113.

Soulé, Michael, and Bruce Wilcox, eds. 1980. *Conservation Biology: An Ecological and Evolutionary Approach.* Sunderland, MA: Sinauer Associates.

Spurr, Stephen. 1963. "The Value of Wilderness to Science." Pages 57–64 in François Leydet, ed., *Tomorrow's Wilderness.* San Francisco: Sierra Club.

Sumner, Francis. 1920. "The Need for a More Serious Effort to Rescue a Few Fragments of Vanishing Nature," *Scientific Monthly* 10: 236–248.
Sumner, Francis. 1921. "The Responsibility of the Biologist in the Matter of Preserving Natural Conditions," *Science* 54: 39–43.
US Code. 1982. 16, Sec. 1604 (g) (3) (B).
US Council on Environmental Quality. 1981. *The Global 2000 Report to the President.* 3 vols. Washington, DC: US Council on Environmental Quality.
US National Park Service. 1980. *State of the Parks—1980.* Washington, DC: US Department of Interior.
US Statutes at Large, 1934. 48: 817.
Whitaker, R. H. 1972. *Communities and Ecosystems.* New York: MacMillan.
Worster, Donald. 1994. *Nature's Economy.* New York: Cambridge University Press.
Wright, George, Ben Thompson, and Joseph Dixon. 1933. *Fauna of the National Parks of the United States: A Preliminary Survey of Faunal Relations in National Parks.* Washington, DC: Government Printing Office.
Yaffee, Stephen. 1982. *Prohibitive Policy: Implementing the Federal Endangered Species Act.* Cambridge, MA: MIT Press.

Jack Turner

In Wildness Is the Preservation of the World (1991)

HANGING FROM THE CEILING of the visitors' center at Point Reyes National Seashore are plaques bearing famous quotations about the value of the natural world. The one from Thoreau reads: "In wilderness is the preservation of the world." This, of course, is a mistake. Henry didn't say "wilderness"; he said "wildness." But the mistake has become a cliché suitable for T-shirts and bumper stickers. So when a recent *Newsweek* article on wolf reintroduction says, "'In wilderness,' Thoreau wrote, 'is the preservation of the world,'" I am not surprised. Confusing wilderness and wildness strikes me increasingly as a Freudian slip. It serves a repressive function, the avoidance of conflict, in this case the inherent tension between wilderness as property and wildness as quality.

Last year at a symposium on nature writing in the Thoreau tradition, William Kittredge was candid enough to admit that

> For decades I misread Thoreau. I assumed he was saying wilderness.... Maybe I didn't want Thoreau to have said wildness, I couldn't figure out what he meant.

I believe that mistaking wilderness for wildness is one cause of our increasing failure to preserve the earth, and that Kittredge's honesty pinpoints the key issue: we aren't sure what Thoreau meant by wildness, nor are we sure what we mean by wildness or why we should preserve it. I don't know either, so what follows is not an explication of Henry's famous saying but more of a prelude or a prolegomena to the issue.

This saying, perhaps Thoreau's most famous, is from his essay "Walking." Along with *Walden* and two other essays, "Resistance to Civil Government" (unfortunately called "Civil Disobedience" most of the time) and "Life without Principle," it expresses the radical heart of Thoreau's life's work, and since he revised the essay just before his death, we may assume it accurately represents his thoughts on wildness.

The most notable thing about "Walking" is that it virtually ignores our current concerns with the preservation of habitats and species. His saying no doubt includes these things—he says "all good things are wild and free"—but Thoreau mainly talks about human beings, their literature, their myths, their history, their work and leisure, and of course their walking. He says, for instance,

> Give me for my friends and neighbors wild men, not tame ones. The wildness of the savage is but a faint symbol of the awful ferity with which good men and lovers meet.

And listen to the essay's opening lines:

> I wish to speak a word for Nature, for absolute freedom and wildness, as contrasted with a freedom and culture merely civil,—to regard man as an inhabitant, or a part and parcel of Nature, rather than a member of society.

Absolute freedom. Absolute wildness. Human beings as inhabitants of absolute freedom, absolute wildness. This is not the usual environmental rhetoric, and I agree with Kittredge: most of us simply don't know what Thoreau means.

Nor should we be surprised, for most people no longer have much experience of wild nature. But language and communication are social phenomena, and without common, shared experience, meaning is impossible. I would go so far as saying that in many inner cities, here and in the developing world, people no longer have the concept of nature. As a New York wit has it, Nature is something I pass through between my hotel and my taxi. As the population grows, the cause of preservation will become increasingly desperate.

What is equally unsettling is this: those people who have led a life of intimate contact with nature at its wildest—a buckaroo working the Owyhee country, a halibut fisherman plying the currents of the Gulf of Alaska, an Eskimo whale hunter, a rancher tending a small cow-calf operation, a logger with a chainsaw—are perceived as the enemies of preservation. The friends of preservation, on the other hand, are often city folk who depend on vacations in wilderness areas and national parks for their (necessarily) limited experience of wildness. The difference in degree of experience of wildness, the dichotomy of friends-enemies of preservation, and the notorious inability of these two groups to communicate shared values indicates the depth of our muddle about wilderness and wildness, and suggests again the increasingly desperate nature of our struggle.

We presume that the experience of wildness goes with wilderness (though the presumption ignores elements of our life that can also plausibly be thought of as wild: sex, dreams, rage). However, since wilderness is a place, and wildness a quality, we can always ask: How wild is our wilderness, and how wild is our experience there? My answer is, not very. There are many reasons, some of them widely acknowledged, and I will pass over them briefly. But there is one reason that is not widely accepted, a reason that is offensive to many minds, but that goes to the heart of Thoreau's opening lines: human beings are no longer residents of wild nature, hence we no longer consider ourselves part of a biological order.

"A little pure wildness is the one great present want," wrote John Muir. It is still true. Why isn't our wilderness wild and why is there so little experience of wildness there? Well, first of all the wilderness that most people visit (excluding Alaska and Canada) is too small—in space and time. Like all experience, the experience of the wild can be a taste or a feast, and feasts take time.

About one-third of our designated wilderness units are less than 10,000 acres, about four miles long on each side. An easy stroll. Some units, usually islands, are less than 100 acres. I have been told that Point Reyes now has "wilderness zones" measuring several hundred yards, a point at which the word becomes meaningless. For comparison, recall that Disney World is 27,000 acres. *Disney World is nearly three times larger than a third of our wilderness areas.*

Even our largest wilderness units are small. Only 4 percent are larger than 500,000 acres, an area 27 miles on a side. And since many follow the ridges of mountain ranges, they are elongated to the point that they can be crossed in a single day by a strong hiker. True, some are adjacent to other wilderness areas and remote BLM lands and national parks, but once you have visited the Amazon, Alaska, the Northwest Territories, or the Himalayas, our local wilderness seems very small indeed.

Without sufficient space and time, the experience of wildness is diminished or simply doesn't exist. Many people agree with Aldo Leopold: it should take a couple of weeks to pack across a true wilderness. It's a simple law: the farther you are from a road, and the longer you are out, the wilder your time. Two weeks is the minimum; a month is better. Until then the mind remains saturated with human concerns and blind to the concerns of the natural world. Until then the body remains bound to metronomic clocks and ignorant of natural biological rhythms, and the wilderness traveler remains ignorant of "forces more fundamental and more calming than the mechanical overlay they have so diligently clamped down on themselves" (Michael Young, *The Metronomic Society*).

Small wilderness units usually lack predators—sometimes simply a function of small size, sometimes a function of artificial borders created according to economic and political rather than ecological criteria. The result is the same: the wilderness is tamed. Predators are perhaps our most accessible experience of the wild. To come upon a grizzly track is to experience the wild in a most intimate, carnal way—an experience marked by gross alterations in attention, perception, body language, body chemistry, and emotion.

The tameness of wilderness is exacerbated by our current model for appropriate human use of the wild—the intensive and commercial recreation that requires trail systems, bridges, signs for direction and distance, back-country rangers, and rescue operations. These in turn create additional commercial activities that further diminish wildness—maps, guidebooks, guiding services, advertising, photography books, instructional films—all of which diminish discovery and surprise and independence and the unknown, the very qualities that make a place wild. Each of these reductions functions like the loss of a predator. It tames and domesticates the wilderness and eliminates wild experience.

The smallness, the artificial boundaries, the loss of predators, and the intensive recreational use all lead to replacing biological methods of growth and interaction with artificial methods of control of plants, animals, humans, and events. Thus, animal populations are managed by controlled hunting, wild fires are suppressed, plants sprayed, and humans treated in a manner best described by the word "surveillance."

The wild then becomes a "problem" to be "solved" by further human intervention—scientific studies, state and federal laws, judicial decisions, political compromise, and administrative and bureaucratic procedures. Once this intervention begins, it never ends; it spirals into further and further human intrusion, rendering wilderness increasingly evaluated, managed, regulated, and controlled. That is, tamed.

Nibble by nibble, decision by decision, animal by animal, we have diminished the wildness of our wilderness. This hasn't just happened; *we have done it.* Thus diminished, wilderness becomes an area, a special unit of property treated like an historic relic or ruin—a valuable remnant. It becomes a place of vacations (a word related to vacant, empty). Humans are strangers there, foreigners to an experience that once grounded their most sacred beliefs and values. The wilderness as relic leads necessarily to tourism, and tourism in the wilderness becomes the primary mode of experiencing a diminished wild. (Here, incidentally, is the reason for the undercurrent of sadness and nostalgia that marks most of our nature writing. Sadness and nostalgia are the emotions that accompany our return to a former residence.)

Wilderness as relic converts places into commodities. We should be concerned, for all tourism is a form of commerce and is to some degree destructive. Virtually everyone in the Nature Business feeds (literally) on wilderness as a commodity. We are enthralled with our ability to make a living with this exchange, but we tend to ignore the practical consequences to wilderness preservation and to ourselves. Preserving relics of tame wilderness and reducing experience there to tourism is not a free lunch. Compared with residency in a wild biological order, where the experience of wildness is part of everyday life, wilderness tourism is pathetic. It has some very bad consequences, and I want to mention several of them.

First, wilderness tourism ignores, perhaps even caricatures, the experience that decisively marked the founders of wilderness preservation and deep ecology—and I am thinking here of Thoreau, Muir, Leopold, and

Murie. The kind of wildness they experienced has become very rare—an endangered experience. As a result, we no longer understand the roots of our own cause.

To read the works of these men and then to look at an issue of, say, *Sierra,* is to experience a severe disorientation. The founders had something we lack, something Thoreau called "Indian Wisdom." For much of their lives, these men lived in and studied nature before it became a "wilderness area."

Thoreau's knowledge of lands surrounding Concord was so vast some of the town's children believed that, like God, Henry had created it all. His knowledge of flora was so precise that a rare fern species not seen for a hundred years was recently rediscovered by examining his surveying notes. His essay on the succession of forest trees is one of the seminal essays of modern ecology. Muir made original contributions to the study of glaciers.

The works of Leopold and Murie are the classics in their fields. To a considerable degree their lives were devotions to wild nature. Without such devotion there is no reason to believe there will be Thoreau's epiphanies on Katadin, Muir's mystical identification with trees, or Leopold's thinking like a mountain.

This is quite different from wilderness tourism, which is devoted to fun. We hunt for fun, fish for fun, climb for fun, ski for fun, and hike for fun. This is the grim harvest of the "fun hog" philosophy that drove the wilderness recreation boom for the past three decades. Despite the poetics and philosophical rhetoric of environmentalists, there is little evidence that either the spiritual or scientific concerns of Muir, Thoreau, Leopold, and Murie (or the scientific concerns of conservation biologists) have trickled down to most wilderness users.

Given the ignorance and arrogance of the fun hog, it is understandable that those who feel they will lose by increased wilderness designation—farmers, ranchers, loggers, commercial fisherman—are often enraged. Instead of a clash of needs, the preservation of the wild appears to be a clash of work vs. recreation. Lacking a deeper experience of wildness and access to the lore, myth, metaphor, and ritual necessary to share that experience, there is no communication, no vision that might shatter the current dead-end of wilderness debate.

Both groups exploit the wild, the first by consumption, the second by alteration into a playpen. Either way, the quality of wildness is destroyed. Until we face that fact we will remain stuck. Meanwhile, the worship of wilderness designation becomes idolatry—the confusion of a symbol with its essence.

Second, with wilderness tourism we lose our most effective weapon for preserving wild nature: emotional identification. At the bedrock level, what drives both reform environmentalism and deep ecology is the practical problem of how to *compel* human beings to respect and care for wild nature. The tradition of Thoreau and Muir says that the best way to do it is raw, visceral contact with wild nature. True residency in the wild brings identification and a generalized NIMBY response that extends sympathy to all the wild world. This is one of the most obvious lessons of primary cultures. Without such identification, solutions become abstract, impotent, and impractical. Right now impracticality and impotence dominate environmental thought.

For example, giving trees and animals moral rights analogous to the rights of humans has bogged down in a morass of value theory, and the aesthetic campaign to preserve the wild has done as much harm as good. It suggests (especially in a nation of relativists) that preservation is a matter of taste, a preference no more compelling than the choice between vanilla and chocolate. It leads to tedious arguments that begin with statements like "Who are you to say we shouldn't have snowmobiles in the Teton wilderness . . . ?", on the model of "Who are you to say I shouldn't eat chocolate?" This, in turn, leads inevitably to questions of egalitarianism and elitism, hence, directly into the dismal swamp of politics, which Henry said in "Walking" is the most alarming of man's affairs. Politicians are invariably people of the polis—city slickers. Yes, even Dave Foreman and Dave Brower are city slickers, tourists in the wilderness, and it shows.

Philosophers have been no more helpful. Deep ecologists are desperately trying to replace the philosophical foundations of the mechanical model of the world with philosophical foundations of an organic model. Unfortunately, these new foundations are not at all obvious to the lay public, and they are even less obvious to professional philosophers. The search for foundations—for science, mathematics, logic, or the social sciences—has been the curse of rationalism from Descartes to the present, and the foundations of deep ecology will not exorcise that curse. Foundationalism is a kind of "pseudo-rationality" (to quote Otto Neurath) and the wild world would be better off without it. Worse yet, explications of these foundations rely on some of the most obscure ruminations of Spinoza, Whitehead, and Heidegger. This bodes ill for condors and rain forests.

All these things are reasonable (sort of), but as Hume saw clearly, *reason alone is insufficient to move the will.* We should repeat this to ourselves everyday like a mantra. Reason has not compelled us to respect and care for wild nature, and we have no basis for the belief that it will in the future. Philosophical arguments, moralizing, aesthetics, legislation, and abstract philosophies are notoriously incapable of compelling human beings. That was the lesson of Prohibition.

Third, wilderness tourism results in little art, literature, poetry, myth, or lore for most of our wild places. In "Walking," Thoreau says, "The West is preparing to add its fables to those of the East. The valleys of the Ganges, the Nile, and the Rhine having yielded their crop, it remains to be seen what the valleys of the Amazon, the Platte, the Orinoco, the St. Lawrence, and the Mississippi will produce." Well, nearly 150 years later it still remains to be seen.

If you ask for the art, literature, lore, myth, and fable of, say, the upper Snake River, I would answer we are working on it, but it might be awhile, because art that takes as its subject a place is created by people who live in that place. This is true of both wilderness and civilization. Joyce grew up in Dublin, Atget lived in Paris, Adams lived in Yosemite, Beston lived on Cape Cod. Many of our best writers on wilderness—Ed Abbey, Gary Snyder, Doug Peacock—worked as fire lookouts for the U.S. Forest Service. (There is probably a Ph.D thesis here: "The Importance of Fire Lookouts in the Development of Western Nature Literature.") If our access to wilderness is limited to tourism, we have no reason to expect a literature and lore of the wild.

And yet, most of us, when we think about it, realize that after our own direct experience of wildness, art and literature, myth and lore have contributed most to our love of wild places, animals, plants, even, perhaps, to our love of human wildness. For here is the language we so desperately lack, the medium so necessary to communicate a shared vision. Mere concepts and abstractions will not do because that which needs to be shared is beyond concepts and abstractions.

Fourth, without residency in the wild there will be no phenology of wild places, and this will be very unfortunate, for phenology, as Paul Shepard has reminded us, is the study of the mature naturalist—the gate through which nature becomes personal. Leopold published phenological studies of two counties in Wisconsin. Thoreau dedicated the last years of his life to studying the mysterious comings and goings of the natural world. Phenology requires a complete immersion in place over time, so the attention, the senses, and the mind can scrutinize and discern widely—the dates of arrivals and departures, the births, the flourishings, the decays and the deaths of wild things, their successions, synchronicities, dependencies, reciprocities, and cycles—the lived life of the earth.

To be absorbed in this life is to merge with larger patterns. Here ecology is not studied but felt. You know these truths the way you know hot from cold; they are immune from doubt and argument. Here is the intimate knowledge conservation biology often seeks to rediscover, the common wisdom of primary peoples. (And again, it is still the common wisdom of many people who actually work in nature.) This will not emerge from tourism in a relict wilderness.

So we are left with the vital importance of residency in wild nature, and a visceral knowledge of that wildness, as the most practical means of preserving the wild. What we need now is a new tradition of the wild that teaches us how human beings live best by living in and studying the wild without taming it or destroying it. Such a tradition of the wild existed; it is as old as the Pleistocene. Before the Neolithic, human beings were always living in, traveling through, and using lands we now call wilderness; they knew it intimately, they respected it, they cared for it. It is the tradition of the people that populated all of the wilderness of North America, a tradition that influenced Taoism and informed major Chinese and Japanese poetic traditions. It is the tradition that emerged again with Emerson and Thoreau (who once asked "Why study the Greeks and Romans? Why not study the American Indian?") It is common in our own myths. On the night before his final battle, King Arthur muses that

> Merlyn had taught him about animals so that the single species might learn by looking at the thousands. He remembered the belligerent ants, who claimed their boundaries, and the pacific geese, who did not. He remembered his lesson from the badger. He remembered Lyo-lyok and the island which they had seen on their migration, where all those puffins, razorbills, guillemots and kittiwakes had lived together peacefully, preserving their own kinds of civilization without war—because they claimed no boundaries.

In short, it is a tradition that could again compel respect, care, and love for wild nature in a way that philosophical foundations, aesthetics, moral theory, and politics cannot compel. It is a tradition we need to recreate for ourselves, borrowing when necessary from native cultures, but making it new—a wild tradition of our own.

A wild bunch is forming—an eclectic tribe returning to the wild to study, learn, and express. From them shall come the lore, myth, literature, art, and ritual we so require. Frank Craighead and John Haines are among the elders of this tribe. And there is also Richard Nelson on his island, Doug Peacock with his grizzlies, Terry Tempest Williams and her beloved birds, Hanna Hinchman with her illuminated journals, Gary Nabhan with his seeds, Dolores LaChapelle and her rituals, Gary Snyder and his poems—all new teachers of the wild.

The presence of these teachers is not sufficient, however. It will not help us if this tradition is created for us, to be read about in yet another book. To affect the self, the self must live the life of the wild, mold a particular form of human character—a form of life. Relics will not do. Tourism will not do.

Out there is the great feeding mass we call the earth. We are, to go back to Thoreau's opening lines, part and parcel of the earth, part and parcel of its cycles, successions, and dependencies. We incorporate and are incorporated in ways not requiring legal papers. We are creator and created, terrorist and hostage, victim and executioner, a guest of honor and a part of the feast. We inhabit a biological order that is terrifying in its identity and reciprocity. It is expressed by Black Elk and his Wakan-Tanka, Lao Tzu and his Tao, the Mahayana Net of Indra, the ecologists' food chain. It is a vision hidden from the urban mind. This vision could inform everything from the most private spiritual matters to the gross facts of nourishment and death. The only interesting question is how can we live it here and now, in this place, in these times.

The moose incorporates the willow, taking the life of the willow into its own life, making the wildness of the willow reincarnate. I kill the moose, its body feeds the willow where it dies, it feeds my body, and in feeding my body the willow and the moose feed the one billion bacteria that inhabit three inches of my colon; they feed the one million spirochetes that live in my mouth; they feed the microscopic brontosauruslike mites that live by devouring the goo on my eyelashes.

From the willow I take branches to make the basket that carries the wortleberries and huckleberries to my home. (And from the willow I learn of salicin and now use salicylic acid to make the aspirin that fights the headache that comes from thinking about all this.) When I die, my friends and family bury me, and I feed the willow and the berry bushes that feed the Clark's Nutcracker that spreads the nuts that feed the grizzly; I feed the children of the moose that fed me and my body's inhabitants, and that will, in turn, feed my children. This great feeding body is the world, parts and parcels that evolved together, mutually, relating in the endless dance of evolution. We are all the dust of old stars. We are the form wildness bred to become conscious of its Self. Nothing less.

In "Walking," Thoreau called this identity and reciprocity a "Sympathy with Intelligence," a "novel and grand surprise in a sudden revelation . . . ," a "lighting up of the mist by the sun." Who shall develop such a sympathy? Where shall we find a sudden revelation? How shall we experience an absence of boundaries, a world mutual and shared, an accord and an understanding grounded in feeling? When shall we know a compassion, an allegiance, and a loyalty that is infinite?

The Taittiriya Upanishad declares:

> O wonderful! O wonderful! O wonderful!
> I am food! I am food! I am food!
> I eat food! I eat food! I eat food!
> My name never dies, never dies, never dies!
> I was born, first in the first of the worlds,
> Earlier than the gods,
> In the belly of what has no death!
> Whoever gives me away has helped me the most!
> I, who am food, eat the eater of food!
> I have overcome this world!
>
> He who knows this shines like the sun.
> Such are the laws of mystery!

Gary Paul Nabhan

Cultural Parallax in Viewing North American Habitats (1995)

A DEBATE IS RAGING with regard to the "nature" of the North American continent—in particular, the extent to which habitats have been managed, diversified, or degraded over the last ten thousand years of human occupation (Gómez-Pompa and Kaus 1992). This debate has at its heart three issues: whether the "natural condition of the land" by definition excludes human management; whether officially designated wilderness areas in the United States should be free of hunting, gathering, and vegetation management by Native Americans or other people; and whether traditional management by indigenous peoples is any more "benign" or "ecologically sensitive" than that imposed by resource managers trained in the use of modern Western scientific principles, methods, and technologies.

The debate, then, is not about the human mental "construction" of nature so much as it is about the physical "reconstructions" of habitats by humans and to what extent these are perceived as "natural" or "ecological." The debate is not merely an academic dispute. It involves hunters, gatherers, ranchers, farmers, and political activists from a variety of cultures, not just "Western scientists" versus "indigenous scientists." The outcome will no doubt shape the destiny of officially designated wilderness areas in national parks and forests throughout North America (Gómez-Pompa and Kaus 1992; Flores et al. 1990).

Consider, for example, the declaration of the 1963 Leopold Report to the U.S. Secretary of Interior: that each large national park should maintain or recreate "a vignette of primitive America," seeking to restore "conditions as they prevailed when the area was first visited by the white man"—as if those conditions were synonymous with "pristine" or "untrammeled" wilderness (Anderson and Nabhan 1991:27). Such a declaration either implies that pre-Columbian Native Americans had no impact on the areas now found within the U.S. National Park System or that indigenous management of vegetation and wildlife as it was done in pre-Columbian times is compatible with and essential to "wilderness quality." For Native Americans with historic ties to land, water, and biota within parks, this latter interpretation provides them a platform for being *co-managers,* not merely harvesters of certain traditionally utilized resources, as currently sanctioned by the National Park Service (1987).

On one side of the debate are those who argue that Native Americans have had a negligible impact on their homelands and left large areas untouched. That is to say, these original human inhabitants did little to actively manage or influence wildlife populations one way or another. An early proponent of this view was John Muir: "Indians walked softly and hurt the landscape hardly more than the birds and squirrels, and their brush and bark huts last hardly longer than those of wood rats, while their enduring monuments, excepting those wrought on the forests by fires they made to improve their hunting grounds, vanish in a few centuries." Yet the Yosemite landscapes he knew so well are now known to have been dramatically shaped by Native American management practices (Anderson and Nabhan 1991:27).

Some proponents of this perspective even deny Muir's exception that controlled burns had a significant impact. Native Americans, they claim, would have no interest in managing the forests even if they were capable of it (Clar 1959:7): "It would be difficult to find a reason why the Indians [of California] should care one way or another if the forest burned. It is quite something else again to contend that the Indians used fire systematically to 'improve' the forest. Improve it for what purpose? . . . Yet this fantastic idea has been and still is put forth time and again."

A second stance on this side of the debate contends that Native American spirituality kept all members of indigenous communities from harming habitats or the biota within them. Leslie Silko (1987:86), who is of Laguna Pueblo descent, has argued that "survival depended upon the harmony and cooperation, not only among human beings, but among all things—the animate and inanimate. . . . As long as good family relations [between all beings] are maintained . . . the Earth's children will survive." The implication is that Native Americans practiced a spirituality "earthly enough" to restrain any tendencies toward overharvesting or toward depletion of diversity through homogenizing habitat mosaics (Anderson 1993). As Max Oelschlaeger (1991:17) has assumed, "*harmony with* rather than *exploitation of* the natural world was a guiding principle for the Paleolithic mind and remains a cardinal commitment among modern aborigines."

The other side of this argument contends that Native Americans and other indigenous peoples have rapaciously exterminated wildlife within their reach and that their farming, hunting, and gathering techniques were often ecologically ill suited for the habitats in which they were practiced. Kent Redford (1985) and others have taken such a stance to play devil's advocate with the romantic notion of "the ecological noble savage." Award-winning science writer Jared Diamond (1993:268) has also tried to dispel what he sees as a myth of native peoples as "environmentally minded paragons of conservation, living in a Golden Age of harmony with nature, in which living things were revered, harvested only as needed, and carefully monitored to avoid depletion of breeding stocks."

Diamond (1993:263, 268) claims that in thirty years of visiting native peoples on the three islands of New Guinea, he has failed to come across a single example of indigenous New Guineans showing friendly responses to wild animals or consciously managing habitats to enhance wildlife populations: "New Guineans kill those animals that their technology permits them to kill," inevitably depleting or exterminating more susceptible species. His claim that all indigenous New Guinean cultures respond in the same manner to nature is astonishing when one considers that about one thousand of the world's remaining languages are spoken in New Guinea and that this cultural/linguistic diversity would presumably encode many distinctive cultural responses to the flora and fauna.

Yet to my knowledge, Diamond himself has never objectively field-tested his game depletion hypothesis as Vickers (1988) has done among Amazonian peoples, where it was demonstrated over several years that native hunters would switch to less desirable prey before locally extirpating rare game species. Redford and Robinson (1987) have also documented that indigenous South American hunters take a wider variety of game species than neighboring nonindigenous South American colonists, who are more likely to use degraded habitats near larger settlements, thereby further impoverishing the abundance and diversity of wildlife. Diamond has never demonstrated that he has systematically asked indigenous consultants about less obvious techniques by which hunters and gatherers may influence habitat quality and wildlife abundance, following research protocols such as those outlined by Blackburn and Anderson (1993:24). And yet, repeatedly, Diamond (1986, 1992, and elsewhere) has used the hypotheses of the Pleistocene overkill, and later selective cutting of fir and spruce in Chaco Canyon by Anasazi city-state dwellers, to indict all Native American hunters, gatherers, and farmers as exterminators of wildlife and aggravators of soil erosion.

This dismissal of enormous historic and cultural differences is at the heart of the problem inherent in most discussions of "the American Indian view of nature" and assessments of the pre-Columbian condition of North American habitats. To assume that even the Hopi and their Navajo neighbors think of, speak of, and treat nature in the same manner is simply wrong. Yet individuals from two hundred different language groups from three historically and culturally distinct colonizations of the continent are commonly lumped under the catchall terms "American Indian" or "Native American."

Even within one mutually intelligible language group, such as the Piman-speaking O'odham, there are considerable differences in what taboos they honor with respect to dangerous or symbolically powerful animals. While the River Pima do not allow themselves to eat badgers, bears, quail, or certain reptiles for fear of "staying sickness," these taboos are relaxed or even dismissed by other Piman groups who live in more marginal habitats where game is less abundant (Rea 1981; Nabhan and St. Antoine 1993). An animal such as the black bear—which is never eaten by one Piman community because it is considered to be one of the "people"—is routinely hunted by another Piman-speaking group which prizes its skin and pit-roasts its meat—an act that would be regarded much like cannibalism in the former group (Rea 1981; personal communication). Moreover, contemporary Pima families do not necessarily adhere to all the taboos that were formerly paramount to all other cultural rules which granted someone Piman identity.

Despite such diversity within and between North American cultures, it is still quite common to read statements implying a uniform "American Indian view of nature"—as if all the diverse cultural relations with particular habitats on the continent can be swept under one all-encompassing rug. Whether one is prejudiced toward the notion of Native Americans as extirpators of species or assumes that most have been negligible or respectful harvesters, there is a shared assumption that all Native Americans have viewed and used the flora and fauna in the same ways. This assumption is both erroneous and counterproductive in that it undermines any respect for the realities of cultural diversity. And yet it continues to permeate land-use policies, environmental philosophies, and even park management plans. It does not grant *any* cultures—indigenous or otherwise—the capacity to evolve, to diverge from one another, or to learn about their local environments through time.

This distortion of the relationships between human cultures and the rest of the natural world is what I call "cultural parallax of the wilderness concept." If you remember your photography or astronomy lessons, *parallax* is the apparent displacement of an observed object due to the difference between two points of view. For example, consider the difference between the view of an object as seen through a camera lens and the view through a separate viewfinder. A cultural parallax, then, might be considered to be the difference in views between those who are actively participating in the dynamics of the habitats within their home range and those who view those habitats as "landscapes" from the outside. As Leslie Silko (1987:84) has suggested: "So long as human consciousness remains *within* the hills, canyons, cliffs, and the plants, clouds, and sky, the term *landscape,* as it has entered the English language, is misleading. 'A portion of territory the eye can comprehend in a single view' does not correctly describe the relationship between a human being and his or her surroundings."

Adherents of the romantic notion of landscape claim that the most pristine and therefore most favorable condition of the American continent worthy of reconstruction is that which prevailed at the moment of European colonization. As William Denevan (1992) has amply documented, the continent was perhaps most intensively managed by Native Americans for the several centuries prior to Columbus's arrival in the West Indies. Because European diseases decimated native populations through the Americas over the following hundred and fifty years, the early European colonists saw only vestiges of these managed habitats, if they recognized them as managed at all (Cronon 1983; Denevan 1992).

And yet, among many ecologists, including Daniel Botkin (1990:195), "the idea is to create natural areas that appear as they did when first viewed by European explorers. In the Americas, this would be the landscape of the seventeenth century. . . . If natural means simply *before human intervention,* then all these habitats could be claimed as natural." Thus Botkin equates the periods prior to European colonization with those prior to *human intervention* in the landscape and assumes that all habitats were equally pristine at that time. By this logic, either the pre-Columbian inhabitants of North America were not human or they did not significantly interact with the biota of the areas where they resided.

Human influences on North American habitats began at least 9,200 years prior to the period Botkin pinpoints—when newly arrived "colonists" came down from the Bering Strait into ice-free country (Janzen and Martin 1982; Martin 1986). Regardless of how major a role humans played in the Pleistocene extinctions, the loss of 73 percent of the North American genera of terrestrial mammals weighing one hundred pounds or more precipitated major changes in vegetation and wildlife abundance. By Paul Martin's (1986) criteria, North American wilderness areas have been lacking "completeness" for over ten millennia and would require the introduction of large herbivores from other continents to simulate the "natural conditions" comparable to those under which vegetation cover evolved over the hundreds of thousands of years prior to these extinctions.

It has always amazed me that many of the same scholars who are willing to grant pre-Columbian cultures of the Americas more ecological wisdom than recent European colonists still deny the possibility that these cultures could have played a role in these faunal extinctions, as if that wisdom did not take centuries to accumulate. Do they believe that the pre-Columbian cultures of North America became "instant natives" incapable of overtaxing any resources in their newfound homeland—an incapability that few European cultures have achieved since arriving in the Americas five centuries ago? As Michael Soulé (1991:746) has pointed out, "the most destructive cultures, environmentally, appear to be those that are colonizing uninhabited territory and those that are in a stage of rapid cultural (often technological) transition." My point is simply this: it may take time for any culture to become truly "native," if that term is to imply any sensitivity to the ecological constraints of its home ground.

I am not arguing that many indigenous American cultures did not develop increasing sensitivity to the plant and animal populations most vulnerable to depletion within their home ranges. To the contrary, I would like to suggest that all of pre-Columbian North America was not pristine wilderness for the very reason that many indigenous cultures actively managed habitats and plant populations within their home ranges as a response to earlier episodes of overexploitation. There is now abundant evidence that hundreds of thousands of acres in various bioregions of North America were actively managed by indigenous cultures (Anderson and Nabhan 1991; Denevan 1992; Fish et al. 1985). This does not mean that the entire continent was a Garden of Eden cultivated by Native Americans, as Hecht and Posey have erroneously implied for the Amazon (Parker 1992). Many large areas of the North American continent remained beyond the influence of human cultures, and should remain so. Nevertheless, it is clear that the degree to which North American plant populations were consciously managed—and conserved—by local cultural traditions has been routinely underestimated.

Hohokam farmers, for example, constructed over seventeen hundred miles of prehistoric irrigation canals along the Salt River in the Phoenix basin and intensively cultivated and irrigated floodplains along the Santa Cruz and Gila rivers, as well as on intermittent watercourses for one hundred and fifty miles south and west of these perennial streams (Nabhan 1989). At the same time, they cultivated agave relatives of the tequila plant over hundreds of square miles of upland slopes and terraces beyond where modern agricultural techniques allow crops to be cultivated today (Fish et al. 1985). Nevertheless, the discovery of native domesticated agaves being grown on a large scale in the Sonoran Desert has been made only within the last decade, despite more than a half century of intensive archaeological investigation in the region. Earlier archaeologists had simply never imagined that pre-Columbian cultures in North America could have cultivated perennial crops on such a scale away from riverine irrigation sources.

In the deserts of southern California, indigenous communities transplanted and managed palms for their fruits and fiber in artificial oases, some of them apparently beyond the "natural distribution" of the California fan palm. Control burns were part of their management of these habitats, and such deliberate use of fire created artificial savannas in regions as widely separated as the California Sierra and the Carolinas. In the Yosemite area, where John Muir claimed that "Indians walked softy and hurt the landscape hardly more than the birds and squirrels," Anderson's (1993) reconstructions of Miwok subsistence ecology demonstrate that the very habitat mosaic he attempted to preserve as wilderness was in fact the cumulative result of Miwok burning, pruning, and selective harvesting over the course of centuries.

So what Muir called wilderness, the Miwok called home; the parallax is apparent again. Is it not odd that after ten to fourteen thousand years of indigenous cultures making their homes in North America, Europeans moved in and hardly noticed that the place looked "lived-in"? There are perhaps two explanations for this failure. One response, as historians William Cronon (1983), William Denevan (1992), and Henry Dobyns (1983) have suggested, is that many previously managed landscapes had been left abandoned between the time when European-introduced diseases spread through the Americas and the time when Europeans actually set foot in second-growth forests, shrub-invaded savannas, or defaunated deserts. The second explanation is that Europeans were so intent on taking possession of these lands and developing them in their own manner that they hardly paid attention to signs that the land had already been managed on a different scale and level of intensity.

It was easier for Europeans to assume possession of a land they considered to be virgin or at least unworked and uninhabited by people of their equal. Columbus himself had set out to discover unspoiled lands where the seeds of Christianity—a faith that was being corrupted in Europe, he felt—could be transplanted. In 1502, well after his own men had unleashed European weeds, diseases, and weapons on the inhabitants of the Americas, Columbus wrote to Pope Alexander VI claiming that he had personally visited the Garden of Eden on his voyages to the New World.

Even those who have condemned Europeans for the effects of their ecological imperialism on indigenous American cultures too often frame their concern as "conquistadors raping a virgin land." As "subjects of rape," American lands and their resident human populations are simply reduced to the role of passive victims, incapable of any resilience or dynamic response to deal capably in any way with such invasions. And so we are often left hearing the truism, "Before the White Man came, North America was essentially a wilderness where the few Indian inhabitants lived in constant harmony with nature"—even though four to twelve million people speaking two hundred languages variously burned, pruned, hunted, hacked, cleared, irrigated, and planted in an astonishing diversity of habitats for centuries (Denevan 1992; Anderson and Nabhan 1991). And we are supposed to believe, as well, that they all lived in some static homeostasis with all the various plants and animals they encountered.

As Daniel Botkin (1990) has convincingly argued, few predator/prey or plant/animal relationships have maintained any long-term homeostasis even where humans are not present, let alone where they are. Although there is little evidence to indicate that indigenous cultures regionally extirpated any rare plants or small vertebrate species, there are intriguing signs that certain prehistoric populations depleted local firewood sources and certain slow-growing fiber plants such as yuccas (Minnis 1981). Because different cultures used native species at different intensities, and each species has a different growth rate and relative abundance, it is impossible to generalize about the conservation of all resources in all places. Nevertheless, numerous localized efforts to sustain or enhance the abundance of certain useful plants have been well documented (Anderson 1993; Anderson and Nabhan 1991). It remains unclear whether by favoring certain useful plants over others, plant diversity increased or decreased in particular areas. To my knowledge, no study adequately addresses indigenous peoples' local effects on biological *diversity* (as opposed to their effects on the *abundance* of key resources).

It can no longer be denied that some cultures had specific conservation practices to sustain plant populations of economic or symbolic importance to their communities. In the case of my O'odham neighbors in southern Arizona, there have been efforts to protect rare plants from overharvesting near sacred sites, to transplant individuals to more protected sites, and to conserve caches of seeds in caves to ensure future supplies (Nabhan 1989). Landscape photographer Mark Klett (1990:73) has written that too often wilderness in the European American tradition is "an entity defined by our absence [as if] the landscape does best without our presence." I find in O'odham oral literature an interesting counterpoint to this notion. The O'odham term for wildness, *doajkam,* is etymologically tied to terms for health, wholeness, and liveliness (Mathiot 1973). While it seems wildness is positively valued as an ideal by which to measure other conditions, the O'odham also feel that certain plants, animals, and habitats "degenerate" if not properly cared for. Thus their failure to take care of a horse or a crop may allow it to go feral, but this degenerated feral state is different from being truly wild. Similarly, their lack of attention to O'odham fields, watersheds, and associated ceremonies may keep the rains from providing sufficient moisture to sustain both wild and cultivated species. Many O'odham express humility in the face of unpredictable rains or game animals, but they still feel a measure of responsibility in making good use of what does come their way.

In short, the O'odham elders I know best still behave as active participants in the desert without assuming that they are ultimately "in control" of it. This, in essence, is the difference between participating in *untrammeled* wilderness (as defined by the U.S. Wilderness Act) and attempting to tame lands through manipulative management. (A *trammel* is a device which shackles, hobbles, cages, or confines an animal, breaking its spirit and capacity to roam.) What may look like uninhabited wilderness to outsiders is a habitat in which the O'odham actively participate. They do not define the desert as it was derived from the Old French *desertus,* "a place abandoned or left wasted." Their term for the desert, *tohono,* can be etymologically understood as a "bright and shining place," and they have long called themselves the Tohono O'odham: the people belonging to that place. They share that place with a variety of plants and animals, a broad range of which still inhabit their oral literature (Nabhan and St. Antoine 1993).

Within their Sonoran Desert homeland, many O'odham people still learn certain traditional land management scripts encoded in their own Piman language, which they then put into practice in particular settings, each in their own peculiar way. What concerns me is that their indigenous language is now being replaced by English. Their indigenous science of desert is being eclipsed by more frequent exposure to Western science. And their internal or etic sense of what it is to be O'odham is being replaced by the mass media's presentation of what it is to be (generically) an American Indian. While I will be among the first to admit that change is presumably inherent to all natural and cultural phenomena, I am not convinced that these three changes are necessarily desirable. In virtually every culture I know of on this continent, similar changes are occurring with blinding speed. Both nature and culture are being rapidly redefined, not so much by what we learn from our immediate surroundings, as by what we learn through the airwaves.

Let me highlight what Sara St. Antoine and I recently learned while interviewing fifty-two children from four different cultures, all of them living in the Sonoran Desert (Nabhan and St. Antoine 1993). Essentially we learned that with regard to knowledge about the natural world, intergenerational differences within cultures are becoming as great as the gaps between cultures. While showing a booklet of drawings of *common* desert plants and animals to O'odham children and their grandparents, for example, we realized that the children knew only a third of the names for these desert organisms in their native language that their grandparents knew. With the loss of those names, we wonder how much culturally encoded knowledge is lost as well. With over half the two hundred native languages on this continent falling out of use at an accelerating rate, a great diversity of perspectives on the structure and value of nature are surely being lost. And culture-specific land management practices are being lost as well.

One driving force in this loss of knowledge about the natural world is that children today spend more time in classrooms and in front of the television than they do directly interacting with their natural surroundings. The vast majority of the children we interviewed are now gaining most of their knowledge about other organisms vicariously: 77 percent of the Mexican children, 61 percent of the Anglo children, 60 percent of the Yaqui children, and 35 percent of the O'odham children told us they had seen more animals on television and in the movies than they had personally seen in the wild.

An even more telling measure of the lack of primary contact with their immediate nonhuman surroundings is this: a significant portion of kids today have never gone off alone, away from human habitations, to spend more than a half hour by themselves in a "natural" setting. None of the six Yaqui children responded that they had; nor had 58 percent of the O'odham, 53 percent of the Anglos, and 71 percent of the Mexican children. We also found that many children today have never been involved in collecting, carrying around, or playing with the feathers, bones, butterflies, or stones they find near their homes. Of those interviewed, 60 percent of the Yaqui children, 46 percent of the Anglos, 44 percent of the Mexicans, and 35 percent of the O'odham had never gathered such natural treasures. Such a paucity of contact with the natural world would have been unimaginable even a century ago, but it will become the norm as more than 38 percent of the children born after the year 2000 are destined to live in cities with more than a million other inhabitants. While few cities are entirely devoid of open spaces, manufactured toys and prefabricated electronic images have rapidly replaced natural objects as common playthings.

However varied the views of the natural world held by the myriad ethnic groups which have inhabited this continent, many of them are now converging on a new view—not so much one of experienced participants dynamically involved with their local environment as one in which they too may feel as though they are outside the frame looking in. Because only a small percentage of humankind has any direct, daily engagement with other species of animals and plants in their habitats, we have arrived at a new era in which ecological illiteracy is the norm. I cannot help concluding that we will soon be losing the many ways in which cultural diversity may have formerly enriched the biological diversity of various habitats of this continent. I can only hope that our children will pay more attention to this warning from Mary Midgley (1978:246) than we have:

> Man is not adapted to live in a mirror-lined box, generating his own electric light and sending for selected images from outside when he needs them. Darkness and bad smell are all that can come from that. We need a vast world, and it must be a world that does not need us; a world constantly capable of surprising us, a world we did not program, since only such a world is the proper object of wonder.

BIBLIOGRAPHY

Anderson, Kat, and Gary Paul Nabhan. "Gardeners in Eden." *Wilderness* 55(194) (1991):27–30.

Anderson, M. Kathleen. "The Experimental Approach to the Assessment of the Potential Ecological Effects of Horticultural Practices by Indigenous Peoples on California Wildlands." Berkeley: University of California Ph.D. dissertation, 1993.

Blackburn, Thomas C., and Kat Anderson. "Introduction: Making the Domesticated Environment." In Thomas C. Blackburn and Kat Anderson (eds.), *Before the Wilderness: Environment Management by Native Californians.* Menlo Park: Ballena Press, 1993.

Botkin, Daniel. *Discordant Harmonies.* Oxford: Oxford University Press, 1990.

Clar, C. R. *California Government and Forestry from Spanish Days Until the Creation of the Department of Natural Resources in 1927*. Sacramento: California Division of Forestry, 1959.

Cronon, William. *Changes in the Land: Indians, Colonists, and the Ecology of New England*. New York: Hill & Wang, 1983.

Denevan, William M. "The Pristine Myth: The Landscape of the Americas in 1492." *Annals of the Association of American Geographers* 82(3) (1992):369–385. [Included in this volume.]

Diamond, Jared. "The Environmentalist Myth: Archaeology." *Nature* 324 (1986):19–20.

———. *The Third Chimpanzee*. New York: HarperCollins, 1992.

———. "New Guineans and Their Natural World." In Stephen Kellert and Edward O. Wilson (eds.), *The Biophilia Hypothesis*. Washington, D.C.: Island Press, 1993.

Dobyns, Henry F. *Their Numbers Become Thinned: Native American Population Dynamics in Eastern North America*. Knoxville: University of Tennessee Press, 1983.

Fish, Suzanne K., Paul R. Fish, Charles Miksicek, and John Madsen. "Prehistoric Agave Cultivation in Southern Arizona." *Desert Plants* 7(2) (1985):107–112.

Flores, Mike, Fernando Valentine, and Gary Paul Nabhan. "Managing Cultural Resources in Sonoran Desert Biosphere Reserves." *Cultural Survival Quarterly* 14(4) (1990):26–30.

Gómez-Pompa, Arturo, and Andrea Kaus. "Taming the Wilderness Myth." *BioScience* 42 (1992):271–279. [Included in this volume.]

Janzen, Daniel H., and Paul S. Martin. "Neotropical Anachronisms: Fruits the Gomphotheres Ate." *Science* 215 (1982):19–27.

Klett, Mark. "The Legacy of Ansel Adams." *Aperture* 120 (1990):72–73.

Martin, Paul S. "Refuting Late Pleistocene Extinction Models." In D. K. Eliot (ed.), *Dynamics of Extinction*. New York: Wiley, 1986.

Mathiot, Madeleine. *A Dictionary of Papago Usage*. Language Science Monograph 8(1). Bloomington: Indiana University Publications, 1973.

Midgley, Mary. *Beast and Man*. Ithaca: Cornell University Press, 1978.

Minnis, Paul S. "Economic and Organizational Responses to Food Stress by Non-stratified Societies: A Prehistoric Example." Ann Arbor: University of Michigan Ph.D. dissertation, 1981.

Nabhan, Gary Paul. *Enduring Seeds*. San Francisco: North Point, 1989.

Nabhan, Gary Paul, and Sara St. Antoine. "The Loss of Floral and Faunal Story: The Extinction of Experience." In Stephen R. Kellert and Edward O. Wilson (eds.), *The Biophilia Hypothesis*. Washington, D.C.: Island Press, 1993.

National Park Service. "Revised Code of Federal Regulation for the National Park Service." *Federal Register* 52(14) (1987):2457–2458.

Oelschlaeger, Max. *The Idea of Wilderness*. New Haven: Yale University Press, 1991.

Parker, Eugene. "Forest Islands and Kayapo Resource Management in Amazonia: A Reappraisal of the *Apete.*" *American Anthropologist* 94 (1992):406–427.
Rea, Amadeo R. "Resource Utilization and Food Taboos of Sonoran Desert Peoples." *Journal of Ethnobiology* 2 (1981):69–83.
Redford, Kent H. "The Ecologically Noble Savage." *Orion* 9(3) (1985):24–29.
Redford, Kent H., and John G. Robinson. "The Game of Choice: Patterns of Indian and Colonist Hunting in the Neotropics." *American Anthropologist* 89(3) (1987): 650–667.
Silko, Leslie M. "Landscape, History, and the Pueblo Imagination." In Daniel Halpern, ed., *On Nature.* San Francisco: North Point, 1987.
Soulé Michael E. "Conservation: Tactics for a Constant Crisis." *Science* 253 (1991): 744–750.

Gary Snyder

The Rediscovery of Turtle Island (1995)

For John Wesley Powell, watershed visionary, and for Wallace Stegner

I

We human beings of the developed societies have once more been expelled from a garden—the formal garden of Euro-American humanism and its assumptions of human superiority, priority, uniqueness, and dominance. We have been thrown back into that other garden with all the other animals and fungi and insects, where we can no longer be sure we are so privileged. The walls between "nature" and "culture" begin to crumble as we enter a posthuman era. Darwinian insights force occidental people, often unwillingly, to acknowledge their literal kinship with critters.

Ecological science investigates the interconnections of organisms and their constant transactions with energy and matter. Human societies come into being along with the rest of nature. There is no name yet for a humanistic scholarship that embraces the nonhuman. I suggest (in a spirit of pagan play) we call it "panhumanism."

Environmental activists, ecological scientists, and panhumanists are still in the process of reevaluating how to think about, how to create policy with, nature. The professional resource managers of the Forest Service and the Bureau of Land Management have been driven (partly by people of conscience within their own ranks) into rethinking their old utilitarian view of the vast lands in their charge. This is a time of lively confluence, as scientists, self-taught ecosystem experts from the communities, land management agency experts, and a new breed of ecologically aware loggers and ranchers (a few, but growing) are beginning to get together.

In the more rarefied world of ecological and social theory, the confluence is rockier. Nature writing, environmental history, and ecological philosophy have become subjects of study in the humanities. There are, however, still a few otherwise humane historians and philosophers who unreflectingly assume that the natural world is primarily a building-supply yard for human projects. That is what the Occident has said and thought for a couple thousand years.

Right now there are two sets of ideas circling about each other. One group, which we could call the "Savers," places value on extensive preservation of wilderness areas and argues for the importance of the original condition of nature. This view has been tied to the idea that the mature condition of an ecosystem is a stable and diverse state technically called "climax." The other position holds that nature is constantly changing, that human agency has altered things to the point that there is no "natural condition" left, that there is no reason to value climax (or "fitness") over any other succession phase, and that human beings are not only part of nature but that they are also dominant over nature and should keep on using and changing it. They can be called the "Users." The Savers' view is attributed to the Sierra Club and other leading national organizations, to various "radical environmentalists," and to many environmental thinkers and writers. The Users' view, which has a few supporters in the biological sciences, has already become a favorite of the World Bank and those developers who are vexed by the problems associated with legislation that requires protection for creatures whose time and space are running out. It has been quickly seized on by the industry-sponsored pseudopopulist-flavored "Wise Use" movement.

Different as they are, both groups reflect the instrumentalist view of nature that has long been a mainstay of occidental thought. The Savers' idea of freezing some parts of nature into an icon of "pristine, uninhabited wilderness" is also to treat nature like a commodity, kept in a golden cage. Some preservationists have been insensitive to the plight of indigenous peoples whose home grounds were turned into protected wildlife preserves or parks, or to the plight of local workers and farmers who lose jobs as logging and grazing policies change.

The Users, in turn, are both pseudopopulist and multinational. On the local level they claim to speak for communities and workers (whose dilemma is real enough), but a little probing discloses industry funding. On the global scale their backers line up with huge forces of governments and corporations, with NAFTA and GATT, and raise the specter of further destruction of local communities. Their organizations are staffed by the sort of professionals whom Wendell Berry calls "hired itinerant vandals."

Postmodern theoreticians and critics have recently ventured into nature politics. Many of them have sided with the Users—they like to argue that nature is part of history, that human beings are part of nature, that there is little in the natural world that has not already been altered by human agency, that in any case our idea of "nature" is a projection of our social condition and that there is no sense in trying to preserve a theoretical wild. However, to say that the natural world is subject to continual change, that nature is shaped by history, or that our idea of reality is a self-serving illusion is not new. These positions still fail to come to grips with the question of how to deal with the pain and distress of real beings, plants and animals, as real as suffering humanity; and how to preserve natural variety. The need to protect worldwide biodiversity may be economically difficult and socially controversial, but there are strong scientific and practical arguments in support of it, and it is for many of us a profound ethical issue.

Hominids have obviously had some effect on the natural world, going back for half a million or more years. So we can totally drop the use of the word *pristine* in regard to nature as meaning "untouched by human agency." "Pristine" should now be understood as meaning "virtually" pristine. Almost any apparently untouched natural environment has in fact experienced some tiny degree of human impact. Historically there were huge preagricultural environments where the human impact, rather like deer or cougar activities, was normally almost invisible to any but a tracker's eye. The greatest single preagricultural human effect on wild nature, yet to be fully grasped, was the deliberate use of fire. In some cases human-caused fire seemed to mimic natural process, as with deliberate use of fire by native Californians. Alvar Núñez "Cabeza de Vaca," in his early-sixteenth-century walk across what is now Texas and the Southwest, found well-worn trails everywhere. But the fact still remains that there were great numbers of species, vast grasslands, fertile wetlands, and extensive forests in mosaics of all different stages in the preindustrial world. Barry Commoner has said that the greatest destruction of the world environment—by far—has taken place since 1950.

Furthermore, there is no "original condition" that once altered can never be redeemed. Original nature can be understood in terms of the myth of the "pool of Artemis"—the pool hidden in the forest that Artemis, goddess of wild things, visits to renew her virginity. The wild has—nay, *is*—a kind of hip, renewable virginity.

We are still laying the groundwork for a "culture of nature." The critique of the Judeo-Christian-Cartesian view of nature (by which complex of views all developed nations excuse themselves for their drastically destructive treatment of the landscape) is well under way. Some of us would hope to resume, reevaluate, re-create, and bring into line with complex science that old view that holds the whole phenomenal world to be our own being: multicentered, "alive" in its own manner, and effortlessly self-organizing in its own chaotic way. Elements of this view are found in a wide range of ancient vernacular philosophies, and it turns up in a variety of more sophisticated but still tentative forms in recent thought. It offers a third way, not caught up in the dualisms of body and mind, spirit and matter, or culture and nature. It is a noninstrumentalist view that extends intrinsic value to the nonhuman natural world.

Scouting parties are now following a skein of old tracks, aiming to cross and explore beyond the occidental (and postmodern) divide. I am going to lay out the case history of one of these probes. It's a potentially new story for the North American identity. It has already been in the making for more than thirty years. I call it "the rediscovery of Turtle Island."

II

In January 1969 I attended a gathering of Native American activists in southern California. Hundreds of people had come from all over the West. After sundown we went out to a gravelly wash that came down from the desert mountains. Drums were set up, a fire started, and for most of the night we sang the pantribal songs called "forty-nines." The night conversations circled around the idea of a native-inspired cultural and ecological renaissance for all of North America. I first heard this continent called "Turtle Island" there by a man who said his work was to be a messenger. He had his dark brown long hair tied in a Navajo men's knot, and he wore dusty khakis. He said that Turtle Island was the term that the people were coming to, a new name to help us build the future of North America. I asked him whom or where it came from. He said, "There are many creation myths with Turtle, East Coast and West Coast. But also you can just hear it."

I had recently returned to the West Coast from a ten-year residence in Japan. It was instantly illuminating to hear this continent renamed "Turtle Island." The realignments that conversation suggested were rich and complex. I was reminded that the indigenous people here have a long history of subtle and effective ways of working with their home grounds. They have had an exuberant variety of cultures and economies and some distinctive social forms (such as communal households) that were found throughout the hemisphere. They sometimes fought with each other, but usually with a deep sense of mutual respect. Within each of their various forms of religious life lay a powerful spiritual teaching on the matter of human and natural relationships, and for some individuals a practice of self-realization that came with trying to see through nonhuman eyes. The landscape was intimately known, and the very idea of community and kinship embraced and included the huge populations of wild beings. Much of the truth of Native American history and culture has been obscured by the self-serving histories that were written on behalf of the conquerors, the present dominant society.

This gathering took place one year before the first Earth Day. As I reentered American life during the spring of 1969, I saw the use of the term "Turtle Island" spread through the fugitive Native American newsletters and other communications. I became aware that there was a notable groundswell of white people, too, who were seeing their life in the Western Hemisphere in a new way. Many whites figured that the best they could do on behalf of Turtle Island was to work for the environment, reinhabit the urban or rural margins, learn the landscape, and give support to Native Americans when asked. By 1970 I had moved with my family to the Sierra Nevada and was developing a forest homestead north of the South Yuba River. Many others entered the mountains and hills of the Pacific slope with virtually identical intentions, from the San Diego backcountry north into British Columbia. They had begun the reinhabitory move.

Through the early seventies I worked with my local forest community, but made regular trips to the cities, and was out on long swings around the country reading poems or leading workshops—many in urban areas. Our new sense of the Western Hemisphere permeated everything we did. So I called the book of poems I wrote from that period *Turtle Island* (New York: New Directions, 1974). The introduction says:

> Turtle Island—the old-new name for the continent, based on many creation myths of the people who have been living here for millennia, and reapplied by some of them to "North America" in recent years. Also, an idea found worldwide, of the earth, or cosmos even, sustained by a great turtle or serpent-of-eternity.
>
> A name: that we may see ourselves more accurately on this continent of watersheds and life communities—plant zones, physiographic provinces, culture areas, following natural boundaries. The "U.S.A." and its states and counties are arbitrary and inaccurate impositions on what is really here.
>
> The poems speak of place, and the energy pathways that sustain life. Each living being is a swirl in the flow, a formal turbulence, a "song." The land, the planet itself, is also a living being—at another pace. Anglos, black people, Chicanos, and others beached up on these shores all share such views at the deepest levels of their old cultural traditions—African, Asian, or European. Hark again to those roots, to see our ancient solidarity, and then to the work of being together on Turtle Island.

Following the publication of these poems, I began to hear back from a lot of people—many in Canada—who were remaking a North American life. Many other writers got into this sort of work each on his or her own—a brilliant and cranky bunch that included Jerry Rothenberg and his translation of Native American song and story into powerful little poem events; Peter Blue Cloud with his evocation of Coyote in a contemporary context; Dennis Tedlock, who offered a storyteller's representation of Zuni oral narrative in English; Ed Abbey, calling for a passionate commitment to the wild; Leslie Silko in her shivery novel *Ceremony;* Simon Ortiz in his early poems and stories—and many more.

A lot of this followed on the heels of the back-to-the-land movement and the diaspora of longhairs and dropout graduate students to rural places in the early seventies. There are thousands of people from those days still making a culture: being teachers, plumbers, chair and cabinet makers, contractors and carpenters, poets in the schools, auto mechanics, geographic information computer consultants, registered foresters, professional storytellers, wildlife workers, river guides, mountain guides, architects, or organic gardeners. Many have simultaneously mastered grassroots politics and the intricacies of public lands policies. Such people can be found tucked away in the cities, too.

The first wave of writers mentioned left some strong legacies: Rothenberg, Tedlock, and Dell Hymes gave us the field of ethnopoetics (the basis for truly appreciating multicultural literature); Leslie Silko and Simon Ortiz opened the way for a distinguished and diverse body of new American Indian writing; Ed Abbey's eco-warrior spirit led toward the emergence of the radical environmental group Earth First!, which (in splitting) later generated the Wildlands Project. Some of my own writings contributed to the inclusion of Buddhist ethics and lumber industry work life in the mix, and writers as different as Wes Jackson, Wendell Berry, and Gary Paul Nabhan opened the way for a serious discussion of place, nature in place, and community. The Native American movement has become a serious player in the national debate, and the environmental movement has become (in some cases) big and controversial politics. Although the counterculture has faded and blended in, its fundamental concerns remain a serious part of the dialogue.

A key question is that of our ethical obligations to the nonhuman world. The very notion rattles the foundations of occidental thought. Native American religious beliefs, although not identical coast to coast, are overwhelmingly in support of a full and sensitive acknowledgment of the subjecthood—the intrinsic value—of nature. This in no way backs off from an unflinching awareness of the painful side of wild nature, of acknowledging how everything is being eaten alive. The twentieth-century syncretism of the "Turtle Island view" gathers ideas from Buddhism and Taoism and from the lively details of worldwide animism and paganism. There is no imposition of ideas of progress or order on the natural world—Buddhism teaches impermanence, suffering, compassion, and wisdom. Buddhist teachings go on to say that the true source of compassion and ethical behavior is paradoxically none other than one's own realization of the insubstantial and ephemeral nature of everything. Much of animism and paganism celebrates the actual, with its inevitable pain and death, and affirms the beauty of the process. Add contemporary ecosystem theory and environmental history to this, and you get a sense of what's at work.

Conservation biology, deep ecology, and other new disciplines are given a community constituency and real grounding by the bioregional movement. Bioregionalism calls for commitment to this continent *place by place,* in terms of biogeographical regions and watersheds. It calls us to see our country in terms of its landforms, plant life, weather patterns, and seasonal changes—its whole natural history before the net of political jurisdictions was cast over it. People are challenged to become "reinhabitory"—that is, to become people who are learning to live and think "as if" they were totally engaged with their place for the long future. This doesn't mean some return to a primitive lifestyle or utopian provincialism; it simply implies an engagement with community and a search for the sustainable sophisticated mix of economic practices that would enable people to live regionally and yet learn from and contribute to a planetary society. (Some of the best bioregional work is being done in cities, as people try to restore both human and ecological neighborhoods.) Such people are, regardless of national or ethnic backgrounds, in the process of becoming something deeper than "American (or Mexican or Canadian) citizens"—they are becoming natives of Turtle Island.

Now in the nineties the term "Turtle Island" continues, modestly, to extend its sway. There is a Turtle Island Office that moves around the country with its newsletter; it acts as a national information center for the many bioregional groups that every other year hold a "Turtle Island Congress." Participants come from Canada and Mexico as well as the United States. The use of the term is now standard in a number of Native American periodicals and circles. There is even a "Turtle Island String Quartet" based in San Francisco. In the winter of 1992 I practically convinced the director of the Centro de Estudios Norteamericanos at the Universidad de Alcalá in Madrid to change his department's name to "Estudios de la Isla de Tortuga." He much enjoyed the idea of the shift. We agreed: speak of the United States, and you are talking two centuries of basically English-speaking affairs; speak of "America" and you invoke five centuries of Euro-American schemes in the Western Hemisphere; speak of "Turtle Island" and a vast past, an open future, and all the life communities of plants, humans, and critters come into focus.

III

The Nisenan and Maidu, indigenous people who live on the east side of the Sacramento Valley and into the northern Sierra foothills, tell a creation story that goes something like this:

> Coyote and Earthmaker were blowing around in the swirl of things. Coyote finally had enough of this aimlessness and said, "Earthmaker, find us a world!"
>
> Earthmaker tried to get out of it, tried to excuse himself, because he knew that a world can only mean trouble. But Coyote nagged him into trying. So leaning over the surface of the vast waters, Earthmaker called up Turtle. After a long time Turtle surfaced, and Earthmaker said, "Turtle, can you get me a bit of mud? Coyote wants a world."
>
> "A world," said Turtle. "Why bother? Oh, well." And down she dived. She went down and down and down, to the bottom of the sea. She took a great gob of mud, and started swimming toward the surface. As she spiraled and paddled upward, the streaming water washed the mud from the sides of her mouth, from the back of her mouth—and by the time she reached the surface (the trip took six years), nothing was left but one grain of dirt between the tips of her beak.
>
> "That'll be enough!" said Earthmaker, taking it in his hands and giving it a pat like a tortilla. Suddenly Coyote and Earthmaker were standing on a piece of ground as big as a tarp. Then Earthmaker stamped his feet, and they were standing on a flat wide plain of mud. The ocean was gone. They stood on the land.

And then Coyote began to want trees and plants, and scenery, and the story goes on to tell how Coyote imagined landscapes that then came forth, and how he started naming the animals and plants as they appeared. "I'll call you skunk because you look like skunk." And the landscapes Coyote imagined are there today.

My children grew up with this as their first creation story. When they later heard the Bible story, they said, "That's a lot like Coyote and Earthmaker." But the Nisenan story gave them their own immediate landscape, complete with details, and the characters were animals from their own world.

Mythopoetic play can be part of what jump-starts long-range social change. But what about the short term? There are some immediate outcomes worth mentioning: a new era of community interaction with public lands has begun. In California a new set of ecosystem-based government/

community joint-management discussions are beginning to take place. Some of the most vital environmental politics is being done by watershed or ecosystem-based groups. "Ecosystem management" by definition includes private landowners in the mix. In my corner of the northern Sierra, we are practicing being a "human-inhabited wildlife corridor"—an area that functions as a biological connector—and are coming to certain agreed-on practices that will enhance wildlife survival even as dozens of households continue to live here. Such neighborhood agreements would be one key to preserving wildlife diversity in most Third World countries.

Ultimately we can all lay claim to the term *native* and the songs and dances, the beads and feathers, and the profound responsibilities that go with it. We are all indigenous to this planet, this mosaic of wild gardens we are being called by nature and history to reinhabit in good spirit. Part of that responsibility is to choose a place. To restore the land one must live and work in a place. To work in a place is to work with others. People who work together in a place become a community, and a community, in time, grows a culture. To work on behalf of the wild is to restore culture.

Val Plumwood

Wilderness Skepticism and Wilderness Dualism (1998)

MULTIPLE MEANINGS OF WILDERNESS IN AUSTRALIA

On my wall wherever I go I put a poster of a place I know I probably will never reach, but whose contemplation gives me intense delight. It shows a curving bay of white sand edged with the intricate lacework of many lines of breakers. Behind this bay of shining, surging waters, the forest-clad contours of the Ironbound Range rise steeply; light filtering through overhanging clouds perfectly illuminates the foreground headlands, whose dark, textured green indicates low thickets of human-impermeable tea-tree. Shadowed in the background, the deep-folded rain forests of beech and leatherwood, botanically linked to those of New Zealand and Chile, speak of the vastness of geological time, when these continents were fused into one great southern continent. This uttermost place, shaped by the immense forces of the earth, carved by the glaciers, by the tearing plates and the great swells rolling in off the stormy Southern Ocean, seems to lie well beyond the limits of any imperium. Beyond, to the south, lies only Antarctica; to the north, fifty miles over the trackless, forbidding Arthur Range, is the nearest road; to the west and east, bay follows bay along this fretted coastline, accessible only to the best-motivated travelers prepared for many days of wet walking.

This place is Prion Bay, on Tasmania's south coast track, part of the magnificent World Heritage area which makes up the southwest part of this heart-shaped island. I walked in near to Prion Bay once, but was persuaded, by a solo woman I met coming the other way making her seventh trip over the track, that no one should attempt a solo walk in this area unless they already knew the terrain. I turned back with some sadness, for I had no available companions for the trip, which was a long-held goal. But the beauty I did see there, as well as the beauty I did not, returns to me often, giving comfort and hope in a world which otherwise seems to offer little of either. The poster reminds me of what is beyond, of the limit and boundary of the mysterious wild Other, of what will not be penetrated and controlled. It also reminds me of an intimate and physical bond of knowledge with the Earth, through a form of conversation with its great, laboriously inscribed body, which can only be entered into through the answering effort of our *human* bodies as we walk within it. In these places travelers seek the wisdom of the land, carry survival on their backs, and measure themselves as limited and only half-hardy animals. Here you can still come to spend what could be some of the most intensely alive but most humbling days of your life.

Two hundred years ago, you might have seen other women making their way along this coast. The hardy Ninene people who built their dwellings on the remote western tip of this coast were also travellers in this place, making a regular trading passage through along the south coast to meet up with the folk from the southeast coast (Ryan 1996). We might imagine Ninene women prior to invasion walking this country with a strong and confident step. As prodigious divers and swimmers, gatherers of lobster and shellfish, and hunters of seals and other animals (Bonwick 1870), they played a major role in supplying their community's food. The visiting French scientist Labillardiere, from d'Entrecasteaux's expedition, was a follower of Rousseau, and recorded in his diary of 22 January 1793 his disapproval of this role and of the local tribe's failure to conform to Rousseau's natural gender arrangements (Ryan 1996). Women, he felt, should take up their natural role as dependent caregivers, properly concerned with children and the family shelter, while men went off to hunt food for their families. The land too was active and insurgent in a way the visiting Frenchman could not imagine. All this was before the removal of the Tasmanians in 1833 to the remote island prison camps designed to force them into a cultivating mode of life, the camps in which they weakened and died, calling perhaps in their forbidden languages and looking toward their homes.

A modern European walker acquainted only with the official history of the Aboriginal people of Tasmania might be tempted to conclude that they no longer moved through their country because it was a place that challenged human ability to survive. But we who know history from the Other's side know that the official history of this genocidal episode in the advance of Western civilization is, as usual, a fake. "The Last Tasmanians," slated from the first as a dying race of savage Nature, were victims of Culture, and not of Nature, as nineteenth-century science tried to claim. They were victims of a virulently imperial form of culture, one that held *them* to be Nature, not fully human inhabitants of a space empty of culture, in which they lived as "strangers to every principle of social order" in the words of Peron's 1801 journal entry from Maria Island on the southeast coast. Aborigines did not cultivate the land in European style, and therefore lived, in the estimation of the European visitors, as mere "beasts of the forest" or as "parasites" upon the land. Whether noble, as in the French estimate, or unclothed and unformed, as in the British, they were "children of nature."

European imagination did not begin to comprehend the mutual nourishment of the land and the people that made up the country. Nor did the intruders entertain the possibility that people here might have more subtle practices of nourishing and managing the land than the kind of cultivation they themselves practiced, although they had recognized such elsewhere in their colonial expansion, notably in New Zealand. In the thought which identifies the truly human with the sphere of culture as cultivation, those who inhabit the oppositional sphere they take to be wild land and forest must be less than fully human. Rousseau's counter-cultural thought did not take these "children of nature" to be less than human, and located the truly human in ennobled nature rather than in artificial culture. But that reversal did not prevent Rousseau and his followers from conceiving this sphere of nature in highly problematic, culture-specific, and gender-specific terms. The contrast between the traditional view of humanity as oppositional to nature and the counter-cultural reversal of Rousseau sets a certain pattern which this essay will explore.

The absence of the Aboriginal people of southwest Tasmania is a convenient one for a contemporary ecological concept of wilderness which continues to construct it as an absence of the human, a conception rather too close in some respects to the traditional European concept of wilderness. In the colonial version of this traditional European concept, wilderness is a place waiting to receive, to be filled, a place with no desire or fully human history other than what Western culture imposes upon it. It is a transparent vessel waiting to be filled with the projects of human labor and cultivation, a site both passive and chaotic, open for the settler effort and improvement that will make it, at last, "productive." Aboriginal critics of wilderness have correctly identified this "feminized" concept of wilderness-as-*emptiness* as one of the modern disguises and affirmations of the *terra nullius* concept (Bayet 1993; Langton 1996; Rose 1996), which envisions indigenous people as part of nature rather than culture.

As it was employed in the framework of initial colonization, the wilderness was a place which lacked occupants who were fully human, where humanity was identified in the ethnocentric and androcentric terms of "civilized" humans and their agricultural works. But when the colonizing human group, the biosphere people, the bringers of light, equated their own absence in the lands they sought to conquer with darkness and emptiness, they denied and erased the humanity and cultural agency of the lands' prior human occupants. They denied and erased also the presence and ecological agency of all other *nonhuman occupants,* both in their imposition of regimes of property and cultivation, and in their colonizing depiction of Europe and its biota as the center in relation to which all Others are defined as a lesser form of existence. This oppositional dynamic remains buried within the new green conception of wilderness and, even though wilderness is now conceived as valuable rather than valueless, defeats efforts to remove Eurocentric and anthropocentric meanings. To escape this colonizing dynamic defining Otherness as vacancy, we need to recognize various prior presences. A conceptual precondition for this, I argue, is to move from an "empty" feminized to a "full" feminist conception of the wild Other as potential presence.

WILDERNESS AS OPPOSITIONAL TO ESTABLISHED ORDER

The traditional Western concept of wilderness was a negative one, delineating a sphere of alterity defined by various nature/culture oppositions, from the Judaic ones which opposed wilderness to the sacred space of the garden, to the Greek and Roman ones which opposed wilderness to rational "civilization," *cives* or civic order associated with the beginning of urban life (Harrison 1992). The wilderness is what is left behind by rational civilization and ordered cultivation; it is the supposedly irrational and chaotic sphere represented by the primeval forest, the dangerous shadow place on the other side of the boundary of order, the haunt of "the wild man," of "barbarians" and beasts (Duerr 1985). The resulting concept of rationality and civic order is defined in opposition to the wild land and its deity—Dionysius, the abandoned god who comes from afar to loose chaos and unbind form and social order.

Greek thought makes a place for this wild, irrational power, but it is a place of danger and disruption, the oppositional space of the Other, especially associated, as in the myth of Pentheus, with the irrationality attributed to women and with erotic power. If such concepts of civilization and human identity are defined by exclusion of the wilderness and the woman, both become symbols of the disorder and unreason which must be transcended or swept away for civilized order and true humanity to prevail. In the traditional story, both Western men and women have drawn their identity from the garden in opposition to the wilderness and the barbarian Other, but it is woman's irrepressible alliance with the Other, the wild, which both spoils the garden and is the principal danger to *cives*. Woman is the Other within, who must be confined and prevented, according to the thought of Hegel and Rousseau, from contaminating the public sphere with a dangerous Otherness that threatens its impartiality and justice. Settler elements dissatisfied with this rational order associated with Europe, and its confining and oppositional conception of humanity, have often looked, like Rousseau, to the wilderness, in a reversal of the traditional gesture of rejection, for intimations of another kind of identity. But they also, like Rousseau, fail to question sufficiently the problematic deriving from its traditional oppositional meaning, exemplified by its gender-coding.

If Prion Bay and its surrounds are now revered by some as wilderness, that status is both recent, hard-won, and precarious, a counter to the traditions of Western civilization dominant until very recently. The British soldiers, convicts, and settlers who invaded this coastline two hundred years ago held few doubts about the danger of wilderness, or the need to drive it back, along with its Aboriginal inhabitants. For the convict unfortunates, this was the *howling* wilderness, the Hell of the Roaring Forties, a place that challenged human survival. The imperial mission to civilize the wilderness flourished in Tasmania, and managed to extinguish the largest marsupial carnivore, the Tasmanian Tiger, by the 1930s. Right up until the 1980s, the main developer of the area, the Tasmanian Hydro Electricity Commission (the HEC, or "Hydro"), was a powerful and explicit advocate of the official mission to dominate and subdue the untamed parts of the island. The southwest was spared because thousands of people, mainly young people, who rejected the traditional treatment of wilderness, put their bodies on the line in front of the bulldozers, and because of a wider cultural revolution which tried to give a liberating meaning to wilderness, to revalue natural and native communities as an essential and positive element in our lives.

That revolution of reversal remains seriously flawed and incomplete, in conception as well as execution. But in the Australian case, the struggle for wilderness reservation included both new, imported elements and older, more local ones. The idea that the settler wilderness movement is inauthentic derives some plausibility from the heavy reliance of some recent sections of it on American wilderness theory, much of which, especially Nash 1982 and 1989, is highly U.S.-centered. But the charge of inauthenticity depends on accepting Nash's claim that wilderness is a uniquely American gift to the world. This claim ignores the almost simultaneous development of a parallel dynamic in many other neo-European contexts, especially in the recently ex-colonial world. In this dynamic, indigenous landscapes and biota are given value and protected in reaction to the colonizers' devaluation, and in an attempt to create a national natural heritage to rival European cultural heritage. Although recent justificatory theory was borrowed heavily from the American movement, the practice of wilderness reservation and the practice of valuing and protecting indigenous biota and landscape has had both genuinely local and genuinely subversive aspects, enabling nature and land for the first time to be seen in parallel terms as colonized and in need of defense.

The first stages of the Australian movement involved the development of a settler landscape sensibility in opposition to a dominant Eurocentric perspective which judged the radical difference of our native biotic communities as a lack or absence, and scorned the original landscape as a jumble of "rubbish" to be erased and reinscribed with a properly ordered European character. The "wilderness" landscape of invasion was conceived as untamed and unredeemed land as yet unmixed with the European-style labor which would silence its ancient tongue and teach it the gentlemanly language of property. This new speech said very little of the shape of the land itself; it spoke first of pounds and pence, then of dollars and cents. But the land was also made to speak of office, as surveyors inscribed a grid of powerful names from the colonial bureaucracy across the continent, which for the colonizer was a *tabula rasa* empty of both the naming of prior occupants and anything of value in indigenous nature.

The early colonizers' devaluation of the Australian biota and landscape was resisted by later settler generations as part of the reworking of the garden and wilderness myth which accompanied the formation of a separate national identity. The anticolonial impulse to take pride in indigenous landscapes and protect indigenous biotic communities began to flower in the late nineteenth and early twentieth centuries. In the Australian wilderness movement, and especially in South-West Tasmania in the 1980s, this anticolonial impulse was directed toward the land, fusing with the newer and larger environmental struggle in Western culture to recognize the long-denied centrality of the earth story, written with such power on the South-West landscape.

I do not suggest that we should turn our backs on this struggle, in which I participated, and abandon places like Prion Bay to the forces of development. But wilderness advocates need to recognize that understandings of wilderness can be dualistic and colonizing, as well as liberating and subversive. Unfortunately, the Australian movement to protect such places from disturbance replaced the imperial concept that denigrated wilderness by an American-centered concept that honored it, but did not rethink its construction as absence, emptiness, or virginity. This honorific concept has now been opened to deep criticism from the environmental justice movement and from a number of ecologists and philosophers. Cultural and philosophical critics speak of the erasures implicit in the "virgin" concept of wilderness, of its retention of the basic Western duality of culture and nature, while ecologists speak of the virgin's complicity with and vulnerability to the continuing environmental degradation of non-pristine lands.

It is increasingly clear that modern environmentalism's dominant conception of wilderness as human absence has not escaped the colonizing and dualistic dynamic inherent in the imperial conception of wilderness as the polar opposite to civilization, as unoccupied, empty land awaiting "discovery," the advent of human history. In contemporary environmental "virgin" accounts, the dualistic understanding of wilderness is merely reversed. The "virgin" conception retains masculinist understandings of human identity as oppositional to and separate from the earth, still conceiving wilderness as the feminized Other to be filled, perhaps no longer by civilizing works of engineering, but by a dominating knowledge of recreational, spiritual, or scientific conquest.

These multiple erasures of the Other are serious problems in the popular concept of wilderness, but they do not, I think, support wilderness skepticism, that is, the position that we should abandon the concepts of wilderness and nature entirely, and with it any attempt to expand wildernesses or to protect places like Prion Bay from humanization. That would be necessary only if the reversal "virgin" metaphor and its analogues were the only way to think about wilderness and the only foundation for managing our relationships to wilderness. We should not abandon concepts of nature and wilderness, I shall argue, but we need to create new, non-colonizing understandings and situate them within the context of a renewed, radical ecology committed to healing the nature/culture split and ending the war on the Other.

GENDER AND THE "VIRGIN" CONCEPT

The virginity metaphor, which has been important to both traditional and counter-cultural meanings of wilderness, is highly problematic, both for wilderness and for non-wilderness nature. The metaphor draws its strength and importance from the very problematic way the land is treated in the dominant social order. The virginity metaphor assumes alternatives in which untouched land is the exceptional option to a normal developmental process which results, eventually, in violence, the complete imposition of a different intentional order which erases the original biotic communities and ecological flows. In a context where rape is the normal form of a sexual encounter, virginity must acquire considerable, unwonted, importance. But purist definitions of wilderness as virginal can obscure possibilities for rejecting these destructive processes at other points than that of initial contact, and can actually work against wilderness reservation by creating a high redefinition only a few areas of land can honestly meet, can unduly limit ambitions for land reservation, and can create unrealistic expectations for the self-sufficiency of wilderness land (Woods 1998; Nelson 1996; Soulé 1995; Callicott 1991a). The virgin concept forces a dualistic conception of land as either totally untouched or "not really nature." It is also problematic in its wider implications for gender.

Carolyn Merchant (1996) has traced a Western recovery narrative which encodes gender at the heart of the meanings and practices of the land, and which make various uses of the virgin metaphor. In the traditional meaning the virgin, original Eve, is the compliant, pristine bride of Adam, land ripe for consummation, for development. After her disruption of the male oneness of the Garden, woman reverts to Fallen Eve, sinful woman as abandoned, disorderly wasteland. Fallen Eve can be redeemed and given positive value through (male) human agency and science (Merchant 1996; Ruether 1975), which render her again the productive, fruitful wife and mother, her fertility controlled and turned to the nurture of her husband's children as wild land gives way to the developed, improved garden.

In the context of such a recovery narrative, I suggest, the Romantic and counter-cultural concept of wilderness can be read as assigning a new and different role to the concept of virginity which makes possible the reversal of the traditional meanings of wilderness and establishes a subversive, positive value for wild nature. This new concept reconceives wilderness as the sacred virgin, sacred nature which has the power to redeem (Ruether 1975), especially the power to redeem fallen culture (as in Thoreau's "in Wildness is the preservation of the World"). The metaphor of virginity played a crucial role in the environmental reconception of wilderness in positive terms, mobilizing the recovery narrative and the feminine gendering of nature in a way which enables nature to be seen as sacred rather than in more traditional terms, as fallen Eve.

The representation of wilderness in the symbolic terms of sacred virginity thus provides a solution to the problem of giving wild nature a positive connotation. But the reconception of wilderness nature in this way as the symbolic sacred feminine generates a further problem: the traditional female gendering of nature, delineated by historians such as Merchant (1981), normally matches a devalued femininity to a nature conceived as a similarly devalued symbolic sphere. Wild land may be redeemed by reconception as the symbolic feminine, but actual women themselves do not change their devalued status, nor are they associated empirically or normatively with the newly redeemed wild lands.[1] The neo-European founding fathers of nineteenth- and twentieth-century environmentalism who invoked the sacred virginity of wilderness nature had no ambitions toward disturbing the wider sphere of hierarchical gender relations, and claimed the sacred wilderness as their own sphere.

Despite the persistence of a symbolic gendering in which nature is feminine, the social and normative gender pattern set up by these revisions of nature's status in wilderness is therefore quite complex. At the social and normative level, the gendering of nature seems to exhibit the same kind of pattern feminist sociologists have found in the case of gender movements in relation to occupational status:[2] when nature is a lower sphere, women (and other devalued groups) are in it, but when nature is elevated, women (and other devalued groups) are out of it. The symbolic gendering of wilderness in female terms as virgin can therefore be complemented by an interpretation and actual treatment of wilderness as a place which is defined in part by the exclusion of actual flesh-and-blood women. This social and normative exclusion is achieved especially via a gender pattern which marks the wilderness as a site of (masculine) transcendence defined in contrast to (feminine) immanence.

If the traditional concept treated wilderness nature as the feminine object of conquest, as a primitive, disordered, or empty space ready to receive the imprint of a masculinist civilization which has seen itself as the pinnacle of humanity, masculinist metaphors of conquest persist in other guises in the reversed wilderness concept. Thus the reverse concept of wilderness is figured in different but still-oppressive ways which proclaim that women, who have been Other to the dominant concept of civilization, remain so in its reversal. Various modern and counter-cultural accounts of wilderness supplement the traditional culture/nature contrast which figures nature as a symbolic feminine Other to be dominated by transcendent masculine culture as part of a new, cross-cutting domestic/wild contrast, which figures the wild (nature) as a masculine sphere of transcendence in opposition to the domestic sphere (culture) as feminine.[3] If feminine immanence is expressed in limitation to the domestic, masculine transcendence is expressed in escape to a superior realm of true spiritual or adventurous experience in a nature defined against an inferiorized, familiar sphere of dullness and dailiness. This symbolic "cross-dressing" of wilderness, like the virgin metaphor, enables a reversal of the traditional contrast in valorizing nature over culture, but retains oppressive and exclusionary gender meanings. These meanings are reflected in the colonial conquest of wilderness and the counter-cultural wilderness movement, both of which have had a strongly masculinist thrust.

The androcentrism of much of the wilderness movement is notorious among walking women and women who are grassroots political workers for wilderness (Dann and Lynch 1989). The normative exclusion of women is achieved not only by treating wilderness as the domain of a masculinity defined as transcendent in opposition to the domestic, but also as the site of a masculinity defined in terms of physical toughness. In familiar wilderness imagery, it is the male body which is invoked, which tests its mettle and proves its manhood by conquering or challenging the land, in the "moral equivalent of war" (Marshall 1930, Olson 1938). Some early walking clubs, such as the Mountain Trails Club in Australia, marked this gendering formally in the exclusion of women members (Barnes and Wells 1985), but exclusions do not need to be formal. That human body which defines itself symbolically through conquest of the wilderness, through its opportunities for adventure, challenge, and fraternal bonding, and in relation to whose capacities for physical achievement wilderness is sometimes itself defined, is preeminently the male body.

But whether wilderness is the feminized Other to be conquered or rescued, or the wild sphere of transcendence of domestic confinement, actual women are erased from this territory, in fantasy if not in fact. In both colonizing and counter-cultural wilderness tropes, the wilderness is "no place for a woman" (Schaffer 1988), unless she is properly subsumed by male escorts, guides, rescuers or protectors. Increasingly, women are subverting or defying this gendering of nature and creating a place for themselves in the wilderness, although all-women parties or solo woman venturers are still greeted by the chorus "You're not doing it on your own!" from people who wouldn't dream of saying anything similar to male or mixed parties (Dann and Lynch 1989).

But not all the limitations come from outside; self-mutilation and cultural limitation to the domestic is an important part of oppressive feminized constructions of self.[4] "Most girls fence off their wilder selves by puberty. We may start out mouthy, loud, happy and brave, and then we leave the wilderness and shrink ourselves to fit the station wagon or office desk," writes Susan Campbell (1995) in an article on women walkers on the Appalachian Trail which notes that few through-walkers are women. As Iris Young shows, many otherwise "liberated" Western women still have to allow themselves, as well as be allowed by others, the full freedom of their bodies and the freedom of the wild.[5] Creating a place for women in and with wilderness implies giving a new, subversive meaning to the female body, to wilderness knowledge (Fullagar and Hailestone 1996), and to our relationship to the Otherness of wilderness.

OPPRESSIVE MEANINGS OF WILDERNESS AND INDIGENOUS ERASURES

It is now established that much of the neo-European land now conceived as wilderness has an extensive indigenous history, and that considerable parts of it at least were subject to various practices of management or alteration. If "nature" is taken in this dominant "virgin" sense, to mean the sphere of the nonhuman, an absence of the human, it would seem that such areas must be considered the work of culture rather than nature. It is obviously and objectionably Eurocentric to conceive of such places as free of any human influence or presence, where land takes the form of virginal, pristine nature. In the American case, historians have mostly excused the failure to recognize this indigenous human history as a genuine mistake of fact, pointing to the disastrous decline in indigenous populations following diseases introduced by Europeans in the sixteenth century (no mention of warfare), which they think caused much of the land to return to a wilderness-like condition in subsequent centuries (Cronon 1983, 1995a, b; Crosby 1986; Denevan 1992).

In the Australian case, where the whole continent was declared "empty" (*terra nullius*) shortly after first contact with Aboriginal people and with no thorough reconnaissance of the land or their relationship to it, such explanations are much too kind. It is necessary to adopt a more sinister and philosophical explanation for this "mistake." It was the product of the perceptual and conceptual distortions of a colonizing worldview that treated the non-European cultures it encountered as inferior and primitive, and justified its land seizure by viewing them as children of savage nature, semi-animal wanderers with neither ties to the land nor capacity to affect its ecological structure. What is involved here is no simple classificatory or ecological error, but the familiar dynamic of colonization which prepares the way for appropriation by representing the Other as inessential and deviant, outside the colonizer's standard that defines the norm. Denial of relationship to or dependency on this Other often involves a perceptual politics in which their labor, skill, and technology is unrecognized, is beneath notice, and is classified into the subordinated sphere of alterity termed "nature," which status also prepares the way for appropriation.

Thus naturalized, the colonized and their trace in the land can be forgotten or subsumed as the inchoate background to "civilization," as the Other whose prior occupancy and whose dispossession and murder need never be spoken or admitted. I think, with Gary Nabhan, that the "disease" explanation proposed in the American case may be a little disingenuous, given that the Miwok Indians were actually being removed from their land in Muir's time so that it could be conformed to the image of wilderness as human absence (Hecht and Cockburn 1990; Nabhan 1995; Kothari et al. 1996). In America, the representation of the Other as part of the sphere of nature or as a "passive parasite on nature," rather than as a maker of culture and remaker of nature, may have played a role in the perception of American territory as pristine, although perhaps not such an explicit one as it did in Australia.[6] But whichever type of explanation we adopt for historical erasures, it cannot excuse contemporary erasures of indigenous history in wilderness land, or continued support for the exile of indigenous inhabitants.[7]

There are practical ways of countering these erasures of indigenous cultures implicit in the "virgin" concept, without abandoning nature reserve or wilderness. In the Australian case, efforts are increasingly made to recognize historical indigenous sites and uses within reserved lands, in management, publicity (where this is appropriate), and in consensual arrangements for use (Harris 1994). In the Australian context, countering the erasure of indigenous people in the presentation and history of land reserved for nature is vital if the concept is not to be, and be discredited as, irrevocably racist. We must deeply regret past and oppose present erasures and removals of indigenous peoples (Langton 1996; Kothari et al. 1996). But it seems to me to be a serious and potentially disastrous mistake to think that the erasure of indigenous culture and its impacts on land took place *only* in land now designated as wilderness, or that it is only or primarily the nature reserve system which now should bear the responsibility of recognizing and rectifying the consequences of indigenous land seizure.

If the *terra nullius* claim provided the foundation of all land takeover in Australia and some parts of Africa, the moral responsibility for meeting the aspirations of indigenous peoples in societies where European land takeover was so based would seem to fall equally on all subsequent land uses and users. There is clearly some likely convergence between nature reserve interests and indigenous land claims, in that places where land claims can most easily be established are usually those most recently occupied by settlers. This land is both more likely to have retained both its biotic communities and its ties with indigenous communities, and will be of significance for both conservation and indigenous groups. In this context, both cooperation and conflict between these communities have emerged in Australia, based both on this competition and on the different economic relationships settler and indigenous communities bear to wilderness, on the one hand, oriented toward recreation and on the other toward gathering (Dann and Lynch 1989).

Eurocentric interests and perceptions intensify the potential for conflict between these communities and between indigenous land claims and the nature reserve system. Thus the channelling of Aboriginal land aspirations into wilderness areas and the nature reserve system is also fed by several more problematic sources. First, Aboriginal people are stereotyped by much of the settler population as "natural" conservationists who can relieve (often without adequate support) settler society of the responsibility of nature conservation. Although indigenous culture undoubtedly has much to offer concepts of conservation, stereotyping of this kind is rejected by many Aboriginal people as limiting the development of their culture and as a reverse form of culture erasure (Bayet 1994; Nabhan 1995; Langton 1996).

Second, channelling indigenous land claims into wilderness and the nature reserve system is beneficial to powerful political interests such as the mining and pastoral elites who control much of the other relevant land over which claims might be made. Third, land kept for nature is often perceived as "empty" or "vacant," rather than as having an owner in nature and as in use by and for nature. In the context of a larger program of recognising native title on pastoral land, losses to some nature reserve land from indigenous land claims can be offset by the new opportunities regional agreements with indigenous groups provide for reworking and enlarging conservation on former pastoral land (Lark 1990; Guy 1994; Walker 1994; Harris 1994). These opportunities could be lost without more generous political recognition of native title on non-wilderness land, and, at the conceptual level, more flexible concepts of wild nature which do not conceive it primarily in terms of human absence.

In this context of potential conflict, some interpreters and supporters of indigenous traditions deny all application to concepts of nature and wilderness, as incompatible with due recognition of indigenous influence (Rose 1996). Similar claims are made by Marcia Langton (1996), who states that there is no such thing as a "natural" landscape. Rose offers an inspiring vision of a humanity so thoroughly at one with the land, so thoroughly engaged in nourishing practices, that there is little need for a concept of "virgin" or "wild" land set apart from oppositional human influence. But it is one thing to discard concepts of nature as land apart from the human in the context where nourishing human land practices are the norm, and an entirely different thing in a context where destructive practices—"a multiplicity of devastations," to use Rose's phrase—are the norm. From the fact that indigenous communities did not need protective concepts that designated land set apart from human influence, it does not follow that settler communities whose practices are oppositional rather than protective do not need them, at least as interim protective measures until their practice becomes more nourishing. Settler and indigenous communities are radically different in their approach to land and require different kinds of protective systems and concepts.

THE ECOLOGICAL DUALISM OF THE VIRGIN CONCEPT

The problems with the reverse concept of wilderness don't end with racism and sexism. A powerful case can be made that, as third-world critics such as Guha (1989) and Kothari et al. (1996), charge, the concept of wilderness has been coopted politically and made part of an approach to environmental problems which is deeply dualistic, colonizing, and politically elitist. Wilderness "set asides" (from the course of ordinary life), together with green consumerism, have been dual parts of a mitigation strategy which has made it possible to act as if environmental problems could be satisfactorily resolved with minor adjustments within a liberal political and economic system which is fundamentally both unjust and hostile to nature (Plumwood 1995).

I think wilderness lovers who hope to rework the wilderness concept in the context of a radical ecology can concede that the "set aside" approach has been emphasized to such exclusion by conservative elements in the green establishment precisely because revering nature in distant wilderness does not force us to reconsider nature in our daily relationships to ordinary land and to economic life. Before his Earth First! days, arch-wilderness advocate Dave Foreman was a "card-carrying" Republican, who campaigned actively for the election of arch-conservative Barry Goldwater to President of the United States in 1964. "Wilderness, like religion and morality, is fine for weekends and holidays, but during the working week it may in no way inform business as usual," writes Tom Birch (1990). But it is important to note that the inadequacy of such a strategy of equating environmental protection with wilderness preservation does not tell against wilderness protection as such, but only against the dualistic limitations of the strategic context in which that protection has often been situated.

Instead of advocating wilderness as part of a far-reaching, sustainable social restructuring, the conservative ecological strategy of wilderness promotes a dualism of land which is akin to that of the "good woman/bad woman" dualism promoted by machismo cultures and institutions in the case of women: on one side, in the spotlight, an idealized, "unspoiled," virgin nature stands on a pedestal, while in the background darkness, on the other side, waits the common "whore" whose disrespectful, exploitative, uncaring, everyday use makes it possible for the virgin to keep her pedestal status. This kind of dualism is implicit in the exclusionary reverence for wilderness as "virgin" land, which is to be accorded a degree of respect as "nature" which common and ordinary land is excluded from; as Marina Warner (1983) notes of the cult of the Virgin Mary, the extreme emphasis on her purity and virginity and its identification with true womanhood implies a subtle denigration of all other women.

In the same way that the dualistic cult which confines reverence and respect to the pure Virgin did not support a culture of respect for ordinary women, a dualistic wilderness cult which confines respect and the status of "nature" to pure virgin land does not support a culture of respect for ordinary land or for nature in the context of everyday life. This buried understanding in terms of the duality of the virgin and the whore, the cathedral and the factory, the wilderness and the *maquiladora,* helps to explain the exclusions and contradictions third-world critics like Guha (1989) point to in the treatment of wilderness in America and in the strategies of much of the Western wilderness movement. However, if a wilderness, as a virgin area of land and its retinue of biotic communities, can be associated with a great variety of cultural constructions and stories about the meaning of its virginity—and this dualistic construction is only one among many—the deficiencies identified in Guha's critique must be those of the dualist protection strategy, and not of the land itself or of protection strategies in general. And we should bear in mind that the overall teleology of this duplicitous cultural construction is to enable unhindered exploitation of most land, and most women, while professing an exaggerated reverence for a few idealized ecological and feminine exceptions.

Botanist Arturo Gómez-Pompa and anthropologist Andrea Kaus (1992) write of the problems this duality creates for resource management policies, which "on the utilitarian side, are permeated with an acceptance of destructive practices, generated from a belief that mitigating measures can halt or reverse environmental depletion and degradation ... [while the other side assumes] that setting aside so-called pristine tracts of land will automatically preserve their biological integrity." The first, utilitarian assumption denies the nature, the wildness that exists in areas we think of as "culture," while the second conceives areas under the rubric of wilderness as self-sufficient, having no dependency on culture for maintenance or support (including, often, financial support). This second assumption ignores the fact that many wildernesses can only now survive if they are supported by an appropriate system of culture which fosters their protection and respect, as well as the dependence of wilderness land on a healthy biospheric condition and hence on healthy non-wilderness land and environmentally sustainable practices in everyday life. The sharply polarized, neat division between the land of nature and the land of culture that the conventional picture of wilderness has tended to presuppose can be maintained neither at the level of ecological management practices nor that of sustainability.

POLARIZED STRUCTURE AND WILDERNESS DUALISM

If these arguments do not show that we should give up on wilderness preservation, in the way skeptics such as Guha seem to think,[8] they do show that we have to give up the dualistic concept of wilderness, which is unhelpful, unviable, and in need of replacement. But we need to start the process of reworking our concepts of wilderness with a careful and precise analysis of what this dualism involves, how it accounts for the problems in wilderness I have outlined, and what it implies for wilderness otherness. Infection by dualism has occurred because the problematic wilderness concept has been based on a reversal, in the classic Romantic style, which replaces the pejorative "emptiness" concept by the closely related but honorific "virgin" concept. Such a reversal retains at its heart the idea of human identity, of culture, and of *cives,* as outside, apart from, and oppositional to nature (Plumwood 1993a). In this reversal, the discontinuity and polarity of these basic concepts have been insufficiently brought into question in the revaluation of the excluded group from positive to negative.

The problems of sexism, like those of elitism and racism, arise from this continued dualistic construction of the wilderness concept, and its situation in the context of dualistic and gendered domestic/wild and culture/nature contrasts. Thus it is the conception of wilderness as the opposite pole to culture which erases indigenous influence, locates it as the site of masculinist transcendence, the site of elite strategies which deny the honorable title of nature to everyday land, and the locus of ecological concepts which are unable to recognize interweavings of nature and culture. The virgin concept suggests, and is usually accompanied by, an account of nature as the absence of human influence. But such a definition allows no significant difference to be marked between the slightly altered ecosystem and the landscape which is totally reconstructed, a thoroughly human product; both are equally non-virginal, "spoilt" (Soulé 1995). From an activist and ecological perspective, such distinctions are crucial. The account of nature in terms of human absence leads to a polarizing assumption of a radical discontinuity between nature and culture which prevents us from seeing wilderness as the extreme end of a spectrum of mixtures of nature and culture, of humanized and wild land (Birch 1990).

These polarizing assumptions are typical of a dualistic conceptual scheme which constructs separate realms between colonizer and colonized through a hyperseparation which writes out the basis of continuity, and which homogenizes and stereotypes the two groups. To overcome this dualism we need to reclaim the ground of continuity, to recognize both the culture which has been denied in the sphere conceived as pure nature, and to recognize the nature which has been denied in the sphere conceived as pure culture. The traditionally dualistic wilderness concept delegitimates both, denying the legitimacy or possibility of the hybrids and boundary crossings which break up the neatly regimented polarity of nature and culture, and which enable wilderness reserves to be understood as part of a continuum. Thus Holmes Rolston III (1991) defends wilderness by reasserting the profound discontinuity between humans and animals, and between Western culture and primitive cultures, the same dualism that the dominant tradition has used to inferiorize nature and to set humans outside it.[9] The now-honorable title of "real nature" must be reserved for "pristine" nature alone, since according to Rolston "it is a fallacy to think that a nature allegedly improved by humans is anymore real nature at all." A position confining nature to the pristine form cannot recognize the continuum of nature and culture, the dependency of certain forms of nature on forms of culture, nor the dependency of culture on the healthy operation of non-pristine, biospheric nature.

The dualistic conception of wilderness is thus part of a background nature/culture dualism which implies the "cordoning off" of nature, its conceptual confinement to pristine situations which entirely lack, or have rendered invisible, cultural influence. But it is precisely this concept of nature which is so problematic from a larger environmental perspective, for it blocks recognition of the embeddedness of culture in nature. It is incompatible with the kind of thoroughgoing environmentalism that aims to recognize *everywhere* in our lives what has been systematically denied and backgrounded as part of this nature/culture dualism—our dependency on the active agency and "labor" of nature and biospheric processes, even in our joint creations, and the limits this imposes on the development of human culture. While we are unable to give that recognition, we will remain at a stage of moral intercourse with nature that corresponds to the Aristotelian theory of reproduction, believing that we are self-sufficient achievers and that nature is no more than a nurse. Wilderness has a great potential to subvert that denial, as one of our best teachers of the limits of the self, and of the agency and wonder of the Other.

Whether or not the interpolation of a human story with that of nature is an improvement or an impoverishment,[10] the result of the conceptual confinement of nature to pristine nature as wilderness is the endorsement of a highly problematic polarity between human/cultural and natural/wilderness spheres. This implies a solution to the problem of respecting nature via wilderness which is of necessity exclusionary, unable to be extended to all areas of land. If only sacred Virgins are revered as truly women, those women we encounter in more profane contexts do not deserve our respect. If nature proper is found only in places without any human influence, there is no way we can recognize the importance of nature or respect its limits in our daily lives, except through elite or exceptional practices of nature escape. If nature is normally "somewhere else," we do not need to be sensitive to its operations in our local environments of urban, working, and domestic life. Of course, as a wilderness lover, I am not advocating that the disrespect extended to ordinary land should be extended to virgin land, but rather the opposite, that the respect presently confined to virgin land should be extended to nature in all our contexts of life.

SKEPTICISM, COUNTER-ERASURES, AND OSCILLATING HEGEMONIES

There is considerable confusion about the implications of these arguments and about how to go about breaking down the polarization of nature and culture which the critique of dualism discerns in both the traditional and reversed wilderness concepts. Recent critical discussion has opened the field for several different kinds of skepticism about concepts of wilderness and nature, associated with different groups and conflicting motivations. The appeal of the skeptical case rests in part upon confusion between these different varieties of skepticism. We need to distinguish a class of anthropocentric reasons for rejecting concepts of nature and wilderness from another class of objections to these concepts based on the oppositional and dualist meanings they are usually given.

Although wilderness critics often assume that the critique of dualism in concepts of wilderness supports wilderness skepticism, I argue the contrary case, that a careful analysis of what is involved in dualism shows that it does not support generalized nature skepticism. A reworked concept of wilderness, I shall show, still has a valid application. It is a feature of dualistic hegemonies that they produce oscillating reductions, in which now one, now the other, pole of the dualism is favored in reductive solutions which often do little more than create a pendulum motion. Thus, in the current discussion there has been a tendency to try to resolve the nature/culture dualism inherent in the conventional understanding of wilderness by a reduction to one or other of the poles of nature or culture, either reducing wilderness to culture, or reducing human activity which might disturb wilderness to nature.[11] Wilderness critics especially, and even some wilderness defenders, seem to think that the problems generated by the dualism of nature and culture show that there is no valid distinction between nature and culture, or that we must indiscriminately applaud hybridity. I think this approach is unnecessary and unhelpful, and that we can give up the dualist or oppositional way of distinguishing between nature and culture without giving up all concepts of nature and wilderness.

Some wilderness skeptics intentionally embrace the anthropocentrism of the dominant tradition, which wilderness reservation aimed to counter, dismissing the devastation of nature and the biodiversity crisis.[12] Thus Guha's list of problems systematically omits those involving the destruction of nature itself, and does not even mention biodiversity loss, or any associated problems of human dominance. Although Guha is right to identify the neglect of the welfare of sanctuary neighbours in India as a serious omission (Kothari 1996), he does not suggest, for example, that the tragic slaughter and extermination of wild elephant populations might be a matter of environmental concern, or that as a humanitarian concern it might be a matter for the peace movement. At this point Guha does not so much *argue* for the position that anthropocentrism is irrelevant, as he claims, as *assume* it in his problem focus, which is highly anthropocentric and exclusionary.[13]

More commonly encountered in the West are philosophical forms of anthropocentric wilderness skepticism. Their methodology usually involves equating the human with culture to yield a human version of solipsism. Solipsistic techniques include reducing nature to culture and prioritizing culture in any mixture of nature and culture. Anthropocentric wilderness skeptics' distinctive claims about human creation of wilderness based on these techniques emerge as absurdly inflated when measured against a place like Prion Bay. For Schaffer (1988, p. 89), nature is a mere "neutral surface inscribed by a network of meanings," a void available for inscription of our stories; it is quite passive, having no role or agency in confirming or constraining our meanings, and has no stories of its own to tell. Chaloupka and Cawley (1993, p. 4) suggest that nature is an artifact of language, which opens on the void, and the term "wilderness" is no more than an attempt to create the Other we choose to encounter. For Cronon (1995a; 1995b, p. 69), too, wilderness is "quite profoundly a human creation," another product of our civilization.[14]

Although the assumptions on which they are based are rarely exposed, these skeptical convictions seem to draw on anthropocentric arguments akin to those for individual solipsism, employing similarly stretched senses or low redefinitions of various common concepts. In the background here we can often glimpse a version of the argument that because the concept of wilderness is a human construct, wilderness itself must be a human construct, an argument which, stated baldly, reveals itself as a use-mention conflation. The wilderness is said to be a "product of civilization," on the grounds that we conceive it in terms of the concepts of our civilization; these grounds would support the same overweening conclusion about the planet Venus, the Sun, or indeed, about anything else we can think of. Sometimes the ground of the claim that wilderness is our creation is that we now have the capacity to destroy it, which we must resist in order for it to exist. This argument seems to rest on a stretched sense of creation which confuses the capacity to create with the capacity to destroy. A comparable argument in the human case would allow one to claim that other people were one's creation because one refrained from murdering them.

The idea that wilderness is our creation because it now shows some human influence rests on prioritizing the human or cultural element in mixtures of nature and culture; a comparable argument in the human case would license the claim that someone was our creation and lacked all autonomy because they had taken some sort of influence from us. But such cultural reduction, which is often associated with certain forms of postmodernism, would abolish conceptual conditions for sensitivity to nature's limits, and to the variation and interweaving of the human and the natural stories which an ecological consciousness aims to foster.[15] These arguments and stretched senses systematically overstate the human contribution and understate nature's contribution, testifying to the growing success of the project of human insulation and self-enclosure. Those postmodernists who employ them may think of themselves as in opposition to the dominant tradition, but are in fact at one with its dualizing approach in continuing to represent the Other, nature, as an absence or void, and to demote its agency.

Indigenous theorists and supporters often have a rather different basis for nature skepticism, one which looks superficially similar to the solipsistic one, but which points to a very different problem. Thus Rose denies even to places "of peace, natural beauty, and spiritual presence" that have escaped settler alteration the appellation of natural landscape, on the grounds that they were subject to prior indigenous influence (Rose 1996, p. 18).[16] There are a number of ways we can understand the concept of nature. The assumption in this passage is that nature is to be understood in the polarized way that is part of nature/culture dualism, as pristine nature, denying any prior human influence or presence in the land. It is certainly understandable that Aboriginal advocates would strongly reject the pristine concept defined in terms of nature/culture dualism, because it is the one under which Aboriginal people were denied full humanity for failing to evidence European-style culture. It is not so clear why we should take this to be the only concept of nature available. The "virgin" usage is one reading of the highly ambiguous concept of "nature," dominant especially in the colonial past. But, as I have suggested, the polarity of virgin nature is highly suspect, and the dualistic reading yields a concept of nature which is incompatible with and unhelpful for many everyday usages, as when we speak of nature in our daily lives, on the farm or in the suburbs.

Rose seems to intend to reject the dualized concept of nature prevalent in the West, but one problem with the skeptical way of putting the point is that statements such as "there is no such thing as a 'natural' landscape," or the claim that "the country is fashioned" by human influence, would normally be understood by people *in the Western anthropocentric tradition* in the same way as those of the solipsistic skeptic, as licensing human self-enclosure and the reduction of agency to human cultural agency. But this is clearly not what Rose intends. Rose's reading of the Aboriginal concept of "country" does not permit such a reduction to culture or to human agency because on her account "country" is highly agentic: "country has its own life, its own imperatives, of which humans are only one aspect. It is not up to humans to take supreme control, or to define the ultimate values of country. Aboriginal relationships to land link people to ecosystems 'rather than giving then dominion over them.'" (Rose 1996, pp. 10–11).

Rose's understanding of the concept of "country" and of human identity is clearly far from being a solipsistic one, and it is the anthropocentrism as well as the oppositionality of the virgin concept which seems to ground the kind of rejection of the concept of nature she makes. Western "nature lovers" are unlikely to be able to assuage the anxiety and narcissism which underlies solipsistic skepticism, but there are conceptual resources available which would enable them to move closer to the indigenous critique of the virgin concept of nature. They have an independent motive also for moving to less oppositional and dualized concepts, deriving from the difficulties these concepts create for environmental understandings and activism.

This nature/culture dualism distorts the way we can think about land, obliging us to view it as either pure nature or as a cultural product, not nature at all, in much the same way that it distorts the way we can think about gender. Polarization and homogenization lead us to classify wilderness, Prion Bay for example, as pure nature, in ways that obscure its continuity with and dependency on culture, and erase the human stories interwoven with it, especially those of its indigenous people. On the other side, conceiving a place according to the opposite homogenized pole of culture has the same distorting result, because we cannot adequately recognize the unique interwoven pattern of nature and culture which makes up the story of a place, and makes each place unique. Recovering the lost ground of continuity that dualistic conception has hidden from us allows us to conceive the field in more continuous and less regimented ways, recognizing nature in what has been seen as pure culture and culture in what has been seen as pure nature.

If the dualistic form of the distinction is a problem for environmentalism, so is the attempt to eliminate the distinction itself. The attempt to eliminate the distinction between nature and culture assumes that we can manage our relations with nature in the same way that we can manage human institutions, either treating nature in the fashion of deconstructive ecology, as simply malleable, as subject to unlimited and rapid reshaping (Worster 1995) and as imposing no limitation or constraint of necessity upon us; or alternatively, as in the case of naturalistic reductionism, failing to recognize the role of human politics and social choice in human social-ecological relations (Soper 1996). Without some distinction between nature and culture, or between humans and nature, it becomes very difficult to present any defense against the total humanization of the world, or to achieve the recognition of the presence and labor of nature which must be a major goal of any thoroughgoing environmental movement. For that, we need sensitivity to the interplay of self and other, and to the interweaving and interdependence of nature/culture narratives in the land. But we need not and should not construct the distinction as a binary opposition, as the Western dualism of nature and culture has done.

The concept of nature is a politically underdetermined one, to be sure, like the concept of woman, and its boundary and meaning must be constantly problematized. But we should reject any form of nature skepticism which discounts earth's agency and invites us to fall back once more on the dominant Western perspective in which nature is the inessential Other, the invisible and unconsidered background "nurse" to human history, with no destiny of its own to fulfill and no story of its own to tell. The skeptical stance risks a dangerous reversion to a *biological* version of *terra nullius*. Where Eurocentrism has encouraged the denial of indigenous presence, co-creation of and agency in the land, we need to be reminded of the human narratives interwoven with the larger creation narratives of the land, and of contemporary claims arising from them. But given the anthropocentrism of the Western tradition, we also need to be reminded of the powerful creation narrative a place like Prion Bay still unfolds before us, the continuing story of the earth itself.

DUALISM AND THE ERASURE OF NATURE

Fortunately, it is unnecessary to adopt these costly reductive routes to the resolution of the dualism: much feminist analysis has rejected the concept of indistinguishability as the solution to the problem of self and other (Grimshaw 1986; Plumwood 1993a), and has given a more precise account of dualism which makes it unnecessary. The idea that eliminating dualism between nature and culture implies eliminating distinction between nature and culture involves a misinterpretation of dualism, conflating the nature/culture polarity which underlies many of the problems discussed above, with nature/culture differentiation (Plumwood 1993a). Dualism creates a polarity, and a polarity involves very much more than a distinction.

A dualistic polarity is composed of two major elements, one of radical separation, and one of homogenization, which together create the typical hyperseparated structure characteristic of dualism, described by Marilyn Frye (1983, p. 32) in the case of gender dualism in the following terms: "To make domination seem natural, it will help if it seems to all concerned that the two groups are very different from each other and . . . that within each group, the members are very like one another. The appearance of the naturalness of dominance of men and subordination of women is supported by the appearance that . . . men are very like other men and very unlike women, and women are very like other women and very unlike men."[17] Dualistic forms of separation involve, then, both a radical exclusion, or maximal separation and distancing between the relevant groups, and an internal regimentation and homogenization of the polarized groups. The reformation of the dualized self/other relation then involves rejecting the radical separation and reclaiming both the ground of continuity that has been denied in radical exclusion, and also the difference that has been denied in the internal regimentation to fit the simplified nature or culture model. Countering the backgrounding and incorporation of the other which usually accompanies this polarised structure is an important element in countering dualism.

It is in the context of such a non-reductionist account that we can come to understand how wilderness reservation, properly understood, can be one part of an appropriate strategy of response to the colonization of nature. The equation by the dominant humans (Europeans) of their absence with emptiness erased indigenous humans, but the erasures in their concept of wilderness did not stop there. A more inclusive account of colonization enables us to see that another major erasure in the traditional concept of wilderness as a European absence is the erasure of nature and of the land itself. Both erasures are implicit in Locke's recipe for annexing new-world "nature" as European property, as the mixing of labor with land which is assumed to be *terra nullius,* unowned and empty, as well as limitless in extent. Locke's formula justified property acquisition by extending back into the European past the imaginary recipe which supposedly justified ongoing European annexation of the new world. This recipe provided some crucial parts of the philosophical technology for European incursions into the fresh "unowned" lands which were to become the neo-Europes. Locke's formula did this by displaying a recipe for a form of appropriation in which the lands of others could be represented as pure nature, containing neither any European-style labor which needed to be recognized, nor any other trace of culture. This points again toward denial as a further neglected reason why America's pioneers were so open to seeing as a wilderness the land they wished to appropriate. Our current perplexity over wilderness is in part a legacy of this colonial problematic.

Implicit in Locke's account also then is an assumption of the emptiness and nullity of nature itself, which serves as the foundation for those other erasures. Locke's recipe assumes an erasure of nature's agency which closely parallels Aristotle's erasure of women's agency in his award of the reproductive ownership of the child to the father, whom he saw as the only active agent in a reproductive situation which we now conceive as normally involving joint and mutual agency. In just the same way that the Aristotelian father could claim the child as his, Europeans could claim the earth and its products as theirs, by denying the Other's agency and treating the Other as inessential. And everywhere in the creation of the neo-Europes, it was not only indigenous people who suffered this denial and its consequences—murder, dispossession and displacement—as inessential Others in the colonization process, but also the indigenous biota, which was colonized and displaced in very much the same terms, and with many of the same results (Crosby 1986). Many indigenous species and biotic communities, like indigenous peoples, are pushed to the margins, and now struggle to survive, but the process continues.

In this process of colonization, nature has been represented according to the same colonizing conception of the inessential Other as indigenous peoples and, before them, women; indeed, in both cases colonization involved explicitly assigning these the status of nature. But some aspects of the colonizing construction of Otherness, like the dualism which it maps partially, have also carried over to the modern ecological wilderness concept. Since Western culture is the world's major colonizing culture (Crosby 1986), it should not be surprising that we have given our basic concepts of Otherness a deeply colonizing meaning. To find a more adequate conception of the wild Other we encounter in wilderness, which can locate the concept of wilderness within such a larger anticolonial critique, demands an effort toward decolonizing the mind and a movement toward a nondualistic conception of the Other.

FROM THE FEMINIZED TO THE FEMINIST OTHER

In these problems we can see how the ghosts of the old wilderness concept have returned to haunt the new ecological one. But not all the ghosts have been dispelled yet. Another serious problem in the new wilderness concept arises from a feature of its structure which it shares with the old and which parallels and extends its Eurocentrism and androcentrism. The modern concept of wilderness still defines the Other as "empty," as human absence, and therefore *is profoundly anthropocentric.* In fact, the emptiness concept of wilderness is anthropocentric for just the same reasons that the *virgin* concept as understood in patriarchy is androcentric. In patriarchy the virginity of a woman is defined and emphasized in relation to male penetration, and hence in relation to a certain kind of male absence. But, as Marilyn Frye reminds us, there is another way of being a virgin: *"The word 'virgin' did not originally mean a woman who was untouched, but a free woman, one not betrothed, not married, not bound to, not possessed by any man. It meant a female who is sexually and hence socially her own person"* (Frye 1992, p. 133). This, I think, is the sense of otherness and virginity that we need to liberate the concept of wilderness. It is not the erotic metaphor itself nor its gendering of nature in female terms which is problematic, but its genderization in terms of the feminized Other—that is, in ways which model human/nature relationships in terms of erasure, subordination, and self-imposition.

The anthropocentrism of the emptiness concept emerges from further consideration of the dualist problematic. The polarized structure I have discussed above is often thought of as being all there is to dualism. But in fact, this is normally only part of the problem, and in most cases we have also to deal with a polarity which is part of a wider hegemonic centrism. In hegemonic centrism, the dualized Other is defined in various ways in relation to a self conceived as center, usually as a lack or absence of the self or its supposed qualities. The colonizer defines the Other as background to his foreground, as deficiency, as inessential in contrast to himself as essential. As we have seen, this definition provides the basis for the common denial of dependency in relation to nature and other colonized groups and for subsuming the Other's labor. The colonizer also claims the right to define the Other in relation to his needs, fails to acknowledge the Other's boundary and limits, and conceives them in a variety of other ways which both assimilate and instrumentalize them (Plumwood 1993a).

The practice of setting land aside for nature through wilderness reservation is open to liberatory meanings which resist colonizing constructions of otherness: for example, respecting wilderness nature as autonomous and as self-defining is a way to resist instrumental and assimilating ways of construing the wild Other. Other familiar ways of reconceiving wilderness, for example as only a more subtle kind of resource for the self, or as some disguised version of the self, aim to take back these liberatory meanings of otherness and to reconquer the Other's space by imposing the framework of otherness meanings generated by the colonizing self (Plumwood 1993a). The conception of wilderness as emptiness is, however, not so open to liberatory reinterpretation. The idea of the Other as a *terra nullius,* as a vacuum, as an absence of self, a space inviting occupation, is the other side of the dualistic formation of a colonizing self primed to fill that vacuum with itself and its works.

As an absence or lack of self, the Other is seen as having a secondary form of existence, defined in relation to the center, who is conceived as primary, subsuming, and self-sufficient (or as Simone de Beauvoir put it, as Absolute). So both the traditional conception and its reversal, the virginal conception of wilderness, define it as a space which is empty of something, as an absence of the human, and conceive it as an absence which we must refer back to the human as center to define. That is, nature's qualities are specified entirely in relation to a human center, and as a lack of that center. The familiar, conventional logic of Otherness prepares us to see nothing to be concerned about here, to treat the difference between presence and absence as academic, a source of jokes about being half-full and half-empty.[18] But there is a major divergence, at this point, between different conceptions of the Other, which can partly be expressed as the difference between perceiving the Other as an absence or emptiness versus perceiving the Other as another center, as a fullness or *presence.*

Thus in the Eurocentric framework, the colonized Other is thought of as empty, primitive, as an absent, deficient, or early stage of civilization. This is also an expression of hegemonic centrism, since the function of this concept of otherness is not to identify an independent center or form of being but to construct a foil for the center, delineated by the Western concept of reason or that of *cives.* Indigenes are conceived as noble if civilization is rejected as corrupt, as in counter-cultural Rousseau, and as low savages when it is conceived traditionally, as a noble and light-bringing enterprise. Of course such a conception of the Other as a deficiency of self must be rejected in any genuinely anticolonial framework, and this is part of what is conveyed, usually rather unclearly, by postmodernist talk of recognizing difference. The experience of the Other as a void or an absence is a prelude to invasion and instrumentalization, whereas the experience of the Other as a presence is the prelude to dialogue.

In the same way, an androcentric framework represents the feminized Other as an absence of or deficiency in relation to the defining center, as in the concept of virginity or as in Aristotle's conception of woman as a deficient man. As Irigaray (1984) points out, modern androcentrism conceives woman, perhaps with slightly more subtlety, according to the housewife model, not as occupying a space herself but as enclosing a space for another. But only a feminized Other, deeply colonized and self-abdicating, could accept a definition of the self entirely in relation to someone else as a center or Absolute, as an absence. Feminists have rejected and exposed the traditional idea that a woman can properly be defined entirely in relation to a husband or another subsuming male as center, for example as someone's wife (Mrs. X), daughter (Miss X), or as someone's "relict" or widow.

Hegemonic forms of definition can also involve general terms, definition not in relation to a particular man but to man in general: "virgin" is one of these. These concepts involve hegemonic centrism not because they involve relationality per se, but because they systematically recognize and distribute this relationality in asymmetrical and hegemonic ways, treating one party as background or support for the other as foreground, or in the "relict" or "virgin" case, as what the center leaves behind or omits to vanquish. Feminists consider such hegemonic ways of introducing or speaking of women to be deeply sexist and insulting in contemporary society, although they are still fairly normal. Feminist consciousness has successfully and widely disrupted these sorts of hegemonic namings and substituted, both in theoretical work and more widely in practice, a different conception of otherness based on a logic of alterity which treats woman as an independent center or presence.

In the same terms, we can see that wilderness in the *full* sense of the wild Other cannot properly be specified as an *absence* of the human; rather it is the *presence* of the Other, the presence of the long-evolving biotic communities and animal species which reside there, the presence of ancient biospheric forces and of the unique combination of them which has shaped that particular, unique place. The dominant Otherness concept of Western culture distorts our conceptions of wilderness and relationship to nature in the same way that it distorts these other relationships of colonization, misdirecting us back upon our tracks and returning us obsessively to our human reference point as center.[19] It is not the absence of humans that we seek in our wilderness quest, as the definitions suggest: you could experience that in your bedroom, or in a place which is totally ecologically devastated, such as a nuclear test site, or in outer space.

In a place like Prion Bay, the dominant experience is not that of the absence of other humans. It is the experience of the *presence of nature,* the company of vast, multiple, and prior presences. This presence may often be obscured for us by (the wrong sort of) human company, which is why the experience of solo traveling in wilderness can be such a powerful one. We may or may not choose to travel alone, but a quest for human absence is an entirely different quest from the quest for the company of nature at large which is at the heart of "the wilderness experience."[20] It is partly this confusion which is behind the stereotype of the wilderness or nature lover as the self-contained, misanthropic, and masculinist self, the Lockean atom drawn away by the "siren song of escape" (Vance 1995) from human society—just as he was drawn to it by the prospect of self-advantage. The definition of wilderness in terms of human absence erases nature as a presence, and even though other parts of the standard definitions of wilderness may try to put nature back *in* again as a presence, the result of the whole is incoherence.[21]

My example of Prion Bay illustrates some ways to understand the wild earth not in terms of an absence of the human, but in terms of a positive presence of place and earth community. We may think of Prion Bay as wild or free for many reasons: as a piece of the earth's body which challenges our survival and measures us as humans, as a place we have to come to know by the physical effort of our own bodies; as a wild presence beyond our remaking which still tells, without our interruption, a history far older than our own, a story we go there to hear; a place which is self-defining, which is for itself and whatever our human occupation does not leave room for.

Such a different conception of the wilderness Other also has implications for how we humans can position ourselves as lovers of wilderness land, and what kind of knowledge we seek in the wilderness. What differentiates the wilderness quest from mere exercise, of the monological sort that can be performed in the gym, is precisely that it has a content, is dialogical, has an orientation to knowledge of another set of presences.

To move to treating the presence of nature and not the absence of humans as the real meaning of wilderness is an important step in several ways. First, we avoid the anthropocentrism of once again placing ourselves at the center, defining the Other once again as our foil, in our absence as much as in our presence. We can then begin to foreground conceptually not ourselves but the wild Others, the free animal species and wild, uncolonized biotic communities, the great ancestral earth forces and their children, the landforms, that we should regard as the real occupants and users of the wilderness. Second, by conceiving wilderness not negatively as a sphere of emptiness but positively as a sphere of presence and freedom for these earth Others, we can open the way conceptually for a non-oppositional account of the relationship between humans and the wild Other.

The dominant account of wilderness as human absence is oppositional because it is only where a human society identifies itself as incompatible with, outside of, or oppositional to nature that the absence of the human is needed to guarantee the presence of nature. That is, in taking for granted that where humans are is where nature will not be, we implicitly assume a paradigm of destruction, the driving out or subjection of nature, rather than one of responsiveness and mutuality. It is little wonder that many indigenous people to whose lives such an oppositional concept is foreign find something deeply problematic in such a concept of wilderness, even apart from their own erasure. If in the present oppositional state of Western society there is a serious problem about its human presences and the assumption of destructiveness is mostly realistic, ensuring the absence of Westernized humans except in transient form may often be, as Birch says, the best we can do for the time being to ensure the presence of nature. But we should not so circumscribe our concepts that we *define* nature as human absence, for to do so is to make any alternative to our present oppositional condition unthinkable.

Third, once we have demoted human absence to being at most a contingent and culturally variable mark of the presence of wild Others, rather than its defining characteristic, we are at liberty to perceive, encounter, and even consort with these Others in contexts where humans are not absent. Defining our wilderness experience as a quest for the presence of wild nature, not the absence of humans, creates conceptual space for the interwoven continuum of nature and culture, and for that recognition of the presence of the wild and of the labor of nature we need to make in all our life contexts, both in wilderness and in places closer to home. It is this recognition that should be the aim of a green society and a green economy. And this may also be what we need to help us end the opposition between culture and nature, the garden and the wilderness, and to come to recognize ourselves at last as at home in both.

NOTES

1. There is no contradiction here, for as Alison Jaggar (1991) notes, something may be counted as symbolically feminine but still be normatively and empirically associated with men rather than women. However, there is a dissonance.

2. See, for example, Tomaskovic-Devy 1993.

3. For a discussion of a similarly gendered domestic/wild distinction in New Guinea, see McCormack and Strathern 1980.

4. As Iris Young's comments indicate, Rousseau remains relevant: "young children of both sexes categorically assert that girls are more likely to get hurt than boys are, and that girls ought to remain close to home, while boys can roam and explore" (Young 1990, p. 154).

5. In Young's view: "Women in sexist society are physically handicapped. Insofar as we learn to live out our existence in accordance with the definition that patriarchal culture assigns to us, we are physically inhibited, confined, positioned, and objectified" (1990, p. 153).

6. However, as Nabhan 1995 shows, such perceptions of indigenes as part of nature rather than culture and as incapable of changing the land appear quite frequently and innocently in modern American ecological writing, and would no doubt have been even more common at earlier times.

7. The erasure of indigenous influence is only one, if perhaps the most glaring, of the erasures of the "virgin" wilderness concept. Also erased is the negative influence of settler users. The presence and impact of the modern adventure tourist is somehow "written out" of focus in much of the land called wilderness. "Hike the many trails through a virgin land" says a hotel brochure, not only propounding but also profiting from this contradiction. The modern subject somehow manages

to be both in and out of this virginal fantasy, appearing by wilderness convention as a disembodied observer (perhaps as the camera eye) in a landscape whose virginity is somehow forever magically renewed, despite the hotel, the campground, the comfort stations and the ever-widening trails which bear witness to the pounding feet. (This conception of the visitor as a disembodied subject makes an implicit appeal to the idea of the body as a "transparent mediator of projects" discussed in Young [1990] "Pregnant Embodiment"). If in the indigenous case the virginity claim can be used to remove awkward questions about prior ownership, in the case of the Western adventure tourist it can be used to remove awkward questions about interference, limits, and responsibility for ongoing damage.

8. See also Bayet 1994.

9. On this human/animal dualism, see Benton 1993.

10. The debate about whether human activity can improve wilderness or not is highly context sensitive. In the case of Prion Bay, there are a few limited human activities, such as track improvements and the genetic resurrection of the Tasmanian Tiger, which might constitute improvements to the area as a wilderness, but it could be argued that these are directed to limiting or reversing the effects of culture. It is not so easy, however, to condemn human improvements such as tree-planting on Saharan or savannah edge land subject to severe processes of desertification, whether or not the land involved is seen as wilderness. No *general* answer to the question of improvement is available. If the virgin is not invariably ennobled by the attentions of civilization or technology, as the traditional view held, she is not invariably defiled by them either.

11. A third, but less common reduction position aims at the counter-reduction of culture to nature, reconceiving humans themselves and all their cultures as already part of nature, such that their activities cannot really change the status of virgin nature (Callicott 1991a; 1991b). Thus, it may be said that all land has both a human cultural history and a history of its development in nature, and that these cannot and need not be separated. But the claim that as humans we are part of nature does not support the claim that everything humans do is natural without the further and problematic assumption that we are not also part of the special, included or intersecting sphere of culture. In short, it requires the assumption that humans are an *indistinguishable* part of nature. The naturalistic reductionism suggested here recalls the problems of sociobiology and of the Malthusian ecologists, which discount the role of the social in human ecological impacts and fail to recognize the difference between the ethical and the ecological (Benton 1994). The result of treating everything as nature is that we obscure the basis for understanding the difference between anthropogenic from non-anthropogenic elements in country, and for understanding social relations and possibilities for change in societies. (See Soper 1996.)

12. Similarly, some argue that wilderness status for places like Prion Bay makes it "land kept artificially frozen in time" (Vance 1995), a virgin kept unwisely and

artificially from her natural consummation by humanity. Rather, wilderness status can enable the land to develop in its own way, through the continued operation of the forces which have changed it so powerfully in the past.

13. Guha has two remaining arguments to show that the critique of anthropocentrism is irrelevant: first, he assumes that wilderness preservation is offered as a sufficient solution for environmental problems. But although wilderness protection, in the sense of the protection of large areas of land for nature, whether virginal or not, and the protection of biotic communities in crisis, is by no means *sufficient* as an answer to environmental problems, it is certainly an *essential* part of any short- to medium-term solution, which should include a spectrum of protection options (Callicott 1991a). Guha's second argument assumes similarly that those who take seriously the critique of anthropocentrism believe it to provide a monocausal focus which undertakes to explain and resolve all current environmental and even social problems, including environmental justice problems. Although some deep ecologists seem to have such a reductionist focus (for example, Fox 1991), there are alternative non-reductionist standpoints (for example some varieties of ecofeminism) in which anthropocentrism is seen as just one form, although a vital and indispensable form for any environmental analysis, of multiple and linked forms of oppression and colonization.

14. If this means that only a particular civilization could have conceived wilderness as its antithesis, that may be true, but it does not follow that it has no resources for other ways to conceive it. On the face of it, Cronon's position appears contradictory: wilderness is our creation, a product of civilization, but is somehow also Other, deeply and irreducibly. Something more in the way of explanation seems to be needed here.

15. But, contrary to the indiscriminate and unreferenced condemnation of postmodernism in Soulé and Lease 1995, it is incorrect to attribute such strong cultural reductionism to all theorists who work in the deconstructive or social justice mode, and some, for example Haraway (1991), have explicitly rejected such a reduction and stress an approach which recognizes the agency of nature. Such highly generalized objections to postmodernism result from failing to distinguish between weak and strong forms of social constructionism (Fraser 1997).

16. It has been claimed that "all of Australia is an Aboriginal artefact" (Bayet 1994), and that there is no place "where the country was not once fashioned and kept productive by Aboriginal people's land management practices" (Rose 1996, p. 18). These strong claims about indigenous influence are applied universally and uniformly, with no exceptions or differential emphases, to all areas of the Australian continent. In a different cultural context, Nabhan (1995) makes a much more modest claim for the more densely populated North American continent, finding it possible to challenge the erasure of indigenous peoples in North American ecology while not denying that "many large areas of the North American continent remained beyond the influence of human cultures and should remain so." The import of Bayet's statement that humans cannot be "removed from the landscape,

whether it's sustainable or not" is unclear. If we read the statement as saying that all areas of land must have human occupants, whether or not their occupation is damaging, it seems highly problematic, especially for areas which have had minor Aboriginal ecological influence or occupation, or where this has been highly localized, such as the ecologically crucial but fragile alpine areas, now subject to potentially damaging occupation claims by high-country stock grazers. Certainly, country needs care, and carers, but sometimes this is best exercised by minimizing permanent presence, as in the high country.

17. For a more detailed discussion of these points, see Plumwood 1993a and Plumwood 1996. This polarized structure itself is often thought of as comprising dualism. But in fact we usually have to deal not only with a dualism but with a wider hegemonic centrism.

18. In fact this difference does not show up at the extensional level, where any rough-and-ready coincidence will do, but is important at the intentional level which establishes meanings. On the logic of presence see Plumwood 1993a and 1993b.

19. For more details on the logic of anthropocentrism, see Plumwood 1996.

20. Of course, there are those who wish to reposition the self at the (hegemonic) center of this experience too, via some form of "spiritual instrumentalism," but I think they have missed the point (Thompson 1990; Grey 1993).

21. Thus the first of the four conditions for wilderness given in the 1964 U.S. Wilderness Act specify nature negatively in terms of human absence, and the fourth condition which specifies it positively is presented as secondary and optional (the area "*may also* contain ecological, geological, or other features of scientific, scenic or historic value"). See the Wilderness Act, Public Law 88–577, 88th Congress, September 3, 1964. [Included in this volume.]

BIBLIOGRAPHY

Barnes, Marj, and Grahame Wells. 1985. "Myles Dunphy, Father of Conservation Dies," *National Parks Journal* (NSW) 29(1): 7–8.

Bayet, Fabienne. 1994. "Overturning the Doctrine: Indigenous People and Wilderness," *Social Alternatives* 13(2) (July): 27–32. [Included in this volume.]

Benton, Ted. 1993. *Natural Relations.* London: Verso.

———. 1994. "Biology and Social Theory in the Environmental Debate," in M. Redclift and T. Benton, eds., *Social Theory and the Global Environment.* London: Routledge, pp. 28–50.

Birch, Thomas H. 1990. "The Incarceration of Wildness: Wilderness Areas as Prisons," *Environmental Ethics* 12: 3–26. [Included in this volume.]

Bonwick, James. 1870. *The Last of the Tasmanians.* London.

Callicott, J. Baird. 1991a. "The Wilderness Idea Revisited: The Sustainable Development Alternative," *The Environmental Professional* 13: 235–47. [Included in this volume.]

———. 1991b. "That Good Old-Time Wilderness Religion," *The Environmental Professional* 13: 378–79. [Included in this volume.]

Campbell, Susan. 1995. "Hiking Alone?: Women on the Appalachian Trail," *News and Observer,* Raleigh, N. C., 13/9.

Chaloupka, William, and R. McGreggor Cawley. 1993. "The Great Wild Hope: Nature, Environmentalism, and the Open Secret," in Jane Bennett and William Chaloupka, eds., *In the Nature of Things: Language, Politics and the Environment.* Minneapolis: University of Minnesota Press, pp. 3–23.

Cronon, William. 1983. *Changes in the Land: Indians, Colonists and the Ecology of New England.* New York: Hill and Wang.

———. 1995a. "The Trouble with Wilderness," *New York Times Magazine,* August 13.

———. 1995b. "The Trouble with Wilderness, or, Getting Back to the Wrong Nature," in William Cronon, ed., *Uncommon Ground: Toward Reinventing Nature.* New York, W. W. Norton, pp. 69–90. [Included in this volume.]

Crosby, Alfred W. 1986. *Ecological Imperialism: The Biological Expansion of Europe.* Cambridge: Cambridge University Press.

Dann, Christine, and Pip Lynch. 1989. *Wilderness Women: Stories of New Zealand Women at Home in Wilderness.* Auckland, N. Z.: Penguin.

Denevan, William M. 1992. "The Pristine Myth: The Landscape of the Americas in 1492," *Annals of the Association of American Geographers* 82: 369–85. [Included in this volume.]

Duerr, H. P. 1985. *Dreamtime: Concerning the Boundary Between Wilderness and Civilization.* Felicitas Goodman, trans. Oxford: Basil Blackwell Ltd.

Fox, Warwick. 1991. *Toward a Transpersonal Ecology.* Boston: Shambhala.

Fraser, Nancy. 1997. *Justus Interruptus.* London: Routledge.

Frye, Marilyn. 1983. *The Politics of Reality.* Trumansberg: Crossing Press.

———. 1992. "Willful Virgin *or* Do You Have to be a Lesbian to be a Feminist?," in *Willful Virgin.* Freedom, California: Crossing Press, pp. 123–37.

Fullagar, Simone, and Susan Hailestone. 1996. "Shifting the Ground: Women and Outdoor Education," *Social Alternatives.* 15(2) (April): 23–27.

Gómez-Pompa, Arturo, and Andrea Kaus. 1992. "Taming the Wilderness Myth," *BioScience* 42(4): 271–79. [Included in this volume.]

Graber, David M. 1995. "Resolute Biocentrism: The Dilemma of Wilderness in National Parks," in M. Soulé and G. Lease, eds., *Reinventing Nature: Responses to Postmodern Deconstruction.* Washington, D.C.: Island Press, pp. 123–36.

Grey, William. 1993. "Anthropocentrism and Deep Ecology," *Australasian Journal of Philosophy* 71: 463–75.

Grimshaw, Jean. 1986. *Feminist Philosophers.* Brighton, Eng.: Wheatsheaf.

Guha, Ramachandra. 1989. "Radical American Environmentalism and Wilderness Preservation: A Third World Critique," *Environmental Ethics* 11: 71–83. [Included in this volume.]

Guy, Kevin. 1994. "'Sanctuaries' and Environmental Justice," *Chain Reaction* 71: 28–30.

Haraway, Donna. 1991. "Situated Knowledges," in *Simians, Cyborgs and Women: The Reinvention of Nature.* London: Free Association Books.

Harris, Alastair, ed. 1994. "A Good Idea Waiting to Happen: Regional Agreements in Australia." Proceedings from Cairns Workshop, July 1994, Cape York Land Council.

Harrison, Robert Pogue. 1992. *Forests: The Shadows of Civilization.* Chicago: University of Chicago Press.

Hecht, Susanna, and Alexander Cockburn. 1990. *The Fate of the Forest.* Harmondsworth, Eng.: Penguin.

Irigaray, Luce. 1984. *The Ethics of Sexual Difference,* Carolyn Sheaffer Jones, trans. Manuscript. See also Luce Irigaray, *An Ethics of Sexual Difference [Ethique de la différence sexuelle].* Carolyn Burke and Gillian C. Gill, trans. Ithaca, N.Y.: Cornell University Press.

Jaggar, Alison. 1991. "Feminist Ethics: Projects, Problems, Prospects," in Claudia Card, ed., *Feminist Ethics.* Lawrence, Kans.: University of Kansas Press, pp. 78–106.

Kothari, Ashish, Saloni Suri, and Neena Singh. 1996. "People and Protected Areas: Rethinking Conservation in India," *The Ecologist* 25: 188–94.

Langton, Marcia. 1996. "What Do We Mean by Wilderness? Wilderness and Terra Nullius in Australian Art," *The Sydney Papers,* (The Sydney Institute) 8(1): 10–31.

Lark, Jon. 1990. "Is Wilderness a Land Rights Issue?," *Chain Reaction* 61: 29–31.

McCormack, Carol P., and Marilyn Strathern, eds. 1980. *Nature, Culture and Gender.* Cambridge: Cambridge University Press.

Marshall, Robert. 1930. "The Problem of the Wilderness," *The Scientific Monthly* 30: 141–48. [Included in this volume.]

Merchant, Carolyn. 1981. *The Death of Nature.* London: Wildwood House.

———. 1995. "Reinventing Eden: Western Culture as a Recovery Narrative," in W. Cronon, ed., *Uncommon Ground: Toward Reinventing Nature.* New York: W. W. Norton, pp. 132–70.

———. 1996. *Earthcare: Women and the Environment.* New York: Routledge.

Nabhan, Gary Paul. 1995. "Cultural Parallax in Viewing North American Habitats," in M. Soulé and G. Lease, eds. *Reinventing Nature? Responses to Postmodern Deconstruction.* Washington, D.C.: Island Press, pp. 87–101. [Included in this volume.]

Nash, Roderick. 1982. *Wilderness and the American Mind,* 3rd edition. New Haven: Yale University Press.

———. 1989. *The Rights of Nature.* Madison: University of Wisconsin Press.

Nelson, Michael P. 1996. "Rethinking Wilderness: The Need for a New Idea of Wilderness," *Philosophy in the Contemporary World* 3(2): 6–9.

Olson, Sigurd. 1938. "Why Wilderness?," *American Forests* 55: 394–430. [Included in this volume.]

Oodgeroo (Kath Walker). 1988. *The Rainbow Serpent.* Canberra: Australian Government Publishing Service.

Plumwood, Val. 1993a. *Feminism and the Mastery of Nature.* London: Routledge.

———. 1993b. "The Politics of Reason," *Australasian Journal of Philosophy* 71: 436–62.

———. 1995. "Has Democracy Failed Ecology?: An Ecofeminist Perspective," *Environmental Politics,* special issue on Ecology and Democracy, 4(4): 134–68.

———. 1996. "Androcentrism and Anthrocentrism: Parallels and Politics," *Ethics and the Environment* 1(2): 119–52.

Rolston, Holmes, III. 1991. "The Wilderness Idea Reaffirmed," *The Environmental Professional* 13: 370–77. [Included in this volume.]

Rose, Deborah Bird. 1996. *Nourishing Terrains: Australian Aboriginal Views of Landscape and Wilderness.* Canberra: Australian Heritage Comission.

Ruether, Rosemary Radford. 1975. *New Woman New Earth.* New York. Seabury Press.

Ryan, Lyndall. 1996. *Aboriginal Tasmanians,* 2nd ed. Sydney: Allen and Unwin (1st ed., St. Lucia: University of Queensland Press, 1981).

Schaffer, Kay. 1988. *Women and the Bush.* Cambridge: Cambridge University Press.

Snyder, Gary. 1955. *A Place in Space.* Washington, D.C.: Counterpoint Press.

Soper, Kate. 1996. "Nature/'nature,'" in G. Robertson et al., eds., *Future Natural.* London: Routledge, pp. 22–35.

Soulé, Michael E. 1995. "The Social Siege of Nature," in M. Soulé and G. Lease, eds., *Reinventing Nature? Responses to Postmodern Deconstruction.* Washington, D.C.: Island Press, pp. 137–70.

Thompson, Janna. 1990. "A Refutation of Environmental Ethics," *Environmental Ethics* 12: 121–46.

Tomaskovic-Devy, Donald. 1993. *Gender and Racial Inequality at Work.* Ames, Ia.: University of Iowa Press.

Vance, Linda. 1995. "Ecofeminism and Wilderness," manuscript.

Walker, Cam. 1994. "Moving Together: Toward Green-Black Alliances," *Chain Reaction* 71: 20–21.

Warner, Marina. 1983. *Alone of All Her Sex: The Myth and Cult of the Virgin Mary.* New York: Vintage.

Woods, Mark. 1998. "Federal Wilderness Preservation in the United States: The Preservation of Wilderness?" [Included in this volume.]

Worster, Donald. 1995. "Nature and the Disorder of History," in M. Soulé and G. Lease, eds., *Reinventing Nature? Responses to Postmodern Deconstruction.* Washington, D.C.: Island Press, pp. 65–85.

Young, Iris Marion. 1990. "Pregnant Embodiment" and "Throwing Like a Girl," in *Throwing Like a Girl and Other Essays in Feminist Philosophy and Social Theory.* Bloomington: Indiana University Press.

INDEX